Springer Monographs in Mathematics

This series publishes advanced monographs giving well-written presentations of the "state-of-the-art" in fields of mathematical research that have acquired the maturity needed for such a treatment. They are sufficiently self-contained to be accessible to more than just the intimate specialists of the subject, and sufficiently comprehensive to remain valuable references for many years. Besides the current state of knowledge in its field, an SMM volume should ideally describe its relevance to and interaction with neighbouring fields of mathematics, and give pointers to future directions of research.

Günter Harder

Cohomology of Arithmetic Groups

Springer

Günter Harder
Max Planck Institute for Mathematics
Bonn, Germany

Mathematical Institute
University of Bonn
Bonn, Germany

ISSN 1439-7382 ISSN 2196-9922 (electronic)
Springer Monographs in Mathematics
ISBN 978-3-032-10377-2 ISBN 978-3-032-10378-9 (eBook)
https://doi.org/10.1007/978-3-032-10378-9

This Springer imprint is published by the registered company Springer Nature Switzerland AG
The registered company address is: Gewerbestrasse 11, 6330 Cham, Switzerland

Preface

During the years 1980-2000 I gave various advanced courses on number theory, algebraic geometry and on the cohomology of arithmetic groups at the University of Bonn. I prepared some very informal notes for my students, I wanted the necessary prerequisites to be available for them in one place.

At some point I had the feeling that these notes could serve as the basis for a book on the subject of *Cohomology of Arithmetic Groups.*

The cohomology groups of arithmetic groups are sheaf cohomology groups. Hence I provided some basic material on sheaves, homological algebra and on the cohomology of sheaves in these lectures. The theory of sheaves and sheaf cohomology are ubiquitous in algebraic geometry and in other branches of mathematics. So I also gave lectures on algebraic geometry and turned my notes into the two volumes [48], [49].

In a sense this present volume is now part 1 of volume III. At many occasions I refer only to these two volumes, even if there are more original references.

The subject has applications to number theory - actually it is part of number theory. The central theme in this book is the relationship between special values of L-functions and the integral structure of the cohomology as a module under the Hecke algebra. We prove rationality results for special values of L-functions (Manin, Shimura and many others). Furthermore we discuss the influence of these special values on the structure of the cohomology as Hecke modules. These special values tell us something about the denominators of the Eisenstein classes. Theorem 5.1.2 is a culminating result in this book.

However, Theorem 5.1.2 can be stated in elementary terms, we do not need any analysis in its formulation. For a given (small) value of the parameter, it can be verified by an algorithm, this is done in Chapter 3. But for a proof we need some analysis and we enter the realm of automorphic forms.

But I am convinced - and there is a lot of evidence for it - that Theorem 5.1.2 is a special case of a much larger class of (conjectural) assertions about L-values and denominators of Eisenstein classes. This is briefly treated in the last chapter. Our method of proof fails badly in the more general cases. But in principle we can write algorithms to confirm this conjecture for given values of the parameters, and I think these conjectures are interesting enough to be checked by some experiments.

In Section 3.3.8 we will see that denominators imply congruences between eigenvalues of Hecke operators acting on the cohomology of different arithmetic groups. It was extremely important for me that these conjectures on congruences could be verified in some finite number of cases by experimental calculations [24]. The experimental support for these congruences had great influence on the content of this book. But these computations confirmed only the congruences for a finite number of Hecke operators, not the denominators.

In chapter 5 we also expound the influence of the denominators on the structure of the Galois group. We discuss in detail a very specific example, namely the prime $\ell = 691$, which divides the denominator of the Eisenstein class in $H^1(\Gamma\backslash\mathbb{H}, \mathcal{M}_{10})$. We construct an extension $K \supset \mathbb{Q}(\zeta_\ell) \supset \mathbb{Q}$ which is unramified outside of ℓ and where $\mathrm{Gal}(K/\mathbb{Q}(\zeta_\ell))$ is nilpotent of degree ℓ^3, but not abelian.

I want to express my gratitude to my colleagues and friends C. Kaiser and J. Schwermer for their help in the final editing of the book. Especially I want to thank C. Kaiser who helped me to recover the proof of Theorem 5.1.2.

And now a final remark. It is not clear whether I will be able to finish Part 2. I have a preliminary version of Part 2 and I will put this version on the archive of the Max-Planck-Institute for Mathematics:

`https://archive.mpim-bonn.mpg.de/id/eprint/5157/`

Günter Harder

Max Planck Institute for Mathematics, Bonn, and
Mathematical Institute, University Bonn, Germany

Bonn, May 2025

Note

Günter Harder finished writing this book at the end of May 2025. He died on 10 June after a long illness.

Harder was a highly esteemed scientist and an inspiring teacher. This last book presents some of the fundamental questions of his scientific work to a broad readership. Along with many other students and colleagues, we are deeply appreciative of his numerous contributions to our field, and we note his passing with much regret.

Joachim Schwermer

Department of Mathematics, University Vienna, Austria

Bonn, June 2025

About the book

The objects of concern in this book are the sheaf-cohomology groups of locally symmetric spaces attached to arithmetically defined groups Γ, contained in an algebraic group. The sheaves are associated to Γ-modules over the ring of integers or finitely generated extension rings thereof. If the modules originate from a representation of the underlying algebraic group, then there is an action of a Hecke algebra on these cohomology groups. Accordingly, they decompose into indecomposable pieces which are of great interest in number theory and algebraic geometry.

Upon extending the coefficient systems to complex vector spaces, an individual indecomposable component corresponds to a space of automorphic forms, thereby giving rise to an attached L-function. For example, the Ramanujan Delta function provides the first case of such a correspondence. Up to division by a carefully chosen period, these L-functions take rational (algebraic) values at certain critical arguments. In various examples, it is discussed how the numerator and denominator ideals of these normalised values shed some light on the integral structure of the cohomology groups as a module under the Hecke algebra. In particular, results concern the denominator of cohomology classes which are represented by Eisenstein series, analytically constructed beforehand. In very special cases, values of the Riemann zeta-function play a decisive role. Within the discussion of these number-theoretic aspects of the cohomology groups, questions of a computational nature unfold. These may lead, by means of experiments, to a better understanding of the general integral structure of these groups.

This introduction to the cohomology of arithmetic groups and the associated theory of automorphic forms and special values of L-functions focuses on number theoretic aspects and questions. It is intended for graduate students and researchers in the field of arithmetic as well as in automorphic forms and differential geometry.

Contents

Introduction

An arithmetic group Γ is a discrete subgroup of a Lie group $G(\mathbb{R}) \subset \mathrm{Gl}_n(\mathbb{R})$ whose matrix entries satisfy certain rationality and integrality conditions. The most basic example of such a group is the group $\mathrm{Sl}_n(\mathbb{Z}) \subset \mathrm{Sl}_n(\mathbb{R})$. More generally, we can start from an algebraic subgroup $G/\mathbb{Q} \subset \mathrm{Gl}_n/\mathbb{Q}$, for instance, the orthogonal group associated to a quadratic form. Then we get arithmetic groups $\Gamma \subset G(\mathbb{Q}) \subset G(\mathbb{R})$ if we impose certain integrality conditions on the matrix coefficients of the elements of Γ.

It is my intention to keep the exposition as elementary as possible, and for the text to be readable by graduate students. We will need some background material from algebraic topology and homological algebra (cohomology and homology of groups, spectral sequences, sheaf cohomology). This material is expounded in the first four chapters in [48], and of course it can be found in many other textbooks.

For the Lie groups $\mathrm{Sl}_2(\mathbb{R})$ and $\mathrm{Sl}_2(\mathbb{C})$ and their arithmetic subgroups $\mathrm{Sl}_2(\mathbb{Z})$ and $\mathrm{Sl}_2(\mathbb{Z}[\sqrt{-1}])$ these prerequisite concepts are easy to explain, and we will do so in this book. For instance, if $\Gamma = \mathrm{Sl}_2(\mathbb{Z})$ or, more generally, a congruence subgroup of finite index, then the associated symmetric space is the upper half plane $\mathbb{H} = \{z \in \mathbb{C} \mid \Im(z) = y > 0\} = \mathrm{Sl}_2(\mathbb{R})/\mathrm{SO}(2)$. The quotient space $\Gamma\backslash\mathbb{H}$ is a punctured Riemann surface. In this special case we have the Γ-module $\mathcal{M}_n = \{\sum a_\nu X^\nu Y^{n-\nu} \mid a_\nu \in \mathbb{Z}\}$. We will study the cohomology groups $H^1(\Gamma\backslash\mathbb{H}, \tilde{\mathcal{M}}_n)$ and their module structure under the Hecke algebra in detail. We will prove some very specific results for these cohomology groups.

In some of the chapters we also need results and concepts from the theory of algebraic groups, the theory of symmetric spaces, arithmetic groups, and reduction theory for arithmetic groups. Furthermore we need results from the theory of representations of real semi-simple groups.

This material is somewhat more advanced, but in the first four chapters all these concepts and results will be explained in terms of special examples, especially the sections on the general reduction theory and the Borel-Serre compactification (Section (1.2.7)) could be skipped in a first reading.

For any Γ-module $\mathcal{M}$ we can define the cohomology groups $H^\bullet(\Gamma, \mathcal{M}) := \bigoplus_q H^q(\Gamma, \mathcal{M})$. These cohomology groups are abelian groups, which are defined in terms of homological algebra. They are the derived functors of the functor $\mathcal{M} \to \mathcal{M}^\Gamma$(= invariants under Γ). We are mainly interested in the cohomology of a very special class of Γ-modules. In fact, we consider rational representations $\rho : G/\mathbb{Q} \to \mathrm{Gl}(\mathcal{M}_\mathbb{Q})$, where $\mathcal{M}_\mathbb{Q}$ is a finite dimensional $\mathbb{Q}$-vector space. Then we can find finitely generated $\mathbb{Z}$-modules $\mathcal{M}$ such that $\mathcal{M}_\mathbb{Q} = \mathcal{M} \otimes_\mathbb{Z} \mathbb{Q}$ which are Γ-invariant and hence Γ-modules.

Let $K_\infty \subset G(\mathbb{R})$ be a maximal compact subgroup in a semi-simple group $G(\mathbb{R})$, for example $\mathrm{SO}(n) \subset \mathrm{Sl}_n(\mathbb{R})$. The quotient $X = G(\mathbb{R})/K_\infty$ is a symmetric space, it carries a Riemannian metric which is $G(\mathbb{R})$-invariant under the left action, and it is connected. It is diffeomorphic to $\mathbb{R}^d$, hence contractible.

Our arithmetic group Γ acts properly discontinuously on X, and we can form the quotient $\Gamma\backslash X$. This quotient is an orbifold. We can pass to a suitable subgroup of finite index $\Gamma' \subset \Gamma$ such that Γ' has no non-trivial elements of finite order (i.e. Γ' is torsion-free). Then $\Gamma'\backslash X$ is a Riemannian manifold, it is a so-called locally symmetric space. The map $\Gamma'\backslash X \to \Gamma\backslash X$ is a finite covering with some ramifications. If Γ has elements of finite order then $\Gamma\backslash X$ is only a Riemannian orbifold. These spaces $\Gamma\backslash X$ provide a very interesting class of spaces, which are of interest to differential geometers, mathematicians interested in analysis on manifolds, and topologists. But they are in a sense of arithmetic origin and therefore they are also of interest to number theorists.

Our Γ-module $\mathcal{M}$ endows the space $\Gamma\backslash X$ with a sheaf $\tilde{\mathcal{M}}$ with stalks in finitely generated abelian groups. If Γ is torsion free then $\tilde{\mathcal{M}}$ is a locally constant sheaf, or, in other words, a local system.

We introduce the sheaf cohomology groups

$$H^\bullet(\Gamma\backslash X, \tilde{\mathcal{M}}) = \bigoplus_q H^q(\Gamma\backslash X, \tilde{\mathcal{M}}).$$

These cohomology groups are "essentially" the same as the above group cohomology groups, and the two versions of cohomology become equal if X is connected and Γ is torsion-free. We will see that these cohomology groups are finitely generated $\mathbb{Z}$-modules.

We have some additional structure on these cohomology groups. In general the quotient space $\Gamma\backslash X$ is not compact. We have the Borel-Serre compactification $i : \Gamma\backslash X \hookrightarrow \Gamma\backslash \bar{X}$, where i is a homotopy equivalence, and $\Gamma\backslash \bar{X}$ is a manifold (orbifold) with corners. The difference set $\partial(\Gamma\backslash X) := (\Gamma\backslash \bar{X}) \setminus (\Gamma\backslash X)$ is the boundary of the Borel-Serre compactification. Moreover we will construct a "punctured tubular neighbourhood" $\overset{\bullet}{\mathcal{N}}(\Gamma\backslash X) \subset \Gamma\backslash X$ of "infinity" (see (1.2.7)). We introduce the cohomology with compact supports $H_c^\bullet(\Gamma\backslash X, \tilde{\mathcal{M}})$ and get the fundamental long exact sequence

$$\begin{aligned} \dots \to H_c^q(\Gamma\backslash X, \tilde{\mathcal{M}}) &\xrightarrow{i_c} H^q(\Gamma\backslash X, \tilde{\mathcal{M}}) \\ &\xrightarrow{r} H^q(\overset{\bullet}{\mathcal{N}}(\Gamma\backslash X), \tilde{\mathcal{M}}) \xrightarrow{\delta} H_c^{q+1}(\Gamma\backslash X, \tilde{\mathcal{M}}) \to \dots . \end{aligned} \tag{1}$$

We define the "inner cohomology" to be

$$H_!^q(\Gamma\backslash X, \tilde{\mathcal{M}}) := \ker(r) = \mathrm{Im}(i_c).$$

A second structural ingredient is the Hecke algebra. We define an action of a big algebra of operators - the Hecke algebra - acting on all these cohomology groups. This action commutes with arrows in the fundamental exact sequence.

We denote this algebra by $\mathcal{H}$ (or $\mathcal{H}_\Gamma$), and it originates from the structure of the arithmetic group Γ. The group Γ has many subgroups Γ' of finite index, for which we can construct two arrows

$$\Gamma'\backslash X \underset{p_2}{\overset{p_1}{\rightrightarrows}} \Gamma\backslash X. \tag{2}$$

Such a pair of arrows is also called a correspondence on $\Gamma\backslash X$. A correspondence, together with a suitable map $u : p_1^*(\tilde{\mathcal{M}}) \to p_2^*(\tilde{\mathcal{M}})$, induces an endomorphism in the cohomology. These endomorphisms act on all the modules in the exact sequence above and are compatible with the arrows. This algebra is a restricted tensor product of local algebras $\mathcal{H}_\Gamma = \bigotimes' \mathcal{H}_p$ where p runs over the set of primes. The local algebras $\mathcal{H}_p$ are commutative for almost all primes p.

In part 1 of this book we sometimes treat only seemingly simpler cases, such as $\Gamma = \mathrm{Sl}_2(\mathbb{Z})$ or $\mathrm{Sl}_2(\mathbb{Z}[\sqrt{-1}])$, and local systems which come from representations of the algebraic group schemes. This means that we consider cohomologies which are unramified everywhere. This has the advantage that the Hecke algebra is commutative, but the main results for these cases are essentially as interesting as the general results.

The basic objects of interest in this book are the various cohomology groups, which appear in the fundamental exact sequence, together with the action of the Hecke algebra $\mathcal{H}$ on them.

As the cohomology of modules under the Hecke-algebra can be decomposed into smaller pieces, the smaller pieces being $\pi_f = \otimes'_p \pi_p$, where for almost all primes the local factor π_p is simply a homomorphism $\pi_p : \mathcal{H}_p \to \bar{\mathbb{Q}}$.

We attach a (cohomological) L-function

$$L(\pi_f, s) = \prod_p L(\pi_p, s),$$

where $L(\pi_p, s)$ are local Euler factors, which are rational functions in the variable p^{-s}. The infinite product converges locally uniformly in some half plane, say $\Re(s) > c_0$, and one hopes that these infinite products have some meromorphic continuation into the entire complex s-plane. This hope has been confirmed in some cases, the easiest one being the Riemann ζ-function.

These L-functions are of course transcendental functions but they are of a cohomological nature - there is always a certain set of critical arguments $\mathbb{Z}_{\mathrm{crit}}(\mathcal{M})$ such that (conjecturally) the following holds: there are numbers $\Omega_{\epsilon(\nu)}(\pi_f), l(\nu)$, where $\epsilon(\nu)$ only depends on the parity of ν and $l(\nu)$ is an integer which can be easily computed, such that for $\nu \in \mathbb{Z}_{\mathrm{crit}}(\mathcal{M})$

$$L^*(\pi_f, \nu) := \frac{\pi^{l(\nu)}}{\Omega_{\epsilon(\nu)}(\pi_f)} \, L(\pi_f, \nu) \in \mathbb{Q}(\pi_f),$$

where $\mathbb{Q}(\pi_f)$ is the field generated by the values $\pi_p(h)$. These numbers $\Omega_{\epsilon(\nu)}$ are the (automorphic) periods, they are only defined modulo $K^\times = \mathbb{Q}(\pi_f)^\times$. A simple example for this rationality result is treated in section 4.2.6 and equation (4.135).

Of course everybody knows the special case of the values of the Riemann ζ-function at even positive integers and odd negative integers.

Now we come to a governing question in this book: Do these values tell us something interesting about the structure of the cohomology? Here we have to take into account that the periods are only well-defined modulo $K^\times$. But in many cases we are able to give a cohomological interpretation of the normalised values $L^*(\pi_f,\nu)$. For instance, these numbers can be given as an evaluation of a cohomology class on a homology class. Then we use the integral structure to pin the periods down up to a unit in $\mathcal{O}_K^\times$ or at least in $\mathcal{O}_{K,S}^\times$ (where S is a finite set of inverted primes). Hence the principal ideals generated by these normalised values are well-defined. In Chapter 4 we make serious efforts to get such a normalisation of these periods in the case $\Gamma = \mathrm{Sl}_2(\mathbb{Z})$.

At various places I formulated the expectation that in certain situations the numerator of these ideals is strongly related to the denominator of an Eisenstein class, for instance, it divides the denominator ([51],[54]) of an Eisenstein class attached to π_f.

For the simplest group $\Gamma = \mathrm{Sl}_2(\mathbb{Z})$ we discuss this issue in detail in Chapters 3 and 5 in a very special case. In Chapter 3 we treat mainly the experimental aspect. We present an algorithm which was written with the help of H. Gangl. It allows us to verify the above assertion in a given case, provided the parameters are small enough. This algorithm works for the group $\Gamma = \mathrm{Sl}_2(\mathbb{Z})$ and provides some interesting experimental data, which we discuss in Chapter 3.

In Chapter 4 we provide the tools from analysis needed for this subject. We explain the necessary results from the representation theory of the real Lie groups $\mathrm{Sl}_2(\mathbb{R})$ and $\mathrm{Sl}_2(\mathbb{C})$. We introduce the notion of Harish-Chandra modules and compute their cohomology, we present the required results from the theory of automorphic forms, and we review the theory of Whittaker-models. We use these Whittaker-models to prove rationality results for the normalised L-values (5.1.1 Manin and Shimura).

Finally we briefly explain how we get the Eisenstein cohomology classes by a process of infinite summation, and how we can use this to prove rationality results of the above kind for the normalised L-values. I propose to call this approach cohomological Langlands-Shahidi method (see [60]).

In Chapter 5 we apply these results to number theory, where we use our analytic tools and prove Theorem 5.1.2 which tells us that the above expectation holds under the given special conditions.

Finally at the end of Chapter 5 we discuss some further applications of these results to number theory. We have to accept some even more advanced results from arithmetic geometry. We concentrate on the case that $\Gamma \subset \mathrm{Sl}_2(\mathbb{Z})$ and we will use the fact that - with a grain of salt - the quotient $\Gamma\backslash\mathbb{H}$ is the set of $\mathbb{C}$-valued points of the moduli space $M_r/\mathbb{Z}[\frac{1}{r}]$ of elliptic curves "with level r structure". I tried to explain this in the subsection "Elliptic curves with level structure" in [48]. We have the famous comparison theorem which says that

$$H^1_{et}(M_1 \times_{\mathbb{Z}} \bar{\mathbb{Q}}, \tilde{\mathcal{M}} \otimes \mathbb{Q}_\ell) \xrightarrow{\sim} H^1(\mathrm{Sl}_2(\mathbb{Z})\backslash\mathbb{H}, \tilde{\mathcal{M}} \otimes \mathbb{Q}_\ell).$$

For any - not too small - prime ℓ we furthermore have an action of the Galois group $\mathrm{Gal}(\bar{\mathbb{Q}}/\mathbb{Q})$ on the ℓ-adic étale cohomology groups $H^1_{et}(M_1 \times_{\mathbb{Z}} \bar{\mathbb{Q}}, \tilde{\mathcal{M}} \otimes \mathbb{Z}_\ell)$ and hence also on $H^1(\mathrm{Sl}_2(\mathbb{Z})\backslash\mathbb{H}, \tilde{\mathcal{M}} \otimes \mathbb{Z}_\ell)$. This action commutes with the action of the Hecke algebra and this action is unramified outside ℓ.

Information about the structure of the cohomology groups as Hecke modules provides information about the structure of the Galois group $\mathrm{Gal}(\bar{\mathbb{Q}}/\mathbb{Q})$. For instance, we may consider the action of the Galois group on the finite-dimensional $\mathbb{F}_\ell$-vector space $H^1_{et}(M_1 \times_{\mathbb{Z}} \bar{\mathbb{Q}}, \tilde{\mathcal{M}} \otimes \mathbb{F}_\ell)$. Hence we constructed a finite Galois-module which is unramified outside ℓ (see 5.2.4). Here the nonvanishing of the denominator of the Eisenstein class allows to reprove the theorem of Herbrand-Ribet ([31], [94]) (see 5.2.4). Since we do not only look at the cuspidal part of the cohomology our results are slightly stronger.

There is an abundant number of cases where we can formulate the conjecture relating normalised L-values to denominators of Eisenstein classes. We have definitely a few more cases where our argument above works and we can prove at least part of the conjecture. But it seems to be clear that our methods above fail in the general case, for instance, it seems to fail in [51] and [54].

At this point we should try to verify the denominator conjecture experimentally and follow the path we took in Chapter 3. I raised this question for many years but it seems to be rather difficult to write such a code. A perhaps promising case is treated in [54].

In Section 3.3.8 we see that for primes ℓ which divide such a denominator, we get congruences between eigenvalues of Hecke operators acting on the cohomology of different groups, the most famous case is the congruence $\tau(p) \equiv p^{11} + 1 \mod 691$. In many cases people have verified these congruences for a certain small set of primes p and sometimes even for all primes (see [1],[2] [10],[11],[37],[28]). But these authors only check the congruences and not the denominators.

There is a hint why the conjecture should be true. I explain this briefly in Section 5 of [51] and I discuss this in more detail in some of the unpublished manuscripts. In the manuscript Mix-Mot.pdf I construct objects which I call Anderson–Tate mixed motives, and the hope is that these A-T-motives are not exotic. If we believe in a generalised Hodge conjecture for mixed Tate motives then we can also believe that exotic Tate motives do not exist. If we accept this then we can compare the Betti-de Rham extension class with the ℓ-adic extension class, and this gives information about the denominator of an Eisenstein class (see [51] Section 5 and [54]). We see that it is of interest to accumulate experimental data which support the conjecture. If we find an example for which the conjecture turns out to be false, then this has serious consequences for the concept of mixed motives. Then we falsify certain expectations, for instance, some kind of extended Hodge conjecture.

In Chapter 6 we develop the analytic tools for the computation of the cohomology. Here we do not use the language of adeles. We assume that the Γ-module $\mathcal{M}$ is a $\mathbb{C}$-vector space being obtained from a rational representation of the underlying algebraic group. In this case one may interpret the sheaf $\tilde{\mathcal{M}}$ as the sheaf of locally constant sections in a flat bundle, and this implies that the cohomology is computable from the de-Rham complex associated to this flat bundle. We could even go one step further and introduce a Laplace operator so that we get some kind of Hodge-theory, and we can express the cohomology in terms of harmonic forms. Here we encounter serious difficulties since the quotient space $\Gamma\backslash X$ is not compact. But we will proceed in a slightly different way. Instead of doing analysis on $\Gamma\backslash X$, we work on $\mathcal{C}_\infty(\Gamma\backslash G(\mathbb{R}))$. This space is a module under the group $G(\mathbb{R})$, which acts by right translations, but we rather consider it as a module under the Lie algebra $\mathfrak{g}$ of $G(\mathbb{R})$ on which the group K_∞ also acts - it is a $(\mathfrak{g}, K)$-module.

Since our module $\mathcal{M}$ comes from a rational representation of the underlying group G, we may replace the de-Rham complex by another complex

$$H^\bullet(\mathfrak{g}, K_\infty, \mathcal{C}_\infty(\Gamma\backslash G(\mathbb{R})) \otimes \tilde{\mathcal{M}}),$$

where this complex computes the $(\mathfrak{g}, K)$-cohomology. The general principle will be to "decompose" the $(\mathfrak{g}, K)$-module $\mathcal{C}_\infty(\Gamma\backslash G(\mathbb{R}))$ into irreducible submodules and therefore to compute the cohomology as the sum of the contributions of the individual submodules. This is a group-theoretic version of the classical approach of Hodge-theory. Again we have to overcome two difficulties - the first one being that the quotient $\Gamma\backslash G(\mathbb{R})$ is not compact, and hence the above decomposition does not make sense.

The second problem is that we have to understand the irreducible $(\mathfrak{g}, K)$-modules and their cohomology.

The first problem is of an analytical nature, and we will give some indication as to how this can be solved by passing to certain subspaces of the cohomology - the cuspidal, or better, the inner cohomology. The central result is Theorem 6.2.1. This result is a partial generalisation of the theorem of Eichler-Shimura, and it describes the cuspidal part of the cohomology in terms of irreducible representations occurring in the space of cusp forms. Thus, it contains some information regarding the contribution of the discrete spectrum to the cohomology.

We shall also state some general results concerning the second problem, and we briefly recall the construction of the irreducible modules with non-trivial $(\mathfrak{g}, K)$-cohomology.

We discuss some consequences of Theorem 6.2.1. It implies some vanishing theorems in cohomology, it implies that the inner cohomology is a semi-simple module for the Hecke-algebra, and it helps to understand the K-theory of algebraic number fields.

Bonn, May 2025

Chapter 1

Basic Notions and Definitions

1.1 Affine algebraic groups over $\mathbb{Q}$.

A linear algebraic group $G/\mathbb{Q}$ is a subgroup $G \subset \mathrm{Gl}_n$, which is defined as the set of common zeroes of a set of polynomials in the matrix coefficients, where in addition these polynomials have coefficients in $\mathbb{Q}$. Of course we cannot take just any set of polynomials, the set has to be somewhat special before its common zeroes form a group. The following examples will clarify what I mean:

1.) The group Gl_n is an algebraic group itself, the set of equations is empty. It has the subgroup $\mathrm{Sl}_n \subset \mathrm{Gl}_n$, which is defined by the polynomial equation

$$\mathrm{Sl}_n = \{x \in \mathrm{Gl}_n \mid \det(x) = 1\}$$

2.) Another example is given by the orthogonal group of a quadratic form

$$f(x_1, \ldots, x_n) = \sum_{i=1}^{n} a_i x_i^2$$

where $a_i \in \mathbb{Q}$ and all $a_i \neq 0$ (this is actually not necessary for the following). We look at all matrices

$$\alpha = \begin{pmatrix} a_{11} & \ldots & a_{1n} \\ \vdots & & \vdots \\ a_{n1} & \ldots & a_{nn} \end{pmatrix}, a_{ij} \in \mathbb{Q},$$

which leave this form invariant, i.e.

$$f(\alpha \underline{x}) = f(\underline{x})$$

for all vectors $\underline{x} = (x_1, \ldots, x_n)$. It is clear that it suffices to check this equality for the unit vectors $e_i = (0, 0, \ldots, 1, 0, 0)$ and their sums. This defines a set of polynomial equations for the coefficients a_{ij} of α. These matrices α form a

G. Harder, *Cohomology of Arithmetic Groups*,
Springer Monographs in Mathematics, https://doi.org/10.1007/978-3-032-10378-9_1

group, which is the linear algebraic group $O(f)$, and if we require in addition that $\det(\alpha) = 1$ then we get the special orthogonal group $SO(f)$. We see that the coefficients of these polynomials lie in $\mathbb{Q}$ and therefore the above groups are defined over $\mathbb{Q}$ and we write $SO(f)/\mathbb{Q} \subset \mathrm{Gl}_n/\mathbb{Q}$.

3.) Instead of taking a quadratic form — which is equivalent to taking a symmetric bilinear form — we could take an alternating bilinear form

$$\begin{aligned}\langle \underline{x}, \underline{y} \rangle :=& \langle x_1, \ldots, x_{2n}, y_1, \ldots y_{2n} \rangle = \\ & \sum_{i=1}^{n} (x_i y_{i+n} - x_{i+n} y_i \rangle = f \langle \underline{x}, \underline{y} \rangle.\end{aligned}$$

This form defines the symplectic group:

$$\mathrm{Sp}_n = \left\{ \alpha \in \mathrm{Gl}_{2n} \mid \langle \alpha \underline{x}, \alpha \underline{y} \rangle = \langle \underline{x}, \underline{y} \rangle \right\}.$$

Important remark: The reader may have observed that we did not specify a field (or a ring) from which we take the entries of the matrices. This is done intentionally, because we may take the entries from any commutative ring R which contains the rational numbers $\mathbb{Q}$ and for which $1 \in \mathbb{Q}$ is the identity element (this means that R is a $\mathbb{Q}$-algebra). In other words: For any algebraic group $G/\mathbb{Q} \subset \mathrm{Gl}_n$ and any commutative $\mathbb{Q}$-algebra R we may define

$$G(R) \subset \mathrm{Gl}_n(R)$$

as the group of those matrices whose coefficients satisfy the required polynomial equations. Adopting this point of view we can say that -

A linear algebraic group $G/\mathbb{Q}$ defines a functor from the category of commutative $\mathbb{Q}$-algebras into the category of groups.

4.) Another example is obtained by the so-called *restriction of scalars.* Let us assume we have a finite extension $K/\mathbb{Q}$, then we consider the vector space $V = K^n$. This vector space may also be considered as a $\mathbb{Q}$-vector space V_0 of dimension $n[K : \mathbb{Q}] = N$. We consider the group

$$\mathrm{Gl}_N/\mathbb{Q}.$$

Our original structure as a K-vector space may be considered as a map

$$\Theta : K \longrightarrow \mathrm{End}_{\mathbb{Q}}(V_0),$$

and the group $\mathrm{Gl}_n(K)$ is then the subgroup of elements in $\mathrm{Gl}_N(\mathbb{Q})$ which commute with all the elements of $\Theta(x)$, $x \in K$. Hence we define the subgroup

$$G/\mathbb{Q} = R_{K/\mathbb{Q}}(\mathrm{Gl}_n) = \{\alpha \in \mathrm{Gl}_N \mid \alpha \text{ commutes with } \Theta(K)\}. \qquad (1.1)$$

Then $G(\mathbb{Q}) = \mathrm{Gl}_n(K)$. For any $\mathbb{Q}$-algebra R we get

$$G(R) = \mathrm{Gl}_n(K \otimes_{\mathbb{Q}} R).$$

This can also be applied to an algebraic subgroup $H/K \hookrightarrow \mathrm{Gl}_n/K$, i.e. a subgroup that is defined by polynomial equations with coefficients in K.

Our definition of a linear algebraic group is a little bit provisional. If we consider for instance the two linear algebraic groups

$$G_1/\mathbb{Q} = \left\{\begin{pmatrix} 1 & x \\ 0 & 1 \end{pmatrix}\right\} \subset \mathrm{Gl}_2$$
$$G_2/\mathbb{Q} = \left\{\begin{pmatrix} 1 & 0 & x \\ 0 & 1 & 0 \\ 0 & 0 & 1 \end{pmatrix}\right\} \subset \mathrm{Gl}_3,$$

then we would like to say that these two groups are isomorphic. They are two different "realizations" of the *additive group* $G_a/\mathbb{Q}$. We see that the same linear algebraic group may be realised in different ways as a subgroup of different Gl_N's.

Of course there is a concept of linear algebraic group which does not rely on embeddings. The understanding of this concept requires a little bit of affine algebraic geometry. The drawback of our definition here is that we cannot define morphisms between linear algebraic groups. Especially when we do not know that they are isomorphic.

We assert to the reader that the general theory implies that a morphism between two algebraic groups is the same thing as a morphism between the two functors from $\mathbb{Q}$-algebras to groups. To pin down a linear algebraic group it is enough to give this functor. For instance, we have the *multiplicative group* $\mathbb{G}_m/\mathbb{Q}$ given by the multiplicative functor

$$R \longrightarrow R^\times$$

and the *additive group* $\mathcal{G}_a/\mathbb{Q}$ given by $R \to R^+$.

We can realise (represent is the right term) the group $\mathbb{G}_m/\mathbb{Q}$ as

$$\mathcal{G}_m/\mathbb{Q} = \left\{\begin{pmatrix} t & 0 \\ 0 & t^{-1} \end{pmatrix}\right\} \subset \mathrm{Gl}_2.$$

1.1.1 Affine group schemes

We say just a few words concerning the systematic development of the theory of linear algebraic groups. This is not directly used in the next few chapters but it will be useful later. If k is any field then $\bar{k}$ will be its algebraic closure.

For any commutative field k with identity 1_k, an affine k-algebra is a finitely generated commutative algebra A/k with identity 1_A and a homomorphism $j : k \to A$ such that $j(1_k) = 1_A$. In other words

$$A = k[x_1, x_2, \ldots, x_n] = k[X_1, X_2, \ldots, X_n]/I,$$

where the X_i are independent variables and where I is a finitely generated ideal of polynomials in $k[X_1, \ldots, X_n]$. Such an affine k-algebra defines a functor from the category of k-algebras to the category of sets, namely $B \mapsto \mathrm{Hom}_k(A, B)$.

The structure of an *affine k-group scheme* consists of the following data:

a) A k-homomorphism $m : A \to A \otimes_k A$ (the comultiplication),
b) A k-valued point $e_A : A \to k$ (the identity element),
c) An inverse $inv : A \to A$,

which satisfy the following requirement: For any k-algebra B our homomorphism m induces a map

$$^t m : \mathrm{Hom}_k(A \otimes_k A, B) = \mathrm{Hom}_k(A, B) \times \mathrm{Hom}_k(A, B) \to \mathrm{Hom}_k(A, B),$$

and we require that this map induces a group structure on $\mathrm{Hom}_k(A, B)$. We also require that the k-valued point e_A is the identity and that inv yields the inverse.

We leave it to the reader to figure out what this means for m, e, inv, and especially what does associativity mean (Hint: Choose $B = A$).

An affine k-algebra A together with such a collection m, e, inv is called an *affine k-group scheme* $G/k = (A, m, e, inv)$. The k-algebra A is the coordinate ring, or the ring of *regular functions* of the group scheme. We will denote it by $A(G)$. The group of B/k-valued points will be denoted by $G(B) = \mathrm{Hom}_k(A(G), B)$. For $g \in G(B)$ and $f \in A(G) \otimes_k B$ we write $g(f) =: f(g)$, i.e. we evaluate the regular function at the point $g \in G(B)$.

Examples: The multiplicative k-group $\mathcal{G}_m$ has the coordinate ring

$$A(\mathcal{G}_m) = k[t, t^{-1} \mid m(t) = t \otimes t, e(t) = 1, \mathrm{inv}(t) = t^{-1}]$$

and the coordinate ring of the additive k-group $\mathcal{G}_a$ is $A(\mathcal{G}_a) = k[x]$ and $m(x) = x \otimes 1 + 1 \otimes x, e(x) = 0, \mathrm{inv}(x) = -x$.

The group scheme Gl_n/k has the coordinate ring

$$A = k[\dots, x_{i,j}, \dots, y]/(\det(x_{i,j})y - 1|1 \leq i, j \leq n),$$

and the comultiplication is given by

$$m(x_{i,j}) = \sum_{\nu=1}^{n} x_{i,\nu} \otimes x_{\nu,j}. \tag{1.2}$$

Now it becomes clear what a homomorphism $\phi : G \to H$ between affine group schemes should be: It is a homomorphism $^t\phi$ between the affine k -algebras $A(H)$ and $A(G)$ which is compatible with the respective maps in a),b),c). A homomorphism $\phi : G \to H$ is injective if the homomorphism $^t\phi : A(H) \to A(G)$ is surjective. This means that $A(G) = A(H)/I$ where I is an ideal which satisfies $m(I) \subset A(H) \otimes I + I \otimes A(H)$. We say that G is a subgroup scheme of H. We write $G/k \subset H/k$.

Now it is clear what an action G/k on a scheme X/k should be. It is morphism $m : G \times X \to X$ which satisfies the obvious rules $m(e, x) = x$ and $m(g_1, m(g_2, x)) = m(g_1 g_2, x)$.

If $H/k \subset G/k$ we have the action of H/k on G/k by left translations. One can define the left (resp. right) quotients $H\backslash G$ (resp. G/H) they are schemes over k. We will encounter these construction later, and it will always be obvious what this quotient will be. But in general $H\backslash G$ will not be affine anymore. If $H/k \subset G/k$ is a normal subgroup then G/H is again an affine scheme.

A *rational representation* of G/k is a homomorphism of group schemes $\rho : G/k \to \mathrm{Gl}_n/k$ for some n.

If for instance V/k is a vector space of dimension n, then we can define the group scheme $\mathrm{Gl}(V)$, if we choose a k-basis on V, then we can identify

$\mathrm{Gl}(V)/k = \mathrm{Gl}_n/k$. If G/k is any affine group scheme, we say that V/k is a G-module if we have a homomorphism $\rho : G/k \to \mathrm{Gl}(V)$. Hence we know that for any k-algebra B/k we have a homomorphism $\rho(B) : G(B) \to \mathrm{Gl}(V \otimes_k B)$. Of course this is functorial in B/k, i.e. a homomorphism $\psi : B/k \to B'/k$ induces a homomorphism $G(B) \to G(B')$.

We may also consider actions of G/k on vector spaces W/k which are not of finite dimension, but here we require a certain finiteness condition. As before we have an action

$$\rho_B : G(B) \times (W \otimes B) \to W \otimes B \tag{1.3}$$

which is functorial in B/k. But now we assume in addition that for any $w \in W$ there is a finite set of elements $w_1, w_2, \ldots, w_d \in W$, such that for any B/k and any $g \in G(B)$

$$\rho_B(g)w = \sum_{i=1}^{d} w_i \otimes b_i(g) \text{ with } b_i \in A(G).$$

It suffices to check this for the "universal" element $\mathrm{Id} \in \mathrm{Hom}_k(A(G), A(G)) = G(A(G))$, this means we have to find $w_1, w_2, \ldots, w_d \in W$ such that

$$\rho_{A(G)}(\mathrm{Id})w = \sum_{i=1}^{d} w_i \otimes h_i \text{ with } h_i \in A(G).$$

This implies of course that the k-subspace $W' = \sum k w_i$ which is generated by these w_i is invariant under the action ρ and it contains w. Hence we see that our k-vector space W is a union of finite dimensional subspaces which are invariant under the action of G/k.

Therefore we say that a vector space W/k with an action of G/k is a *G-module* if it satisfies the above finiteness condition. The category of these modules will be called Mod_G.

The ring of regular functions $A(G)$ is a $(G \times_k G)$-module: For $(g_1, g_2) \in (G \times_k G)(B) = G(B) \times G(B)$ and $f \in A(G \otimes_k B)$, $x \in G(B)$ the action is defined by

$$\rho(g_1, g_2)f(x) = f(g_1^{-1} x g_2).$$

We have to verify the finiteness condition. To do this we write a formula for $\rho(g_1, g_2)f \in A(G) \otimes B$. We have the comultiplication $m : A(G) \to A(G) \otimes_k A(G)$, we apply it to the first factor on the right hand side and get $m_{1,2} \circ m : A(G) \to A(G) \otimes_k A(G) \otimes_k A(G)$. Then

$$m_{1,2} \circ m(f) = \sum_{\mu} h'_\mu \otimes h_\mu \otimes h''_\mu.$$

Then by definition

$$\rho(g_1, g_2)f = \sum_{\mu} h_\mu \otimes \mathrm{inv}(h'_\mu)(g_1) h''_\mu(g_2),$$

and this says that $\rho(g_1, g_2)f$ lies in the submodule generated by the h_μ.

Of course we may restrict the action to each of the two factors, we simply choose $g_1 = e$ - we get the action by right translations - or we choose $g_2 = e$ which gives the action by left translations.

We say that G/k is *absolutely reduced* if $A(G) \otimes \bar{k}$ does not have non zero nilpotent elements. It is not difficult to show that for an absolutely reduced affine group scheme we can find a collection of elements $e_0, e_1, \ldots, e_r \in A(G) \otimes \bar{k}$ such that $e_i^2 = e_i \; \forall i, e_i e_j = 0 \; \forall i \neq j$ such that $1_A = \sum_i e_i$ and such that the subalgebras $A(G) \otimes \bar{k} e_i$ are integral. Then there is exactly one element (say $e_0 \in A(G)$) such that $e(e_0) = 1$. Then $A(G)e_0$ is a subgroup scheme and it is called the *connected component of the identity* (See for instance [49], Chap. 7, 7.2)

A group scheme G/k is *connected* if its affine algebra $A(G) = A(G)e_0$ is integral, i.e. it does not have zero divisors.

Base change

If we have a field $L \supset k$ and a linear group G/k then the group $G/L = G \times_k L$ is the group over L where we forget that the coefficients of the equations are contained in k. The group $G \times_k L$ is the *base extension* from G/k to L.

1.1.2 Tori, their character module

A special class of algebraic groups is given by *tori*. We briefly recall the results of T. Ono [89].

An algebraic group T/k over a field k is called a *split torus* if it is isomorphic to a product of multiplicative groups $\mathbb{G}_m/k$-s,

$$T/k \xrightarrow{\sim} \mathbb{G}_m^d/k.$$

The algebraic group T/k is called a *torus* if it becomes a split torus after a suitable finite extension of the ground field, i.e we have $T \times_k L \xrightarrow{\sim} \mathbb{G}_m^r/L$.

If we take an arbitrary separable finite field extension L/k we may consider the functor

$$R \to (L \otimes_k R)^\times.$$

It is not hard to see that this functor can be represented by an algebraic group over k, which is denoted by $R_{L/k}(\mathbb{G}_m/L)$ and called the *Weil restriction* of $\mathbb{G}_m/L$. We propose the notation

$$R_{L/k}(\mathbb{G}_m/L) =: \mathbb{G}_m^{L/k}. \tag{1.4}$$

The reader should try to prove that for a finite extension $\tilde{L}/L$ which is normal over k we have

$$\mathbb{G}_m^{L/k} \times_k \tilde{L} \xrightarrow{\sim} (\mathbb{G}_m/\tilde{L})^{[L:k]}$$

and this shows that $\mathbb{G}_m^{L/k}$ is a torus .

A torus T/k is called *anisotropic* if is does not contain a non-trivial split torus. Any torus C/k contains a maximal split torus S/k and a maximal anisotropic torus C_1/k. The multiplication induces a map

$$m : S \times C_1 \to C,$$

which is a surjective (in the sense of algebraic groups) homomorphism whose kernel is a finite algebraic group. We call such a map an *isogeny* and we write that $C = S \cdot C_1$, we say that C is the product of S and C_1 up to isogeny.

We give an example. Our torus $R_{L/k}(\mathbb{G}_m/L)$ contains $\mathbb{G}_m/k$ as a subtorus: For any commutative ring R containing k we have $R^\times = \mathbb{G}_m(R) \subset (R \otimes_k L)^\times$. On the other hand we have the norm map $N_{L/k} : (R \otimes_k L)^\times \to R^\times$, and the kernel defines a subgroup

$$R^{(1)}_{L/k}(\mathbb{G}_m/L) \subset R_{L/k}(\mathbb{G}_m/L),$$

and it is clear that

$$m : \mathbb{G}_m/k \times R^{(1)}_{L/k}(\mathbb{G}_m/L) \to R_{L/k}(\mathbb{G}_m/L)$$

has a finite kernel which is the finite algebraic group of $[L : k]$-th roots of unity.

For any torus T we define the character module as the group of homomorphisms

$$X^*(T) = \mathrm{Hom}(T \times_k k_s, \mathbb{G}_m \times_k k_s).. \tag{1.5}$$

If the torus is split, i.e. $T = \mathbb{G}_m^r$ then $X^*(T) = \mathbb{Z}^r$ and the identification is given by $(n_1, n_2, \dots, n_r) \mapsto \{(x_1, x_2, \dots, x_r) \mapsto x_1^{n_1} x_2^{n_2} \dots x_r^{n_r}\}$. We write the group structure on $X^*(T)$ additively, which means that $(\gamma_1 + \gamma_2)(x) = \gamma_1(x)\gamma_2(x)$.

It is a theorem that for any torus T/k we can find a finite, separable, normal extension L/k such that $T \times_k L$ splits. Then it is easy to see that we have an action of the Galois group $\mathrm{Gal}(L/k)$ on $X^*(T) \cong \mathbb{Z}^r$. If we have two tori T_1/K, T_2/K that split over L then

$$\mathrm{Hom}_k(T_1, T_2) \xrightarrow{\sim} \mathrm{Hom}_{\mathrm{Gal}(L/k)}(X^*(T_2 \times_k L), X^*(T_1 \times_k L)). \tag{1.6}$$

To any $\mathrm{Gal}(L/k)$-action on $\mathbb{Z}^n$ we can find a torus T/k which splits over L and which realises this action.

A homomorphism $\phi : T_1/k \to T_2/k$ is called an *isogeny* if $\dim(T_1) = \dim(T_2)$ and if ${}^t\phi : X^*(T_2) \to X^*(T_1)$ is injective. Then the kernel $\ker(\psi)$ is a finite group scheme of multiplicative type. If $Y \subset X^*(T_1)$ is a submodule of finite index the $Y = X^*(T_2)$ and the inclusion provides an isogeny $\psi : T_1 \to T_2$. The quotient $X^*(T_1)/Y$ is a finite constant group scheme and $\ker(\psi)$ the *dual* of this quotient.

We also define the *cocharacter module* $\mathrm{Hom}(\mathbb{G}_m, T)$. If the torus $T/k = \mathbb{G}_m^r$ then every cocharacter is of the form $x \mapsto (x^{n_1}, x^{n_2}, \dots, x^{n_r})$. It is clear that we have a pairing

$$\langle\, , \,\rangle : X_*(T) \times X^*(T) \to \mathbb{Z} \text{ which is defined by } \gamma(\chi(t)) = t^{\langle \chi, \gamma \rangle}$$

A very prominent torus is the torus $\mathbb{S}^1/\mathbb{R}$, this being the one dimensional torus whose character module $X^*(\mathbb{S}^1 \times \mathbb{C}) = \mathbb{Z}$ and the complex conjugation acts by -1.

1.1.3 Semi-simple groups, reductive groups

An important class of linear algebraic groups is formed by the *semi-simple* and the *reductive* groups (for a general reference [106]). We do not want to give the precise definition here. Roughly, a linear group is *reductive* if it is connected and if $G \times_k \bar{k}$ does not contain a non-trivial normal subgroup which is isomorphic to a product of groups of type G_a. A group is called *semi-simple*, if it is reductive and does not contain a non-trivial torus in its centre.

A semi-simple group G/k is simple, if it does not contain any normal subgroup of dimension > 0. Any semi-simple group is up to isogeny a product of simple groups. Any semi-simple group G/k contains a maximal torus $T/k \subset G/\mathbb{Q}$ which is equal to its own centraliser. A semi-simple group is split if it contains a split maximal torus T_0/k, i.e. a maximal torus which is split. If $T/k \subset G/k$ is any (maximal) torus, then there is a finite extension $L/\mathbb{Q}$ such that $T \times_{\mathbb{Q}} L$ is split, and hence $G \times_{\mathbb{Q}} L$ is also split.

For example, the groups Sl_n, Sp_n are (split) semi-simple, and the groups $\mathrm{SO}(f)$ are semi-simple provided $n \geq 3$ (see next subsection 1.1.6). The groups Gl_n and especially the multiplicative group $\mathrm{Gl}_1/\mathbb{Q} = \mathbb{G}_m/\mathbb{Q}$ are reductive. Any reductive group $G/\mathbb{Q}$ (or over any field of characteristic zero) has a central torus $C/\mathbb{Q}$, and this central torus contains a maximal split torus S. The derived group $G^{(1)}/\mathbb{Q}$ is semi-simple, and we get an isogeny

$$m : G^{(1)} \times C_1 \times S \to G$$

or briefly $G = G^{(1)} \cdot C_1 \cdot S$.

If for instance $G = R_{L/\mathbb{Q}}(\mathrm{Gl}_n/L)$ then $G^{(1)} = R_{L/\mathbb{Q}}(\mathrm{Sl}_n/L)$ and $C = R_{L/\mathbb{Q}}(\mathbb{G}_m/L)$ and this yields the product decomposition up to isogeny

$$G = G^{(1)} \cdot R^{(1)}_{L/\mathbb{Q}}(\mathbb{G}_m/L) \cdot \mathbb{G}_m. \tag{1.7}$$

For $\mathrm{Gl}_n/\mathbb{Q}$ the central torus is the group $\mathbb{G}_m/\mathbb{Q}$. The center of $\mathrm{Sl}_n/\mathbb{Q}$ is the finite group scheme μ_n of n-th roots of unity. The coordinate ring of μ_n is the finite algebra $A(\mu_n) = \mathbb{Q}[t]/(t^n - 1)$. Of course we may replace $\mathbb{Q}$ by any commutative ring R.

We can form the quotient group scheme

$$\mathrm{PGL}_n/\mathbb{Q} = (\mathrm{Gl}_n/\mathbb{G}_m)/\mathbb{Q} \xrightarrow{\sim} (\mathrm{Sl}_n/\mathbb{Q})/\mu_n \tag{1.8}$$

this quotient is also called the *adjoint* group of $\mathrm{Gl}_n/\mathbb{Q}$ and $\mathrm{Sl}_n/\mathbb{Q}$.

We could certainly drop the assumption that a reductive group should be connected, instead we could demand that the connected component of the identity is reductive (semi-simple).

Another important class of semi-simple groups is given by the *quasisplit* groups (see also section 1.1.8). A group $G/\mathbb{Q}$ is called quasisplit if it contains a *Borel subgroup* $B/\mathbb{Q} \subset G/\mathbb{Q}$. A Borel subgroup $B/\mathbb{Q}$ is a maximal connected solvable subgroup, which contains a maximal torus $T/\mathbb{Q} \subset B/\mathbb{Q}$, with this torus also being a maximal torus in $G/\mathbb{Q}$. Then $B = U \rtimes T$ is the semidirect product of this torus and the *unipotent radical* $U/\mathbb{Q}$. We will discuss a special example which is of great relevance for our subject.

Let $L/\mathbb{Q}$ be a quadratic extension, let us denote the non-trivial automorphism by $a \mapsto \bar{a}$. Let V/L be a finite dimensional vector space together with a hermitian form $h : V \times_L V \to L$, i.e.

$$h(v,w) = \overline{h(w,v)};\ h(\lambda u + \mu v, w) = \lambda h(u,w) + \mu h(v,w)\ \forall u,v,w \in V, \lambda, \mu \in L.$$

Furthermore we assume that h is non-degenerate, i.e. for any $v \in V, v \neq 0$ we find a $w \in V$ such that $h(v,w) \neq 0$. Then we can define the group $\mathrm{SU}(h)/\mathbb{Q}$: For any commutative $\mathbb{Q}$-algebra R we define

$$\mathrm{SU}(h)(R) = \{g \in R_{L/k}\mathrm{Sl}(V)(R) \mid h(gv, gw) = h(v,w) \text{ and } \det(g) = 1\}. \quad (1.9)$$

Then $\mathrm{SU}(h)/\mathbb{Q}$ is a semi-simple group over $\mathbb{Q}$. We can also define the unitary group $\mathrm{U}(h)/\mathbb{Q}$ where we drop the condition that the determinant equals one, and the group of hermitian similitudes $\mathrm{GU}(h)$ where

$$\begin{aligned} \mathrm{GU}(h)(R) = \{g \in \mathrm{Gl}(V \otimes_{\mathbb{Q}} R) \mid h(gv, gw) &= d(g)h(v,w) \\ &\forall v,w \in V \otimes_{\mathbb{Q}} R\}. \end{aligned} \quad (1.10)$$

Here $d : \mathrm{GU}(h) \to \mathbb{G}_m/\mathbb{Q}$ is a homomorphism, and the kernel of d is the group $\mathrm{U}(h)$.

We consider the special case where

$$V = Le_1 \oplus \cdots \oplus Le_n \oplus (Le_0) \oplus Lf_n \oplus \cdots \oplus Lf_1$$

with the convention that the summand Le_0 is left out if $\dim_L V$ is even. The hermitian scalar product is given by

$$h_1(e_i, f_i) = h_1(f_i, e_i) = 1,\ \forall i = 1, \dots, n,\ \ (h_1(e_0, e_0) = 1)$$

and all other scalar products are equal to zero. Then $\mathrm{SU}(h_1)$ is a quasi-split semi-simple group over $\mathbb{Q}$: The elements $t \in \mathrm{Sl}(V)$ for which

$$t = \begin{cases} t : e_i \mapsto t_i e_i \\ t : f_i \mapsto \bar{t_i}^{-1} f_i \\ (t : e_0 \mapsto t_0 e_0 \text{ with } t_0 \bar{t_0} = 1) \end{cases}$$

are the $\mathbb{Q}$-valued points of a maximal torus $T_1/\mathbb{Q} \subset \mathrm{SU}(h_1)$. The vector space V/L comes with a natural flag

$$\begin{aligned} \mathcal{F} := \{0\} &\subset Le_1 \subset \cdots \subset Le_1 \oplus \cdots \oplus Le_n \\ &\subset (Le_1 \oplus \dots Le_n \oplus Le_0) \subset (Le_1 \oplus \dots Le_n \oplus Le_0 \oplus Lf_n) \\ &\subset \cdots \subset (Le_1 \oplus \cdots \oplus Le_n \oplus Le_0 \oplus Lf_n \oplus \cdots \oplus Lf_1) \subset V \end{aligned} \quad (1.11)$$

Now the subgroup $B_1/\mathbb{Q} \subset \mathrm{SU}(h_1)/\mathbb{Q}$ which fixes $\mathcal{F}$ is a maximal solvable subgroup in $\mathrm{SU}(h_1)$.

1.1.4 The Lie-algebra

We need some basic facts about the Lie-algebras of algebraic groups.

For any algebraic group G/k we can consider its group of points with values in $k[\epsilon] = k[X]/(X^2)$. We have the homomorphism $k[\epsilon] \to k$ sending ϵ to zero and hence we get an exact sequence

$$0 \to \mathfrak{g} \to G(k[\epsilon]) \to G(k) \to 1.$$

The kernel $\mathfrak{g}$ is a k-vector space, and if the characteristic of k is zero, then its dimension is equal to the dimension of G/k. It is denoted by $\mathfrak{g} = \mathrm{Lie}(G)$.

Let us consider the example of the group $G = SO(f)$, where $f : V \times V \to k$ is a non-degenerate symmetric bilinear form. In this case an element in $G(k[\epsilon])$ is of the form $\mathrm{Id} + \epsilon A$, with $A \in \mathrm{End}(V)$ for which

$$f((\mathrm{Id} + \epsilon A)v, (\mathrm{Id} + \epsilon A)w) = f(v, w)$$

for all $v, w \in V$. Taking into account that $\epsilon^2 = 0$ we get

$$\epsilon(f(Av, w) + f(v, Aw)) = 0,$$

i.e. A is skew symmetric with respect to the form, and $\mathfrak{g}$ is the k-vector space of skew endomorphisms. If we give V a basis and if $f = \sum x_i^2$ with respect to this basis then this means the matrix of A is skew symmetric.

If we consider $G = \mathrm{Gl}_n/k$ then $\mathfrak{g} = M_n(k)$, here the Lie-bracket is given by

$$(A, B) \mapsto AB - BA. \tag{1.12}$$

We have some kind of a standard basis for our Lie algebra, namely

$$\mathfrak{g} = \bigoplus_{i=1}^{n} kH_i \oplus \bigoplus_{i,j,i \neq j} kE_{i,j} \tag{1.13}$$

where H_i (resp. $E_{i,j}$) are the matrices

$$H_i = \begin{pmatrix} 0 & 0 & \dots & 0 & 0 & 0 \\ 0 & 0 & \dots & 0 & 0 & 0 \\ 0 & 0 & \ddots & & 0 & 0 \\ 0 & 0 & 0 & 1 & 0 & 0 \\ 0 & 0 & 0 & \dots & \ddots & 0 \\ 0 & 0 & 0 & 0 & 0 & 0 \end{pmatrix} \text{ resp. } E_{i,j} = \begin{pmatrix} 0 & 0 & \dots & 0 & 0 & 0 \\ 0 & 0 & \dots & 0 & 0 & 0 \\ 0 & 0 & \ddots & & 1 & 0 \\ 0 & 0 & 0 & 0 & 0 & 0 \\ 0 & 0 & 0 & \dots & \ddots & 0 \\ 0 & 0 & 0 & 0 & 0 & 0 \end{pmatrix}$$

and the only non-zero entries (=1) are at (i, i) on the diagonal (resp. at (i, j) off the diagonal).

For the group Sl_n/k the Lie-algebra is $\mathfrak{g}^{(0)} = \{A \in M_n(k) | \mathrm{tr}(A) = 0\}$, and again we have a standard basis

$$\mathfrak{g}^{(0)} = \bigoplus_{i=1}^{n-1} k(H_i - H_{i+1}) \oplus \bigoplus_{i,j,i \neq j} kE_{i,j} \tag{1.14}$$

If $\rho : G \to \mathrm{Gl}(V)$ is a rational representation of our group G/k then it is clear from our considerations above that we have a "derivative" of this representation

$$d\rho : \mathfrak{g} = \mathrm{Lie}(G/k) \to \mathrm{Lie}(\mathrm{Gl}(V)) = \mathrm{End}(V), \tag{1.15}$$

which moreover is k-linear.

Every group scheme G/k has a very special representation, this is the *adjoint representation.* We observe that the group acts on itself by conjugation, this is the morphism

$$\mathrm{Inn} : G \times_k G \to G,$$

which on R valued points is given by

$$\mathrm{Inn} : (g_1, g) \mapsto g_1 g {g_1}^{-1}.$$

This action clearly induces

$$\mathrm{Ad} : G/k \to \mathrm{Gl}(\mathfrak{g})$$

This is the *adjoint representation* of the group. It has a derivative

$$D_{\mathrm{Ad}} = \mathrm{ad} : \mathfrak{g} \to \mathrm{End}(\mathfrak{g}).$$

This is the *adjoint representation* of the Lie-algebra.

We introduce the following notation: For $T_1, T_2 \in \mathfrak{g}$ we put

$$[T_1, T_2] := \mathrm{ad}(T_1)(T_2).$$

Now we can state the famous and fundamental result

Theorem 1.1.1. *The map $\mathfrak{g} \times \mathfrak{g} \longrightarrow \mathfrak{g}$, $(T_1, T_2) \mapsto [T_1, T_2]$ is bilinear and antisymmetric. It induces the structure of a Lie-algebra on $\mathfrak{g}$, i.e. we have the Jacobi identity*

$$[T_1, [T_2, T_3]] + [T_2, [T_3, T_1]] + [T_3, [T_1, T_2]] = 0.$$

We do not prove this here. In the case $G/k = \mathrm{Gl}(V)$ and $T_1, T_2 \in \mathrm{Lie}(\mathrm{Gl}(V)) = \mathrm{End}(V)$ we have $[T_1, T_2] = T_1T_2 - T_2T_1$ and in this case the Jacobi identity is a well-known identity.

On any Lie algebra $\mathfrak{g}$ we have a distinguished symmetric bilinear form (the *Killing form*)

$$B : \mathfrak{g} \times \mathfrak{g} \to k \tag{1.16}$$

which is defined by the rule $B(T_1, T_2) = \mathrm{tr}(\mathrm{ad}(T_1) \circ \mathrm{ad}(T_2))$.

A simple computation shows that for the examples $\mathfrak{g} = \mathrm{Lie}(\mathrm{Gl}_n)$ and $\mathfrak{g}^{(0)} = \mathrm{Lie}(\mathrm{Sl}_n)$ we have

$$B(T_1, T_2) = 2n\mathrm{tr}(T_1T_2) - 2\mathrm{tr}(T_1)\mathrm{tr}(T_2). \tag{1.17}$$

We observe that in case one of the T_i is central, i.e. $T_i = u\mathrm{Id}$ we have $B(T_1, T_2) = 0$. In the case of $\mathfrak{g}^{(0)}$ the second term is zero.

It is well-known that a linear algebraic group is semi-simple if and only if the Killing form B on its Lie algebra is non-degenerate. [22]

1.1.5 Affine group schemes over $\mathbb{Z}$

In our book [49] section 7.5.6 and on p. 50,51 we discuss briefly the general notion of a group scheme over an arbitrary base scheme S. An affine group scheme over $G/\mathbb{Z}$ is a finitely generated $\mathbb{Z}$-algebra $A(G)$ together with a comultiplication $m : A(G) \to A(G) \otimes A(G)$. For any $\mathbb{Z}$ -algebra B (commutative and with identity) the comultiplication m induces a multiplication on the B-valued points

$$^t m : \mathrm{Hom}_{\mathrm{alg}}(A, B) \times \mathrm{Hom}_{\mathrm{alg}}(A, B) \to \mathrm{Hom}_{\mathrm{alg}}(A, B)$$

and the requirement is that this multiplication defines a group structure on $G(B) = \mathrm{Hom}_{\mathrm{alg}}(A, B)$. In educated language : $G/\mathbb{Z}$ is a functor from the category of affine schemes into the category of groups.

For instance we can define the group scheme $\mathrm{Gl}_n/\mathbb{Z}$. The affine algebra is

$$A(\mathrm{Gl}_n) = \mathbb{Z}[X_{11}, X_{12}, \ldots, X_{1n}, X_{21}, \ldots, X_{nn}, Y]/(Y \cdot \det(X_{ij} - 1)$$

Then the group $\mathrm{Gl}_n(\mathbb{Z})$ of $\mathbb{Z}$-valued points of $\mathrm{Gl}_n/\mathbb{Z}$ is our group $Gl_n(\mathbb{Z})$.

If $G/\mathbb{Q} \subset \mathrm{Gl}_n/\mathbb{Q}$ is a subgroup, then its affine algebra is $A(G) = A(\mathrm{Gl}_n) \otimes \mathbb{Q}/I$, where I is an ideal in $A(\mathrm{Gl}_n) \otimes \mathbb{Q}$. Since $G/\mathbb{Q}$ is a subgroup this ideal must satisfy

$$m_{\mathrm{Gl}_n}(I) \subset A(\mathrm{Gl}_n) \otimes \mathbb{Q} \otimes I + I \otimes A(\mathrm{Gl}_n) \otimes \mathbb{Q}.$$

Let $J = A(\mathrm{Gl}_n) \cap I$, then it is easy to check that the comultiplication of $A(\mathrm{Gl}_n)$ satisfies

$$m_{\mathrm{Gl}_n}(J) \subset A(\mathrm{Gl}_n) \otimes J + J \otimes A(\mathrm{Gl}_n)$$

and this tells us that m_{Gl_n} induces a comultiplication

$$m : A(\mathrm{Gl}_n)/J \to A(\mathrm{Gl}_n)/J \otimes A(\mathrm{Gl}_n)/J$$

which provides a group scheme structure. This means that we have extended the group scheme $G/\mathbb{Q}$ to a group scheme $\mathcal{G}/\mathbb{Z}$., its affine algebra $A(\mathcal{G}) = A(\mathrm{Gl}_n)/J$. This extension depends on the choice of the embedding into $\mathrm{Gl}_n/\mathbb{Q}$ and it is called the *flat extension*. Then the base extension $\mathcal{G} \times_{\mathbb{Z}} \mathbb{Q} = G/\mathbb{Q}$, this base extension is called the *generic fiber* of $\mathcal{G}/\mathbb{Z}$.

There is the special class of semi-simple or reductive group schemes. Roughly speaking an affine group scheme $G/\mathbb{Z}$ is semi-simple (resp. reductive) , if its generic fiber $G \times_{\mathbb{Z}} \mathbb{Q}$ is semi-simple (resp. reductive) and if for all primes p the group scheme $G \times_{\mathbb{Z}} \mathbb{F}_p$ (the reduction $\bmod p$) is a semi-simple ((resp. reductive)) group scheme over $\mathbb{F}_p$.

Of course the simplest example of a semi-simple (resp. reductive) group (scheme) over $\mathbb{Z}$ is the group $\mathrm{Sl}_n/\mathbb{Z}$ (resp. $\mathrm{Gl}_n/\mathbb{Z}$).

In the following section we discuss split forms of the so called classical groups over $\mathrm{Spec}(\mathbb{Z})$

1.1.6 The classical groups and their realisation as split semi-simple group schemes over $\mathrm{Spec}(\mathbb{Z})$

We will not discuss the general notion of a semi-simple group scheme over a base S, instead we will discuss the examples of classical groups and explain the main structure theorems in examples.

The group scheme $\mathrm{Sl}_n/\mathrm{Spec}(\mathbb{Z})$

We consider a free module M of rank n over $\mathrm{Spec}(\mathbb{Z})$. We define the group scheme $\mathrm{Sl}(M)/\mathrm{Spec}(\mathbb{Z})$, and for any commutative $\mathbb{Z}$-algebra R we have

$$\mathrm{Sl}(M)(R) = \mathrm{Sl}(M \otimes_{\mathbb{Z}} R).$$

This is clearly a semi-simple group scheme over $\mathrm{Spec}(\mathbb{Z})$ because

a) The group scheme is smooth over $\mathrm{Spec}(\mathbb{Z})$.
b) For any field k - which is of course a $\mathbb{Z}$-algebra - we have

$$\mathrm{Sl}(M) \times_{\mathrm{Spec}(\mathbb{Z})} \mathrm{Spec}(k) = \mathrm{Sl}(M \otimes_{\mathbb{Z}} k)/\mathrm{Spec}(k),$$

and its center is a finite group scheme.

If we fix a basis $e_1, e_2, \dots, e_n$ then we get a split maximal torus $T/\mathrm{Spec}(\mathbb{Z})$ This is the sub-group scheme which fixes the lines $\mathbb{Z}e_i$. With respect to this basis we have

$$T(R) = \left\{ \begin{pmatrix} t_1 & 0 & \dots & 0 \\ 0 & t_2 & \dots & 0 \\ 0 & 0 & \ddots & 0 \\ 0 & 0 & 0 & t_n \end{pmatrix} \mid t_i \in R^\times, \prod_i t_i = 1 \right\}.$$

With respect to this torus $T/\mathrm{Spec}(\mathbb{Z})$ we define root subgroups. These are smooth subgroup schemes $U \subset G$ which are isomorphic to the additive group scheme $\mathcal{G}_a/\mathrm{Spec}(\mathbb{Z})$ and which are normalized by T. It is clear that these root subgroups are given by

$$\tau_{ij} : \mathcal{G}_a \to \mathrm{Sl}(M)$$

$$\tau_{ij} : x \mapsto \begin{pmatrix} 1 & 0 & \dots & 0 & 0 \\ 0 & 1 & \dots & 0 & 0 \\ 0 & 0 & \ddots & x & 0 \\ 0 & 0 & 0 & \ddots & 0 \\ 0 & 0 & 0 & 0 & 1 \end{pmatrix},$$

where the entry x on the right is placed in the i-th row and j-th column. Let us denote the image by U_{ij}. Then we get the relation

$$t\tau_{ij}(x)t^{-1} = \tau_{ij}((t_i/t_j)x, t \in T(R).$$

(If I write such a relation then I always mean that $t, x, \dots$ are elements in $T(R), \mathcal{G}_a(R) \dots$ for some unspecified $\mathbb{Z}$-algebra R).

The root system

The characters

$$\alpha_{ij} : T \to G_m$$

$$\alpha_{ij} : \begin{pmatrix} t_1 & 0 & \dots & 0 \\ 0 & t_2 & \dots & 0 \\ & & & \\ 0 & 0 & \ddots & 0 \\ 0 & 0 & 0 & t_n \end{pmatrix} \to t_i/t_j$$

form the set Δ of roots in the character module of the torus. We may select a subset of positive roots

$$\Delta^+ = \{\alpha_{ij} \mid i < j\}.$$

Then the torus T and the $U_{\alpha_{ij}}$ with $\alpha_{ij} \subset \Delta^+$ stabilize the flag

$$\mathcal{F} = (0) \subset \mathbb{Z}e_1 \subset \mathbb{Z}e_1 \oplus \mathbb{Z}e_2 \subset \cdots \subset M. \tag{1.18}$$

The stabiliser of the flag is a smooth sub-group scheme $B/\mathrm{Spec}(\mathbb{Z})$. It turns out that B is a maximal solvable sub-group scheme. These maximal sub-group schemes are called Borel subgroups.

It is clear that the morphism

$$T \times \prod_{\alpha_{ij}, i<j} U_{\alpha_{ij}} \to B,$$

which is induced by the multiplication, is an isomorphism of schemes.

The set Δ^+ of positive roots contains the subset $\pi \subset \Delta$ of simple roots $\alpha_{i,i+1}(t) := t_i/t_{i+1}$. For simplicity we write $\alpha_i := \alpha_{i,i+1}$. Every positive root can be written as a sum of simple roots with positive coefficients.

We consider the normalizer $N(T) \subset \mathrm{Sl}_n$, which acts by permutations on the set of submodules $\mathbb{Z}e_i$. The quotient $N(T)/T = W$ is the Weyl group, in this case it is isomorphic to the symmetric group S_n. It is easy to see that we have a positive-definite, symmetric, W-invariant bilinear form on $X^*(T)$ which is given by

$$\langle \alpha_i, \alpha_i \rangle = 2, \langle \alpha_i, \alpha_{i+1} \rangle = -1, \text{ and } \langle \alpha_i, \alpha_j \rangle = 0 \text{ if } |i-j| > 1.$$

All these data about the set of roots and simple roots are encoded in the Dynkin diagram

to be denoted A_{n-1}, where the dots are labelled in the following order

$$A_{n-1} := \alpha_1 \quad - \quad \alpha_2 \quad - \cdots - \quad \alpha_{n-1}.$$

The flag variety

It is not so difficult to see that the flags (see 1.18) form a projective scheme $\mathrm{Gr}/\mathrm{Spec}(\mathbb{Z})$. From this it follows: For any Dedekind ring A and its quotient field K we have

$$\mathrm{Gr}(K) = \mathrm{Gr}(A).$$

If A is even a discrete valuation ring then we can show easily that the group $\mathrm{Gl}_n(A)$ acts transitively on $\mathrm{Gr}(A)$.

The whole point is that results of this type are true for arbitrary split semi-simple groups $\mathcal{G}/\mathrm{Spec}(\mathbb{Z})$. This is not so easy to explain and also much more difficult to prove. But we have the series of so called classical groups and for those these results are again easy to see. The main problem in the general approach is that we have to start from an abstract definition of a semi-simple group and not from a group which is given to us in a rather explicit way like Sl_n or the classical groups.

The group scheme $\mathrm{Sp}_g/\mathrm{Spec}(\mathbb{Z})$

Now we choose again a free $\mathbb{Z}$-module M but we assume that we have a non-degenerate alternating pairing

$$\langle\, ,\, \rangle : M \times M \to \mathbb{Z}$$

where non-degenerate means: If $x \in M$ and $\langle x, M\rangle \subset a\mathbb{Z}$ with some integer $a > 1$, then $x = ay$ with $y \in M$. It is well-known and also very easy to prove that M is of even rank, say $2g$, and that we can find a basis

$$\{e_1, \ldots, e_g, f_g, \ldots, f_1\}$$

such that $\langle e_i, f_i\rangle = -\langle f_i, e_i\rangle = 1$ and all other values of the pairing on basis elements are zero.

The automorphism group scheme $G = \mathrm{Aut}((M, \langle\, .\, \rangle))$ is the symplectic group $\mathrm{Sp}_g/\mathrm{Spec}(\mathbb{Z})$. Again it is easy to find out how a maximal torus must look like. With respect to our basis we can take

$$T = \left\{ \begin{pmatrix} t_1 & 0 & & & \cdots & 0 \\ 0 & \ddots & & & \cdots & 0 \\ 0 & 0 & t_g & & & 0 \\ 0 & 0 & 0 & t_g^{-1} & \cdots & \\ 0 & & & & \ddots & 0 \\ & 0 & & & & t_1^{-1} \end{pmatrix} \right\} \tag{1.19}$$

We can say that the torus is the stabiliser of the ordered collection of rank 2 submodules $\mathbb{Z}e_i, \mathbb{Z}f_i$. We can define a Borel subgroup $B/\mathbb{Z}$ which is the stabiliser of the flag

$$\mathcal{F} = (0) \subset \mathbb{Z}e_1 \subset \cdots \subset \mathbb{Z}e_1 \cdots \oplus \ldots \mathbb{Z}e_g \subset \mathbb{Z}e_1 \cdots \oplus \ldots \mathbb{Z}e_g \oplus \mathbb{Z}f_g \subset M$$

A flag starts with isotropic subspaces until we reach half the rank of the module. But then this lower part of the flag determines the upper half, because it is given by the orthogonal complements of the members in the lower half.

Again we can define the root subgroups (with respect to T)

$$\tau_\alpha : \mathcal{G}_a \xrightarrow{\sim} U_\alpha \subset \mathcal{G}$$

which are normalized by T. As before we have the relation

$$t\tau(x)t^{-1} = \tau(\alpha(t)x),$$

where $\alpha \in \Delta \subset X^*(T)$.

Now it is not quite so easy to write down what these root subgroups are. We give the simple positive roots in the case $g = 2$: We have the maximal torus

$$T = \left\{ \begin{pmatrix} t_1 & 0 & 0 & 0 \\ 0 & t_2 & 0 & 0 \\ 0 & 0 & t_2^{-1} & 0 \\ 0 & 0 & 0 & t_1^{-1} \end{pmatrix} \right\},$$

and we want to find one-parameter subgroups $U_\alpha \subset \mathcal{G}$ which stabilize the flag.

The one parameter subgroups corresponding to the simple roots are

$$\tau_{\alpha_1} : x \mapsto \{e_1 \mapsto e_1, e_2 \mapsto e_2 + xe_1, f_2 \mapsto f_2, f_1 \mapsto f_1 - xf_2\},$$

$$\tau_{\alpha_2} : y \mapsto \{e_1 \mapsto e_1, e_2 \mapsto e_2, f_2 \mapsto f_2 + ye_2, f_1 \mapsto f_1\}$$

where $\alpha_1(t) = t_1/t_2, \alpha_2(t) = t_2^2$. The unipotent radical is

$$\left\{ \begin{pmatrix} 1 & x & v & u \\ 0 & 1 & y & v - xy \\ 0 & 0 & 1 & -x \\ 0 & 0 & 0 & 1 \end{pmatrix} \right\} \tag{1.20}$$

For later reference: The dominant fundamental weights are $\gamma_1 =!/2\alpha_2+\alpha_1, \gamma_2 = \alpha_1 + \alpha_2$

From here it is not difficult to see that for all values of g the simple roots are $\alpha_i(t) = t_i/t_{i+1}$ with $1 \leq i < g$ and $a_g(t) = t_g^2$. Again we define the Weyl group W as above, and we then have a W invariant positive-definite, symmetric bilinear form on $X^*(T)$ and for this form we have

$$\langle \alpha_i, \alpha_i \rangle = 2 \text{ for } i < g \text{ and } \langle \alpha_g, \alpha_g \rangle = 4,$$
$$\langle \alpha_i, \alpha_{i+1} \rangle = -1 \text{ if } i < g - 1 \text{ and } \langle \alpha_{g-1}, \alpha_g \rangle = -2,$$

and all other values of the pairing between simple roots are zero. See, for example, [22]. Again these data are encoded for $g \geq 3$ in the Dynkin diagram

,

to be denoted C_g, where the dots are labelled in the following order

$$C_g := \alpha_1 \quad - \quad \alpha_2 \quad - \cdots - <= \quad \alpha_g$$

As before it is not so difficult to show that the flags form a smooth projective scheme $\mathcal{X}/\mathrm{Spec}(\mathbb{Z})$ (see also [49] V.2.4.3), as well as to show that for any discrete valuation ring A the group $\mathcal{G}(A)$ acts transitively on $\mathcal{X}(A) = \mathcal{X}(K)$. It is also easy to verify the statements in 1.1.

The group scheme $\mathrm{SO}(n,n)/\mathrm{Spec}(\mathbb{Z})$

We can play the same game with quadratic forms. Let M be together with its basis as above, but we now replace g with n. Furthermore we take the quadratic form $F : M \to \mathbb{Z}$ which is defined by

$$F(x_1e_1 \cdots + x_ne_n + y_nf_n + \cdots + y_1f_1) = \sum x_iy_i.$$

We define the group scheme of isomorphisms but in addition we require the determinant to be equal to one. Hence

$$\mathrm{SO}(n,n)/\mathrm{Spec}(\mathbb{Z}) = \mathrm{Aut}(M, F, \det = 1).$$

The maximal torus and the flags look pretty much the same as in the previous case. But the set of roots looks different. For $n = 2$ the torus T and the unipotent radical U are given by

$$T = \left\{ \begin{pmatrix} t_1 & 0 & 0 & 0 \\ 0 & t_2 & 0 & 0 \\ 0 & 0 & t_2^{-1} & 0 \\ 0 & 0 & 0 & t_1^{-1} \end{pmatrix} \right\}, \; U = \left\{ \begin{pmatrix} 1 & x & y & -xy \\ 0 & 1 & 0 & -y \\ 0 & 0 & 1 & -x \\ 0 & 0 & 0 & 1 \end{pmatrix} \right\}.$$

The system of positive roots consists of two roots $\alpha_1(t) = t_1/t_2, \alpha_2(t) = t_1t_2$. This is the Dynkin diagram $A_1 \times A_1$ hence there exists a homomorphism (isogeny) between group schemes over $\mathrm{Spec}(\mathbb{Z})$:

$$\mathrm{Sl}_2 \times \mathrm{Sl}_2 \to \mathrm{SO}(2,2).$$

It is an amusing exercise to write down this isogeny.

Another even more interesting exercise is the computation of the roots for the torus (here $n = 3$)

$$T = \left\{ \begin{pmatrix} t_1 & 0 & 0 & 0 & 0 & 0 \\ 0 & t_2 & 0 & 0 & 0 & 0 \\ 0 & 0 & t_3 & 0 & 0 & 0 \\ 0 & 0 & 0 & t_3^{-1} & 0 & 0 \\ 0 & 0 & 0 & 0 & t_2^{-1} & 0 \\ 0 & 0 & 0 & 0 & 0 & t_1^{-1} \end{pmatrix} \right\}. \tag{1.21}$$

In this case we have the three root subgroups

$$\tau_{\alpha_1} : x \mapsto \begin{pmatrix} 1 & x & 0 & 0 & 0 & 0 \\ 0 & 1 & 0 & 0 & 0 & 0 \\ 0 & 0 & 1 & 0 & 0 & 0 \\ 0 & 0 & 0 & 1 & 0 & 0 \\ 0 & 0 & 0 & 0 & 1 & -x \\ 0 & 0 & 0 & 0 & 0 & 1 \end{pmatrix}, \; \tau_{\alpha_2} : x \mapsto \begin{pmatrix} 1 & 0 & 0 & 0 & 0 & 0 \\ 0 & 1 & x & 0 & 0 & 0 \\ 0 & 0 & 1 & 0 & 0 & 0 \\ 0 & 0 & 0 & 1 & -x & 0 \\ 0 & 0 & 0 & 0 & 1 & 0 \\ 0 & 0 & 0 & 0 & 0 & 1 \end{pmatrix}$$

and

$$\tau_{\alpha_3} : x \mapsto \begin{pmatrix} 1 & 0 & 0 & 0 & 0 & 0 \\ 0 & 1 & 0 & x & 0 & 0 \\ 0 & 0 & 1 & 0 & -x & 0 \\ 0 & 0 & 0 & 1 & 0 & 0 \\ 0 & 0 & 0 & 0 & 1 & 0 \\ 0 & 0 & 0 & 0 & 0 & 1 \end{pmatrix}$$

where

$$\alpha_1(t) = t_1/t_2,\ \alpha_2(t) = t_2/t_3,\ \alpha_3(t) = t_2t_3$$

Here is an excercise for the reader: Use the result of this computation to show that we have an isogeny

$$\mathrm{Sl}_4 \to \mathrm{SO}(3,3).$$

How can we give a linear algebra interpretation of these isogenies?

If we now consider the maximal torus (1.19) and put (1.21) into the middle then we see that the simple roots are

$$\alpha_i(t) = t_i/t_{i+1} \text{ for } i = 1, \dots n-1, \text{ and } \alpha_n(t) = t_{n-1}t_n$$

which gives us for any $n \geq 4$ the Dynkin diagram

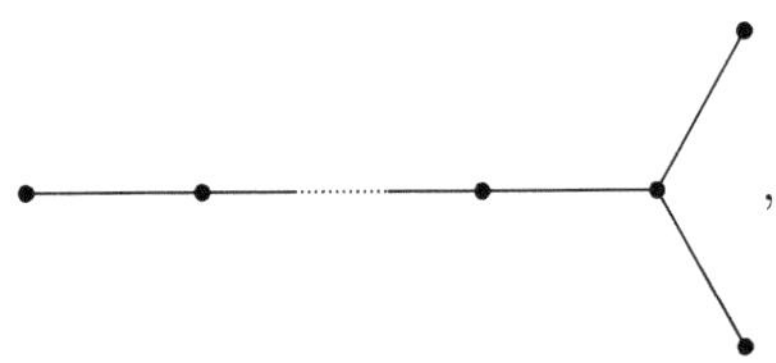

$$\alpha_{n-1}$$

$$D_n := \alpha_1 \quad - \quad \alpha_2 \quad - \cdots - \alpha_{n-2}$$

$$\alpha_n$$

The group scheme $\mathrm{SO}(n+1,n)/\mathrm{Spec}(\mathbb{Z})$

Of course we can also consider quadratic forms in an odd number of variables. We take a free $\mathbb{Z}$-module of rank $2n+1$ with a basis

$$\{e_1, \dots, e_n, h, f_n, \dots, f_1\}.$$

On this module we consider the quadratic form

$$F : M \to \mathbb{Z}$$

$$F\left(\sum x_i e_i + zh + \sum y_i f_i\right) = \sum x_i y_i + z^2.$$

From this quadratic form we get the bilinear form

$$B(u,v) = F(u+v) - F(u) - F(v).$$

We have the relation

$$2F(u) = B(u,u),$$

hence we can reconstruct the quadratic form from the bilinear form if we extend $\mathbb{Z}$ to a larger ring where 2 is invertible.

We consider the automorphism scheme

$$\mathcal{G}/\mathrm{Spec}(\mathbb{Z}) = \mathrm{SO}(n+1,n)/\mathrm{Spec}(\mathbb{Z}) = \mathrm{Aut}(M, F, det = 1)/\mathrm{Spec}(\mathbb{Z}),$$

and I claim that this is indeed a semi-simple group scheme over $\mathrm{Spec}(\mathbb{Z})$. To see this I strongly recommend to discuss the case $n=1$.

We have of course the maximal torus

$$T = \left\{ \begin{pmatrix} t & 0 & 0 \\ 0 & 1 & 0 \\ 0 & 0 & t^{-1} \end{pmatrix} \right\}.$$

It is also the stabilizer of the collection of three subspaces $\mathbb{Z}e, \mathbb{Z}h, \mathbb{Z}f$, where here we used the determinant condition.

Now one has to discuss the root subgroups with respect to this torus. From this we can derive that we have an isogeny

$$\mathrm{Sl}_2 \to \mathrm{SO}(2,1)$$

It is also interesting to look at the case $n=2$. In this case we can compare the root systems of Sp_2 and $\mathrm{SO}(3,2)$ as they are isomorphic. Now it is a general theorem in the theory of split semi-simple group schemes that the root system determines the group scheme up to isogeny. Hence we should be able to construct an isogeny between Sp_2 and $\mathrm{SO}(3,2)$. Who can do it?

For an arbitrary value of n, $n \geq 2$, we get the Dynkin diagram

,

$$B_n := \alpha_1 \quad - \quad \alpha_2 \quad - \cdots - => \quad \alpha_n$$

The element w_0

Finally we will take a short look at the automorphism groups of the Dynkin diagrams.

For the Dynkin diagram of type A_{n-1} the group of automorphisms is trivial if $n=1$ and for $n>1$ the group of automorphism is $\mathbb{Z}/2\mathbb{Z}$, the non-trivial element ϵ exchanges the roots α_i and α_{n-1-i}.

For the diagrams B_n, C_n the automorphism group is trivial,

For the Dynkin diagram D_n and $n>4$ the group of automorphisms is $\mathbb{Z}/2\mathbb{Z}$ and the non-trivial automorphism ϵ fixes the simple roots α_i with $1 \leq i \leq n-2$ and interchanges α_{n-1} and α_n.

For $n=4$ the automorphism group is the symmetric group S_3 and it acts by permutations on the three simple roots $\alpha_1, \alpha_3, \alpha_4$.

The Weyl group $W = N(T)/T$ acts simply transitively on the set of Borel subgroups $B' \supset T$. Hence there is a unique element $w_0 \in W$ which sends our Borel subgroup B into its opposite B^-. This is the Borel subgroup whose simple roots are the roots $-\alpha_i$.

If the automorphism group of the Dynkin diagram is trivial we have $w_0 = -1$, i.e. it acts by multiplication by -1 on $X^*(T)$.

For the diagram A_n and $n>1$ the element w_0 equals $-\epsilon$ and therefore is not equal to -1.

For the diagram D_n the element w_0 equals -1 if n is even and is equal to $-\epsilon$ if n is odd.

The element w_0 will play an important role later in this book.

The fundamental and the dominant weights

The Weyl group $W = N(T)/T$ acts by conjugation on the character module $X^*(T)$, and this module carries a positive-definite symmetric bilinear form $\langle\ ,\ \rangle : X^*(T) \times X^*(T) \to \mathbb{Q}$ which is invariant under this action. The *Weyl group* is generated by the reflections

$$s_{\alpha_i} : \gamma \mapsto \gamma - \frac{2\,\langle\gamma, \alpha_i\rangle}{\langle\alpha_i, \alpha_i\rangle}\alpha_i,$$

and this implies that $\frac{2\langle\gamma,\alpha_i\rangle}{\langle\alpha_i,\alpha_i\rangle} \in \mathbb{Z}$. Of course we have $\alpha_i \in X^*(T)$ for all simple roots, with the sub lattice $X^*(T^{(\mathrm{ad})}) = \bigoplus \mathbb{Z}\alpha_i$ being of finite index in $X^*(T)$. To this sub lattice belongs an isogeny $\psi : T \to T^{(\mathrm{ad})}$. The kernel $\ker(\psi)$ is the centre of our group scheme $G/\mathbb{Z}$ and the quotient $G/\ker(\psi) = G^{(\mathrm{ad})}$ is the *adjoint group*.

In $X^*(T) \otimes \mathbb{Q}$ we have the elements γ_i which are defined by

$$\frac{2\langle\gamma_i, \alpha_i\rangle}{\langle\alpha_j, \alpha_j\rangle} = \delta_{i,j}.$$

These elements are the *fundamental weights.* The lattice $\bigoplus \mathbb{Z}\gamma_i$ contains $X^*(T)$ as a sublattice of finite index. It provides a torus $T^{(\mathrm{sc})}$ and an isogeny $\psi_1 : T^{(\mathrm{sc})} \to T$. This torus is the maximal torus in a semi-simple group scheme $G^{(\mathrm{sc})}/\mathbb{Z}$ which admits an isogeny

$$\psi_1 : G^{(\mathrm{sc})} \to G$$

whose kernel is $\ker(\psi_1) \subset T^{(\mathrm{sc})}$. The group $G^{(\mathrm{sc})}/\mathbb{Z}$ is the *simply connected cover* of $G/\mathrm{Spec}(\mathbb{Z})$.

The *dominant weights* are the weights $\lambda = \sum n_i\gamma_i$ with all $n_i \geq 0$. They parametrise the irreducible representation of G. We come to this in Chapter 6.

The abstract group $G^{(\mathrm{sc})}(k)$

We want to show that for a split semi-simple group the abstract group $G^{(\mathrm{sc})}(k)$ is generated by the groups $U_\alpha(k)$.

For any root α we can consider the two root subgroups U_α and $U_{-\alpha}$. It is easy to see - at least in our examples above - that these root subgroups generate a subgroup $H_\alpha \subset G$, this being the smallest subgroup which contains $U_\alpha, U_{-\alpha}$. This subgroup is either PSl_2 or Sl_2. Then $T^{(\alpha)} = H_\alpha \cap T$ is a maximal torus in H_α.

If our group $G = G^{(\mathrm{sc})}$ is simply connected then $H_\alpha = \mathrm{Sl}_2$ and we define the coroot $\alpha^\vee \in X_*(T^{(\mathrm{sc})})$ by $\alpha^\vee : \mathbb{G}_m \to T^{(\alpha)}$ and $< \alpha^\vee, \alpha >= 2$. We have the relation $\langle\alpha^\vee, \gamma_j\rangle = \delta_{i,j}$ and this implies that the $\alpha_i^\vee$ form a basis of $X_*(T^{(\mathrm{sc})})$. This in turn implies that the map given by multiplication

$$m : \prod_i \alpha_i^\vee(\mathbb{G}_m) \xrightarrow{\sim} T^{(\mathrm{sc})}$$

is an isomorphism.

Now it is easy to see that for any field k the abstract group $\mathrm{Sl}_2(k)$ is generated by the two root subgroups $U_\alpha(k), U_{-\alpha}(k)$. Combined with the observation above this implies that $T^{(\mathrm{sc})}(k)$ is contained in the subgroup which is generated by the subgroups $U_{\alpha_i}(k), U_{-\alpha_i}(k)$. Now we recall the Bruhat decomposition. The unipotent radical U_+ of B is equal to the product $U_+ = \prod_{\alpha \in \Delta^+} U_\alpha$ and the same holds for $U_- = \prod_{\alpha \in \Delta^-} U_\alpha$. The Bruhat decomposition tells us that the multiplication $m : U_- \times T^{(\mathrm{sc})} \times U_+ \to G$ provides an isomorphism of the left hand side with a Zariski-open $\mathcal{V} \subset G$ (this is the Big Cell). This means that we get a bijection

$$U_-(k) \times T^{(\mathrm{sc})}(k) \times U_+(k) \xrightarrow{\sim} \mathcal{V}(k).$$

Our previous arguments imply that $\mathcal{V}(k)$ lies in the subgroup generated by the $U_\alpha(k)$. But then it is clear that $G^{(\mathrm{sc})}(k)$ is generated by the $U_\alpha(k)$.

1.1.7 k-forms of algebraic groups

For the following concepts and results on Galois cohomology we also refer to [90] and [103].

Exercise: 1) *Consider the following two quadratic forms over $\mathbb{Q}$:*

$$f(x,y,z) = x^2 + y^2 - z^2 \;,\; f_1(x,y,z) = x^2 + y^2 - 3z^2.$$

Prove that the first form is isotropic. This means there exists a vector $(a,b,c) \in \mathbb{Q}^3 \setminus \{0\}$ with

$$f(a,b,c) = 0.$$

Show that the second form is anisotropic, i.e. it has no such vector.

2) Prove that the two linear algebraic group $G/\mathbb{Q} = \mathrm{SO}(f)/\mathbb{Q}$ and $G_1/\mathbb{Q} = \mathrm{SO}(f_1)/\mathbb{Q}$ cannot be isomorphic. (Hint: This is not so easy since we did not define when two groups are isomorphic.)

Here is some advice: In general we call an element $e \neq u \in G(\mathbb{Q})$ unipotent if it is unipotent in $\mathrm{Gl}_n(\mathbb{Q})$ where we consider $G/\mathbb{Q} \hookrightarrow \mathrm{Gl}_n/\mathbb{Q}$. It turns out that this notion of unipotence does not depend on the embedding.

Now it is possible to show that our first group $G(\mathbb{Q}) = \mathrm{SO}(f)(\mathbb{Q})$ has unipotent elements, and $G_1(\mathbb{Q})$ does not. Hence these two groups cannot be isomorphic.

3) Prove that the two algebraic groups $G \times_{\mathbb{Q}} \mathbb{R}$ and $G_1 \times_{\mathbb{Q}} \mathbb{R}$ are isomorphic, and therefore the two groups $G(\mathbb{R})$ and $G_1(\mathbb{R})$ are isomorphic.

In this example we see, that we may have two groups G/k and G_1/k which are not isomorphic but which become isomorphic over some extension L/k. Then we say that the groups are *k-forms* of each other. To determine the different k-forms of a given group G/k is sometimes difficult. One has to employ the concepts of Galois cohomology. For a separable normal extension L/k we have the almost tautological description

$$G(k) = \{g \in G(L) | \sigma(g) = g \text{ for all elements } \sigma \in \mathrm{Gal}(L/k)\}.$$

Now we consider the functor $\mathrm{Aut}(G)$: it attaches to any field extension L/k the group of automorphisms $\mathrm{Aut}(G)(L)$ of the algebraic group $G \times_k L$. The Galois group $\mathrm{Gal}(L/k)$ acts upon $\mathrm{Aut}(G)(L)$, and we denote this action by

$h \mapsto \sigma(h) = h^\sigma$. Note that this notation gives us the rule $h^{(\sigma\tau)} = (h^\tau)^\sigma$. A 1-cocycle of $\mathrm{Gal}(L/k)$ with values in $\mathrm{Aut}(G)$ is a map $c : \sigma \mapsto c_\sigma \in \mathrm{Aut}(G)(L)$ which satisfies the cocycle rule

$$c_{\sigma\tau} = c_\sigma c_\tau^\sigma. \tag{1.22}$$

Now we define a new action of $\mathrm{Gal}(L/k)$ on $G(L)$: an element σ acts by

$$g \mapsto c_\sigma(g^\sigma)$$

We define a new algebraic group G_1/k: for any extension E/k we have an action of $\mathrm{Gal}(L/k)$ on $E \otimes_k L$ and we put

$$G_1(E) = \{g \in G(E \otimes_k L) | g = c_\sigma(g^\sigma)\}. \tag{1.23}$$

For the trivial cocycle $\sigma \mapsto 1$ - this gives us back the original group.

It is plausible and in fact not very difficult to show that $E \to G_1(E)$ is in fact represented by an algebraic group G_1/k. This group is clearly a k-form of G/k.

We define an equivalence relation on the set of cocycles: we say that

$$\{\sigma \mapsto c_\sigma\} \sim \{\sigma \mapsto c'_\sigma\}$$

if and only if we can find an $a \in G(L)$ such that

$$c'_\sigma = a^{-1} c_\sigma a^\sigma \text{ for all } \sigma \in \mathrm{Gal}(L/k).$$

We define $H^1(L/k, \mathrm{Aut}(G))$ as the set of 1-cocycles modulo this equivalence relation. If we have a larger normal separable extension $L' \supset L \supset k$ then we get an inclusion $H^1(L/k, \mathrm{Aut}(G)) \hookrightarrow H^1(L'/k, \mathrm{Aut}(G))$. If $\bar{k}_s$ is a separable closure of k we can form the direct limit over all finite extensions $k \subset L \subset \bar{k}_s$ and put

$$H^1(\bar{k}_s/k, \mathrm{Aut}(G)) = \lim_{\to} H^1(L/k,\ \mathrm{Aut}(G))$$

This set is isomorphic to the set of isomorphism classes of k-forms of G/k.

If L/k is a cyclic extension and if $\sigma \in \mathrm{Gal}(L/k)$ is a generator, then a cocycle $c : \mathrm{Gal}(L/k) \to \mathrm{Aut}(G)(L)$ is determined by its value $g = c(\sigma) \in \mathrm{Aut}(G)(L)$. But we have to satisfy the cocycle relation. We have a useful little lemma:

Lemma 1.1.1. *The assigment $\sigma \mapsto c(\sigma) = g$ provides a 1-cocycle if and only if*

$$\mathrm{Norm}(g) = g g^\sigma \cdots g^{\sigma^{n-1}} = \mathrm{Id}$$

and

$$H^1(\mathrm{Gal}(L/k), \mathrm{Aut}(G)(L)) = \{g \in \mathrm{Aut}(G)(L) |\ \mathrm{Norm}(g) = \mathrm{Id}\}/\{hgh^{-\sigma} \sim g\}.$$

Proof. Straightforward calculation. □

For any algebraic group G/k we considered the adjoint group $\mathrm{Ad}(G)$, which is the quotient of G/k by its center. It can be shown, that this is again an algebraic group over k. It is clear that we have an embedding

$$\mathrm{Ad}(G) \to\ \mathrm{Aut}(G)$$

which for any $g \in \mathrm{Ad}(G)(L)$ is given by

$$g \mapsto \{x \mapsto g^{-1}xg\}.$$

A k-form G_1/k of a group G/k is called an *inner k-form* if it is in the image of $H^1(\bar{k}_s/k, \mathrm{Ad}(G)) \to H^1(\bar{k}_s/k, \mathrm{Aut}(G))$.

We may apply the same concepts in a slightly different situation. A k-algebra $\mathcal{D}$ over a field k is called a *central simple algebra*, if it has a unit element $\neq 0$, if it is finite dimensional over k, if its centre is k (embedded via the unit element) and if it has no non-trivial two sided ideals. It is a classical theorem, that such an algebra over a separably closed field k_s is isomorphic to a full matrix algebra $M_n(k_s)$ of (n,n) matrices with entries in k_s. Hence we can say that over an arbitrary field k, any central simple algebra is a k-form of $M_n(k)$ and its dimension is n^2. The famous Skolem-Noether theorem tells us that all automorphisms of $M_n(k)$ are inner automorphisms, i.e. of the form $x \mapsto axa^{-1}$ with $a \in M_n(k)^\times = \mathrm{Gl}_n(k)$. The centre $\mathbb{G}_m$ gives the trivial automorphisms, hence $\mathrm{Aut}(M_n(k)) = \mathrm{PGL}_n(k)$. Hence we see that

The set of isomorphism classes of k-forms $\mathcal{D}$ of M_n/k is isomorphic to $H^1(k_s/k, \mathrm{PGL}_n)$

Then it is also clear that the multiplicative group $\mathcal{D}^\times$ is equal to $G_1(k)$ where G_1/k is a k-form of Gl_n/k. It is also clear that the determinant map $\det : \mathrm{Gl}_n \to \mathbb{G}_m$ provides the reduced norm map

$$\mathrm{Nred} : \mathcal{D}^\times \to \mathbb{G}_m,$$

and the kernel $G_1^{(1)}/k$ of this reduced norm is a k-form of Sl_n/k.

We call a semi-simple group G/k *anisotropic* if it does not contain a non-trivial k-split torus.

We give some indication of how to construct anisotropic groups over $k = \mathbb{Q}$ (or even over any number field). We choose a cyclic extension $L/\mathbb{Q}$ of degree n and we pick a number $a \in \mathbb{Q}^\times$, and let $A(a) \in \mathrm{Gl}_n(\mathbb{Q})$ be the following matrix

$$A(a) = \begin{pmatrix} 0 & 1 & 0 & \dots & 0 \\ 0 & 0 & 1 & \dots & 0 \\ & & \vdots & \ddots & \\ 0 & 0 & \dots & 0 & 1 \\ a & 0 & 0 & 0 & 0 \end{pmatrix}. \tag{1.24}$$

Let $\sigma \in \mathrm{Gal}(L/\mathbb{Q})$ be a generator, then $\sigma^\nu \mapsto A(a)^\nu \mod \mathbb{G}_m$ is a homomorphism from $\mathrm{Gal}(L/\mathbb{Q})$ to $\mathrm{PGL}_n(\mathbb{Q})$ and since $A(a) \in \mathrm{Gl}_n(\mathbb{Q})$ this is also a 1-cocycle $c : \mathrm{Gal}(L/K) \to \mathrm{PGL}_n(\mathbb{Q})$. Therefore it defines a cohomology class $[A(a)] \in H^1(L/\mathbb{Q}, \mathrm{Ad}(\mathrm{Gl}_n))$, and hence an inner $\mathbb{Q}$-form $G/\mathbb{Q}$ of $\mathrm{Gl}_n/\mathbb{Q}$. In Galois cohomology we have the boundary map ([103])

$$\delta : H^1(L/\mathbb{Q}, \mathrm{Ad}(\mathrm{Gl}_n)) \to H^2(L/\mathbb{Q}, \mathbb{G}_m) = \mathbb{Q}^\times/N_{L/\mathbb{Q}}(L^\times)$$

and it is clear that $\delta([A(a)]) = a \in \mathbb{Q}^\times/N_{L/\mathbb{Q}}(L^\times)$. It is well-known that the resulting $\mathbb{Q}$-form $G^{(1)}/\mathbb{Q}$ of $\mathrm{Sl}_n/\mathbb{Q}$ is anisotropic if and only if the class $a \in \mathbb{Q}^\times/N_{L/\mathbb{Q}}(L^\times)$ is an element of order n. We know from algebraic number theory that there are infinitely many primes p which are inert, i.e. p is unramified in L and the prime ideal (p) stays prime in the ring of integers $\mathcal{O}_L$. Then it is easy to see that the order of $p \in \mathbb{Q}^\times/N_{L/\mathbb{Q}}(L^\times)$ is n. Hence we see that the set of isomorphism classes of anisotropic $\mathbb{Q}$-forms of $\mathrm{Gl}_n/\mathbb{Q}$ is abundant.

The central simple algebra $\mathcal{D}$ defined by the class $[A(a)]$ can be described explicitly as a cyclic algebra:

It contains the field $L/\mathbb{Q}$ as a maximal commutative subalgebra and it is generated by L and another element $a_\sigma \in \mathcal{D}$ which satisfies the following relations

$$\forall x \in L \text{ we have } a_\sigma x \alpha_\sigma^{-1} = \sigma(x)\ ;\ a_\sigma^n = a.$$

It is easy to see that for $a = 1$ the central simple algebra is equal to the endomorphism ring of the $\mathbb{Q}$ vector space $L/\mathbb{Q}$ (this is the linear independence of the elements σ^ν in $\mathrm{End}(L/\mathbb{Q})$).

I want to give an example: we consider the algebraic group $\mathrm{Gl}_2/\mathbb{Q}$ where we pick two integers $a, b \neq 0$, which, for simplicity, we assume that b is not a square. Then we have the quadratic extension $L = \mathbb{Q}(\sqrt{b})$, and let σ be its non-trivial automorphism. The element $\begin{pmatrix} 0 & 1 \\ a & 0 \end{pmatrix}$ defines the inner automorphism

$$\mathrm{Ad}(\begin{pmatrix} 0 & 1 \\ a & 0 \end{pmatrix}) : g \mapsto \begin{pmatrix} 0 & 1 \\ a & 0 \end{pmatrix} g \begin{pmatrix} 0 & 1 \\ a & 0 \end{pmatrix}^{-1}$$

of the group Gl_2. Then $\sigma \mapsto \mathrm{Ad}\left(\begin{pmatrix} 0 & 1 \\ a & 0 \end{pmatrix}\right)$ and $\mathrm{Id}_{\mathrm{Gal}(L/k)} \mapsto \mathrm{Id}_{\mathrm{Aut}(\mathrm{Gl}_2)(L)}$ is a 1-cocycle, and we get an inner $\mathbb{Q}$-form of our group.

Now we can easily see that the group of rational points of our above group $G(a,b)(\mathbb{Q})$ is the multiplicative group of a central simple algebra $D(a,b)/\mathbb{Q}$. To get this algebra we consider the algebra $M_2(L)$ of (2,2)-matrices over L. We define

$$D(a,b) = \left\{ x \in M_2(L) | x = \mathrm{Ad}\left(\begin{pmatrix} 0 & 1 \\ a & 0 \end{pmatrix}\right)(x^\sigma) \right\}. \tag{1.25}$$

We have an embedding of the field L into this algebra, which is given by $u \mapsto \begin{pmatrix} u & 0 \\ 0 & u^\sigma \end{pmatrix}$. Let u_b be the image of $\sqrt{b}$ under this map. We also have the element $u_a = \begin{pmatrix} 0 & a \\ 1 & 0 \end{pmatrix}$ in this algebra.

Now I leave it as an exercise to the reader that as a $\mathbb{Q}$ vector space

$$D(a,b) = \mathbb{Q} \oplus \mathbb{Q}u_b \oplus \mathbb{Q}u_a \oplus \mathbb{Q}u_a u_b$$

We have the relation $u_a^2 = a, u_b^2 = b, u_a u_b = -u_b u_a$.

Of course we should ask ourselves: When is $D(a,b)$ split, i.e. isomorphic to $M_2(\mathbb{Q})$? To answer this question we consider the norm homomorphism Nred, which is defined by the assignment

$$x + yu_b + zu_a + wu_a u_b \xrightarrow{\mathrm{Nred}}$$
$$(x + yu_b + zu_a + wu_a u_b)(x - yu_b - zu_a - wu_a u_b) = x^2 - y^2 b - z^2 a + w^2 ab.$$

It is easy to see that $D(a,b)$ splits if and only if we can find a non-zero element whose norm is zero. If we do this over $\mathbb{R}$ as the base field and if we take $a=-1, b=-1$ then we get the Hamiltonian quaternions.

We may also look at the p-adic completions $\mathbb{Q}_p$ of our field. Then it is not difficult to see that $D(a,b)$ splits over $\mathbb{Q}_p$ if $p \neq 2$ and $p \nmid ab$. Hence it is clear that there is only a finite number of primes p for which $D(a,b)$ does not split. If we consider $\mathbb{R}$ as the completion at the infinite place, and the $\mathbb{Q}_p$ as the completions at the finite places, then we have that:

The algebra $D(a,b)$ splits if and only if it splits at all places. The number of places where it does not split is always even.

The first assertion is the so called Hasse-Minkowski principle, the second assertion is essentially equivalent to the quadratic reciprocity law.

1.1.8 Quasisplit $\mathbb{Q}$-forms

We recall that a semi-simple group $G/\mathbb{Q}$ is quasisplit if it contains a Borel subgroup $B/\mathbb{Q}$. This Borel subgroup contains its unipotent radical $U/\mathbb{Q}$ and a maximal torus $T/\mathbb{Q}$. Two such maximal tori $T/\mathbb{Q}, T_1/\mathbb{Q}$ are conjugate by an element $u \in U(\mathbb{Q})$. Let $G_0/\mathbb{Q}$ be a split group which is a $\mathbb{Q}$-form of $G/\mathbb{Q}$. We pick a maximal split torus $T_0/\mathbb{Q}$ and a Borel $B_0/\mathbb{Q} \supset T_0/\mathbb{Q}$. Then we see that the triple $(G,B,T)/\mathbb{Q}$ is a $\mathbb{Q}$-form of $(G_0,B_0,T_0)/\mathbb{Q}$. Hence it can by constructed from a 1-cocycle representing a cohomology class $\xi \in H^1(\bar{\mathbb{Q}}/\mathbb{Q}, \mathrm{Aut}((G_0,B_0,T_0)))$, where of course $\mathrm{Aut}((G_0,B_0,T_0))$ is the subgroup of $\mathrm{Aut}(G_0)$ which fixes (T_0,B_0). Obviously we have an exact sequence

$$1 \to T_0^{(\mathrm{ad})} \to \mathrm{Aut}((G_0,B_0,T_0)) \to \mathrm{Autext}((G_0,B_0,T_0)) \to 1, \tag{1.26}$$

here $\mathrm{Autext}((G_0,B_0,T_0))$ is the very "small" group of automorphisms of the Dynkin diagram Φ. This is also the subgroup of $\mathrm{Aut}(X^*(T_0))$ which leaves the set Δ^+ of positive roots invariant. We could say $\mathrm{Autext}((G_0,B_0,T_0)) = \mathrm{Aut}(X^*(T_0),\Delta^+)$.

It is well-known - and easy to see in the examples of classical groups - that this sequence has a section $s_0 : \mathrm{Autext}((G_0,B_0,T_0)) \to \mathrm{Aut}((G_0,B_0,T_0))$ and this gives us a map in Galois cohomology

$$\begin{aligned} s_0^\bullet &: H^1(\bar{\mathbb{Q}}/\mathbb{Q}, \mathrm{Autext}((G_0,B_0,T_0))) \\ &= \mathrm{Hom}(\mathrm{Gal}(\bar{\mathbb{Q}}/\mathbb{Q}), \mathrm{Autext}((G_0,B_0,T_0)) \to H^1(\bar{\mathbb{Q}}/\mathbb{Q}, \mathrm{Autext}(\mathrm{G}_0). \end{aligned} \tag{1.27}$$

Hence we see that the isomorphism classes of quasisplit $\mathbb{Q}$-forms of $G_0/\mathbb{Q}$ are given by homomorphisms $\psi : \mathrm{Gal}(\bar{\mathbb{Q}}/\mathbb{Q}) \to \mathrm{Autext}((G_0))$. The maximal torus $T/\mathbb{Q} \subset B/\mathbb{Q}$ is not split (unless $G/\mathbb{Q}$ is split). Hence there is a finite normal extension $F_0/\mathbb{Q}$ such that $T \times_{\mathbb{Q}} F_0$ splits, and we assume that $F_0/\mathbb{Q}$ is minimal, i.e. $\mathrm{Gal}(F_0\mathbb{Q}) \subset \mathrm{Aut}(X^*(T \times_{\mathbb{Q}} F_0),\Delta^+)$. We see that a quasisplit form of $G_0/\mathbb{Q}$ is given by a finite normal extension $F_0/\mathbb{Q}$ and an injection $\psi : \mathrm{Gal}(F_0/\mathbb{Q}) \hookrightarrow \mathrm{Aut}(X^*(T_0),\Delta^+)$.

In the special case $G_0/\mathbb{Q} = \mathrm{Sl}_n/\mathbb{Q}$ with $T_0/\mathbb{Q}$, $B_0/\mathbb{Q}$ being the standard diagonal torus and the standard Borel subgroup of upper triangular matrices, this looks as follows: we have the element

$$w_0 = \begin{pmatrix} 0 & 0 & 0 & \dots & 1 \\ 0 & 0 & \dots & 1 & 0 \\ & & \vdots & \ddots & \\ 0 & 1 & \dots & 0 & 0 \\ \pm 1 & 0 & 0 & 0 & 0 \end{pmatrix} \in \mathrm{Sl}_n(\mathbb{Q}), \tag{1.28}$$

this element w_0 conjugates B_0 into its opposite B_0^- - the group of lower triangular matrices. The standard Cartan involution $\Theta : g \to {}^t g^{-1}$ does the same and therefore the composition $\mathrm{Ad}(w_0) \circ \Theta$ is an automorphism of $G_0/\mathbb{Q}$ which fixes (B_0, T_0). It is an outer automorphism if $n \geq 3$ and gives us the non-trivial element of $\mathrm{Autext}(G_0)$. Hence we get a 1-cocycle, if we choose a quadratic extension $L/\mathbb{Q}$ and send the non-trivial element in $\mathrm{Gal}(L/\mathbb{Q})$ to $\mathrm{Ad}(w_0) \circ \Theta$.

We leave it as an exercise to the reader to show that the $\mathbb{Q}$-form obtained from this cocycle (cohomology class) is isomorphic to the above group $\mathrm{SU}(h)/\mathbb{Q}$. (see 1.34).

An important class of quasisplit groups is given by the groups $G/\mathbb{Q} = R_{F_0/\mathbb{Q}}(G_0)$ where $F_0/\mathbb{Q}$ is a finite extension of $\mathbb{Q}$ and G_0/F_0 is a split group. If $B_0/F_0 \subset G_0$ is a Borel subgroup then $B = R_{F_0/Q}(B_0)$ is a Borel subgroup in $G/\mathbb{Q}$. Let $F \supset F_0$ be a normal closure of F_0, then

$$G \times_{\mathbb{Q}} F = \prod_{\iota : F_0 \to F} G_0 \times_{F_0, \iota} F,$$

where ι runs over the set Σ of maps from F_0 to F. The Galois group acts on the product via the action on Σ and on F on both sides.

1.1.9 Semi-simple groups over $\mathbb{R}$ and symmetric spaces

We need some information concerning the structure of the group $G_\infty = G(\mathbb{R})$ for semi-simple groups over $G/\mathbb{R}$. We will provide this information simply by discussing a series of examples.

Of course the group $G(\mathbb{R})$ is a topological group, actually it is even a Lie group. This means it has a natural structure of a $\mathcal{C}^\infty$-manifold with respect to this structure. Instead of $G(\mathbb{R})$ we will very often write G_∞. Let G_∞^0 be the connected component of the identity in G_∞. It is an open subgroup of finite index. We will discuss the following:

Theorem of E. Cartan: *The group G_∞^0 always contains a maximal compact subgroup $K_\infty \subset G_\infty^0$ and all maximal compact subgroups are conjugate under G_∞^0. The quotient space $X = G_\infty^0/K_\infty$ is again a $\mathcal{C}^\infty$-manifold. It is diffeomorphic to an $\mathbb{R}^N$ and carries a Riemannian metric which is invariant under the operation of G_∞^0 from the left. It has sectional curvature ≤ 0 and therefore any two points can be joined by a unique geodesic. The maximal compact subgroup $K \subset G_\infty^0$ is connected and equal to its own normaliser. Therefore the space X can be viewed as the space of maximal compact subgroups in G_∞^0.*

See for instance [63].

For any maximal compact subgroup $K_x \subset G_\infty$ there exists a unique automorphism Θ_x with $\Theta_x^2 = Id$ such that $K_x = \{g \in G_\infty^0 \mid \Theta_x(g) = g\}$, which is the Cartan involution corresponding to K_x. The Cartan involutions are in one-to-one correspondence with the maximal compact subgroups.

A Cartan involution Θ_x induces an involution, also called Θ_x, on the Lie algebra $\mathfrak{g}_\mathbb{R}$ of G_∞, and we get a decomposition into $\pm$ eigenspaces

$$\mathfrak{g} = \mathfrak{k}_x \oplus \mathfrak{p}_x;\ \mathfrak{k}_x = \{U \in \mathfrak{g} \mid \Theta_x(U) = U\}\ ;\ \mathfrak{p}_x = \{V \in \mathfrak{g} \mid \Theta_x(V) = -V\}$$

where of course $\mathfrak{k}_x$ is the Lie algebra of K_x. The differential of the action of G_∞ on $G(\mathbb{R})/K_x$ provides an isomorphism $D_x : \mathfrak{p}_x \xrightarrow{\sim} T_x^X$ (the tangent space at x). For $V_1, V_2 \in \mathfrak{p}_x$ we have $[V_1, V_2] \in \mathfrak{k}_x$, with the map $R : \mathfrak{p}_x \times \mathfrak{p}_x \to \mathfrak{k}_x$ being the curvature tensor. The $\mathbb{R}$-vector space $\mathfrak{g}_c := \mathfrak{k}_x + \sqrt{-1}\mathfrak{p}_x \subset \mathfrak{g} \otimes_\mathbb{R} \mathbb{C}$ is a Lie algebra, for $U_1 + \sqrt{-1}V_1, U_2 + \sqrt{-1}V_2 \in \mathfrak{g}_c$ we get for the Lie bracket

$$[U_1 + \sqrt{-1}V_1, U_2 + \sqrt{-1}V_2] = [U_1, U_2] - [V_1, V_2] + \sqrt{-1}([U_1, V_2] + [U_2, V_1]) \in \mathfrak{g}_c.$$

To this Lie algebra $\mathfrak{g}_c$ corresponds an algebraic group $G_c/\mathbb{R}$ which is an $\mathbb{R}$-form of $G/\mathbb{R}$, with the group $G_c(\mathbb{R})$ being compact. The group $G_c/\mathbb{R}$ is called the compact dual of $G/\mathbb{R}$. On $G_c/\mathbb{R}$ we have only one Cartan involution $\Theta = \mathrm{Id}$.

This theorem of Cartan is fundamental. We discuss a series of examples:

The groups $\mathrm{Sl}_n(\mathbb{R})$ and $\mathrm{Gl}_n(\mathbb{R})$

The group $\mathrm{Sl}_n(\mathbb{R})$ is connected. If $K \subset \mathrm{Sl}_d(\mathbb{R})$ is a closed compact subgroup, I claim that we can find a positive definite quadratic form $f : \mathbb{R}^n \to \mathbb{R}$ such that $K \subset SO(f, \mathbb{R})$. Since the group $SO(f, \mathbb{R})$ itself is compact, it is maximally compact. Two such forms f_1, f_2 define the same maximal compact subgroup K_∞ if there is a $\lambda > 0$ in $\mathbb{R}$ such that $\lambda f_1 = f_2$. We say that f_1 and f_2 are *conformally equivalent.*

The existence of such a form f follows easily from the existence of the Haar integral. The Haar integral provides a positive invariant integral

$$\int_K : C_c(K) \longrightarrow \mathbb{R}; \varphi \mapsto \int_K \varphi(k)dk$$

where $\int \varphi > 0$ if $\varphi \geq 0$ and not identically zero (positivity), and which is bi-invariant under the action of K, i.e. for all $k_0 \in K$ we have $\int \varphi(kk_0)dk = \int \varphi(k_0k)dk = \int \varphi(k)dk$. To get our form f we start from any positive-definite form f_0 on $\mathbb{R}^n$ and put

$$f(\underline{x}) = \int_K f_0(k\underline{x})dk.$$

A positive-definite quadratic form on $\mathbb{R}^n$ is the same as a symmetric positive-definite bilinear form. Hence the space of positive-definite forms is the same as the space of positive-definite symmetric matrices

$$\tilde{X} = \left\{A = (a_{ij}) \in M_n(\mathbb{R}) \mid A = A^t, A > 0\right\}.$$

Hence we can say that the space of maximal compact subgroups in $\mathrm{Sl}_n(\mathbb{R})$ is given by

$$X = \tilde{X}/\mathbb{R}^*_{>0}.$$

It is easy to see that a maximal compact subgroup $K_\infty \subset \mathrm{Sl}_n(\mathbb{R})$ is equal to its own normalizer (why?). If we view X as the space of positive-definite symmetric matrices with determinant equal to one, then the action of $\mathrm{Sl}_d(\mathbb{R})$ on $X = \mathrm{Sl}_n(\mathbb{R})/K_\infty$ is given by

$$(g, A) \longrightarrow gA\,{}^t g,$$

and if we view it as the space of maximal compact subgroups, then the action is via conjugation.

There is still another interpretation of the points $x \in X$. In our above interpretation a point was a symmetric, positive-definite bilinear form $\langle\ ,\ \rangle_x$ on $\mathbb{R}^n$, up to a homothety. From this we get a transposition $g \mapsto {}^{t_x}g$, which is defined by the rule $\langle gv, u\rangle_x = \langle v, {}^{t_x}gu\rangle_x$ and from this we get the involution

$$\Theta_x : g \mapsto ({}^{t_x}g)^{-1}. \tag{1.29}$$

Then the corresponding maximal compact subgroup is

$$K_x = \{g \in \mathrm{Sl}_n(\mathbb{R}) \mid \Theta_x(g) = g\}. \tag{1.30}$$

This involution Θ_x is a Cartan involution, it induces an involution, also called Θ_x, on the Lie algebra, and it has the property that (see 1.16)

$$(u, v) \mapsto B(u, \Theta_x(v)) =: B_{\Theta_x}(u, v) \tag{1.31}$$

is negative definite. This bilinear form is K_x-invariant. All these Cartan involutions are conjugate to each other.

If we work with $\mathrm{Gl}_n(\mathbb{R})$ instead then we have some freedom to define the symmetric space. In this case we have the non-trivial center $\mathbb{R}^\times$, and it is sometimes useful to define

$$X = \mathrm{Gl}_n(\mathbb{R})/\mathrm{SO}(\mathbb{R}) \cdot \mathbb{R}^\times_{>0}, \tag{1.32}$$

then our symmetric space has two components, and a point is pair (Θ_x, ϵ) where ϵ is an orientation. If we do not divide by $\mathbb{R}^\times_{>0}$, we multiply the Riemannian manifold X by a flat space and we get two copies of the above space $\tilde{X}$.

A Cartan involution on $\mathrm{Gl}_n(\mathbb{R})$ is an involution which induces a Cartan involution on $\mathrm{Sl}_n(\mathbb{R})$ and which is trivial on the center.

Proposition 1.1.1. *The Cartan involutions on* $\mathrm{Gl}_n(\mathbb{R})$ *are in one to one correspondence to the euclidian metrics on* $\mathbb{R}^n$*, up to conformal equivalence.*

Finally we recall the Iwasawa decomposition. Inside $\mathrm{Gl}_n(\mathbb{R})$ we have the standard Borel subgroup $B(\mathbb{R})$ of upper triangular matrices, and it is well-known that

$$\mathrm{Gl}_n(\mathbb{R}) = B(\mathbb{R}) \cdot \mathrm{SO}(\mathbb{R}) \cdot \mathbb{R}^\times_{>0}, \tag{1.33}$$

and hence we see that $B(\mathbb{R})$ acts transitively on X.

The compact dual of $\mathrm{Sl}_n(\mathbb{R})$

If $G/\mathbb{R}$ is a semi-simple group, then $G_c/\mathbb{R}$ is an $\mathbb{R}$-form of $G/\mathbb{R}$. Therefore we find a cohomology class $\xi_c \in H^1(\mathbb{C}/\mathbb{R}, \mathrm{Aut}(G))$ corresponding to G_c. It is clear from the theorem of Cartan how we get a cocycle representing this class: We choose a Cartan involution $\Theta \in \mathrm{Aut}(G)$. The Galois group $\mathrm{Gal}(\mathbb{C}/\mathbb{R})$ is cyclic of order 2. Let $\mathbf{c}$ be the generator (the complex conjugation). Then $\mathbf{c} \mapsto \mathbf{c} \circ \Theta$ yields a 1-cocycle in $C^1(\mathrm{Gal}(\mathbb{C}/\mathbb{R}), \mathrm{Aut}(G)(\mathbb{C}))$ (Lemma 1.1.1), and this 1-cocycle represents the class ξ_c.

This means for the group $\mathrm{Sl}_n/\mathbb{R}$ that

$$G_c(\mathbb{R}) = \{g \in \mathrm{Sl}_n(\mathbb{C}) \mid \mathbf{c}({}^t g^{-1}) = g\}$$

and if we go back to the usual notation and write $\mathbf{c}(g) = \bar{g}$, then we get

$$G_c(\mathbb{R}) = \{g \in \mathrm{Sl}_n(\mathbb{C}) \mid {}^t\bar{g}g = \mathrm{Id}\} = \mathrm{SU}(n). \tag{1.34}$$

Here of course $\mathrm{SU}(n) = \mathrm{SU}(h_c)$ where $h_c(z_1, z_2, \dots, z_n) = \sum_{i=1}^n z_i\bar{z}_i$ is the standard positive-definite hermitian form on $\mathbb{C}^n$.

We know that for $G/\mathbb{R} = \mathrm{Sl}_n/\mathbb{R}$ and $n > 2$ the Cartan involution Θ is the generator of $\mathrm{Aut}(G)/\mathrm{Ad}(G)$ and hence it is clear that ξ_c is not in the image of $H^1(\mathbb{C}/\mathbb{R}, \mathrm{Ad}(G)) \to H^1(\mathbb{C}/(\mathbb{R}, \mathrm{Aut}(G))$. This means that in this case $G_c/\mathbb{R} = \mathrm{SU}(n)/\mathbb{R}$ is not an inner $\mathbb{R}$-form of $\mathrm{Sl}_n/\mathbb{R}$, and in turn this also means that $\mathrm{Sl}_n/\mathbb{R}$ is not an inner $\mathbb{R}$-form of $\mathrm{SU}(n)/\mathbb{R}$.

In this context the following general proposition is of importance:

Proposition 1.1.2. *A semi-simple group scheme $G/\mathbb{R}$ is an inner $\mathbb{R}$-form of its compact dual $G_c/\mathbb{R}$ if and only if one of the following holds:*
a) The Cartan involution Θ of $G/\mathbb{R}$ is an inner automorphism of $G/\mathbb{R}$.
b) The group $G/\mathbb{R}$ has a compact maximal torus $T_c/\mathbb{R} \subset G/\mathbb{R}$.

I propose to call these groups cmT groups, a reductive group is cmT if its semi-simple component is cmT.

The group $\mathrm{Sl}_d(\mathbb{C})$

We now consider the group $G/\mathbb{R}$ whose group of real points is $G(\mathbb{R}) = \mathrm{Sl}_d(\mathbb{C})$ (see 1.1 example 4)).

A completely analogous argument as before shows that the maximal compact subgroups are in one-to-one correspondence to the positive-definite hermitian forms on $\mathbb{C}^n$ (up to multiplication by a scalar). Hence we can identify the space of maximal compact subgroups with the space of positive-definite hermitian matrices

$$X = \left\{A \mid A = {}^t\overline{A}\,,\ A > 0\,,\ \det A = 1\right\}.$$

The action of $\mathrm{Sl}_d(\mathbb{C})$ by conjugation on the maximal compact subgroups becomes

$$A \longrightarrow gA\,{}^t\overline{g}$$

on the space of matrices.

The orthogonal group:

The next example we want to discuss is the orthogonal group of a non-degenerate quadratic form

$$f(x_1,\dots,x_n)=x_1^2+\dots+x_m^2-x_{m+1}^2-\dots-x_n^2,$$

and since at this point we consider only groups over the real numbers, we may assume that our form is of this type. In this case one has the usual notation

$$\mathrm{SO}(f,\mathbb{R})=\mathrm{SO}(m,n-m).$$

Of course we can use the same argument as before and see that for any maximal compact subgroup $K\subset \mathrm{SO}(f,\mathbb{R})$, we may find a positive-definite form ψ

$$\psi:\mathbb{R}^n\longrightarrow\mathbb{R}$$

such that $K=\mathrm{SO}(f,\mathbb{R})\cap\mathrm{SO}(\psi,\mathbb{R})$. But now we cannot take all forms ψ, i.e. only special forms ψ provide a maximal compact subgroup.

We leave it to the reader to verify that any compact subgroup K fixes an orthogonal decomposition $\mathbb{R}^n=V_+\oplus V_-$ where our original form f is positive-definite on V_+ and negative definite on V_-. Then we can take a ψ which is equal to f on V_+ and equal to $-f$ on V_-.

Exercise 3 *a) Let $V/\mathbb{R}$ be a finite dimensional vector space and let f be a symmetric non-degenerate form on V. Let $K\subset\mathrm{SO}(f)$ be a compact subgroup. If f is not definite then the action of K on V is not irreducible.*

b) We can find a K-invariant decomposition $V=V_-\oplus V_+$ such that f is negative-definite on V_- and positive-definite on V_+

In this case the structure of the quotient space $G(\mathbb{R})/K$ is not so easy to understand. We consider the special case of the form

$$x_1^2+\dots+x_n^2-x_{n+1}^2=f(x_1,\dots,x_{n+1}).$$

We consider in $\mathbb{R}^{n+1}$ the open subset

$$X_-=\{v=(x_1,\dots,x_{n+1})\mid f(v)<0\}$$

It is clear that this set has two connected components, one of them is

$$X_-^+=\{v\in X_-\mid x_{n+1}>0\}.$$

Since it is known that $\mathrm{SO}(n,1)$ acts transitively on the vectors of a given length, we find that $\mathrm{SO}(n,1)$ cannot be connected. Let $G_\infty^0\subset\mathrm{SO}(n,1)$ be the subgroup leaving X_-^+ invariant.

Now it is not too difficult to show that for any maximal compact subgroup $K_\infty\subset G_\infty^0$ we can find a ray $\mathbb{R}_{>0}^*\cdot v\subset X_-^{(+)}$ which is fixed by K_∞ (start from $v_0\in X_-^{(+)}$ and show that $R_{>0}^*K_\infty v_0$ is a closed convex cone in $X_-^{(+)}$. It is K_∞-invariant and has a ray which has a "centre of gravity" and this is fixed under K_∞).

For a vector $v=(x_1,\dots,x_{n+1})\in X_-^{(+)}$ we may normalise the coordinate x_{n+1} to be equal to one; then the rays $\mathbb{R}_{>0}^+v$ are in one-to-one correspondence with the points of the ball

$$\overset{\circ}{D}_n=\{(x_1,\dots,x_n)\mid x_1^2+\dots+x_n^2<1\}\subset X_-^{(+)}. \tag{1.35}$$

This tells us that we can identify the set of maximal compact subgroups $K_\infty \subset G^0_\infty$ with the points of this ball. The first conclusion is that $G^0_\infty/K_\infty \simeq D^n$ is topologically a cell (diffeomorphic to $\mathbb{R}^n$). Secondly we see that for a $v \in X^+_-$ we have an orthogonal decompositon with respect to f

$$\mathbb{R}^{n+1} = \langle v\rangle + \langle v\rangle^\perp,$$

and the corresponding maximal compact subgroup is the orthogonal group on $\langle v\rangle^\perp$. The corresponding Cartan involution is given by $\Theta_v : v \mapsto -v$ and Θ_v is the identity on $< v >^\perp$.

The space X^+_- is often called the n-dimensional *hyperbolic space* $\mathbb{H}_n$.

1.1.10 Low dimensional cases

1) We consider the (semi-simple) group $\mathrm{Sl}_2/\mathbb{R}$. It acts on the upper half plane

$$\mathbb{H} = \{z \mid z \in \mathbb{C}, \Im(z) > 0\}$$

by

$$(g, z) \longrightarrow \frac{az+b}{cz+b}, \qquad g = \begin{pmatrix} a & b \\ c & d \end{pmatrix} \in \mathrm{Sl}_2(\mathbb{R}).$$

It is clear that the stabiliser of the point $i \in \mathbb{H}$ is the standard maximal compact subgroup

$$K_\infty = \mathrm{SO}(2) = \left\{ \begin{pmatrix} \cos\varphi & \sin\varphi \\ -\sin\varphi & \cos\varphi \end{pmatrix} \right\}.$$

Since the action is transitive we have $\mathbb{H} = \mathrm{Sl}_2(\mathbb{R})/K_\infty$. But this quotient has also been realized as the space of symmetric positive-definite 2×2 matrices with determinant equal to one

$$Z = \left\{ \begin{pmatrix} y_1 & x_1 \\ x_1 & y_2 \end{pmatrix} \mid y_1 y_2 - x_1^2 = 1, y_1 > 0 \right\}.$$

It is clear how to find an isomorphism between these two explicit realizations. The map

$$\begin{pmatrix} y_1 & x_1 \\ x_1 & y_2 \end{pmatrix} \longrightarrow \frac{i + x_1}{y_2}$$

is compatible with the action of $\mathrm{Sl}_2(\mathbb{R})$ on both sides and sends the identity $\begin{pmatrix} 1 & 0 \\ 0 & 1 \end{pmatrix}$ to the point i.

If we start from a point $z \in \mathbb{H}$ the corresponding metric is as follows: We identify the lattices $\langle 1, z\rangle = \{a + bz \mid a, b \in \mathbb{Z}\} =: \Omega$ to the lattice $\mathbb{Z}^2 \subset \mathbb{R}^2$ by sending $1 \to \binom{1}{0}$ and $z \to \binom{0}{1}$. The standard euclidian metric on $\mathbb{C} = \mathbb{R}^2$ induces a metric on $\Omega \subset \mathbb{C}$, and this metric is transported to $\mathbb{R}^2$ by the identification $\Omega \otimes \mathbb{R} \to \mathbb{R}^2$. Hence the symmetric matrix will be $\begin{pmatrix} 1 & x \\ x & z\bar{z} \end{pmatrix}$.

We may also start from the (reductive) group $\mathrm{Gl}_2(\mathbb{R})$, which has centre $C(\mathbb{R}) = \left\{ \begin{pmatrix} t & 0 \\ 0 & t \end{pmatrix} \right\}$. Let $C(\mathbb{R})^{(0)}$ be the connected component of the identity of $C(\mathbb{R})$. In this case we define $K_\infty = \mathrm{SO}(2) \times C(\mathbb{R})^{(0)}$. Then we can identify the quotient $\mathrm{Gl}_2(\mathbb{R})/K_\infty$ with $\mathbb{H} \cup \mathbb{H}_-$ where $\mathbb{H}_-$ is the lower half plane.

2) The two groups $\mathrm{Sl}_2(\mathbb{R})$ and $\mathrm{PSl}_2(\mathbb{R}) = \mathrm{Sl}_2(\mathbb{R})/\{\pm \mathrm{Id}\}$ give rise to the same symmetric space. The group $\mathrm{PSl}_2(\mathbb{R})$ acts on the space $M_2(\mathbb{R})$ of 2×2-matrices by conjugation (the group $\mathrm{Gl}_2(\mathbb{R})$ acts by conjugation and the centre acts trivially) and leaves invariant the space

$$\{A \in M_2(\mathbb{R}) \mid \mathrm{trace}(A) = 0\} =: M_2^0(\mathbb{R}).$$

On this three-dimensional space we have a quadratic form

$$\begin{aligned} B &: M_2^0(\mathbb{R}) \longrightarrow \mathbb{R} \\ B &: A \mapsto \frac{1}{2} \text{ trace } (A^2) \end{aligned}$$

and with respect to the basis

$$H = \begin{pmatrix} 1 & 0 \\ 0 & -1 \end{pmatrix}, V = \begin{pmatrix} 0 & 1 \\ 1 & 0 \end{pmatrix}, Y = \begin{pmatrix} 0 & 1 \\ -1 & 0 \end{pmatrix}, \tag{1.36}$$

this form is $x_1^2 + x_2^2 - x_3^2$.

Hence we see that $\mathrm{SO}(M_2^0(\mathbb{R}), B) = \mathrm{SO}(2,1)$, and hence we have an isomorphism between $\mathrm{PSl}_2(\mathbb{R})$ and the connected component of the identity $G_\infty^0 \subset \mathrm{SO}(2,1)$. We see that our symmetric space $\mathbb{H} = \mathrm{Sl}_2(\mathbb{R})/K_\infty = \mathrm{PSl}_2(\mathbb{R})/\overline{K}_\infty$ can also be realised as a disc

$$D = \{(x_1, x_2) \mid x_1^2 + x_2^2 < 1\}$$

where we normalised $x_3 = 1$ on $X_-^{(+)}$ as in (1.35) .

The group $\mathrm{Sl}_2(\mathbb{C})$.

Recall that in this case the symmetric space is given by the positive-definite

$$X = \left\{ A = \begin{pmatrix} y_1 & z \\ \overline{z} & y_2 \end{pmatrix} \mid \det(A) = 1, y_1 > 0, y_2 \in \mathbb{R} \right\}.$$

In this case we also have a realization of the symmetric space as an upper half space. We send

$$\begin{pmatrix} y_1 & w \\ \overline{w} & y_2 \end{pmatrix} \longmapsto \left(\frac{w}{y_2}, \frac{1}{y_2} \right) = (z, \zeta) \in \mathbb{C} \times \mathbb{R}_{>0}$$

The inverse of this isomorphism is given by

$$(z, \zeta) \mapsto \begin{pmatrix} \zeta + z\overline{z}/\zeta & z/\zeta \\ \overline{z}/\zeta & 1/\zeta \end{pmatrix}$$

As explained earlier, the action of $\mathrm{Gl}_2(\mathbb{C})$ on the maximal compact subgroups given by conjugation yields the action

$$G(\mathbb{R}) \times X \longrightarrow X,$$

$$(g, A) \longrightarrow gA{}^t\overline{g},$$

on the hermitian matrices. Translating this into the realisation as an upper half space yield the slightly scaring formula

$$G \times (\mathbb{C} \times \mathbb{R}_{>0}) \longrightarrow \mathbb{C} \times \mathbb{R}_{>0},$$
$$(g, (z, \zeta)) \longrightarrow \left(\frac{(az+b)\,\overline{(cz+d)} + a\bar{c}\,\zeta^2}{(cz+d)\,\overline{(cz+d)} + c\bar{c}\,\zeta^2}, \frac{\zeta}{(cz+d)\,\overline{(cz+d)} + c\bar{c}\,\zeta^2} \right).$$

Here X is another model the three dimensional hyperbolic space $\mathbb{H}_3$.

The Riemannian metric

It was already mentioned in the statement of the theorem of Cartan that we always have a G^0_∞-invariant Riemannian metric on X. It is not too difficult to construct such a metric, which in many cases is rather canonical.

In the general case we observe that the maximal compact subgroup is the stabiliser of the point $x_0 = e \cdot K_\infty \in G^0_\infty/K_\infty = X$. Hence it acts on the tangent space of x_0, and we can construct a K_∞-invariant positive definite quadratic form on this tangent space. Then we use the action of G^0_∞ on X to transport this metric to an arbitrary point in X: if $x \in X$ we find a g so that $x = gx_0$, which defines an isomorphism between the tangent space at x_0 and the tangent space at x. Hence we get a quadratic form on the tangent space at x, which will not depend on the choice of $g \in G^0_\infty$. In our examples this metric is always unique up to scalars.

a) In the case of the group $\mathrm{Sl}_n(\mathbb{R})$ we may take as a base point $x_0 \in X$ the identity $\mathrm{Id} \in \mathrm{Sl}_n(\mathbb{R})$. The corresponding maximal compact subgroup is the orthogonal group $\mathrm{SO}(n)$. The tangent space at Id is given by the space

$$\mathrm{Sym}^0_n(\mathbb{R}) = T^X_{\mathrm{Id}}$$

of symmetric matrices with trace zero. On this space we have the form

$$Z \longrightarrow \mathrm{trace}(Z^2),$$

which is positive-definite (a symmetric matrix has real eigenvalues). It is easy to see that the orthogonal group acts on this tangent space by conjugation, hence the form is invariant.

b) A similar argument applies to the group $G_\infty = \mathrm{Sl}_d(\mathbb{C})$. Again the identity Id is a nice positive-definite hermitian matrix. The tangent space consists of the hermitian matrices

$$T^X_{\mathrm{Id}} = \left\{ A \mid A = {}^t\overline{A} \text{ and } \mathrm{tr}(A) = 0 \right\},$$

and the invariant form is given by

$$A \longrightarrow \mathrm{tr}(A\overline{A}^t).$$

c) In the case of the group $G^0_\infty \subset \mathrm{SO}(f)(\mathbb{R})$ where f is the quadratic form

$$f(x_1, \ldots, x_{n+1}) = x_1^2 + \ldots + x_n^2 - x_{n+1}^2,$$

we realised the symmetric space as the open ball

$$\overset{\circ}{D}_n = \{(x_1, \ldots, x_n) \mid x_1^2 + \ldots + x_n^2 < 1\}.$$

The orthogonal group $\mathrm{SO}(n,1)$ is the stabiliser of $0 \in \overset{\circ}{D}_n$, and hence it is clear that the Riemannian metric has to be of the form

$$h(x_1^2 + \ldots + x_n^2)(dx_1^2 + \ldots + dx_n^2)$$

(in the usual notation). A closer look shows that the metric has to be

$$\frac{dx_1^2 + \ldots + dx_n^2}{\sqrt{1 - x_1^2 - \ldots - x_n^2}}.$$

In our two low dimensional spacial examples the metric is easy to determine. For the action of the group $\mathrm{Sl}_2(\mathbb{R})$ on the upper half plane $\mathbb{H}$ we observe that for any point $z_0 = x + iy \in \mathbb{H}$ the tangent vectors $\frac{\partial}{\partial x}|_{z_0}, \frac{\partial}{\partial y}|_{z_0}$ form a basis of the tangent spaces at z_0.

If we take $z_0 = i$ then the stabiliser is the group $\mathrm{SO}(2)$. For

$$e(\varphi) = \begin{pmatrix} \cos\varphi & \sin\varphi \\ -\sin\varphi & \cos\varphi \end{pmatrix}$$

we have

$$\begin{aligned} e(\varphi) \cdot \left(\frac{\partial}{\partial x}|_i\right) &= \cos 2\varphi \cdot \frac{\partial}{\partial x}|_i + \sin 2\varphi \cdot \frac{\partial}{\partial y}|_i, \\ e(\varphi)\left(\frac{\partial}{\partial y}|_i\right) &= -\sin 2\varphi \cdot \frac{\partial}{\partial x}|_i + \cos 2\varphi \cdot \frac{\partial}{\partial y}|_i. \end{aligned}$$

Hence we find that $\frac{\partial}{\partial x}|_i$ and $\frac{\partial}{\partial y}|_i$ have to be orthogonal and of the same length.

Now the matrix

$$\begin{pmatrix} y & x \\ 0 & 1 \end{pmatrix}$$

sends i into the point $z = x + iy$. It sends $\frac{\partial}{\partial x}|_i$ and $\frac{\partial}{\partial y}|_i$ into $y \cdot \frac{\partial}{\partial x}|_z$ and $y \cdot \frac{\partial}{\partial y}|_z$, and hence we must have for our invariant metric

$$\langle\frac{\partial}{\partial x}|_z, \frac{\partial}{\partial y}|_z\rangle = 0\ ;\ \langle\frac{\partial}{\partial x}|_z, \frac{\partial}{\partial x}|_z\rangle = \frac{1}{y^2}\ ;\ \langle\frac{\partial}{\partial y}|_z, \frac{\partial}{\partial y}|_z\rangle = \frac{1}{y^2},$$

and this is in the usual notation the metric

$$ds^2 = \frac{1}{y^2}(dx^2 + dy^2). \tag{1.37}$$

A completely analogous argument yields the metric

$$ds^2 = \frac{1}{\zeta^2}\,(d\zeta^2 + dx^2 + dy^2) \tag{1.38}$$

for the space $\mathbb{H}_3$.

1.2 Arithmetic groups

If we have a linear algebraic group $G/\mathbb{Q} \hookrightarrow \mathrm{Gl}_n$ we may consider the group $\Gamma = G(\mathbb{Q}) \cap \mathrm{Gl}_n(\mathbb{Z})$. This is the prototype of an *arithmetic* group. It has the following fundamental property:

Proposition 1.2.1. *The group Γ is a discrete subgroup of the topological group $G(\mathbb{R})$.*

This is rather easily reduced to the fact that $\mathbb{Z}$ is discrete in $\mathbb{R}$.

There is a slightly more sophisticated view of arithmetic groups. We may consider the flat extension $\mathcal{G}/\mathbb{Z}$ of $G/\mathbb{Q}$ and then it is clear that $\Gamma = \mathcal{G}(\mathbb{Z})$.

Actually our construction provides a big family of arithmetic groups. For any integer $m > 0$ we have the homomorphism of reduction $\mod m$, namely

$$\mathcal{G}\mathbb{Z}) \longrightarrow \mathcal{G}(\mathbb{Z}/m\mathbb{Z}).$$

The kernel $\mathcal{G}(\mathbb{Z})(m)$ of this homomorphism has finite index in $\mathcal{G}(\mathbb{Z})$, and we will call this kernel a *congruence subgroup* of $\Gamma = \mathcal{G}(\mathbb{Z})$. More generally we define

Definition 2.1.: *A subgroup Γ'' of Γ is called a congruence subgroup if we can find an integer m such that*

$$\mathcal{G}(\mathbb{Z})(m) \subset \Gamma'' \subset \Gamma.$$

At this point a small remark is in order. We explained already that a linear algebraic group $G/\mathbb{Q}$ may be embedded in different ways into different groups Gl_n, i.e. we may have

$$G \hookrightarrow GL_{n_1} \quad \text{resp.} \quad G \hookrightarrow GL_{n_1}$$

and this means that we have two possibly different flat extensions $\mathcal{G}_1, \mathcal{G}_2$. Now it is not hard to show that we can find an $m > 0$ such that

$$\mathcal{G}_1(\mathbb{Z}) \cap \mathcal{G}_2(\mathbb{Z}) \supset \mathcal{G}_1(\mathbb{Z})(m) \text{ and } \mathcal{G}_2(\mathbb{Z})(m) \tag{1.39}$$

and this means that $\mathcal{G}_1(\mathbb{Z}) \cap \mathcal{G}_2(\mathbb{Z})$ is a congruence subgroup of $\mathcal{G}_1(\mathbb{Z})$ and $\mathcal{G}_2(\mathbb{Z})$ as well.

Hence we say that $\Gamma \subset G(\mathbb{Q})$ is a congruence subgroup, if we can find a flat extension $\mathcal{G}/\mathbb{Z}$ such that $\Gamma \subset \mathcal{G}(\mathbb{Z})$, is a congruence subgroup. Then the notion of *congruence subgroup* does not depend on the way we realised the group $G/\mathbb{Q}$ as a subgroup in the general linear group.

It is clear from the above considerations that all congruence subgroups are *commensurable* , this means that for two such congruence subgroups Γ_1, Γ_2 the intersection $\Gamma_1 \cap \Gamma_2$ is of finite index in both of them.

Now we may also define the notion of an *arithmetic group* . A subgroup $\Gamma' \subset G(\mathbb{Q})$ is called *arithmetic* if it is commensurable to a congruence subgroup, especially subgroups of finite index in a congruence subgroup are arithmetic subgroups. All arithmetic subgroups are commensurable. Of course all congruence subgroups are arithmetic subgroups.

The most prominent example of an arithmetic group is the group

$$\Gamma = \mathrm{Sl}_2(\mathbb{Z}) \subset \mathrm{Sl}_2(\mathbb{Q}).$$

Another example is obtained as follows. We defined for any number field $K/\mathbb{Q}$ the group

$$G/\mathbb{Q} = R_{K/\mathbb{Q}}(\mathrm{Sl}_d)$$

for which $G(\mathbb{Q}) = \mathrm{Sl}_d(K)$. If $\mathcal{O}_K$ is the ring of integers in K, then $\Gamma = \mathrm{Sl}_d(\mathcal{O}_K)$ (and also $\tilde{\Gamma} = \mathrm{Gl}_n(\mathcal{O}_K)$) is a congruence (and hence arithmetic) subgroup of $G(\mathbb{Q})$.

It is very interesting that the groups $\Gamma = \mathrm{Sl}_2(\mathbb{Z})$ and $\mathrm{Sl}_2(\mathcal{O}_K)$ for imaginary quadratic $K/\mathbb{Q}$ always contain arithmetic subgroups $\Gamma' \subset \Gamma$ which are not congruence subgroups. This means that in general the class of arithmetic subgroups is larger than the class of congruence subgroups, see, for example, [100].

If only the group $G(\mathbb{R})$ is given (as the group of real points of a group $G/\mathbb{R}$ or perhaps only as a Lie group), then the notion of an arithmetic group $\Gamma \subset G(\mathbb{R})$ is not defined. The notion of an arithmetic subgroup $\Gamma \subset G(\mathbb{R})$ requires the choice of a group scheme $G/\mathbb{Q}$ such that the group $G(\mathbb{R})$ is the group of real points of this group over $\mathbb{Q}$. The exercise in 1.1.2. shows that different $\mathbb{Q}$-forms provide different arithmetic groups.

Exercise 2 *If* $\gamma \in \mathrm{Gl}_n(\mathbb{Z})$ *is a nontrivial torsion element and if* $\gamma \equiv \mathrm{Id} \mod m$ *then* $m = 1$ *or* $m = 2$. *In the latter case the element* γ *is of order* 2.

This implies that for $m \geq 3$ *the congruence subgroup* $\mathrm{Gl}_n(\mathbb{Z})(m)$ *of* $\mathrm{Gl}_n(\mathbb{Z})$ *is torsion free.*

This implies of course that any arithmetic group has a subgroup of finite index, which is torsion free.

We may now understand our arithmetic group $\Gamma = G(\mathbb{Q}) \cap \mathrm{Gl}_n(\mathbb{Z})$ as the group $\mathcal{G}(\mathbb{Z})$ of $\mathbb{Z}$-valued points of a group scheme over $\mathbb{Z}$. Since we know what $\mathcal{G}(\mathbb{Z}/m\mathbb{Z})$ is we can define congruence subgroups Γ_H as inverse images of subgroups $H \subset \mathcal{G}(\mathbb{Z}/m\mathbb{Z})$ under the projection $\mathcal{G}(\mathbb{Z}) \to \mathcal{G}(\mathbb{Z}/m\mathbb{Z})$.

There is the special class of semi-simple or reductive group schemes. Roughly speaking, an affine group scheme $G/\mathbb{Z}$ is semi-simple (resp. reductive), if it is smooth, if its generic fiber $G \times_{\mathbb{Z}} \mathbb{Q}$ is semi-simple (resp. reductive) and finally if for all primes p the group scheme $G \times_{\mathbb{Z}} \mathbb{F}_p$ (the reduction $\mod p$) is a semi-simple (resp. reductive) group scheme over $\mathbb{F}_p$.

Of course the simplest example of a semi-simple (resp. reductive) group (scheme) over $\mathbb{Z}$ is the group $\mathrm{Sl}_n/\mathbb{Z}$ (resp. $\mathrm{Gl}_n/\mathbb{Z}$).

We can also construct semi-simple group-schemes by taking flat extensions of orthogonal (resp. symplectic) groups over $\mathbb{Q}$ (cf. Examples 2 and 3 of Section 1.1.5). Here the symmetric (resp. alternating) form has to satisfy certain arithmetic conditions (see chap4.pdf).

1.2.1 Γ-modules

We consider modules $\mathcal{M}$ (i.e. abelian groups) with an action of Γ, and we want to briefly discuss some special classes of such Γ-modules.

The most important classes of Γ-modules are the modules of *arithmetic origin.* To construct such modules we realise our arithmetic group as $\Gamma = G(\mathbb{Q}) \cap \mathrm{Gl}_n(\mathbb{Z})$. Then we take any rational representation $\rho : G/\mathbb{Q} \to \mathrm{Gl}(V)$, where V is a finite dimensional $\mathbb{Q}$ vector space. Now we look for finitely generated submodules $\mathcal{M} \subset V$ such that $\mathcal{M} \otimes \mathbb{Q} = V$ which are invariant under the action of Γ. Such a module is a Γ-module of arithmetic origin.

It is not to difficult to show that given any finitely generated module $\mathcal{M}'$ which is a full sublattice, i.e. $\mathcal{M}' \otimes \mathbb{Q} = V$, we can find a congruence subgroup $\Gamma_1 \subset \Gamma$ such that $\Gamma_1 \mathcal{M}' = \mathcal{M}'$. Then

$$\mathcal{M} = \bigcap_{\gamma \in \Gamma/\Gamma_1} \gamma \mathcal{M}'$$

is a Γ-module of *arithmetic origin.*

A second class of Γ-modules are those of *congruence origin.* To get such a module we simply pick a congruence subgroup $\Gamma(N) \subset \Gamma$ and then we simply look at finitely generated abelian groups V with an action of $\Gamma/\Gamma(N)$ on V.

We get some important examples of Γ-modules of congruence origin if we start from a Γ-module $\mathcal{M}$ of arithmetic origin. Then we choose an integer N and consider the Γ-module $\mathcal{M} \otimes \mathbb{Z}/N\mathbb{Z}$. On this module $\Gamma(N)$ acts trivially, hence this module is a $\Gamma/\Gamma(N)$-module of congruence origin.

We go back to the more sophisticated point of view above, our arithmetic group is now the group $\Gamma = \mathcal{G}(\mathbb{Z})$ of $\mathbb{Z}$-valued points of the flat extension $\mathcal{G}/\mathbb{Z}$.

Now we pick a torsion-free finitely generated module $\mathcal{M}$, and we know what it means that $\mathcal{M}$ is a $\mathcal{G}/\mathbb{Z}$ module: it simply means that for any commutative ring R with identity we have a R-linear action of the $\mathcal{G}(R)$ on the R-module $\mathcal{M} \otimes R$, or in other words we have a homomorphism $\mathcal{G}(R) \to \mathrm{Gl}_B(\mathcal{M} \otimes_{\mathbb{Z}} R)$. Of course we require that this action is functorial in R.

For this book - especially for the first half - the group scheme $\mathrm{Gl}_2/\mathbb{Z}$ plays a dominant role. In this case the irreducible representations of $\mathrm{Gl}_2 \times_{\mathbb{Z}} \mathbb{Q}$ are well-known. We consider the $\mathbb{Q}$ vector space of homogenous polynomials in two variables and of degree n

$$\mathcal{M}_{n,\mathbb{Q}} := \{P(X,Y) = \sum_{\nu=0}^{n} a_\nu X^\nu Y^{n-\nu} \mid a_\nu \in \mathbb{Q}\}.$$

We choose an integer m and define an action of $\mathrm{Gl}_2(\mathbb{Q})$:

$$\begin{pmatrix} a & b \\ c & d \end{pmatrix} P(X,Y) = P(aX+cY, bX+dY) \det\left(\begin{pmatrix} a & b \\ c & d \end{pmatrix}\right)^m, \tag{1.40}$$

which gives us the $\mathrm{Gl}_2/\mathbb{Q}$-module $\mathcal{M}_{n,\mathbb{Q}}[m]$.

But now it is easy to get $\mathrm{Gl}_2/\mathbb{Z}$-modules, we simply define

$$\mathcal{M}_n := \{P(X,Y) = \sum_{\nu=0}^{n} a_\nu X^\nu Y^{n-\nu} \mid a_\nu \in \mathbb{Z}\}, \tag{1.41}$$

and then we define the $\mathrm{Gl}_2/\mathbb{Z}$ modules $\mathcal{M}_n[m]$ by the same formula as above. If n is even we will sometimes work with the module $\mathcal{M}[-\frac{n}{2}]$, because in this case the centre acts trivially.

At this point a small remark is in order. If we look at $\mathcal{M}_n[m]$ only as a $\mathrm{Gl}_2(\mathbb{Z})$-module then the module "knows" what n is, clearly $n = \mathrm{rank}(\mathcal{M}_n) - 1$. But this $\mathrm{Gl}_2(\mathbb{Z})$-module does not "know" what m is. The only information we get is

$$\begin{pmatrix} -1 & 0 \\ 0 & -1 \end{pmatrix} P = (-1)^n P$$

and from this we only get the value of $m \mod 2$. But if we consider $\mathcal{M}_n[m]$ as a module for the group scheme $\mathrm{Gl}_2/\mathbb{Z}$, then the module also knows the value of m, because then we know

$$\begin{pmatrix} \alpha & 0 \\ 0 & \alpha \end{pmatrix} P = \alpha^{n+2m} P,$$

for any $\alpha \in R^\times$. Actually the value of the parameter m does not play a relevant role. If n is even then we often work with the $\mathrm{PGL}_2/\mathbb{Z} = \mathrm{Gl}_2/\mathbb{G}_m$. module $\mathcal{M}_n[-\frac{n}{2}]$.

In Section 4.1.1 we discuss the corresponding situation for groups of type $\mathrm{Gl}_2(\mathbb{Z}[\sqrt{-d}])$.

1.2.2 The locally symmetric spaces

We start from a semi-simple group $G/\mathbb{Q}$. To this group we attached the group of real points $G(\mathbb{R}) = G_\infty$. In G_∞ we have the connected component G_∞^0 of the identity and in this group we choose a maximal compact subgroup K_∞. The quotient space $X = G_\infty/K_\infty$ is a symmetric space which now may have several connected components. On this space we have the action of an arithmetic group $\Gamma \subset G(\mathbb{Q})$.

We have a fundamental fact: *The action of Γ on X is properly discontinuous, i.e. for any point $x \in X$ there exists an open neighbourhood U_x such that for all $\gamma \in \Gamma$ we have*

$$\gamma U_x \cap U_x = \emptyset \quad \text{or} \quad \gamma x = x.$$

Moreover for all $x \in X$ the stabiliser

$$\Gamma_x = \{\gamma \mid \gamma x = x\}$$

is finite.

This is easy to see: If we consider the projection $p : G(\mathbb{R}) \to G(\mathbb{R})/K_\infty = X$, then the inverse image $p^{-1}(U_x)$ of a relatively compact neighbourhood U_x of $x = g_0 K_\infty$ is of the form $V_{g_0} \cdot K_\infty$, where V_{g_0} is a relatively compact neighbourhood of g_0. Hence we look for solutions of the equation

$$\gamma v k = v'k', \gamma \in \Gamma, v, v' \in V_{g_0}, k, k' \in K_\infty.$$

Since Γ is discrete in $G(\mathbb{R})$ there are only finitely many possibilities for γ and they can be ruled out by shrinking U_x with the exception of those γ for which

$\gamma x = x$. If $\gamma x = x$ this means that $\gamma g_0 K_\infty = g_0 K_\infty$ and hence $\gamma \in \Gamma \cap g_0 K_\infty g_0^{-1}$ this intersection is a compact discrete set, hence finite.

If Γ has no torsion then the projection

$$\pi : X \longrightarrow \Gamma\backslash X$$

is locally a $\mathcal{C}^\infty$-diffeomorphism. To any point $x \in \Gamma\backslash X$ and any point $\tilde{x} \in \pi^{-1}(x)$ we find a neighbourhood $U_{\tilde{x}}$ such that

$$\pi : U_{\tilde{x}} \xrightarrow{\sim} U_x.$$

Hence the space $\Gamma\backslash X$ inherits the Riemannian metric, and the quotient space is a *locally symmetric space*. In the following we will denote the dimension of $\Gamma\backslash X$ by d, i.e. $d = \dim(\Gamma\backslash X)$.

If our group Γ has torsion, then a point $\tilde{x} \in X$ may have a nontrivial stabiliser $\Gamma_{\tilde{x}}$. Then it is not difficult to prove that $\tilde{x}$ has a neighbourhood $U_{\tilde{x}}$ which is invariant under $\Gamma_{\tilde{x}}$ and that for all $\tilde{y} \in U_{\tilde{x}}$ the stabiliser $\Gamma_{\tilde{y}} \subset \Gamma_{\tilde{x}}$. This gives us a diagram

$$\begin{array}{ccc} U_{\tilde{x}} & \longrightarrow & \Gamma_{\tilde{x}}\backslash U_{\tilde{x}} = U_x \\ \downarrow & & \downarrow \\ X & \xrightarrow{\pi} & \Gamma\backslash X \end{array}$$

i.e. the point $x \in \Gamma\backslash X$ has a neighbourhood which is the quotient of a neighbourhood $U_{\tilde{x}}$ by a finite group.

We see that the quotient space $\Gamma\backslash X$ may have singularities. These spaces are called *orbifolds*. They have a natural stratification. Any point x defines a Γ conjugacy class $[\Gamma_{\tilde{x}}]$ of finite subgroups $\Gamma_{\tilde{x}} \subset \Gamma$. On the other hand, a conjugacy class $[c]$ of finite subgroups defines the subset (stratum) $\Gamma\backslash X([c])$ of those points $\tilde{x} \in \Gamma\backslash X$ for which $\Gamma_{\tilde{x}} \in [c]$.

These strata are easy to describe. We consider for any $\Gamma_{\tilde{x}} \subset \Gamma$ the fixed point set $X^{\Gamma_{\tilde{x}}}$ and its open subset $(X^{\Gamma_{\tilde{x}}})^{(0)}$ consisting of those points $\tilde{x}'$ with $\Gamma_{\tilde{x}} = \Gamma_{\tilde{x}'}$. If for $\gamma \in \Gamma$ and $\tilde{x}', \tilde{x}'' \in (X^{\Gamma_{\tilde{x}}})^{(0)}$ we have that $\gamma\tilde{x}' = \tilde{x}''$, then necessarily γ normalizes $\Gamma_{\tilde{x}}$. So, putting $\Gamma[\tilde{x}]$ the normalizer of $\Gamma_{\tilde{x}}$ in Γ, we get a locally closed embedding

$$\Gamma[\tilde{x}]\backslash(X^{\Gamma_{\tilde{x}}})^{(0)} \hookrightarrow \Gamma\backslash X.$$

The image of this embedding only depends on the Γ-conjugacy class of $\Gamma_{\tilde{x}}$ and is in fact its attached stratum.

We have an ordering on the set of conjugacy classes - $[c_1] \leq [c_2]$ if for any $H_1 \in [c_1]$ there exists a subgroup $H_2 \in [c_2]$ such that $H_1 \subset H_2$. These strata are not closed, the closure $\overline{\Gamma\backslash X([c])}$ being the union of lower dimensional strata.

If we start investigating the stratification above we immediately hit upon number theoretic problems. Let us pick a prime p and we consider the group $\Gamma = \mathrm{Sl}_{p-1}[\mathbb{Z}]$. The ring of p-th roots of unity $\mathbb{Z}[\zeta_p]$ as a $\mathbb{Z}$-module is free, of rank $p-1$, hence we get an element

$$\zeta_p \in \mathrm{Sl}(\mathbb{Z}[\zeta_p]) \cong \mathrm{Sl}_{p-1}(\mathbb{Z}),$$

and thus a cyclic subgroup of order p. But clearly we have many conjugacy classes of elements of order p in Γ because any ideal $\mathfrak{a}$ is a free $\mathbb{Z}$-module. If we want to understand the conjugacy classes of elements of order p or the conjugacy classes of cyclic subgroups of order p in $\mathrm{Sl}_{p-1}(\mathbb{Z})$, we need to understand the ideal class group. In the next section we will discuss some simple examples.

These quotient spaces $\Gamma\backslash X$ attract the attention of various different kinds of mathematicians. They provide interesting examples of Riemannian manifolds and they are intensively studied from that point of view. On the other hand, number theoretic data enter into their construction. Hence any insight into the structure of these spaces contains number theoretic information. This is the main theme of this book.

It is not difficult to see that any arithmetic group Γ contains a normal congruence subgroup Γ' which does not have torsion. This can be deduced easily from the exercise above. . Hence we see that $\Gamma'\backslash X$ is a Riemannian manifold which is a finite cover of $\Gamma\backslash X$ with covering group Γ/Γ'.

We discuss special examples below.

1.2.3 Low dimensional examples

We consider the action of the group $\Gamma = \mathrm{Sl}_2(\mathbb{Z}) \subset \mathrm{Sl}_2(\mathbb{R})$ on the upper half plane

$$X = \mathbb{H} = \{z \mid \Im(z) = y > 0\} \cong \mathrm{Sl}_2(\mathbb{R})/\mathrm{SO}(2).$$

We want to describe the quotient $\Gamma\backslash\mathbb{H}$, and we construct further down the fundamental domain $\mathcal{F}$.

In Section 1.1.10 we explained that we have a positive-definite euclidian metric on $\Omega\otimes\mathbb{R} = \mathbb{C}$. The action of $\mathrm{Sl}_2(\mathbb{Z})$ on the upper half plane corresponds to changing the basis $1, z$ of Ω into another basis. Without changing the conformal class we can normalise the first basis vector to 1. But we may also achieve that the first basis vector is of shortest length, in other words $\Omega = \langle 1, z\rangle$ where now $|z| \geq 1$.

We still can change the basis by $1 \to 1$ and $z \to z + n$. We still have $|z + n| \geq 1$. Hence we see that this condition implies that we can move z by these translations into the strip $-1/2 \leq \Re(z) \leq 1/2$ and since 1 is still a shortest vector we end up in the classical fundamental domain:

$$\mathcal{F} = \{z \mid -1/2 \leq \Re(z) \leq 1/2, |z| \geq 1\}. \tag{1.42}$$

Two different points $z_1, z_2 \in \mathcal{F}$ are inequivalent under the action of $\mathrm{Sl}_2(\mathbb{Z})$ unless they differ by a translation. i.e.

$$z_1 = -\frac{1}{2} + it \ , \ z_2 = z_1 + 1 = \frac{1}{2} + it,$$

or we have $|z_1| = 1$ and $z_2 = -\frac{1}{z_1}$. Hence the quotient $\mathrm{Sl}_2(\mathbb{Z})\backslash\mathbb{H}$ is given by the following picture: The circles are actually the images of horizontal lines $iy + x$ where $x \in \mathbb{R}$ or $x \in [0, 1]$ in the quotient. The picture is a little bit misleading because it does not reflect the Riemannian metric: the circumference of the circle at level iy is $\frac{1}{y}$.

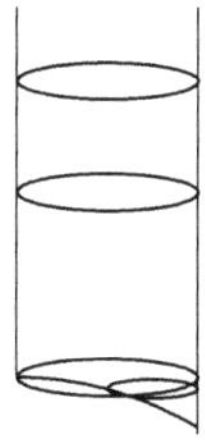

Figure 1.1: The quotient $\Gamma\backslash\mathbb{H}$

It turns out that this quotient is actually a Riemann surface, i.e. the finite stabilisers at i and ρ do not produce singularities. As a Riemann surface the quotient is the complex plane or better the projective line $\mathbb{P}^1(\mathbb{C})$ minus the point at infinity.

It is clear that the points i and $\rho = \frac{1}{2} + \frac{1}{2}\sqrt{-3}$ in the upper half plane are - up to conjugation by an element $\gamma \in \mathrm{Sl}_2(\mathbb{Z})$ - the only points with non-trivial stabiliser. Actually the stabilisers are given by

$$\Gamma_i = \left\langle \begin{pmatrix} 0 & 1 \\ -1 & 0 \end{pmatrix} \right\rangle \quad , \quad \Gamma_\rho = \left\langle \begin{pmatrix} -1 & 1 \\ -1 & 0 \end{pmatrix} \right\rangle .$$

The second example is given by the group $\Gamma = \mathrm{Sl}_2(\mathbb{Z}[i]) \subset \mathrm{Sl}_2(\mathbb{C}) = G_\infty = R_{\mathbb{C}/\mathbb{R}}(\mathrm{Sl}_2/\mathbb{C})(\mathbb{R})$ (see (1.1)). Here we should remember that the choice of G_∞ allows a whole series of arithmetic groups. For any imaginary quadratic extension $K = \mathbb{Q}(\sqrt{-d})$ with $\mathcal{O}_K$ as its ring of integers we may embed K into $\mathbb{C}$ and get

$$\mathrm{Sl}_2(\mathcal{O}_K) = \Gamma \subset G_\infty .$$

If the number d becomes larger then the structure of the group Γ becomes more and more complicated. We only discuss the simplest case $\mathcal{O}_K = \mathbb{Z}[i]$. We will construct a fundamental domain for the action of Γ on the three-dimensional hyperbolic space $\mathbb{H}_3 = \mathbb{C} \times \mathbb{R}_{>0}$.

We identify $\mathbb{H}_3$ with the space of positive-definite hermitian matrices

$$X = \{A \in M_2(\mathbb{C}) \mid A = {}^t\bar{A}, A > 0, \det(A) = 1\}.$$

We consider the lattice

$$\Omega = \mathbb{Z}[i] \cdot \begin{pmatrix} 1 \\ 0 \end{pmatrix} + \mathbb{Z}[i] \cdot \begin{pmatrix} 0 \\ 1 \end{pmatrix}$$

in $\mathbb{C}^2$ and view A as a hermitian metric on $\mathbb{C}^2$ where $\mathbb{C}^2/\Omega$ has volume 1. Let $e_1' = \binom{\alpha}{\beta}$ be a vector of shortest length. We can find a second vector $e_2' = \binom{\gamma}{\delta}$ so that $\det \begin{pmatrix} \alpha & \beta \\ \gamma & \delta \end{pmatrix} = 1$. We consider the vectors $e_2' + \nu e_1'$ where $\nu \in \mathbb{Z}[i]$. We have

$$\langle e_2' + \nu e_1', e_2' + \nu e_1' \rangle_A = \langle e_2', e_2' \rangle_A + \nu \langle e_1', e_2' \rangle_A + \overline{\nu} \langle e_2', e_1' \rangle_A + \nu\overline{\nu} \langle e_1', e_1' \rangle_A .$$

We can choose ν such that

$$-\frac{1}{2}\langle e_1', e_1'\rangle_A \leq \mathrm{Re}\langle e_1', e_2'\rangle_A, \Im\langle e_1', e_2'\rangle_A \leq \frac{1}{2}\langle e_1', e_1'\rangle_A.$$

If we translate this to the action of $\mathrm{Sl}_2(\mathbb{Z}[i])$ on $\mathbb{H}_3$ then we find that every point $x = (z, \zeta) \in \mathbb{H}_3$ is equivalent to a point in the domain

$$\tilde{F} = \{(z, \zeta) \mid -\frac{1}{2} \leq \mathrm{Re}(z), \Im(z) \leq \frac{1}{2}; z\bar{z} + \zeta^2 \geq 1\}.$$

Since we still have the action of the matrix $\begin{pmatrix} i & 0 \\ 0 & -i \end{pmatrix}$ we can find a fundamental domain

$$F = \{(z, \zeta) \mid -\frac{1}{2} \leq \mathrm{Re}(z), \Im(z) \leq \frac{1}{2}; z\bar{z} + \zeta^2 \geq 1 \text{ and } \mathrm{Re}(z) + \Im(z) \geq 0\}.$$

I also want to discuss the extension of our considerations to the case of the reductive group $\mathrm{Gl}_2(\mathbb{C})$. In such a case we have to enlarge the maximal compact subgroup. In this case the group $\tilde{K} = \mathrm{SU}(2) \cdot \mathbb{C}^* = K \cdot \mathbb{C}^*$ is a good choice where $\mathbb{C}^*$ is the centre of $\mathrm{Gl}_2(\mathbb{C})$. Then we get

$$\mathbb{H}_3 = \mathrm{Sl}_2(\mathbb{C})/K = \mathrm{Gl}_2(\mathbb{C})/\tilde{K}$$

i.e. we still get the same symmetric space. But the group $\tilde{\Gamma} = \mathrm{Gl}_2(\mathbb{Z}[i])$ is still larger. We have an exact sequence

$$1 \to \Gamma \to \tilde{\Gamma} \to \{i^\nu\} \to 1.$$

The centre $Z_{\tilde{\Gamma}}$ of $\tilde{\Gamma}$ is given by the matrices $\left\{\begin{pmatrix} i^v & 0 \\ 0 & i^v \end{pmatrix}\right\}$. The centre Z_Γ has index 2 in $Z_{\tilde{\Gamma}}$. Since the centre acts trivially on the symmetric space, the above fundamental domain will be "cut into two halves" by the action of $\tilde{\Gamma}$. the matrices $\begin{pmatrix} i^v & 0 \\ 0 & 1 \end{pmatrix}$ induce a rotation of $\nu \cdot 90°$ around the axis $z = 0$ and therefore it becomes clear that the region

$$F_0 = \{(z, \zeta) \mid 0 \leq \Im(z), \mathrm{Re}(z) \leq \frac{1}{2}, z\bar{z} + \zeta^2 \geq 1\}$$

is a fundamental domain for $\tilde{\Gamma}$.

The translations $z \to z + 1$ and $z \to z + i$ identify the opposite faces of $\tilde{F}$. This induces an identification on F_0, namely

$$\left(\frac{1}{2} + iy, \zeta\right) \longrightarrow \left(-\frac{1}{2} + iy, \zeta\right) \longrightarrow \left(y + \frac{i}{2}, \zeta\right).$$

On the bottom of the domain F_0, namely

$$F_0(1) = \{(z, \zeta) \in F_0 \mid z\bar{z} + \zeta^2 = 1\}$$

we have the further identification

$$(z, \zeta) \longrightarrow (i\bar{z}, \zeta).$$

Hence we see that the quotient space $\tilde{\Gamma}\backslash\mathbb{H}_3$ is given by the following figure.

I want to discuss the fixed points and the stabilisers of the fixed points of $\tilde{\Gamma}$. Before I can do that, I need some simple facts concerning the structure of Gl_2.

The group $\mathrm{Gl}_2(K)$ acts upon the projective space

$$\mathbb{P}^1(K) = K \cup \{\infty\} \ ; \ K(xe_1 + e_2) = x, Ke_1 = \infty.$$

It is quite clear that the action of $g = \begin{pmatrix} \alpha & \beta \\ \gamma & \delta \end{pmatrix} \in \mathrm{Gl}_2(K)$ is given by

$$gx = \frac{\alpha x + \beta}{\gamma x + \delta}.$$

The action of $\mathrm{Gl}_2(K)$ on $\mathbb{P}^1(K)$ is transitive. For a point $x \in \mathbb{P}^1(K)$ the stabiliser B_x is clearly a linear subgroup of Gl_2/K. If $x = \infty$, then this stabiliser is the subgroup

$$B_\infty = \left\{ \begin{pmatrix} a & u \\ 0 & b \end{pmatrix} \right\},$$

and for $x = 0$ we get

$$B_0 = \left\{ \begin{pmatrix} a & 0 \\ u & b \end{pmatrix} \right\}.$$

It is clear that these subgroups B_x are conjugate under the action of $\mathrm{Gl}_2(K)$. They are in fact maximal connected solvable subgroups of Gl_2.

If we have two different points $x_1, x_2 \in \mathbb{P}^1(K)$, then this corresponds to a choice of a basis where the basis vectors are only determined up to scalars. Then the intersection of the two groups $B_{x_1} \cap B_{x_2}$ is a so-called maximal torus. If we choose $x_1 = Ke_1$, $x_2 = Ke_2$, then

$$B_{x_1} \cap B_{x_2} = \left\{ \begin{pmatrix} a & 0 \\ 0 & b \end{pmatrix} \right\}.$$

Any other maximal torus of the form $B_{x_1} \cap B_{x_2}$ is conjugate to T_0 under $\mathrm{Gl}_2(K)$.

Now we assume $K = \mathbb{C}$. We compactify the three dimensional hyperbolic space by adding $\mathbb{P}^1(\mathbb{C})$ at infinity, i.e.

$$\mathbb{H}_3 \hookrightarrow \overline{\mathbb{H}}_3 = \mathbb{H}_3 \cup \mathbb{P}^1(\mathbb{C}) = \mathbb{C} \times \mathbb{R}_{\geq 0} \cup \{\infty\}.$$

Now let us assume that $a \in \mathrm{Gl}_2(\mathbb{C})$ is an element which has a fixed point on $\mathbb{H}_3$ and which is not central. Since it lies in a maximal compact subgroup times $\mathbb{C}^\times$ we see that this element a can be diagonalised

$$a \longrightarrow g_0 a g_0^{-1} = \begin{pmatrix} \alpha & 0 \\ 0 & \beta \end{pmatrix} = a'$$

with $\alpha \neq \beta$ and $|\alpha/\beta| = 1$.

Then it is clear that the fixed point set for a' is the line

$$\mathrm{Fix}\,(a') = \{(0, \zeta) \mid \zeta \in \mathbb{R}_{>0}\},$$

i.e. we do not get an isolated fixed point but a full fixed line.

The element a' has the two fixed points $\infty, 0$ in $\mathbb{P}^1(\mathbb{C})$, and hence it defines the torus $T_0(\mathbb{C})$. Then it is clear that

$$\mathrm{Fix}(a') = \{(0, \zeta) \mid \zeta > 0\} = T_0(\mathbb{C}) \cdot (0, 1),$$

i.e. the fixed point set is an orbit under the action of $T_0(\mathbb{C})$.

1.2.4 Fixed point sets and stabilisers for $\tilde{\Gamma} = \mathrm{Gl}_2(\mathbb{Z}[i])$

If we want to describe the stabilisers up to conjugation, we can focus our attention on F_0.

If we have an element $\gamma \in \tilde{\Gamma}$, γ not central, and if we assume that γ has fixed points on $\mathbb{H}_3$, then we know that γ defines a torus $T_\gamma = \text{centraliser}_{\mathrm{Gl}_2}(\gamma) =$ stabiliser of $x_\gamma, x_{\gamma'} \in \mathbb{P}^1(\mathbb{C})$. This torus is defined over $\mathbb{Q}(i)$, but it is not necessarily diagonalisable over $\mathbb{Q}(i)$, as it may be that the coordinates of $x_\gamma, x_{\gamma'}$ lie in a quadratic extension $F/\mathbb{Q}(i)$. This is the quadratic extension defined by the eigenvalues of γ.

We look at the edges of the fundamental domain F_0. We saw that they consist of connected pieces of the straight lines

$$G_1 = \{(z,\zeta) \mid z = 0\}, G_2 = \{(z,\zeta) \mid z = \frac{1}{2}\}, G_3 = \{(z,\zeta) \mid z = \frac{1+i}{2}\},$$

and of the circles (these circles are euclidean circles and geodesics for the hyperbolic metric)

$$D_1 = \{(z,\zeta) \mid z\bar{z} + \zeta^2 = 1, \Im(z) = \mathrm{Re}(z)\}, D_2 = \{(z,\zeta) \mid z\bar{z} + \zeta^2 = 1, \Im(z) = 0\},$$

$$D_3 = \{(z,\zeta) \mid z\bar{z} + \zeta^2 = 1, \mathrm{Re}(z) = \frac{1}{2}\}.$$

The pair of points $(\infty, (z_0, 0)) \in \mathbb{P}^1(\mathbb{C}) \times \mathbb{P}^1(\mathbb{C})$ has as its stabiliser

$$T_{z_0}(\mathbb{C}) = \left\langle \begin{pmatrix} 1 & z_0 \\ 0 & 1 \end{pmatrix} \begin{pmatrix} \alpha & 0 \\ 0 & \beta \end{pmatrix} \begin{pmatrix} 1 & -z_0 \\ 0 & 1 \end{pmatrix} \right\rangle, z_0 \in \mathbb{C}, \alpha, \beta \in \mathbb{C}^\times$$

where the straight line $\{(z_0, \zeta) \mid \zeta > 0\}$ is an orbit under $T_{z_0}(\mathbb{C})$ and it consists of fixed points for

$$T_{z_0}(\mathbb{C})(1) := \left\{ \begin{pmatrix} \alpha & z_0(\beta - \alpha) \\ 0 & \beta \end{pmatrix} \middle| \alpha/\beta \in S^1 \right\}, \; z_0 \in \mathbb{C}, \alpha\beta \in \mathbb{C}^times$$

We can easily compute the pointwise stabiliser of G_1, G_2, G_3 in $\tilde{\Gamma}$. They are

$$\begin{aligned}
\tilde{\Gamma}_{G_1} &= \left\{ \begin{pmatrix} i^\nu & 0 \\ 0 & i^\mu \end{pmatrix} \right\} = \left\{ \begin{pmatrix} i^\nu & 0 \\ 0 & 1 \end{pmatrix} \right\} \cdot Z_{\tilde{\Gamma}} \\
\tilde{\Gamma}_{G_2} &= \left\{ \begin{pmatrix} i^\nu & \frac{1-i^\nu}{2} \\ 0 & 1 \end{pmatrix} \middle| \frac{1-i^\nu}{2} \in \mathbb{Z}[i] \right\} \cdot Z_{\tilde{\Gamma}} = \left\{ \begin{pmatrix} \pm 1 & \frac{1 \mp 1}{2} \\ 0 & 1 \end{pmatrix} \right\} \cdot Z_{\tilde{\Gamma}} \\
\tilde{\Gamma}_{G_3} &= \left\{ \begin{pmatrix} i^\nu & \frac{(1-i^\nu)(1+i)}{2} \\ 0 & 1 \end{pmatrix} \right\} \cdot Z_{\tilde{\Gamma}},
\end{aligned}$$

where in the last case we have to take into account that $\frac{(1-i^\nu)(1+i)}{2} \in \mathbb{Z}[i]$ for all ν.

Hence modulo the centre $Z_{\tilde{\Gamma}}$ these stabilisers are cyclic groups of order 4, 2 and 4 respectively.

The arcs D_i are also pointwise fixed under the action of certain cyclic groups, namely

$$\begin{aligned} D_1 =& \mathrm{Fix}\left(\begin{pmatrix} 0 & i \\ 1 & 0 \end{pmatrix}\right) \\ D_2 =& \mathrm{Fix}\left(\begin{pmatrix} 0 & 1 \\ 1 & 0 \end{pmatrix}\right) \\ D_3 =& \mathrm{Fix}\left(\begin{pmatrix} 1 & -1 \\ 1 & 0 \end{pmatrix}\right), \end{aligned}$$

and we check easily that these arcs are geodesics joining the following points on the boundary

$$\begin{aligned} &D_1 \text{ runs from } \sqrt{i} \text{ to } -\sqrt{i} \\ &D_2 \text{ runs from } 1 \text{ to } -1 \\ &D_3 \text{ runs from } \rho = e^{\frac{2\pi i}{6}} = e^{\frac{\pi i}{3}} \text{ to } \overline{\rho}. \end{aligned}$$

The corresponding tori are

$$\begin{aligned} T_1 =& \mathrm{Stab}(-\sqrt{i}, \sqrt{i}) = \left\{\begin{pmatrix} \alpha & i\beta \\ \beta & \alpha \end{pmatrix}\right\} \\ T_2 =& \mathrm{Stab}(-1, 1) = \left\{\begin{pmatrix} \alpha & \beta \\ \beta & \alpha \end{pmatrix}\right\} \\ T_3 =& \mathrm{Stab}(\rho, \overline{\rho}) = \left\{\begin{pmatrix} \delta-\beta & \beta \\ -\beta & \delta \end{pmatrix}\right\}. \end{aligned}$$

The torus T_2 splits over $\mathbb{Q}(i)$, the other two tori split over a quadratic extension of $\mathbb{Q}(i)$.

Now it is not difficult anymore to describe the finite stabilisers and the corresponding fixed point sets. If $x \in \mathbb{H}_3$ is such that the stabiliser is bigger than $Z_{\tilde{\Gamma}}$, then we can conjugate x into F_0. It is very easy to see that x cannot lie in the interior of F_0 because then we would get an identification of two points nearby x and hence still in F_0 under $\tilde{\Gamma}$.

If x is on one of the lines D_1, D_2, D_3 or on one of the arcs G_1, G_2, G_3 but not on the intersection of two of them, then the stabiliser Γ_x is equal to $Z_{\tilde{\Gamma}}$ times the cyclic group we attached to the line or the arc earlier. Finally we are left with the three special points

$$\begin{aligned} x_{12} =& D_1 \cap D_2 \cap G_1 = \{(0,1)\} \\ x_{13} =& D_1 \cap D_3 \cap G_3 = \left\{\left(\frac{1+i}{2}, \frac{\sqrt{2}}{2}\right)\right\} \\ x_{23} =& D_2 \cap D_3 \cap G_2 = \left\{\left(\frac{1}{2}, \frac{\sqrt{3}}{2}\right)\right\}. \end{aligned}$$

In this case it is clear that the stabilisers are given modulo central elements by

$$\tilde{\Gamma}_{x_{12}} = \left\langle \begin{pmatrix} 0 & i \\ 1 & 0 \end{pmatrix}, \begin{pmatrix} i & 0 \\ 0 & 1 \end{pmatrix} \right\rangle = D_4$$

$$\tilde{\Gamma}_{x_{13}} = \left\langle \begin{pmatrix} 0 & i \\ 1 & 0 \end{pmatrix}, \begin{pmatrix} 1 & -1 \\ 1 & 0 \end{pmatrix}, \begin{pmatrix} i & 1 \\ 0 & 1 \end{pmatrix} \right\rangle = S_4$$

$$\tilde{\Gamma}_{x_{23}} = \left\langle \begin{pmatrix} 0 & 1 \\ 1 & 0 \end{pmatrix}, \begin{pmatrix} 1 & -1 \\ 1 & 0 \end{pmatrix} \right\rangle = S_3.$$

1.2.5 Compactification of $\Gamma\backslash X$

Our two special low dimensional examples show clearly that the quotient spaces $\Gamma\backslash X$ are not compact in general. There exist various constructions to compactify them.

If, for instance, $\Gamma \subset \mathrm{Sl}_2(\mathbb{Z})$ is a subgroup of finite index, then the quotient $\Gamma\backslash\mathbb{H}$ is a Riemann surface. It can be embedded into a compact Riemann surface by adding a finite number of points. This is a special case of a more general theorem of Satake and Baily-Borel: If the symmetric space X is actually hermitian symmetric (this means it has a complex structure) then we have the structure of a quasi-projective variety on $\Gamma\backslash X$. This is the so-called Baily-Borel compactification.

I will discuss the process of compactification in some more detail for our special low-dimensional examples.

Compactification of $\mathrm{Sl}_2(\mathbb{Z})\backslash\mathbb{H}$ by adding points

Let $\Gamma \subset \mathrm{Sl}_2(\mathbb{Z})$ be any subgroup of finite index. The group Γ acts on the rational projective line $\mathbb{P}^1(\mathbb{Q})$. We add it to the upper half plane and form

$$\overline{\mathbb{H}} = \mathbb{H} \cup \mathbb{P}^1(\mathbb{Q}),$$

and we extend the action of Γ to this space. Since the full group $\mathrm{Sl}_2(\mathbb{Z})$ acts transitively on $\mathbb{P}^1(\mathbb{Q})$ we find that Γ has only finitely many orbits on $\mathbb{P}^1(\mathbb{Q})$.

Now we introduce a topology on $\overline{\mathbb{H}}$. We define a system of neighbourhoods of points $\frac{p}{q} = r \in \mathbb{P}^1(\mathbb{Q})$. We define the Farey circles $S\left(c, \frac{p}{q}\right)$ which touch the real axis in the point $r = p/q$, $(p,q) = 1$, and have the radius $\frac{c}{2q^2}$. For $c = 1$ we get the picture: re Let us denote by $D\left(c, \frac{p}{q}\right) := \bigcup_{c':0<c'\leq c} S\left(c', \frac{p}{q}\right)$ the Farey disks. For $c \to 0$ these Farey disks $D\left(c, \frac{p}{q}\right)$ define a system of neighbourhoods of the point $r = p/q$. The Farey disks at $\infty \in \mathbb{P}^1(\mathbb{Q})$ are given by the regions

$$D(c,\infty) = \{z \mid \Im(z) \geq 1/c\}.$$

It is easy to check that an element $\gamma \in \mathrm{Sl}_2(\mathbb{Z})$ which sends $\infty \in \mathbb{P}^1(\mathbb{Q})$ into the point $r = \frac{p}{q}$ sends $D(c,\infty)$ to $D\left(c, \frac{p}{q}\right)$. These Farey disks $D(c,r)$ do not meet provided we take $c < 1$. The considerations in 1.6.1 imply that the complement

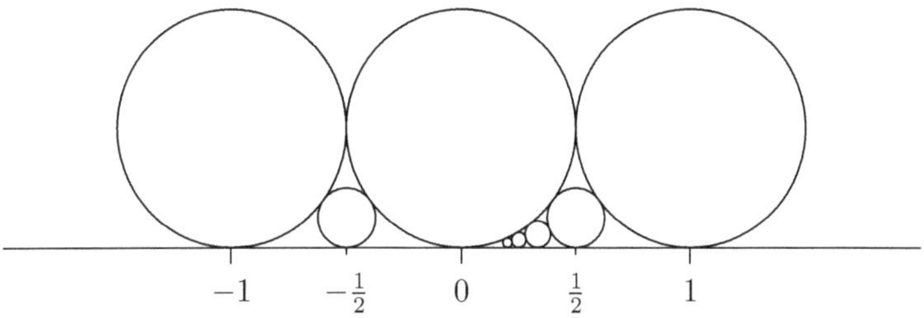

Figure 1.2: The Farey circles

of the union of Farey disks is relatively compact modulo Γ, and since Γ has finitely many orbits on $\mathbb{P}^1(\mathbb{Q})$, we see easily that

$$Y_\Gamma = \Gamma\backslash\overline{\mathbb{H}}$$

is compact (which means of course also Hausdorff).

It is essential that the set of Farey circles $D(c,r)$ and $D(c,\infty)$ is invariant under the action of Γ on the one hand and decomposes into several connected components (which are labeled by the point $r \in \mathbb{P}^1(\mathbb{Q})$) on the other hand. Hence

$$\Gamma\backslash\bigcup_r D(c,r) = \bigcup \Gamma_{r_i}\backslash D(c,r_i)$$

where r_i is a set of representatives for the action of Γ on $\mathbb{P}^1(\mathbb{Q})$ and where Γ_{r_i} is the stabiliser of r_i in Γ.

It is now clear that $\Gamma_{r_i}\backslash D(c,r_i)$ is holomorphically equivalent to a punctured disc and hence the above compactification is obtained by filling the point into this punctured disc, which makes it clear that Y_Γ is a Riemann surface.

1.2.6 The Borel-Serre compactification of $\mathrm{Sl}_2(\mathbb{Z})\backslash\mathbb{H}$

There is another construction of a compactification. We look at the disks $D(c,r)$ and divide them by the action of Γ_r. For any point $y \in S(c',r)-\{r\}$ there exists a unique geodesic joining r and y, passing orthogonally through $S(c',r)$ and hitting the projective line in another point y_∞ ($= -1/4$ in the picture below): If $r=\infty$, then this system of geodesics is given by the vertical lines $\{y\cdot i + x \mid y \in \mathbb{R}\}$. This allows us to write the set

$$D(c,r) - \{r\} = X_{\infty,r} \times [c,0)$$

where $X_{\infty,r} = \mathbb{P}^1(\mathbb{R}) - \{r\}$. The stabiliser Γ_r acts on $D(c,r)$ and on the right hand side of the identification it acts on the first factor, the quotient $\Gamma_r\backslash X_{\infty,r}$ being a circle. Hence we can compactify the quotient

$$\Gamma_r\backslash D(c,r) - \{r\} \hookrightarrow \Gamma_r\backslash X_{\infty,r} \times [c,0].$$

This gives us a second way to compactify $\Gamma\backslash\mathbb{H}$ - we apply this process to a finite set of representatives of $\mathbb{P}^1(\mathbb{Q})$ mod Γ.

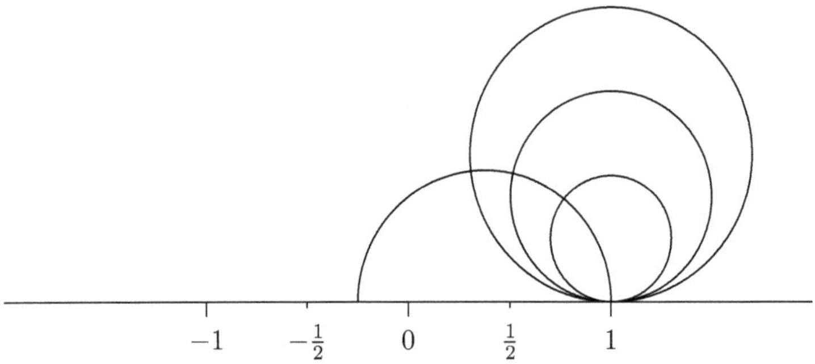

Figure 1.3: The compactification

There is a slightly different way of looking at this. We may form the union

$$\mathbb{H} \cup \bigcup_r X_{\infty,r} = \tilde{\mathbb{H}},$$

and topologize it in such a way that

$$D(c,r) = X_{\infty,r} \times [c,0) \subset X_{\infty,r} \times [c,0]$$

is a local homeomorphism. Then we see that the compactification above is just the quotient $\Gamma\backslash\tilde{\mathbb{H}}$ and the boundary is simply

$$\partial(\Gamma\backslash\bar{\mathbb{H}}) = \Gamma\backslash \bigcup_{r\in\mathbb{P}^1(\mathbb{Q})} X_{\infty,r}.$$

If take $r = \mathbf{i}\infty$. Then the stabilizer is the group $\Gamma_\infty = \begin{pmatrix} \pm1 & n \\ 0 & \pm1 \end{pmatrix}$

This compactification is called the Borel-Serre compactification. Its relation to the Baily-Borel compactification is that the latter is obtained by the former, by collapsing the circles at infinity to a point.

It is quite clear that a similar construction applies to the action of a congruence sub group $\Gamma \subset \mathrm{Sl}_2(\mathbb{Z}[i])$ on the three-dimensional hyperbolic space. The Farey circles will be substituted by spheres $S(c,\alpha)$ which touch the complex plane $\{(z,0) \mid z \in \mathbb{C}\} \subset \overline{\mathbb{H}}_3$ at the point $(0,\alpha), \alpha \in \mathbb{P}^1(\mathbb{Q}(i))$, and for $\alpha = \infty$ the Farey sphere is the horizontal plane $S(\infty,\zeta_0) = \{(z,\zeta_0) \mid z \in \mathbb{C}\}$. An element $\gamma \in \Gamma$ which maps ∞ to α maps $S(\infty,\zeta_0)$ to $S(c,\alpha)$, where $c = 1/\zeta_0$. For a given α we may identify the different spheres if we vary c, and for any point $\alpha \in \mathbb{P}^1(\mathbb{Q}(i))$ we define $X_{\infty,\alpha} = \mathbb{P}^1(\mathbb{C}) \setminus \{\alpha\}$. Again we can identify

$$D(c,\alpha) \setminus \{\alpha\} = X_{\infty,\alpha} \times (0,c] \subset \overline{D(c,\alpha) \setminus \{\alpha\}} = \partial(\mathbb{H}) = X_{\infty,\alpha} \times [0,c].$$

The stabiliser Γ_α acts on $D(c,\alpha) \setminus \{\alpha\}$ and again this yields an action on the first factor. If we choose $\alpha = \infty$ then

$$\Gamma_\infty = \left\{ \begin{pmatrix} \zeta & a \\ 0 & \zeta^{-1} \end{pmatrix} \mid \zeta \text{ power of } i, a \in \mathfrak{a}_\infty \right\}$$

where $\mathfrak{a}_\infty$ is a free rank 2 module in $\mathbb{Z}[i]$. If ζ does not assume the value i then $\Gamma_\infty\backslash X_{\infty,\infty}$ is a two-dimensional torus, a product of two circles. If ζ assumes the value i then $\Gamma_\infty\backslash X_{\infty,\infty}$ is a two dimensional sphere. If $\Gamma = \mathrm{Sl}_2(\mathbb{Z}[i])$ then $\mathfrak{a}_\infty = \mathbb{Z}[i]$. Of course we get the same result for an arbitrary α.

Then we get an action of the group Γ on $\tilde{\mathbb{H}}_3 = \mathbb{H}_3 \cup \bigcup_{\alpha\in\mathbb{P}^1(\mathbb{Q}(i))} \overline{D(c,\alpha)\setminus\{\alpha\}}$ with compact quotient. The set of orbits of Γ on $\mathbb{P}^1(\mathbb{Q}(i))$ is finite, these orbits are called (the) cusps.

1.2.7 The Borel-Serre compactification and the reduction theory of arithmetic groups

This section could be skipped in a first reading. For the particular groups $\mathrm{Sl}_2/\mathbb{Q}$ or $\mathrm{Sl}_2(\mathbb{Q}(\sqrt{-d}))$ this compactification has been discussed in detail in the previous sections. A reader who is interested in the specific applications to number theory, which will be discussed in the following Chapters 2-5, only needs the results from section 1.2.6.

The Borel-Serre compactification works in complete generality for any semi-simple or reductive group $G/\mathbb{Q}$. To explain it, we need the notion of a *parabolic subgroup* of $G/\mathbb{Q}$.

A subgroup $P/\mathbb{Q} \hookrightarrow G/\mathbb{Q}$ is parabolic if the quotient variety $P\backslash G$ is a projective variety. We mentioned already earlier that for the group $\mathrm{Gl}_2/\mathbb{Q}$ we have an action of Gl_2 on the projective line $\mathbb{P}^1$ and the stabilisers B_x of the points $x \in \mathbb{P}^1(\mathbb{Q})$ are the so-called Borel subgroups of $\mathrm{Gl}_2/\mathbb{Q}$. They are maximal solvable subgroups and

$$\mathrm{Gl}_2/B_x = \mathbb{P}^1,$$

hence they are also parabolic.

More generally we get parabolic subgroups of $\mathrm{Gl}_n/\mathbb{Q}$, if we choose a flag on the vector space $V = \mathbb{Q}^n = \mathbb{Q}e_1 \oplus \cdots \oplus \mathbb{Q}e_n$. This is an increasing sequence of subspaces

$$\mathcal{F} := \{0\} = V_0 \subsetneq V_1 \subsetneq V_2 \subsetneq \ldots \subsetneq V_k = V.$$

The stabiliser P of such a flag is always a parabolic subgroup; the quotient space

$$G/P = \text{variety of all flags of the given type},$$

where the type of the flag is the sequence of the dimensions $n_i = \dim V_i$.

These *flag varieties* (the Grassmannians) are smooth projective schemes over $\mathrm{Spec}(\mathbb{Z})$, and this implies that any flag $\mathcal{F}$ is induced by a flag

$$\mathcal{F}_\mathbb{Z} := (0) = L_0 \subsetneq L_1 \subsetneq L_2 \subsetneq \ldots \subset L_k = L = \mathbb{Z}^n \tag{1.43}$$

where $L_i = V_i \cap L$, and of course $L_i \otimes \mathbb{Q} = V_i$. This is an elementary fact which will be used later.

If our group $G/\mathbb{Q}$ is the orthogonal group of a quadratic form

$$f(x_1,\ldots,x_n) = \sum_{i=1}^n a_i x_i^2$$

with $a_i \in \mathbb{Q}^\times$, then we have to replace the flags by sequences of subspaces

$$\mathcal{F}: 0 \subsetneq W_1 \subsetneq W_2 \cdots \qquad \subsetneq W_2^\perp \subsetneq W_1^\perp \subsetneq V,$$

where the W_i are isotropic spaces for the form f, i.e. $f \mid W_i \equiv 0$, and where the $W_i^\perp$ are the orthogonal complements of the subspaces. Again, the stabilisers of these flags are the parabolic subgroups defined over $\mathbb{Q}$. Especially if the form f is anisotropic over $\mathbb{Q}$, i.e. if there is no non-zero vector $\underline{x} \in K^n$ with $f(\underline{x}) = 0$, then the group $G/\mathbb{Q}$ does not have any parabolic subgroup over $\mathbb{Q}$. This is equivalent to the fact that $G(\mathbb{Q})$ does not have unipotent elements.

These parabolic subgroups always have a unipotent radical U_P. It is the subgroup which acts trivially on the successive quotients of the flag. The unipotent radical is a normal subgroup, the quotient $P/U_P = M$ is a reductive group again and is called the *Levi-quotient* of P.

We go back to the group $\mathrm{Gl}_n/\mathbb{Z}$. It contains the standard maximal torus whose R-valued points are

$$T_0(R) = \left\{ \underline{t} = \begin{pmatrix} t_1 & 0 & \dots & 0 \\ 0 & t_2 & \dots & 0 \\ 0 & 0 & \ddots & 0 \\ 0 & 0 & 0 & t_n \end{pmatrix} \quad \middle| \; t_i \in R^\times \right\}. \tag{1.44}$$

It is a subgroup of the Borel subgroup (maximal connected solvable subgroup or minimal parabolic subgroup) whose R-valued points are

$$B_0(R) = \left\{ \underline{b} = \begin{pmatrix} t_1 & u_{1,2} & \dots & u_{1,n} \\ 0 & t_2 & \dots & u_{2,n} \\ 0 & 0 & \ddots & u_{n-1,n} \\ 0 & 0 & 0 & t_n \end{pmatrix} \quad \middle| \; t_i \in R^\times, u_{ij} \in R \right\}, \tag{1.45}$$

and its unipotent radical U_0 consists of those $b \in B_0$ where all the $t_i = 1$. This unipotent radical contains the one dimensional *root subgroups*

$$U_{i,j} = \left\{ \begin{pmatrix} 1 & 0 & \dots & 0 & 0 \\ 0 & 1 & \dots & 0 & 0 \\ 0 & 0 & \ddots & x & 0 \\ 0 & 0 & 0 & \ddots & 0 \\ 0 & 0 & 0 & 0 & 1 \end{pmatrix} \middle| \; x \in R \right\} \tag{1.46}$$

where $i < j$. These one dimensional subgroups are isomorphic to the one dimensional additive group $\mathbb{G}_a$. They are normalized by the torus. For an element $\underline{t} \in T(R)$ and $x_{i,j} \in U_{i,j}(R) \cong R$ we have

$$\underline{t} x_{i,j} \underline{t}^{-1} = (t_i/t_j) x_{i,j}. \tag{1.47}$$

For $i, j = 1, \dots, n, i \neq j$ (resp. $i < j$) the characters $\alpha_{i,j}(\underline{t}) = t_i/t_j$ are called the roots (resp. positive roots) of T_0 in Gl_n. We denote these systems of roots by Δ^{Gl_n} (resp. $\Delta_+^{\mathrm{Gl}_n}$). The one-dimensional subgroups $U_{i,j}, i \neq j$ are called the root subgroups.

Inside the set of positive roots we have the set of simple roots

$$\pi = \pi^{\mathrm{Gl}_n} = \{\alpha_{1,2}, \dots, \alpha_{i,i+1}, \dots, \alpha_{n-1,n}\}. \tag{1.48}$$

If we pass to the semi-simple subgroup $\mathrm{Sl}_n/\mathbb{Q}$ then the torus and the Borel-subgroup has to be replaced by $T_0^{(1)}, B_0^{(1)}$, where we have $\prod_i t_i = 1$. The system of roots does not change, and we have $\pi = \pi^{\mathrm{Gl}_n} = \pi^{\mathrm{Sl}_n}$.

We change the notation slightly: for $i = 1, \dots, n-1$ we define $\alpha_i := \alpha_{i,i+1}$. Then for $i < j$ we get $\alpha_{i,j} = \alpha_i + \dots + \alpha_{j-1}$, and $\pi = \{\alpha_1, \alpha_2, \dots, \alpha_{n-1}\}$.

The Borel subgroup B_0 is the stabiliser of the "complete" flag

$$\{0\} \subset \mathbb{Q}e_1 \subsetneq \mathbb{Q}e_1 \oplus \mathbb{Q}e_2 \subsetneq \dots \subsetneq \mathbb{Q}e_1 \oplus \mathbb{Q}e_2 \oplus \dots \oplus \mathbb{Q}e_n, \tag{1.49}$$

the parabolic subgroups $P_0 \supset B_0$ are the stabilisers of "partial" flags

$$\begin{aligned} &\{0\} \subsetneq \mathbb{Q}e_1 \oplus \dots \oplus \mathbb{Q}e_{n_1} \\ &\subsetneq \mathbb{Q}e_1 \oplus \dots \oplus \mathbb{Q}e_{n_1} \oplus \mathbb{Q}e_{n_1+1} \oplus \dots \oplus \mathbb{Q}e_{n_1+n_2} \subsetneq \dots \subsetneq \mathbb{Q}^n. \end{aligned} \tag{1.50}$$

The parabolic subgroup P_0 also acts on the direct sum of the successive quotients

$$(\mathbb{Q}e_1 \oplus \dots \oplus \mathbb{Q}e_{n_1}) \bigoplus (\mathbb{Q}e_{n_1+1} \oplus \dots \oplus \mathbb{Q}e_{n_1+n_2}) \bigoplus \cdots \tag{1.51}$$

and this yields a homomorphism

$$r_{P_0} : P_0 \to M_0 = \mathrm{Gl}_{n_1} \times \mathrm{Gl}_{n_2} \times \dots \tag{1.52}$$

hence M_0 is the Levi-quotient of P_0. By definition the unipotent radical U_{P_0} of P_0 is the kernel of r_0.

A parabolic subgroup $P_0 \supset B_0$ defines a subset

$$\Delta^{P_0} = \{\alpha_{i,j} \in \Delta^{\mathrm{Gl}_n} \mid U_{i,j} \subset P_0\}$$

and the set decomposes into two sets

$$\Delta^{M_0} = \{\alpha_{i,j} \mid U_{i,j} \text{ and } U_{j,i} \subset \Delta^{P_0}\};\ \Delta^{U_{P_0}} = \Delta^{P_0} \setminus \Delta^{M_0}. \tag{1.53}$$

Intersecting this decomposition with the set π^{Gl_n} yields a disjoint decomposition

$$\pi^{\mathrm{Gl}_n} = \pi^{M_0} \cup \pi^U, \tag{1.54}$$

where $\pi^U = \{\alpha_{n_1}, \alpha_{n_1+n_2}, \dots, \}$. In turn, any such decomposition of π^{Gl_n} yields a well-defined parabolic $P_0 \supset B_0$.

We define the *parabolic rank* of a parabolic subgroup as the number

$$d(P) = \#\pi^U.$$

The proper maximal parabolic subgroups are the ones with $d(P) = 1$.

If we choose another maximal split torus T_1 and a Borel subgroup $B_1 \supset T_1$, then this amounts to the choice of a second ordered basis $v_1, v_2, \dots, v_n$ of V, where the v_i are given up to a non-zero scalar factor. We can find a $g \in \mathrm{Gl}_n(\mathbb{Q})$ which maps $e_1, e_2, \dots, e_n$ to $v_1, v_2, \dots, v_n$, and hence we can conjugate the pair

(B_0, T_0) to (B_1, T_1), and thus also the parabolic subgroups containing B_0 into the parabolic subgroups containing B_1. The conjugating element g also identifies

$$i_{T_0,B_0,T_1,B_1} : X^*(T_0) \xrightarrow{\sim} X^*(T_1),$$

and this identification does not depend on the choice of the conjugating element g. This allows us to identify the two sets of positive simple roots $\pi^{\mathrm{Gl}_n} \subset X^*(T_0)$ and $\pi \subset X^*(T_1)$. Eventually we can speak of the set π of simple roots of Gl_n. Hence we have the fundamental fact:

The $\mathrm{Gl}_n(\mathbb{Q})$ conjugacy classes of parabolic subgroups $P/\mathbb{Q}$ are in one to one correspondence with the subsets $\pi^M \subset \pi$. The number of elements in $\pi \setminus \pi^M = \pi^U$ is called the parabolic rank of P, the set π^U is called the type of P.

We also introduce the notation π' for π^U.

We formulated this result for $\mathrm{Gl}_n/\mathbb{Q}$, but we can replace $\mathbb{Q}$ by any field k and Gl_n by any reductive group G/k. We have to define the system of relative simple positive roots π^G for any G/k (see [B-T]). We also refer to section 1.1.6.

The group G/k itself is also a parabolic subgroup which corresponds to $\pi' = \emptyset$. We decide that we do not like it and hence we consider only proper parabolic subgroups $P \neq G$, i.e. $\pi' \neq \emptyset$. We can define the Grassmann variety $\mathrm{Gr}^{[\pi']}$ of parabolic subgroups of type π'. This is a smooth projective variety and $\mathrm{Gr}^{[\pi']}(\mathbb{Q})$ is the set of parabolic subgroups of type π'.

There is always a unique minimal conjugacy class corresponding to $\pi' = \pi$. In our examples this minimal class is given by the maximal flags, i.e. those flags where the dimension of the subspaces increases by one at each step (until we reach a maximal isotropic space in the case of an orthogonal group). The (proper) maximal parabolic subgroups are those for which $\pi' = \{\alpha_i\}$, i.e. $\pi^{U_{P_i}} = \{\alpha_i\}$.

For any parabolic subgroup $P/\mathbb{Q} \subset G/\mathbb{Q}$ we consider the character module $X^*(P) := \mathrm{Hom}(P/\mathbb{Q}, \mathbb{G}_m)$. Since we do not have any non-trivial homomorphisms from the unipotent radical U_P to $\mathbb{G}_m$ we have $\mathrm{Hom}(P/\mathbb{Q}, \mathbb{G}_m) = \mathrm{Hom}(M_P, \mathbb{G}_m)$.

The reductive quotient $M_P = M_P^{(1)} \cdot Z_P$, where Z_P is the central torus and $M_P^{(1)}$ the semi-simple part (the derived group). The quotient $M_P/M_P^{(1)} = Z'_P$ is a torus and $Z_P \to Z'_P$ is an isogeny. Hence we have

$$\begin{aligned}\mathrm{Hom}(P/\mathbb{Q}, \mathbb{G}_m) \otimes \mathbb{Q} &= \mathrm{Hom}(M_P, \mathbb{G}_m) \otimes \mathbb{Q} \\ &= \mathrm{Hom}(Z_P, \mathbb{G}_m) \otimes \mathbb{Q} = \mathrm{Hom}(Z'_P, \mathbb{G}_m) \otimes \mathbb{Q}.\end{aligned} \tag{1.55}$$

If G is semisimple then for any maximal parabolic subgroup P of type $\pi' = \{\alpha_i\}$, the module $\mathrm{Hom}(P, \mathbb{G}_m) \otimes \mathbb{Q} \subset X^*(T) \otimes \mathbb{Q}$ is of rank one. In general

$$\mathrm{Hom}(P, \mathbb{G}_m) \otimes \mathbb{Q} = \mathbb{Q}\gamma_i \oplus X^*(S_P) \otimes \mathbb{Q}$$

where γ_i is a fundamental dominant weight. In the special case $G = \mathrm{Gl}_n$ we have

$$\gamma_i(t) = \left(\prod_{\nu=1}^{\nu=i} t_\nu \right) \det(t)^{-i/n}. \tag{1.56}$$

If our maximal parabolic subgroup $P/\mathbb{Q}$ is defined as the stabiliser of a flag $0 \subsetneq W \subsetneq V = \mathbb{Q}^n$, then the unipotent radical is $U \cong \mathrm{Hom}(V/W, W)$. An element $y \in P(\mathbb{Q})$ induces linear maps $y_W, y_{V/W}$ and hence $\mathrm{Ad}(y)$ on $U = \mathrm{Hom}(V/W, W)$. We get a character $\gamma_P(y) = \det(\mathrm{Ad}(y)) \in \mathrm{Hom}(P, \mathbb{G}_m)$ which is called the sum of the positive roots. An easy computation shows that

$$n\gamma_i = \gamma_P,\ i = 1, \ldots, n-1. \tag{1.57}$$

We add points at infinity to our symmetric space X: We consider the disjoint union $\cup_{\pi' \neq \emptyset} \mathrm{Gr}^{[\pi']}(\mathbb{Q})$ and form the space

$$\overline{X} = X \cup \bigcup_{\pi' \neq \emptyset} \mathrm{Gr}^{[\pi']}(\mathbb{Q}).$$

This is the analogue of $\mathbb{H} \cup \mathbb{P}^1(\mathbb{Q})$ in our first example, but it is now more complicated because we have several types of Grassmannians, and we have maps (deleting some members in the filtration)

$$r_{\pi_1,\pi_2} :\ \mathrm{Gr}^{[\pi_1]}(\mathbb{Q}) \to \mathrm{Gr}^{[\pi_2]}(\mathbb{Q}) \text{ if } \pi_2 \subset \pi_1.$$

Our first aim is to put a topology on this space such that $\Gamma\backslash\overline{X}$ becomes a compact Hausdorff space.

In our first example we interpreted the Farey discs $D\left(c, \frac{p}{q}\right)$ with $0 < c < 1$ as a system of open neighborhoods of $\frac{p}{q} \in \mathbb{H}$. They are close to the point $\frac{p}{q}$, but far away from any other point in $\mathbb{P}^1(\mathbb{Q})$.

We mimic this concept: For any parabolic $P \in \mathrm{Gr}^{[\pi']}(\mathbb{Q})$ (here we also allow $P = G$) we will define open sets

$$X^P(\underline{c}_{\pi'}, r(\underline{c}_{\pi'})) \subset X \tag{1.58}$$

which depend on certain parameters $\underline{c}_{\pi'}, r(c_{\pi'})$. The points in $X^P(\underline{c}_{\pi'}, r(\underline{c}_{\pi'}))$ should be viewed as the points, which are "very close" to the parabolic subgroup P (controlled by $\underline{c}_{\pi'}$) but "keep a certain distance" (controlled by $r(\underline{c}_{\pi'})$) to the parabolic subgroups $Q \not\supset P$. They are the analogues of the Farey circles. We will see:

a) This system of open sets is invariant under the action $\mathrm{Gl}_n(\mathbb{Z})$

b) For $P = G$ the set $X^G(\{\ \}, r_0)$ is relatively compact modulo the action of $\mathrm{Gl}_n(\mathbb{Z})$.

c) Any subgroup $\Gamma \subset \mathrm{Gl}_n(\mathbb{Z})$ has only finitely many orbits on any $\mathrm{Gr}^{[\pi']}(\mathbb{Q})$.

d) For a suitable choice of the parameters $\underline{c}_{\pi'}$, and $r(c_{\pi'})$ we have

$$X = \bigcup_P X^P(\underline{c}_{\pi'}, r(c_{\pi'})) = X^G(\emptyset, r_0) \cup \bigcup_{P:P \text{ proper}} X^P(\underline{c}_{\pi'}, r(\underline{c}_{\pi'}))$$

and if the parabolic groups P and P_1 are conjugate and $P \neq P_1$ then we have $X^P(\underline{c}_{\pi'}, r(\underline{c}_{\pi'})) \cap X^{P_1}(\underline{c}_{\pi'}, r(\underline{c}_{\pi'})) = \emptyset$.

Let us assume that we have constructed such a system of open sets. Then c) and d) imply that for a given type π' we have

$$\Gamma\backslash(\bigcup_{P:\ \mathrm{type}(P)=\pi'} X^P(c_{\pi'}, r(c_{\pi'}))) = \bigcup \Gamma_{P_i}\backslash X^{P_i}(c_{\pi'}, r(c_{\pi'}))$$

where $\{\dots, P_i, \dots\} = \Sigma(\pi, \Gamma)$ is a set of representatives of $\mathrm{Gr}^{[\pi']}(\mathbb{Q})$ modulo the action of Γ and $\Gamma_{P_i} = \Gamma \cap P_i(\mathbb{Q})$.

This tells us that we have a covering

$$\Gamma\backslash X = \Gamma\backslash X^G(\emptyset, r_0) \cup \bigcup_{\pi'\neq\emptyset} \bigcup_{P\in\Sigma(\pi',\Gamma)} \Gamma_P\backslash X^P(\underline{c}_{\pi'}, r(\underline{c}_{\pi'})). \tag{1.59}$$

The philosophy of reduction theory is that $\Gamma\backslash X^G(\emptyset, r_0)$ is relatively compact and that we have an explicit description of the sets $\Gamma_P\backslash X^P(\underline{c}_{\pi'}, r(\underline{c}_{\pi'}))$ as fiber bundles with nilmanifolds as fibers, over the locally symmetric spaces $\Gamma_M\backslash X^M$.

We give the definition of the sets $X^P(\underline{c}_{\pi'}, r(\underline{c}_{\pi'}))$. We stick to the case that $G = \mathrm{Gl}_n/\mathbb{Q}$ and $\Gamma \subset \Gamma_0 = \mathrm{Gl}_n(\mathbb{Z})$ is a (congruence) subgroup of finite index. We defined the positive-definite bilinear form (see 1.31)

$$\tilde{B}_{\Theta_x} = -\frac{1}{2n} B_{\Theta_x} : \mathfrak{g}_{\mathbb{R}} \times \mathfrak{g}_{\mathbb{R}} \to \mathbb{R},$$

and we have the identification $\mathfrak{g}_{\mathbb{R}} \xrightarrow{\sim} T_e^{\mathcal{G}(\mathbb{R})}$, and hence we get a euclidian metric on the tangent space $T_e^{\mathcal{G}(\mathbb{R})}$ at the identity e. This extends to a left invariant Riemannian metric on $G(\mathbb{R})$, and we denote it by $d_{\Theta_x}s^2$. Hence we get a volume form $d^{\Theta_x}_{\mathrm{vol_H}}$ on any closed subgroup $H(\mathbb{R}) \subset G(\mathbb{R})$.

For any point $x \in X$ and any parabolic subgroup $P/\mathbb{Q}$ with unipotent radical $U/\mathbb{Q}$ we define

$$p(P, x) = \mathrm{vol}_U^{\Theta_x}((\Gamma_0 \cap U(\mathbb{R}))\backslash U(\mathbb{R})) \tag{1.60}$$

The Arakelov-Chevalley scheme $(\mathrm{Gl}_n/\mathbb{Z}, \Theta_0)$

We start from the free lattice $L = \mathbb{Z}e_1 \oplus \mathbb{Z}e_2 \oplus \cdots \oplus \mathbb{Z}e_n$ and we think of $\mathrm{Gl}_n/\mathbb{Z}$ as the scheme of automorphisms of this lattice. If we choose an euclidian metric $\langle\ ,\ \rangle$ on $L \otimes \mathbb{R}$, then we call the pair $(L, \langle\ ,\ \rangle)$ an *Arakelov vector bundle.* From the conformal class of the metric we get a Cartan involution Θ on $\mathrm{Gl}_n(\mathbb{R})$, and the pair $(\mathrm{Gl}_n/\mathbb{Z}, \Theta)$ is an *Arakelov group scheme.*

We may choose the standard euclidian metric with respect to the given basis, i.e. $\langle e_i, e_j\rangle = \delta_{i,j}$. The resulting Cartan involution is the standard one: $\Theta_0 : g \mapsto ({}^t g)^{-1}$. This pair $(\mathrm{Gl}_n/\mathbb{Z}, \Theta_0)$ is called an Arakelov-Chevalley scheme (in a certain sense the integral structure of $\mathrm{Gl}_n/\mathbb{Z}$ and the choice of the Cartan involution are "optimally adapted").

In this case we find for our basis elements in (1.13)

$$B_{\Theta_0}(H_i, H_j) = -2n\delta_{i,j} + 2,\ B_{\Theta_0}(E_{i,j}, E_{k,l}) = -2n\delta_{i,k}\delta_{j,l}, \tag{1.61}$$

hence the $E_{i,j}$ are part of an orthonormal basis.

We propose to call a pair $(L, \langle\ ,\ \rangle_x)$ an Arakelov vector bundle over $\mathrm{Spec}(\mathbb{Z}) \cup \{\infty\}$ and $(\mathrm{Gl}_n, \Theta_x)$ an Arakelov group scheme. The Arakelov vector bundles modulo conformal equivalence are in one-to-one correspondence with the Arakelov group schemes of type Gl_n.

For the Arakelow-Chevalley scheme $(\mathrm{Gl}_n/\mathbb{Z}, \Theta_0)$ (see 1.2.7) we have that $\tilde{B}_{\Theta_0}(E_{i,j}) = 1$. We have by construction

$$U_{i,j}(\mathbb{Z})\backslash U_{i,j}(\mathbb{R}) = \mathbb{R}/\mathbb{Z} \tag{1.62}$$

and under this identification $E_{i,j}$ maps to $\frac{\partial}{\partial x}$. Hence we get

$$\mathrm{vol}^{\Theta_0}_{U_{i,j}}(U_{i,j}(\mathbb{Z})\backslash U_{i,j}(\mathbb{R})) = 1$$

and from this we immediately get:

Proposition 1.2.2. *For any parabolic subgroup P_0 containing the torus T_0 we have*

$$p(P_0, \Theta_0) = 1.$$

Let $(L, \langle\ ,\ \rangle_x)$ be an Arakelow vector bundle and $(\mathrm{Gl}_n, \Theta_x)$ the corresponding Arakelow group scheme (of type Gl_n). Let

$$\mathcal{F}_{\mathbb{Z}} : (0) = \{(0)\} = L_0 \subset L_1 \subset L_2 \subset \cdots \subset L_k = L = \mathbb{Z}^n$$

be a flag and $P/\mathbb{Z}$ the corresponding parabolic subgroup. Then we have the homomorphism

$$r_P : P/\mathrm{Spec}(\mathbb{Z}) \to M/\mathbb{Z} = \prod_{i=1}^{k} \mathrm{Gl}(L_i/L_{i-1}) \tag{1.63}$$

with kernel $U_P/\mathbb{Z}$. The metric $\langle\ ,\ \rangle_x$ on $L \otimes \mathbb{R}$ yields an orthogonal decomposition

$$L \otimes \mathbb{R} = \bigoplus_{i=1}^{k} L_i/L_{i-1} \otimes \mathbb{R}$$

and hence an Arakelow bundle structure $(L_i/L_{i-1}, (\Theta_x)_i)$ for all i, and therefore an Arakelow group scheme structure on $M/\mathbb{Z}$.

Hence we get:

Proposition 1.2.3. *If (Gl_n, Θ) is an Arakelow group scheme, then Θ induces an Arakelow group scheme structure Θ^M on any reductive quotient $M = P/U$.*

Definition: A pair $(\mathrm{Gl}_n/\mathbb{Z}, \Theta)$ is called *stable (resp. semi-stable)* if for any proper parabolic subgroup $P/\mathbb{Q} \subset \mathrm{Gl}_n/\mathbb{Q}$ we have

$$p(P, \Theta) > 1 \text{ (resp. } p(P, \Theta) \geq 1) \tag{1.64}$$

In our example in 1.2.5 the stable points are those outside the union of the closed Farey circles $D(1, \frac{p}{q})$. The stable points in the fundamental domain (1.42) are the points with $\Im(z) \leq 1$. The lattice $\Omega =< 1, z >$ comes with the flag

$B_0 = \{0\} \subset \mathbb{Z} \cdot 1 \subset \Omega$ and clearly $p(B_0, z) = \frac{1}{(}\Im(z)$. Hence we see that the maximum value for $p(B_0, z)$ and $z \in \mathcal{F}$ is assumed at $z = \rho = \frac{1+\sqrt{-3}}{2}$ and this maximum value is

$$p(B_0, \rho) = \frac{2}{\sqrt{3}} \tag{1.65}$$

To get a better understanding of these numbers we have to do some computations with roots and weights. Let us start from an Arakelow vector bundle $(L = \mathbb{Z}^d, \langle\ ,\ \rangle)$ and let us assume that L is equipped with a complete flag

$$\mathcal{F}_0 := \{0\} = L_0 \subset L_1 \subset \cdots \subset L_{d-1} \subset L_d \tag{1.66}$$

which defines a Borel subgroup $B/\mathbb{Z}$. The quotients $(L_i/L_{i-1}, \langle\ ,\ \rangle_i)$ are Arakelow line bundles over $\mathbb{Z}$, or in a less sophisticated language, they are free $\mathbb{Z}$-modules of rank one and the generating vector $\bar{e}_i$ has a length $\sqrt{\langle \bar{e}_i, \bar{e}_i \rangle_i}$. This length is of course also the volume of $(L_i/L_{i-1} \otimes \mathbb{R})/(L_i/L_{i-1})$. The unipotent radical $U/\mathbb{Z} \subset B/\mathbb{Z}$ has a filtration

$$\{(0)\} \subset V_1 \subset \ldots, V_{(n(n-1)/2)-1} \subset V_{n(n-1)/2} = U$$

by normal subgroups, the successive quotients are isomorphic to $\mathbb{G}_a$ and the torus $T = B/U$ acts by a positive root $\alpha_{i,j}$. This is a one-to-one correspondence between the subquotients and the positive roots. Then it is clear: if ν corresponds to (i, j), then

$$(V_\nu/V_{\nu+1}, \Theta_\nu) = (L_i/L_{i-1}, \langle\ ,\ \rangle_i) \otimes (L_j/L_{j-1}, \langle\ ,\ \rangle_j)^{-1}. \tag{1.67}$$

Moreover the quotients $(V_\nu/V_{\nu+1}, \Theta_\nu)$ depend only on the conformal class of $\langle\ ,\ \rangle$ and hence only on the resulting Cartan involution Θ.

The unipotent subgroup $U/\mathbb{Z}$ contains the one parameter subgroup $U_{i,j}/\mathbb{Z}$ and this one parameter subgroup maps isomorphically to $(V_\nu/V_{\nu+1})$. Hence our construction defines the Arakelow line bundle $(U_{i,j}, \Theta_{i,j})$.

If we now define $n_{\alpha_{i,j}}(B, x) = \mathrm{vol}_{U_{i,j}}^{\Theta_{i,j}}(U_{i,j}(\mathbb{R})/U_{i,j}(\mathbb{Z}))$ then it is clear that

$$p(B, x) = \prod_{i<j} n_{\alpha_{i,j}}(B, x). \tag{1.68}$$

If $P \supset B$ then its unipotent radical is $U_P \subset U$ and we defined the set Δ^{U_P} as the set of positive roots for which $U_{i,j} \subset U_P$. Then we have

$$p(P, x) = \prod_{\alpha_{i,j} \in \Delta^{U_P}} n_{\alpha_{i,j}}(B, x). \tag{1.69}$$

Here it is important to notice that the right hand side does not depend on the choice of $B \subset P$.

We follow a convention and put $2\rho_P = \sum_{\alpha_{i,j} \in \Delta^{U_P}} \alpha_{i,j}$ so that ρ_P is half the sum of positive roots in the unipotent radical. Formula (1.57) tells us that for any maximal parabolic subgroup P_{i_0}

$$2\rho_{P_{i_0}} = \sum_{i \le i_0, j \ge i_0+1} \alpha_{i,j} = n\gamma_{i_0}. \tag{1.70}$$

For any $\gamma = \sum \alpha_i \otimes z_i \in X^*(T) \otimes \mathbb{C}$ we define the homomorphism

$$|\gamma| : T(\mathbb{R}) \to \mathbb{C}^\times : |\gamma| : t \to \prod_i |\alpha_i(t)|^{z_i}. \tag{1.71}$$

Since the numbers $n_{\alpha_i}(B,x)$ are positive real numbers, we define:

$$n_\gamma(B,x) = \prod_{i=1}^{n-1} n_{\alpha_i}(B,x)^{z_i}. \tag{1.72}$$

Here we see that the second argument is a Borel-subgroup B. But if the above character $\gamma : B \to \mathbb{G}_m$ extends to a character $\gamma : P \to \mathbb{G}_m$ then we can define

$$n_\gamma(P,x) := n_\gamma(B,x),$$

because this number only depends on P and not on the Borel subgroup $B \subset P$. The characters in $\gamma \in X^*(T)$ for which $|\gamma|$ extend to $P(\mathbb{R})$ are exactly the linear combinations (see (1.74) below) $\gamma = \sum_{\alpha_i \in \pi'} x_i \gamma_i$. The characters being $\gamma_P = \sum_{\alpha_i \in \pi^U} r_i \gamma_i$ where the $r_i > 0$ are rational numbers. Let P_i be the maximal parabolic subgroup of type $\pi \setminus \{\alpha_i\}$ containing P. Then the above formula implies that

$$p(P,x) = \prod_{\alpha_i \in \pi^U} n_{\gamma_i}(P_i,x)^{r_i} = \prod_{\alpha_i \in \pi^U} p_{P_i}(P_i,x)^{\frac{r_i}{n}}, \tag{1.73}$$

and hence we know

The Arakelow scheme $(\mathrm{Gl}_n/\mathbb{Z}, \Theta)$ *is stable if for all maximal parabolic subgroups* $p_{P_i}(P_i,\Theta) = n_{\gamma_i}(P_i,\Theta)^n > 1$.

We need a few more formulas relating roots and weights. For any parabolic subgroup we have the decomposition of the set of simple roots into two parts

$$\pi = \pi^M \cup \pi^{U_P}.$$

This induces a splitting of the character module

$$X^*(T) \otimes \mathbb{Q} = \bigoplus_{\alpha_i \in \pi^M} \mathbb{Q}\alpha_i \oplus \bigoplus_{\alpha_i \in \pi^{U_P}} \mathbb{Q}\gamma_i \tag{1.74}$$

where γ_i is the dominant fundamental weight attached to α_i (see (1.56)).

If now $\alpha_i \in \pi^{U_P}$ then we can project α_i to the second component, with projection

$$\alpha_i^P = \alpha_i + \sum_{\alpha_\nu \in \pi^M} c_{i,\nu}\alpha_\nu. \tag{1.75}$$

Here an elementary - but not completely trivial - computation shows that

$$c_{i,\nu} \geq 0 \tag{1.76}$$

Since $\alpha_i^P \in \bigoplus_{\alpha_i \in \pi^{U_P}} \mathbb{Q}\gamma_i$, these characters extend to $P(\mathbb{R})$ and hence $n_{\alpha_i^P}(P,x)$ is defined.

We state the two fundamental theorems of reduction theory

Theorem 1.2.1. *For any Arakelow group scheme* $(\mathrm{Gl}_n, \Theta_x)$ *we can find a Borel subgroup* $B \subset \mathrm{Gl}_n$ *for which*

$$n_{\alpha_i}(B, \Theta_x) = n_{\alpha_i}(B, x) < \frac{2}{\sqrt{3}} \text{ for all } i = 1, \dots, n-1$$

We call a Borel subgroup B *reduced* if it satisfies the inequalities above.

Theorem 1.2.2. *For any Arakelow group scheme* (Gl_n, Θ) *we can find a unique parabolic subgroup* P *such that for all* $\alpha_i \in \pi^{U_P}$ *we have*

$$n_{\alpha_i^P}(P, \Theta) < 1 \text{ and the reductive quotient } (M, \Theta^M) \text{ is semi-stable.}$$

This parabolic subgroup is called the *canonical destabilizing* group. We denote it by $P(\Theta)$, if (G, Θ) is semi-stable then $P(\Theta) = G$.

The first theorem is due to Minkowski, we recall his argument.

We pick a B, assume there is an index i such that $n_{\alpha_i}(B, \Theta_x) > \frac{2}{\sqrt{3}}$, we look at the corresponding flag and look at the quotient L_{i+1}/L_{i-1}. Then the image of L_i in this quotient has a generator which is not a shortest vector. Hence we can replace L_i by an L_i' whose image in the quotient is generated by a shortest vector. We get a modified flag $\mathcal{F}'$ and a modified Borel subgroup B'. The punchline is that now $p_{P_j}(B', \Theta) = p_{P_j}(B, \Theta)$ for $j \neq i$ but $p_{P_i}(B', \Theta) < p_{P_j}(B, \Theta)$. We can repeat this argument with B' but this process must stop and then we found a reduced subgroup.

The second theorem is proved in [108], [41].

This gives us a dissection of X into the subsets

$$X = \bigcup_{P\text{: parabolic subgroups of } G/\mathbb{Q}} X^{[P]} = \{x \in X \mid P(x) = P\}. \tag{1.77}$$

Clearly $\gamma X^{[P]} = X^{[\gamma P \gamma^{-1}]}$. If we divide by the group Γ then we get

$$\Gamma \backslash X = \bigcup_{P \in \mathrm{Par}(\Gamma)} \Gamma_P \backslash X^{[P]}. \tag{1.78}$$

where Par (Γ) is a set of representatives of Γ-conjugacy classes of parabolic subgroups of $\mathrm{Gl}_n/\mathbb{Q}$. This is a decomposition of $\Gamma\backslash X$ into a disjoint union of subsets. The subset $\Gamma\backslash X^{[\mathrm{Gl}_n]}$ is compact, it is the set of semi-stable pairs (Gl_n, x). The subsets $\Gamma_P\backslash X^{[P]}$ for $P \neq G$ are in a certain sense “open in some directions” and “closed in some other directions”. Therefore this decomposition is not so useful for the study of cohomology groups.

To remedy this we introduce larger subsets. For a real number r, $0 < r < 1$ we define

$$X^{\mathrm{Gl}_n}(r) = \{x \in X \mid n_{\gamma_\alpha}(P(x), x) > r, \text{ for all } \alpha \in \pi'\}. \tag{1.79}$$

It contains the set of semi-stable (Gl_n, x). If we choose $r < 1$ but close to one then some of the elements in $X^{\mathrm{Gl}_n}(r)$ may be unstable but only a "little bit".

Together with the first theorem this has a consequence

Proposition 1.2.4. *The quotient* $X^{\mathrm{Gl}_n}(r) = \Gamma\backslash X^{\mathrm{Gl}_n}(r)$ *is a relatively compact open subset of* $\Gamma\backslash X$. *It contains the set of semi-stable* (Gl_n, x).

Now we start from a parabolic subgroup P and let $M = P/U_P$ be its Levi-quotient. Let $\pi' \subset \pi$ be the type of P, then $\pi = \pi' \cup \pi^M$. our considerations above also apply to $M/\mathbb{Q}$. The group $P(\mathbb{R})$ acts transitively on X and we put (see (1.63))

$$X^M = U_P(\mathbb{R})\backslash X \text{ and let } p_M : X \to X^M \text{ be the projection.} \tag{1.80}$$

Here $X^M = M(\mathbb{R})/K_\infty^M$ where K_∞^M is the image of $P(\mathbb{R}) \cap K_\infty$ in $M(\mathbb{R})$. Let $Z \subset M$ be the center of M, and as usual $M^{(1)}$ the semi-simple component. Then $M_1(\mathbb{R})$ will be the intersection of the kernels of all $|\gamma| : M(\mathbb{R}) \to \mathbb{R}^\times_{>0}$ where $\gamma \in X^*(Z) \otimes \mathbb{Q}$, then the index of $M^{(1)}(\mathbb{R}) \subset M_(\mathbb{R})$ is a power of two.

Then we define $X^{M^{(1)}} = M_1(\mathbb{R})/K_\infty^M$, and if Γ_M is the image of $P(\mathbb{R}) \cap \Gamma$ in $M(\mathbb{R})$ then

$$\Gamma_M\backslash X^M := \Gamma_M\backslash X^{M^{(1)}} \times Z^{(0)}(\mathbb{R}) \tag{1.81}$$

where of course $Z^{(0)}(\mathbb{R})$ is the connected component of the identity of $Z(\mathbb{R})$.
We change the notation and put

$$Z^{(0)}(\mathbb{R}) = A_{\pi'} = \{(\dots, a_\alpha, \dots)_{\alpha \in \pi'} \mid a_\alpha \in \mathbb{R}^\times_{>0}\}.$$

For simple roots $\alpha \in \pi^M$, and a Borel subgroup $\bar{B} \subset M/\mathbb{Q}$, and a point $x_M = p_M(x)$, we can define the numbers $n_\alpha(\bar{B}, x_M)$ essentially in the same way as before and clearly

$$n_\alpha(\bar{B}, x_M) = n_\alpha(B, x)$$

if B is the inverse image of $\bar{B}$.

We have to be a little bit careful with the numbers $p_{\bar{Q}}(\bar{Q}, x^M)$, because for the inverse image Q, the unipotent radical U_Q is larger than $U_{\bar{Q}}$. Therefore we have to look at the dominant fundamental weights $\gamma_\alpha^M \in \bigoplus_{\alpha_i \in \pi^M} \mathbb{Q}\alpha_i$, and formulate the stability condition for x^M in terms of these γ_α^M:

The point x^M *is stable, if for all* $\alpha_i \in \pi^M$ *the inequality* $n_{\gamma^M_{\alpha_i}}(\bar{P}_{\alpha_i}, x^M) > 1$ *holds. Again we denote the destabilizing group by* $P(x^M)$.

Hence we see that for a number $r_M < 1$ we can define regions

$$X^M(r_M) = \{x^M \mid n_{\gamma^M_{\alpha_i}}(\bar{P}_{\alpha_i}, x^M) > r_M \text{ whenever } \bar{P}_{\alpha_i} \supset \bar{P}(x^M)\}. \tag{1.82}$$

For any type π' and numbers $0 < c_{\pi'} < 1$ and $0 < r(\pi') < 1$ we define

$${}^*X^P(c_{\pi'}, r(\pi')) = \{x \mid n_{\alpha^P}(P, x) < c_{\pi'} \text{ for all } \alpha \in \pi'; x_M \in X^M(r(c_{\pi'}))\}$$

Proposition 1.2.5. *For a given $0 < r < 1$, we can find numbers $c_{\pi'}$ such that for any $x \in {}^*X^P(c_{\pi'}, r)$, the destabilising parabolic subgroup $P(x) \subset P$. The same is true in the other direction: for a given $0 < c_{\pi'} < 1$, we can find $r(\pi') < 1$ such that for $x \in {}^*X^P(c_{\pi'}, r(\pi')))$, the destabilising parabolic subgroup $P(x) \subset P$.*

To see this we have to look at the destabilising subgroup $\bar{Q} \subset (M, x_M)$. Its inverse image $Q \subset P$ is a parabolic subgroup of Gl_n. The reductive quotient $(\bar{M}, x_{\bar{M}})$ of Q is semi-stable. We want to show that Q is the destabilising parabolic of (Gl_n, x). We have to show that:

$$n_{\alpha^Q}(Q, x) < 1 \,\forall\, \alpha \in \pi^{U_Q} = \pi^{U_P} \cup \pi^{U_{\bar{Q}}}.$$

For $\alpha \in \pi^{U_{\bar{Q}}}$ this is true by definition. For $\alpha \in \pi^{U_P}$ we have

$$\alpha^P = \alpha + \sum_{\beta \in \pi^M} a_{\alpha,\beta}\beta \text{ and } \alpha^Q = \alpha + \sum_{\beta' \in \pi^{\bar{M}}} a'_{\alpha,\beta'}\beta',$$

where $a_{\alpha,\beta} \geq 0$. The roots $\beta \in \pi^{U_{\bar{Q}}}$ can be expressed in terms of the $\beta^{\bar{Q}} = \beta^Q$:

$$\beta^Q = \beta + \sum_{\beta' \in \pi^{\bar{M}}} a^*_{\beta,\beta'}\beta' \tag{1.83}$$

and hence

$$\alpha^Q = \alpha^P - \sum_{\beta \in \pi^{U_{\bar{Q}}}} a_{\alpha,\beta}\beta^Q + \sum_{\beta' \in \pi^{\bar{M}}} c_{\alpha\beta'}\beta'. \tag{1.84}$$

The last sum is zero because $\alpha^Q, \alpha^P, \beta^Q$ are orthogonal to the module $\oplus_{\beta'}\mathbb{Z}\beta'$.
We get the relation

$$n_{\alpha^Q}(Q, x) = n_{\alpha^P}(P, x) \cdot \prod_{\beta \in \pi^{U_{\bar{Q}}}} n_{\beta^Q}(Q, x)^{-a_{\alpha,\beta}}. \tag{1.85}$$

Now it comes down to show that

$$\begin{gathered} n_{\alpha^P}(P, x) < c_\alpha,\ \forall\, \alpha \in \pi^{U_P} \text{ and } n_{\beta^Q}(Q, x) > r,\ \forall \beta\ \in \pi^{U_{\bar{Q}}} \\ \Longrightarrow n_{\alpha^Q}(P, x) < 1; \forall\, \alpha \in \pi^{U_P}. \end{gathered} \tag{1.86}$$

This is certainly true if either the $c_{\pi'}$ is small enough or if r is sufficiently close to one. In this case we say that $(c_{\pi'}, r(\pi'))$ is *well chosen.*
Therefore we define

$$X^P(c_{\pi'}, r(\pi')) = \{x \in {}^*X^P(c_{\pi'}P, r(\pi')) \mid P(x) \subset P\},$$

we have $X^P(c_{\pi'}, r(\pi')) = {}^*X^P(c_{\pi'}, r(\pi'))$, if $(c_{\pi'}, r(\pi'))$ is well chosen.

We claim that we can find a well chosen list of parameters

$$(\ldots, (c_{\pi'}, r(\pi')), \ldots)_{P\text{: parabolic over } \mathbb{Q}}$$

where $(c_{\pi'}, r(\pi'))$ are well chosen such that we get a covering

$$X = \bigcup_P X^P(c_{\pi'}, r(\pi'))) \tag{1.87}$$

and hence

$$\Gamma\backslash X = \Gamma\backslash \bigcup_P X^P(c_{\pi'}, r(\pi')) = \bigcup_{P\in \mathrm{Par}(\Gamma)} \Gamma_P\backslash X^P(c_{\pi'}, r(\pi')). \tag{1.88}$$

We argue by induction on $\#\pi'$. If $\pi' = \emptyset$ we choose $c_\emptyset$ any number and $0 < r(\emptyset) < 1$. We choose $c_{\{\alpha_i\}}$ such that $r(\emptyset) < c_{\{\alpha_i\}} < 1$ For all α_i We apply the above proposition and find $r_{\{\alpha_i\}} < 1$ such that $c_{\{\alpha_i\}}, r(\{\alpha_i)\}$ are well chosen. This can be repeated. Now we assume that $\#\pi' > 1$. Assume we have well chosen all $c_{\pi'\setminus\{\alpha_i\}}$, and $r(\pi'\setminus\{\alpha_i\}$. Then we chose $c_{\pi'}$ such that for all $1 > (c_{\pi'} > r(\pi'\setminus\{\alpha_i\}$. Finally we choose $r(\pi') < 1$ so that $(c_{\pi'}, r(\pi'))$ is well chosen. This continues until we reach $\pi' = \pi$. Now it is clear that a point $x \in X$ lies in in some $X^P(c_{\pi'}, r(\pi')$ we simply can choose $P = P(x)$

We can choose the numbers $c^+_{\pi'}$, (resp. $r(c_{\pi'})^-$) a tiny bit larger (resp. smaller) and these numbers are still well chosen. The closure $\overline{\Gamma\backslash \bigcup_P X^P(c_{\pi'}, r(c_{\pi'}))}$ is contained in $\Gamma\backslash \bigcup_P X^P(c^+_{\pi'}, r(c^-_{\pi'}))$. We get a second covering by slightly larger open sets. This can be used to produce a partition of unity:

Proposition 1.2.6. *We can find a family of $\mathcal{C}_\infty$-functions $h_P \geq 0$ which satisfies*

a) h_P restricted to $\Gamma_P\backslash X^P(c_{\pi'}, r(c_{\pi'})$ is identically equal to one
b) $\sum_P h_P = 1$ and h_P is zero outside of $\Gamma_P\backslash X^P(c^+_{\pi'}, r(c^+_{\pi'}))$

Proof. well-known □

We consider the evaluation map

$$\begin{gathered} n^{\pi'} : \Gamma_P\backslash X^P(c_{\pi'}, r(c_{\pi'})) \to \textstyle\prod_{\alpha\in\pi'} (0, c_\alpha) \\ x \mapsto (\ldots, n_{\alpha^P}(P, x), \ldots)_{\alpha\in\pi'} \end{gathered} \tag{1.89}$$

and consider the diagram

$$\begin{array}{lccc} p_M = p_{M_1} \times n^{\pi'} : & X^P(c_{\pi'}, r(c_{\pi'}))) & \to & X^{M_1}(r(\pi')) \times A_{\pi'}(c_{\pi'}) \\ & \downarrow & & \downarrow q_1 \times q_a \\ \underline{p}_{M_1} \times n^{\pi'} : & \Gamma_P\backslash X^P(c_{\pi'}, r(c_{\pi'})) & \to & \Gamma_{M_1}\backslash X^{M_1}(r(\pi')) \times A_{\pi'}(c_{\pi'}) \end{array} \tag{1.90}$$

The group $U(\mathbb{R})$ acts by left translations on the fibres of p_M and this action on the fibres is simply transitive, hence the fibres are isomorphic to $U(\mathbb{R})$, they are principal homogenuous $U(\mathbb{R})$ spaces. (See 1.80)

The geodesic action

We construct a distinguished trivialisation of the fibration p_M. We start from the Arakelow-Chevalley scheme $(\mathrm{Sl}(L, <\ ,\ >)$, where $L = \mathbb{Z}e_1, \oplus \cdots \oplus \mathbb{Z}e_i \oplus \mathbb{Z}e_n$ and $< e_i, e_j >= \delta_{i,j}$. We rather look at $(\mathrm{Sl}(L), \Theta_0)$ where Θ_0 is the standard Cartan involution. This Cartan involution determines the split maximal torus $\mathcal{T}$, so that Θ_0 acts by $t \mapsto t^{-1}$ on this torus, and so that for all Borel subgroups $B \supset \mathcal{T}$ and all roots α we have $n_\alpha(B, \Theta_0) = 1$.

Let B_0 be the Borel subgroup of upper triangular matrices. Then for any parabolic subgroup $P \supset B_0$ we have a unique section

$$s_{\Theta_0} : M \to P;\ p_M \circ s_{\Theta_0} = \mathrm{Id}_M,$$

which induces the identity on T. Hence we get $P = s_{\Theta_0}(M) \times U_P$, this parabolic subgroup is determined by its type π'.

(Comment: It is my point here that choosing a Levi factor is sometimes a little bit problematic but here this Levi-factor is uniquely determined by Θ_0.)

Now it is clear that we get a section $s : X^{M_1} \to X$: A point $y \in X^{M_1}$ of the form $g_1\Theta_0^M$ with $g_1 \in M_1(\mathbb{R})$ We send this point y to $s_{\Theta_0}(g_1)\Theta_0$, this is well defined.

We use this trivialisation of the fibration p_M in the top line to construct a "local" trivialisation for the fibration $\underline{p}_M$ in the second line. We pick a point $y = (y_M, a) \in \Gamma_{M_1}\backslash X^{M_1}(r(\pi')) \times A_{\pi'}(c_{\pi'})$. We choose a point $\tilde{y}_M \in q_1^{-1}(y_M)$ in the fibre. We identified the fibre of $p_M^{-1}(\tilde{y}_M, a)$ with $\mathrm{U}(\mathbb{R})$ and hence we get an isomorphism

$$\psi_{\tilde{y}_M} : (p_{M_1} \times n^{\pi'})^{-1}((y_M, a) \xrightarrow{\sim} U(\mathbb{Z})\backslash U(\mathbb{R}).$$

This isomorphism depends on the choice of $\tilde{y}_M$, a second point is of the form $y'_M = \gamma_{M_1}\tilde{y}_M$ and then

$$\psi_{\tilde{y}'_M} = \mathrm{Ad}(s_{\theta_0}\gamma_M) \circ \psi_{\tilde{y}_M} \tag{1.91}$$

This construction extends to a small convex neighborhood V_y of y_M. Let $\tilde{V}_{\tilde{y}_M}$ be the connected component of $q_1^{-1}(V_y)$ which contains $\tilde{y}_M$ Then we the above construction yields an identification

$$\psi_{\tilde{y}_M} : (p_{M_1} \times n^{\pi'})^{-1}(V_y \times A_{\pi'}(c_{\pi'}) \xrightarrow{\sim} U(\mathbb{Z})\backslash U(\mathbb{R}) \times \tilde{V}_{\tilde{y}_M} \times A(c_{\pi'}) \tag{1.92}$$

and the above formula 1.91 also extends to Y_y. We consider the semi group

$$A_{\pi'} = \{\underline{t} \mid \underline{t} = \{\ldots, t_\alpha, \ldots\}_{\alpha\in\pi'}\ ;\ 0 < t_\alpha < 1\} \tag{1.93}$$

and observe that it acts by multiplication on this last factor in the right hand side of (1.92) and hence it also acts on the left hand side and we get an action of this semi group on $\Gamma_P\backslash X^P(c_{\pi'}, r(c_{\pi'})$ This is the geodesic action which has been introduced by Borel and Serre. The geodesic action provides a distinguished identification of any two fibres $(n^{\pi'})^{-1}(a_0) \xrightarrow{\sim} (n^{\pi'})^{-1}(a_1)$ for any two points $a_0, a_1 \in A_{\pi'}(c_{\pi'})$.

We put

$$A_{\pi'}(c_{\pi'}) = \prod_{\alpha\in\pi'} (0, c_{\pi'}) \subset \bar{A}_{\pi'}(c_{\pi'}) = \prod_{\alpha\in\pi'} [0, c_{\pi'}),$$

then we can extend the fibration $n^\pi : \Gamma_P\backslash X^P(c_{\pi'}, r(\pi')) \to A_{\pi'}(c_{\pi'})$ to a locally trivial fibration

$$\bar{n}^{\pi'} : \overline{\Gamma_P\backslash X^P(c_{\pi'}), r(\pi'))} \to \bar{A}_{\pi'}(c_{\pi'}) \tag{1.94}$$

We define the boundary

$$\partial\overline{\Gamma_P\backslash X^P(c_{\pi'}, r(\pi'))} = \overline{\Gamma_P\backslash X^P(c_{\pi'}, r(c_{\pi'}))} \setminus \Gamma_P\backslash X^P(c_\pi, r(\pi'))$$

this is equal to

$$\partial\overline{\Gamma_P\backslash X^P(c_{\pi'}, r(\underline{c}_{\pi'}))} = \Gamma_P\backslash(n^{\pi^{\pi'}})^{-1}(c_0) \times \partial(\prod_{\nu\in\pi_G\setminus\pi}[0, c_\pi])$$

where of course $\partial(\prod_{\alpha\in\pi'}[0, c_\pi]) \subset \prod_{\alpha\in\pi'}[0, c_\pi]$ is the subset where at least one of the coordinates x_α is equal to zero.

We form the disjoint union of these boundaries over the subsets π' and a set of representatives of Γ-conjugacy classes of parabolic subgroups. We introduce an equivalence relation on this disjoint union. If we have two parabolic subgroups $Q \subset P$ then the intersection $X^P(c_{\pi'}, r(\pi')) \cap X^Q(c_{\pi''}, r(\pi''))$ may not be empty. If we now have points

$$x \in \partial\overline{\Gamma_P\backslash X^P(c_{\pi'}, r(\pi')))}, y \in \partial\overline{\Gamma_Q\backslash X^Q(c_{\pi''}, r((\pi''))},$$

then we identify these two points if we have a sequence of points $\{x_n\}_{n\in\mathbb{N}}$ which lies in the intersection $X^P(c_{\pi'}, r(\pi'))) \cap X^Q(c_{\pi''}, r(\pi''))$ and which converges to x in $\overline{X^P(c_{\pi'}, r(\pi'))))}$ and to y in $\overline{X^Q(c_{\pi''}, r(\pi'')}$. A careful inspection shows that this provides an equivalence relation $\sim$, and we define

$$\partial(\Gamma\backslash X) = \bigcup_{\pi', P\in\mathrm{Par}(\Gamma)} \partial\overline{\Gamma_P\backslash X^P(c_\pi, r(\underline{c}_{\pi'}))}/\sim$$

and the Borel-Serre compactification will be the manifold with corners

$$\overline{\Gamma\backslash X} = \Gamma\backslash(X \cup \bigcup_{P:P \text{ proper}} \overline{X^P(\underline{c}_{\pi'}, r(\underline{c}_{\pi'})}). \tag{1.95}$$

We also define the Borel-Serre strat ification of the Borel-Serre boundary, here $\partial_P(\Gamma\backslash\bar{X})(r(c_{\pi'}))$ is the subset of $\overline{\Gamma_P\backslash X^P(\underline{c}_{\pi'}, r(\underline{c}_{\pi'})}$ where $a_\alpha = 0$ for all $\alpha \in \pi'$ and where we take the union over all $r(c_{\pi'}) > 0$. (Under these conditions the parameters $r(\pi')$ become obsolete, $(0, r(\pi))$ is always well chosen.) These strata are of course not closed, the strata $\partial_Q(\Gamma\backslash\bar{X})$ with $Q \subset P$ lie in the closure of these strata.

The projection $p_{M_1} \times n^{\pi'}$ extends to a fibration on the closure and restriction to the boundary yields the fibration with fibre $\Gamma_U\backslash U(\mathbb{R})$:

$$r_P : \partial_P(\Gamma\backslash\bar{X}) \to \Gamma_M\ X^M \times \{(0)\}.$$

We define a "tubular" neighbourhood of the boundary

$$\mathcal{N}(\Gamma\backslash X)(\mathbf{c}) = \Gamma\backslash \bigcup_{P:P\text{proper}} \overline{X^P(\underline{c}_{\pi'}, r(\underline{c}_{\pi'}))} \tag{1.96}$$

where $\mathbf{c}$ stands for the collection of parameters $(c_{\pi'}, r(c_{\pi'}))$. Then we define the *punctured tubular neighbourhood* as

$$\overset{\bullet}{\mathcal{N}}(\Gamma\backslash X)(\mathbf{c}) = \Gamma\backslash \bigcup_{P:P \text{ proper}} X^P(\underline{c}_{\pi'}, r(\underline{c}_{\pi'})) = (\Gamma\backslash X) \cap \mathcal{N}(\Gamma\backslash X). \tag{1.97}$$

Eventually we want to use the above covering as a tool to understand cohomology (see section 2.1.3). For this it is also necessary to understand the intersections

$$X^{P_1}(c_{\pi_1}, r(\underline{c}_{\pi_1})) \cap \cdots \cap X^{P_\nu}(c_{\pi_k}, r(\underline{c}_{\pi_k})). \tag{1.98}$$

Our proposition 1.2.5 implies that for any point x in the intersection, the destabilizing parabolic subgroup $P(x) \subset P_1 \cap \cdots \cap P_k$. Hence we see that the above intersection can only be non empty if $Q = P_1 \cap \cdots \cap P_k$ is a parabolic subgroup. Then $\pi^{U_Q} = \cup_{\nu=1}^k \pi^{U_{P_\nu}}$. Let M be the reductive quotient of Q.

Now we look at the product $\prod_{\alpha\in\pi^{U_Q}} \mathbb{R}^\times_{>0}$ - here it seems to be helpful to identify it - using the logarithm - with $\mathbb{R}^{d_Q}$:

$$\log : \prod_{\alpha\in\pi^{U_Q}} \mathbb{R}^\times_{>0} \xrightarrow{\sim} \mathbb{R}^{d_Q}. \tag{1.99}$$

We consider the map

$$\begin{aligned} N^Q &: X^{P_1}(c_{\pi_1}, r(\pi_1)) \cap \cdots \cap X^{P_k}(c_{\pi_k}, r(\pi_k)) \to \mathbb{R}^{d_Q} \\ N^Q &: x \mapsto (\ldots, -\log(n_{\alpha^Q}(Q,x)), \ldots)_{\alpha^Q\in\pi^{U_Q}} \end{aligned} \tag{1.100}$$

Consider a point $x \in X^{P_\nu}(c_{\pi_\nu}, r\pi_\nu))$. For $\alpha \in \pi^{U_{P_\nu}}$ we have

$$-\log(n_{\alpha^{P_\nu}}(P_\nu, x)) \geq -\log(c_{\pi_\nu})$$

We can express the left hand side $-\log(n_{\alpha^{P_\nu}}(P_\nu, x))$ as a linear combination of the terms $-\log(n_{\alpha^Q}(Q,x))$, with $\alpha \in \pi^{U_Q}$. This means that the root $\alpha \in \pi^{U_{P_\nu}}$ defines a half space $H_\nu^+(\alpha)$ in $\mathbb{R}^{d_Q}$ and $N^Q(x) \in H_\nu^+(\alpha)$ in $\mathbb{R}^{d_Q}$.

Now we assume that x is in the intersection (1.98). For the roots $\alpha \in \pi\setminus\pi^{U_{P_\nu}}$ we have the condition (1.82). For the roots $\alpha \in \pi^{U_Q} \setminus \pi^{U_{P_\nu}}$ this yields

$$-\log(n_{\gamma_\alpha^{M_\nu}}(P_\nu, x)) \leq -\log(r(\pi_\nu)).$$

Therefore we see that the image of N^Q is contained in the intersection of a finite number of half spaces, which are obtained from a finite family of hyperplanes. These hyperplanes depend on the parameters $c_{\pi_\nu}, r(\pi_\nu)$. Let us call this intersection $C(\underline{c}, \underline{r})$, and it is a convex - possibly empty - subset of $\mathbb{R}^{d_Q}$.

We investigate the restriction

$$N^Q : X^{P_1}(c_{\pi_1}, r(\pi_1)) \cap \cdots \cap X^{P_\nu}(c_{\pi_k}, r(\pi_k)) \to C(\underline{c}, \underline{r})$$

We observe that the unipotent radical $U_Q(\mathbb{R})$ acts by left translations on the intersection, we get a diagram

$$\begin{array}{ccc} X^{P_1}(c_{\pi_1}, r(\pi_1)) \cap \cdots \cap X^{P_\nu}(c_{\pi_k}, r(\pi_k)) & \to & C(\underline{c}, \underline{r}) \\ \downarrow p_M & & \\ X^M \times \mathbb{R}^{d_Q} & \to & \mathbb{R}^{d_Q}. \end{array} \tag{1.101}$$

Now it is clear from the definitions that the image of p_M is a set

$$\mathrm{Im}(p_M) = \Omega^M(\underline{c}, \underline{r}) \times C(\underline{c}, \underline{r})$$

where $\Omega^M(\underline{c}, \underline{r}) \subset X^M$ is a subset containing the set $X^{M,st}$ of semi-stable points and is described by certain inequalities as in (1.79). This subset is Γ_M invariant and $\Gamma_M \backslash \Omega^M(\underline{c}, \underline{r})$ is relatively compact.

Hence we see that we have essentially the same situation as before. The map

$$\begin{aligned} p_M : & X^{P_1}(c_{\pi_1}, r(\underline{c}_{\pi_1})) \cap \cdots \cap X^{P_\nu}(c_{\pi_k}, r(\underline{c}_{\pi_k})) \\ & \to \Gamma_M \backslash \Omega^M(\underline{c}, \underline{r}) \times C(\underline{c}, \underline{r}) \end{aligned} \tag{1.102}$$

is a fiber bundle with fiber isomorphic to $\Gamma_{U_Q} \backslash U_Q(\mathbb{R})$.

In the following we refer to the book of S. Helgason [63].

We mention an important property of the sets $X^P(\underline{c}_{\pi'}, r(c_P))$. We assume that our symmetric space X is connected, then it is well known that it is convex, any two points $p, q \in X$ can be joined by a unique geodesic $[p, q]$. We say that a subset $U \subset X$ is convex if for any two points $p, q \in U$ also the geodesic $[p, q] \subset U$.

Proposition 1.2.7. *Let $\Omega \subset \Omega^M(\underline{c}, \underline{r})$ be a convex subset. Then the inverse image $\overset{-1}{p_M}(\Omega \times C(\underline{c}, \underline{r}))$ is a convex subset of $X^{P_1}(c_{\pi_1}, r(\underline{c}_{\pi_1})) \cap \cdots \cap X^{P_\nu}(c_{\pi_k}, r(\underline{c}_{\pi_k}))$*

Proof. The assertion is easily reduced to the following:

Let P be a maximal parabolic subgroup, let M be its reductive quotient, let α be the simple root not in π^M and $\Omega \subset X^{M^{(1)}}$. Then the set for any choice of We choose a $c_\alpha > 0$ and claim that $X^P(c_\alpha, \Omega) = \{x \in X \mid n_{\alpha^P}(P, x) \leq c_\alpha \,; q_M(x) \in \Omega\}$ is convex .

We pick two points $x_1, x_2 \in X^P(c_\alpha, \Omega)$, let $t \mapsto c(t)$ be the unique geodesic in X joining these two points. Assume that this geodesic leaves $X^P(c_\alpha, \Omega)$, then there is a t_0 where $n_{\alpha^P}(P, (c(t_0)) > c_\alpha$ and we may assume that this value is a local maximum. We put $x_0 = c(t_0)$, let $Y_0 = c'(t_0) \in T^X_{x_0}$ be the tangent vector to this geodesic. The action of $G(\mathbb{R})$ on X gives us a surjective map $D_{x_0} : \mathfrak{g}_\mathbb{R} \to T^X_{x_0}$ and this induces an isomorphism $D_{x_0} : \mathfrak{g}_\mathbb{R}/\mathfrak{k}_{x_0} \xrightarrow{\sim} T^X_{x_0}$, here of course $\mathfrak{k}_{x_0}$ is the Lie-algebra of K_{x_0}. We have the Cartan decomposition of the Lie-algebra

$$\mathfrak{g}_\mathbb{R} = \mathfrak{k}_{x_0} \oplus \mathfrak{p}_{x_0} \text{ where } \mathfrak{p}_{x_0} = \{V \in \mathfrak{g}_\mathbb{R} \mid \Theta_{x_0}(V) = -V\} \tag{1.103}$$

and we get the isomorphism $D_{x_0} : \mathfrak{p}_{x_0} \xrightarrow{\sim} T^X_{x_0}$. Starting from our parabolic subgroup P we get a finer decomposition of $\mathfrak{p}_{x_0}$. Let $\mathfrak{P}_{\mathbb{R}}$ be the Lie algebra of $P \times \mathbb{R}$. The intersection $P \times \mathbb{R} \cap \Theta_{x_0}.(P \times \mathbb{R}) = M_{x_0}$. and we get for the Lie algebras $\mathfrak{m}_{x_0}. = \mathfrak{m}^{(0)} \oplus \mathfrak{a}$ and this gives the finer decomposition $\mathfrak{m}_{x_0}. = \mathfrak{k}_{x_M} \oplus \mathfrak{p}^{(M_{x_0}).} \oplus \mathfrak{a}$ and then

$$\mathfrak{p}_{x_0}. = \mathfrak{p}^{(M_{x_0}.)} \oplus \mathfrak{a} \oplus \{V - \Theta_{x_0}(V)\}_{V \in \mathfrak{u}} \tag{1.104}$$

where $V \in \mathfrak{u}_{\mathbb{R}}$ and $\mathfrak{a} = \mathbb{R}Y_A$. The above tangent vector Y_0 is now

$$Y_0 = Y_M + aY_A + (V - \Theta_{x_0}V).$$

The theorem 3.3 in Chapter IV in [63] says that this geodesic is $c(t) = \exp(tY_0) \cdot x_0$. But since we are at a local maximum we also know that $a = 0$ and then a somewhat tedious but elementary computation yields

$$n_{\alpha^P}(P, \exp(tY_0) \cdot x_0) = n_{\alpha^P}(P, x_0)) - B(V, \Theta_{x_0}V)t^2 + O(t^3) \tag{1.105}$$

where $B(\, , \,)$ - normalised by a non zero positive - factor is the Killing form. Then $B(V, \Theta_{x_0}V) < 0$ for $V \neq 0$ and $n_{\alpha^P}(P, \exp(tY_0) \cdot x_0)$ can not assume a local maximum at $t = 0$, this is a contradiction and we have proved that $X^P(c_\alpha, \Omega)$ is convex. (See [26])

□

We formulated the main theorems of reduction theory only for $\Gamma = \mathrm{Sl}_n(\mathbb{Z})$ or $\Gamma = \mathrm{Gl}_n(\mathbb{Z})$, but if one is willing to accept the structure theorems for split semi simple group schemes $\mathcal{G}/\mathbb{Z}$ (see [33]), then it is not very difficult to extend these two theorems to groups $\Gamma = \mathcal{G}(\mathbb{Z})$.

The split semi simple group scheme $\mathcal{G}/\mathbb{Z}$ has a split maximal torus $\mathcal{T}_0/Z$. Then we easily see that we have an involution $g \mapsto \Theta_0(g)$ of this grouo scheme which induces $t \mapsto t^{-1}$ on $\mathcal{T}_0$. Again we have the system of roots $\alpha \in \Delta \subset X^*(\mathcal{T})$ For any root we have the one parameter subgroup $U_\alpha \xrightarrow{\sim} \mathcal{G}_a/\mathbb{Z}$ and Θ_0 maps U_α isomorphically to $U_{-\alpha}$. We can choose a system $\Delta^+ \subset \Delta$ and inside Δ^+ we have the set π of simple roots.

The set Δ^+ determines a Borel subgroup $B \supset \mathcal{T}$, a one parameter subgroup $U_\alpha \subset B$ if and only if $\alpha \in \Delta^+$ The parabolic sub groups containing B correspond one to one to the subsets $\pi' \subset \pi$.

The $\Gamma = \mathcal{G}(\mathbb{Z})$ conjugacy classes of parabolic subgroups $P/\mathbb{Q} \subset \mathcal{G} \times \mathbb{Q}$ are in one to one correspondence with the subsets $\pi' \subset \pi$.

We call such a datum $(\mathcal{G}, T, \Theta_0)$ a Arakelow-Chevalley scheme. A split semi simple group scheme $\mathcal{G}/\mathbb{Z}$ becomes an Arekelow group scheme $(\mathcal{G}, \Theta)$ if we have chosen a Cartan involution Θ The space of these Cartan involutions will be denoted by X, for $x \in X$ we denote the corresponding Cartan involution by Θ_x. If K_x is the group of fixed elements under Θ_x then K_x is maximal compact. Our space X is the symmetric space

$$X = \mathcal{G}(\mathbb{R})/K_{x_0} \text{ where } \Theta_{x_0} = \Theta_0 \tag{1.106}$$

Now we can proceed as before, we can define the numerical invariants $n_\alpha(B, x)$ or $p(P, x)$ and can introduce the notions of stable and semistable Arakelow groups schemes $(\mathcal{G}, \Theta_x)$. Then we can state the two theorems 1.2.1 and 1.2.2.

We still can go one step further. For any finite extension $F/\mathbb{Q}$ and its ring of algebraic integers $\mathcal{O}_F \subset F$ we have the notion of a split semi simple group scheme $\mathcal{G}/\mathcal{O}_F$. For instance we get such a group scheme if we take the base extension of our Chevalley scheme $\mathcal{G} \times_{\mathbb{Z}} \mathcal{O}_F$. Any parabolic subgroup $P/\mathcal{O}_F \subset \mathcal{G}/\mathcal{O}_F$ contains a split maximal torus $\mathcal{T} \subset \mathcal{G}$. Here we have to observe that two such split maximal tori $\mathcal{T}_1, \mathcal{T}_2$ are not necessarily conjugate under $\mathcal{G}(\mathcal{O}_F)$, but the number of such conjugacy classes is finite (finiteness of the class number). But these tori are conjugate under $\mathcal{G}(F)$. and hence we can speak of the type $t(P) = \pi'$ of the parabolic sub group P. The number of $\mathcal{G}(\mathcal{O}_F)$ conjugacy classes of parabolic subgroups is finite.

We can play the same game as before. We consider $F_\infty = F \otimes \mathbb{R} = \prod_{v \in \Sigma} F_v$, where $F_v = \mathbb{R}$ or $\mathbb{C}$. Then $\mathcal{G}(F_\infty) = \prod_{v \in \Sigma} \mathcal{G}(F_v)$ The associated symmetric space $X = X_v$ where X_v is the space of Cartan involutions on $\mathcal{G}(F_v)$. For $x \in X$ let Θ_x be the corresponding Cartan involution, i.e. $x = \Theta_x$. If we put $\Gamma = \mathcal{G}(\mathcal{O}_F)$ then $\Gamma \backslash X$ parametrises the isomorphism classes of Arakelow structures on $\mathcal{G}/\mathcal{O}_F$.

We have the distinguished Arakelow -Chevalley scheme $(\mathcal{G}_0/\mathcal{O}_F, \Theta_0)$ which is simply the base extension of $(\mathcal{G}_0/\mathbb{Z}, \Theta_{x_0})$.

. We can define the numbers $n_\alpha(B, \mathcal{G}, x), p(P, \mathcal{G}, x)$. In contrast to the previous definition we introduce the variable $\mathcal{G}$, i.e. this is the semi simple extension which was the distinguished $\mathcal{G}_0$ in the previuos case. Again we have the notion stability and semi stability. Our distinguished Arakelow -Chevalley scheme $(\mathcal{G}_0/\mathcal{O}_F, \Theta_0)$ comes with a split maximal torus $\mathcal{T}_0$ and for the Borel sub groups $B \supset \mathcal{T}_0$ and the parabolic sub groups $P \supset \mathcal{T}_0$ we will have the values $n_\alpha(B, \mathcal{G}_0, x_0) = 1, p(P, \mathcal{G}_0, x_0) = 1$.

Now we adopt the definition of stability and semi stability, our Arakelow -Chevalley scheme $(\mathcal{G}_0/\mathcal{O}_F, \Theta_0)$ is semi stable.

We can state the generalisation of the two theorems 1.2.1. and 1.2.2.

Theorem 1.2.3. *For any finite extension $F/\mathbb{Q}$ we can find a constant $c_F > 0$ such that for any Arakelow group scheme $(\mathcal{G}/\mathcal{O}_F, \Theta_x)$ we can find a Borel subgroup $B \subset \mathcal{G}$ such that*

$$n_{\alpha_i}(B, \mathcal{G}, x) < c_F \text{ for all } i = 1, \dots, n-1$$

It is not very difficult to show that the space of split semi stable Arakelow group schemes is compact.

Theorem 1.2.4. *For any Arakelow group scheme $(\mathcal{G}/\mathcal{O}_F, \Theta)$ we can find a unique parabolic subgroup P such that for all $\alpha_i \in \pi^{U_P}$ we have*

$$n_{\alpha_i^P}(P, \mathcal{G}, \Theta) < 1 \text{ and the reductive quotient } (M, \Theta^M) \text{ is semi-stable.}$$

For the first theorem we just copy the classical approach, as we did earlier we can reduce it to the case $\mathrm{Sl}_2/\mathbb{Z}$. For the second one we refer to the paper [5] of Kai Behrends.

These issues are also discussed in an unpublished manuscript with U.Stuhler, which hopefully will be available in the near future somewhere.

From these two theorems we can derive general statements concerning the reduction theory of arbitrary semi simple (or reductive) groups G/K over a number field K. We can find a normal extension F/K such that $G \times_K F$ splits. We can extend G/K to a flat group scheme $\mathcal{G}/\mathcal{O}_K$. We consider the two arithmetic groups $\Gamma = \mathcal{G}_{\mathcal{O}_K}$ and $\tilde{\Gamma} = \mathcal{G}_{\mathcal{O}_F}$. We have the two spaces X and $\tilde{X}$ of Cartan involutions on $\mathcal{G}(K \otimes \mathbb{R})$ and $\mathcal{G}(F \otimes \mathbb{R})$ and get a map

$$j : \Gamma\backslash X \to \tilde{\Gamma}\backslash\tilde{X}$$

If Γ and $\tilde{\Gamma}$ are torsion free- a condition we can easily achieve- then then it is not difficult to see that j is a closed embedding. This implies that in general this map j is proper with finite fibres.

Since $\mathcal{G} \times_K F$ is split we also have an extension $\mathcal{G}^{(sp)}/\mathcal{O}_F$ to a semi simple group scheme. This group scheme may not be invariant under the action of the Galois group $\mathrm{Gal}(F/K)$ and hence the arithmetic group $G \times_K F \to \tilde{\Gamma}^{(s)} \subset \mathcal{G}(F)$ may not be invariant under the action of the Galois group. But we can arrange our data such that $\Gamma \subset (\Gamma^{(sc)})^\sigma$ for all $\sigma \in \mathrm{Gal}(F/K)$. We get a family of Galois conjugated diagrams

$$\begin{array}{ccc} \Gamma\backslash X & \xrightarrow{j} & \tilde{\Gamma}\backslash\tilde{X} \\ & \searrow & \downarrow p^\sigma \\ & & (\Gamma^{(sc)})^\sigma\backslash\tilde{X} \end{array} \tag{1.107}$$

Now we look at the numbers $n_{\alpha^P}(B, ((\mathcal{G}^{(sc)})^\sigma, x)$. Since the different conjugates $(\tilde{\Gamma}^{(sc)})^\sigma \in \mathcal{G}(F))$ are commensurable it is clear that we can find a constant $C > 1$ such that for σ and all parabolic subgroups P the inequality

$$\frac{1}{C} < \frac{n_{\alpha^P}(P, \mathcal{G}^{(sc)}, x)}{n_{\alpha^P}(P, (\mathcal{G}^{(sc)})^\sigma, x^\sigma)} < C$$

holds. This has the following consequence: If x is in the image of j and hence $x = x^\sigma$ and if $n_{\alpha^P}(P, \mathcal{G}^{(sc)}, x) << 1$, i.e. is very small, then also $n_{\alpha^{P^\sigma}}(P^\sigma, \mathcal{G}^{(sc)}, x) << 1$. But this implies that $(\mathcal{G}, \Theta_x)$ is unstable, and the destabilising parabolic $P' \subset \bigcap_\sigma P^\sigma$. Then it is clear that $Q^\dagger = \bigcap_\sigma P^\sigma$ is the base extension of a parabolic subgroup Q/K, i.e. $Q \times_K F = Q^\dagger$.

Hence we see: If the image $j(\Gamma\backslash X)$ is not compact then the group $\mathcal{G}/K$ contains a non trivial parabolic subgroup. Since the map j is proper we get the famous result of Borel,Harish-Chandra and Mostow,Tamagawa

Theorem 1.2.5. *([18],[86]): If G/K is a reductive group and $\Gamma \subset G(K)$ is an arithmetic subgroup then*

$$\Gamma\backslash X = \Gamma\backslash G(\mathbb{R})/K_\infty$$

is compact if and only if G/K is anisotropic.

Chapter 2

The Cohomology Groups

2.1 Cohomology of arithmetic groups as cohomology of sheaves on $\Gamma\backslash X$.

We are now in the position to unify — at least for the special case of arithmetic groups — the two cohomology theories from chapter II and chapter IV in [48].

We start from a semi-simple group $G/\mathbb{Q}$, and we choose an arithmetic subgroup $\Gamma \subset G(\mathbb{Q})$. Let $X = G(\mathbb{R})/K_\infty$ as before. A second datum will be a Γ-module $\mathcal{M}$, in principle, this can be any Γ-module.

To such a Γ-module we attach a sheaf $\tilde{\mathcal{M}}$ on $\Gamma\backslash X$. This sheaf has values in the category of abelian groups. For any open subset $U \subset X$ we have to define the group of sections $\tilde{\mathcal{M}}(U)$. We start from the projection

$$\pi : X \longrightarrow \Gamma\backslash X$$

and define

$$\tilde{\mathcal{M}}(U) = \{f : \pi^{-1}(U) \to \mathcal{M} \mid f \text{ is locally constant and } f(\gamma u) = \gamma f(u), \gamma \in \Gamma\}.$$

This is clearly a sheaf. For any point $x \in \Gamma\backslash X$ we can find a neighbourhood V_x with the following property: We choose a point $\tilde{x} \in \pi^{-1}(x)$, then $\tilde{x}$ has a convex $\Gamma_{\tilde{x}}$-invariant neighbourhood $U_{\tilde{x}}$, for which $\gamma U_{\tilde{x}} \cap U_{\tilde{x}} \neq \emptyset \iff \gamma \in \Gamma_{\tilde{x}}$. Then we put $V_x = \Gamma_{\tilde{x}}\backslash U_{\tilde{x}}$. We call such a neighbourhood V_x an *orbiconvex* neighbourhood. It is clear that

$$\tilde{\mathcal{M}}(V_x) = \mathcal{M}^{\Gamma_x} \tag{2.1}$$

Since x has a cofinal system of neighbourhoods of this kind, we see that we get an isomorphism

$$j_{\tilde{x}} : \tilde{\mathcal{M}}(V_x) = \tilde{\mathcal{M}}_x \xrightarrow{\sim} \mathcal{M}^{\Gamma_{\tilde{x}}}.$$

If we are in the special case of Γ without any fixed points, then we can cover $\Gamma\backslash X$ by open sets U so that the retriction $\tilde{\mathcal{M}} \mid U$ is isomorphic to a constant sheaf $\underline{\mathcal{M}}_U$. These sheaves are called *local systems*. In case of fixed points we call them *orbilocal systems*.

G. Harder, *Cohomology of Arithmetic Groups*,
Springer Monographs in Mathematics, https://doi.org/10.1007/978-3-032-10378-9_2

In [48] Chap. 4 we defined the sheaf cohomology groups $H^\bullet(\Gamma\backslash X, \tilde{\mathcal{M}})$ and in this book these cohomology groups are the main objects of interest.

If our $\Gamma-$ module is equipped with an R -module structure-here R is a commutative ring with identity, then the cohomology groups are also R-modules.

The motivations for these constructions are

1) If we have some understanding of the geometry of the quotient space $\Gamma\backslash X$, we gain some insight into the structure of Γ. This will become clear when we discuss the examples in (2.1.4).

2) The cohomology groups $H^\bullet(\Gamma, \mathcal{M})$ are closely related - and in many cases even isomorphic - to the sheaf cohomology groups $H^\bullet(\Gamma\backslash X, \tilde{\mathcal{M}})$. Again the geometry of the spaces $\Gamma\backslash X$ provides tools to compute these cohomology groups in some cases (again 2.1.4).

3) If the Γ-module $\mathcal{M}$ is a $\mathbb{C}$-vector space and obtained from a rational representation of $G/\mathbb{Q}$, then we can apply analytic tools (de Rham cohomology, Hodge theory, see Chapter 6) to get insight into the structure of the groups $H^\bullet(\Gamma, \mathcal{M})$ which are defined in purely algebraic terms.

2.1.1 The relation between $H^\bullet(\Gamma, \mathcal{M})$ and $H^\bullet(\Gamma\backslash X, \tilde{\mathcal{M}})$

For the following we refer to [48]. In this section we assume that X is connected. The functor

$$\mathcal{M} \mapsto H^0(\Gamma\backslash X, \tilde{\mathcal{M}}) = \mathcal{M}^\Gamma.$$

is a functor from the category of Γ-modules to the category **Ab** of abelian groups. The functor $\mathcal{M} \to \mathcal{M}^\Gamma$ is not exact and homological algebra provides the construction of derived functor and this is given by the group cohomology groups $H^\bullet(\Gamma, \mathcal{M})$. ([48] 2.3)

Now we can write this functor as a composition of the two functors $\mathrm{sh}_\Gamma :$ $\mathcal{M} \mapsto \tilde{\mathcal{M}}$ and $\tilde{\mathcal{M}} \mapsto H^0(\Gamma\backslash X, \tilde{\mathcal{M}})$ and we want to apply the composition rule. We extend the functor sh_Γ to the derived functor between the derived categories (See [48, 4.6.4])

$$R\mathrm{sh}_\Gamma : \mathcal{D}(\mathbf{Mod}_\Gamma) \to \mathcal{D}(S_{\Gamma\backslash X}),$$

and define $\mathrm{sh}_\Gamma^\bullet := H^\bullet(R\mathrm{sh}_\Gamma)$ This means $\mathrm{sh}_\Gamma^0(\mathcal{M}) = \tilde{\mathcal{M}}$ and for $q \geq 1$ we get $\mathrm{sh}_\Gamma^q(\mathcal{M})_x = H^q(\Gamma_{\tilde{x}}, \mathcal{M})$ for $\pi(\tilde{x}) = x$.

Now we can apply the composition rule from [48] 4.6.4. For this we have to show that sh_Γ sends injective Γ-modules $\mathcal{I}$ to acyclic sheaves, i.e. for which $H^q(\Gamma\backslash X, \tilde{\mathcal{I}}) = 0$ for $q > 0$.

In 2.24. of [48], we constructed the induced Γ-module $\mathrm{Ind}_{\{1\}}^\Gamma \mathcal{M}$. For any Γ module $\mathcal{M}$ this is the module of all functions $f : \Gamma \to \mathcal{M}$. An element $\gamma_1 \in \Gamma$ acts on this module by $(\gamma_1 f)(\gamma) = f(\gamma\gamma_1)$. The map

$$m \mapsto f_m = \{\gamma \mapsto \gamma m\} \tag{2.2}$$

is an injective Γ-module homomorphism.
We prove that for any such induced module the corresponding sheaf

$$\mathrm{sh}^\Gamma(\mathrm{Ind}_{\{1\}}^\Gamma \mathcal{M}) = \widetilde{\mathrm{Ind}_{\{1\}}^\Gamma \mathcal{M}}$$

is acyclic. We compute this sheaf, for any point $x \in \Gamma\backslash X$ and $\tilde{x} \in \pi^{-1}(x)$. It follows directly from the definitions that

$$\mathrm{sh}^0_\Gamma(\mathcal{M})_x = (\mathrm{Ind}^\Gamma_{\{1\}}\mathcal{M})^\Gamma_{\tilde{x}} = \mathcal{M} \text{ for } q > 0 \ \mathrm{sh}^q_\Gamma(\mathcal{M})_x = H^q(\Gamma_{\tilde{x}}, \mathrm{Ind}^\Gamma_{\{1\}}(\mathcal{M})) = 0$$

in other words the sheaf $\mathrm{sh}^\bullet_\Gamma(\mathrm{Ind}^\Gamma_{\{1\}}\mathcal{M}) = \mathrm{sh}^0_\Gamma(Ind^\Gamma_{\{1\}}\mathcal{M}))$ is the direct image of the constant sheaf $\pi_*(\underline{\mathcal{M}})_X)$.

Now acyclicity is clear. We apply example d) in [48], 4.6.3 to this situation. The fibre of π is a discrete space and hence $R^0\pi_*(\tilde{\mathcal{M}}) = \widetilde{\mathrm{Ind}^\Gamma_{\{1\}}\mathcal{M}}$ and $R^q\pi_*(\underline{\mathcal{M}}_X) = 0$ for $q > 0$. Therefore the spectral sequence yields

$$H^q(X, \underline{\mathcal{M}}_X) = H^q(\Gamma\backslash X, \pi_*(\underline{\mathcal{M}}_X)) = H^q\left(\Gamma\backslash X, \widetilde{\mathrm{Ind}^\Gamma_{\{1\}}\mathcal{M}}\right),$$

and since X is a cell, we see that this is zero for $q \geq 1$.

We apply this to the case that $\mathcal{M} = \mathcal{I}$ is an injective Γ-module. As seen above we can always embed $\mathcal{I} \longrightarrow \mathrm{Ind}^\Gamma_{\{1\}}\mathcal{I}$. But this is now a direct summand; hence it follows from the acyclicity of $\widetilde{\mathrm{Ind}^\Gamma_{\{1\}}\mathcal{I}}$ that also $\tilde{\mathcal{I}}$ must be acyclic.

Now we can apply the composition rule and get a spectral sequence with an E_2 term

$$H^p(\Gamma\backslash X, \mathrm{sh}^q_\Gamma)(\mathcal{M})) \Rightarrow H^n(\Gamma, \mathcal{M})$$

the differential maps $d_2^{p,q} : E_2^{p,q} \to E_2^{p+2,q-1}$. We have the two edge homomorphisms

$$H^n(\Gamma\backslash X, \mathrm{sh}^0_\Gamma(\mathcal{M})) \to H^n(\Gamma, \mathcal{M}),\ H^n(\Gamma, \mathcal{M}) \to H^0(\Gamma\backslash X, \mathrm{sh}^n_\Gamma(\mathcal{M})) \qquad (2.3)$$

which in general are neither injective nor surjective.

We have seen in Section 1.2.1 that - under our assumption that $G/\mathbb{Q}$ is semi-simple - the stabilisers Γ_x are finite. If now $\mathcal{M}$ is an R module and if all primes, which divide the order $\#\Gamma_x$ of some Γ_x, are invertible in R, then we have $\mathrm{sh}^q_\Gamma(\mathcal{M}) = 0$ for $q > 0$. This implies that under these circumstances

$$H^n(\Gamma\backslash X, \mathrm{sh}^0_\Gamma(\mathcal{M})) \xrightarrow{\sim} H^n(\Gamma, \mathcal{M})$$

is an isomorphism.

As already said: In this book our interest are the cohomology groups $H^\bullet(\Gamma\backslash X, \tilde{\mathcal{M}})$. and not so much the $H^\bullet(\Gamma, \mathcal{M})$.

2.1.2 Functorial properties of cohomology

We investigate the functorial properties of the cohomology with respect to the change of Γ. If $\Gamma' \subset \Gamma$ is a subgroup of finite index, then we have the functor

$$\mathbf{Mod}_\Gamma \longrightarrow \mathbf{Mod}_{\Gamma'},$$

which is obtained by restricting the Γ-module structure to Γ'. Since for any Γ-module $\mathcal{M}$ we have $\mathcal{M}^\Gamma \longrightarrow \mathcal{M}^{\Gamma'}$, we obtain a homomorphism

$$\mathrm{res} : H^i(\Gamma, \mathcal{M}) \longrightarrow H^i(\Gamma', \mathcal{M}).$$

We give an interpretation of this homomorphism in terms of sheaf cohomology. We have the diagram

$$\begin{array}{ccc} & X & \\ \pi_{\Gamma'} \swarrow & & \searrow \pi_\Gamma \\ \pi = \pi_{\Gamma,\Gamma'} : \Gamma'\backslash X & \longrightarrow & \Gamma\backslash X \end{array}$$

and a Γ-module $\mathcal{M}$ produces sheaves $\mathrm{sh}^0_\Gamma(\mathcal{M}) = \tilde{\mathcal{M}}$ and $\mathrm{sh}^0_{\Gamma'}(\mathcal{M}) = \tilde{\mathcal{M}}'$ on $\Gamma\backslash X$ and $\Gamma'\backslash X$ respectively. It is clear that we have a homomorphism

$$\pi^*(\tilde{\mathcal{M}}) \longrightarrow \tilde{\mathcal{M}}'.$$

To get this homomorphism we observe that for $y_1 \in \Gamma'\backslash X$ we have $\pi^*_{\Gamma'}(\tilde{\mathcal{M}})_{y_1} = \tilde{\mathcal{M}}_{\pi_1(y_1)}$, and this last module is

$$\{f : \pi^{-1}(\pi_\Gamma(y)) \to \mathcal{M} \mid f(\gamma\tilde{y}) = \gamma f(\tilde{y}) \text{ for all } \gamma \in \Gamma, \tilde{y} \in \pi^{-1}(\pi(y_1))\}.$$

Now

$$\tilde{\mathcal{M}}'_{y_1} = \{f : (\pi')^{-1}(y_1) \to \mathcal{M} \mid f(\gamma'\tilde{y}) = \gamma' f(\tilde{y}) \text{ for all } \gamma \in \Gamma', \tilde{y} \in (\pi')^{-1}(y_1)\},$$

and if we pick a point $\tilde{y} \in (\pi')^{-1}(y_1) \subset \pi^{-1}(\pi_1(y_1))$, then

$$\pi^*(\mathcal{M})_{y_1} \simeq \mathcal{M}^{\Gamma_{\tilde{y}_1}} \subset \tilde{\mathcal{M}}'_{y_1} = \mathcal{M}^{\Gamma'_{\tilde{y}_1}}.$$

Hence we get (or define) our restriction homomorphism as

$$H^i(\Gamma\backslash X, \mathrm{sh}^0_\Gamma(\mathcal{M})) \longrightarrow H^i(\Gamma'\backslash X, \pi^*_{\Gamma'}(\mathrm{sh}^0_\Gamma(\mathcal{M})) \longrightarrow H^i(\Gamma'\backslash X, \mathrm{sh}^0_{\Gamma'}(\mathcal{M})).$$

There is also a map in the opposite direction. Since the fibres of π are discrete, we have

$$H^i(\Gamma'\backslash X, \tilde{\mathcal{M}}) \xrightarrow{\sim} H^i(\Gamma\backslash X, \pi_*(\tilde{\mathcal{M}})).$$

But the same reasoning as in the previous section yields an isomorphism

$$\pi_*(\tilde{\mathcal{M}}) \xrightarrow{\sim} \widetilde{\mathrm{Ind}^\Gamma_{\Gamma'}\mathcal{M}}.$$

Hence we get an isomorphism

$$H^i(\Gamma'\backslash X, \tilde{\mathcal{M}}) \xrightarrow{\sim} H^i(\Gamma\backslash X, \widetilde{\mathrm{Ind}^\Gamma_{\Gamma'}\mathcal{M}})$$

which is well-known as Shapiro's lemma. But we have a Γ-module homomorphism

$$e : \mathrm{Ind}^\Gamma_{\Gamma'}\mathcal{M} \longrightarrow \mathcal{M}$$

which sends an $f \in \mathrm{Ind}^\Gamma_{\Gamma'}\mathcal{M}$ to the sum

$$\mathrm{tr}(f) = \sum \gamma_i^{-1} f(\gamma_i)$$

where the γ_i are representatives for the classes of $\Gamma'\backslash\Gamma$. This homomorphism induces a map in the cohomology, and we get a composition

$$\pi_\bullet : H^i(\Gamma'\backslash X, \tilde{\mathcal{M}}) \longrightarrow H^i(\Gamma\backslash X, \tilde{\mathcal{M}}).$$

It is not difficult to check that

$$\pi_\bullet \circ \pi^\bullet = [\Gamma : \Gamma']\mathrm{Id}$$

2.1.3 How to compute the cohomology groups $H^i(\Gamma\backslash X, \tilde{\mathcal{M}})$

The Čech complex of an orbiconvex Covering

We consider a point $\tilde{x} \in X$ and an open neighbourhood $\tilde{U}_{\tilde{x}} \subset X$. We say that $\tilde{U}_{\tilde{x}}$ is an *orbiconvex* neighbourhood of $\tilde{x}$ if

a) The set $\tilde{U}_{\tilde{x}}$ is convex, i.e. for any two points in $\tilde{x}_1, \tilde{x}_2 \in \tilde{U}_{\tilde{x}}$ the geodesic joining $\tilde{x}_1$ and $\tilde{x}_2$ lies in $\tilde{U}_{\tilde{x}}$.

b) We have $\gamma\tilde{U}_{\tilde{x}} \cap \tilde{U}_{\tilde{x}} = \emptyset$ unless $\gamma\tilde{x} = \tilde{x}$ and in this case we even have $\gamma\tilde{U}_{\tilde{x}} = \tilde{U}_{\tilde{x}}$.

This also means that we require that $\tilde{x}$ is a *central point* in $U_{\tilde{x}}$: The stabiliser of the open set equals the stabiliser $\Gamma_{\tilde{x}.}$ It is well known that any open convex set $V \subset X$ which has a stabiliser Γ_V, also contains a central point $\tilde{x}$, i.e $\Gamma_V = \Gamma_{\tilde{x}}$ We will then also use the notation $V = V_{\tilde{x}}$. and the image $\pi(V)$ will be denoted by V_x where $x = \pi(\tilde{x})$.

A family of orbiconvex neighbourhoods $\{\tilde{U}_{\tilde{x}_i}\}_{i=1,\dots,r}$ ($\tilde{x}_1, \dots, \tilde{x}_r$ set of points) will be called an *orbiconvex covering,* if

$$\bigcup_{i=1}^{r} \bigcup_{\gamma\in\Gamma} \gamma\tilde{U}_{\tilde{x}_i} = X. \tag{2.4}$$

We will show later that we can always find a finite orbiconvex covering of X.

If now $\{\tilde{U}_{\tilde{x}_i}\}_{i=1,\dots,r}$ is an orbiconvex covering we put $U_{x_i} = \pi(\tilde{U}_{\tilde{x}_i})$, and then we get finite covering by open sets

$$\bigcup_{x_i} U_{x_i} = \Gamma\backslash X.$$

We abbreviate and use the usual notation $\mathfrak{U} = \{U_{x_i}\}$ for this orbiconvex covering of $\Gamma\backslash X$. The resulting Čech complex can be written down explicitly, so that it can be implemented on a computer. For this we consider sequences of indices $\underline{i} = \{i_0 < i_1 < \cdots < i_q\}$, the set of such sequences will be denoted by $I^q_<$. We compute the module $\tilde{\mathcal{M}}(U_{\underline{i}}) = \tilde{\mathcal{M}}(U_{x_{i_0}} \cap U_{x_{i_1}} \cap \cdots \cap U_{x_{i_q}})$. The intersection can be empty, in this case the module of sections is trivial and uninteresting. If this intersection is non trivial it may have several connected components. Let $\pi_0(U_{\underline{i}})$ be the set of connected components and for $\eta \in \pi_0(U_{\underline{i}})$ let V_η be the corresponding connected component. Then clearly

$$\tilde{\mathcal{M}}(U_{\underline{i}}) = \bigoplus_{\eta\in\pi_0(U_{\underline{i}})} \tilde{\mathcal{M}}(V_\eta) \tag{2.5}$$

We want to give an explicit description of this module. We choose a point $\tilde{\eta} \in \pi^{-1}(V_\eta)$ we may assume that it is central in its connected component $V_{\tilde{\eta}}$. Moreover we may assume that $\tilde{\eta} \in \tilde{U}_{\tilde{x}_0}$. Then $\tilde{\eta}$ determines an array $\underline{\gamma} = \{e, \gamma_1, \dots, \gamma_\nu, \dots, \gamma_q\}$ such that

$$\tilde{\eta} \in \tilde{U}_{\tilde{\xi}_0} \cap \gamma_1\tilde{U}_{\tilde{x}_{i_1}} \cap \cdots \cap \gamma_\nu\tilde{U}_{\tilde{\xi}_\nu} \cap \cdots \cap \gamma_q\tilde{U}_{\tilde{\xi}_q} = V_{\tilde{\eta}} \tag{2.6}$$

Here the elements $\gamma_\nu \in \Gamma/\Gamma_{\tilde{x}_\nu}$. Let us denote by $\mathcal{F}_\eta$ the set of arrays

$$\underline{\gamma} = \{e, \gamma_1, \ldots, \gamma_\mu, \ldots, \gamma_q\} \in \Gamma \times \prod_{\nu=1}^{q} \Gamma/\Gamma_{\tilde{x}_\nu}$$

which are obtained from $\pi_0(U_{\underline{i}})$, this is the set of arrays for which the intersection in (2.6) is not empty Then we get an identification

$$\pi_0(U_{\underline{i}}) \xrightarrow{\sim} \mathcal{F}_{\underline{i}}. \tag{2.7}$$

The stabiliser of $V_{\tilde{\eta}}$ or of $\tilde{\eta}$ is

$$\Gamma_{\tilde{\eta}} = \Gamma_{\underline{\gamma}} = \Gamma_{\tilde{\xi}_0} \cap \cdots \cap \Gamma_{\gamma_\nu \tilde{\xi}_\nu} \cap \cdots \cap \Gamma_{\gamma_\nu \tilde{x}_{\tilde{\xi}_q}} \tag{2.8}$$

and since the $V_{\tilde{\eta}}$ are orbiconvex they are acyclic, more precisely

$$H^q(U_{\underline{i}}, \tilde{\mathcal{M}}) = \begin{cases} \bigoplus_{\underline{\gamma} \in \mathcal{F}_{\underline{i}}} \mathcal{M}^{\Gamma_{\underline{\gamma}}} & q = 0 \\ 0 & q > 0 \end{cases} \tag{2.9}$$

This implies that the Čech complex (See [48], Chap. 4, 4.5)

$$C^\bullet(\mathfrak{U}, \tilde{\mathcal{M}}) :=\rightarrow \bigoplus_{\underline{i} \in I_<^{q-1}} \tilde{\mathcal{M}}(U_{\underline{i}}) \xrightarrow{d^{q-1}} \bigoplus_{\underline{i} \in I_<^{q}} \tilde{\mathcal{M}}(U_{\underline{i}}) \xrightarrow{d^q} \tag{2.10}$$

computes the cohomology, i.e. the cohomology groups $H^\bullet(\Gamma\backslash X, \tilde{\mathcal{M}})$. are the cohomology groups of this complex. The differentials are given by the formula

$$(d^{q-1}c)_{i_0,\ldots,i_q} = \sum_{\nu=0}^{q} (-1)^\nu c_{i_0,\ldots,\hat{i}_\nu,\ldots,i_q} | U_{i_0,,\ldots,i_q}$$

where the "hat" means that i_ν is deleted from the array. Hence we need to write explicit formulas for the restriction maps

$$\bigoplus_{\underline{\gamma}' \in \mathcal{F}_{\underline{i}^\nu}} \mathcal{M}^{\Gamma_{\underline{\gamma}'}} \rightarrow \bigoplus_{\gamma \in \mathcal{F}_{\underline{i}}} \mathcal{M}^{\Gamma_{\underline{\gamma}}}.$$

For any pair $\underline{\gamma}^\nu = \{e, \gamma_1, \ldots, \hat{\gamma}_\nu, \ldots, \gamma_q\}, \underline{\gamma} = \{e, \gamma_1, \ldots, \gamma_\nu, \ldots, \gamma_q\}$ we have to determine the restriction map $\mathcal{M}^{\Gamma_{\underline{\gamma}^\nu}} \rightarrow \mathcal{M}^{\Gamma_{\underline{\gamma}}}$. If $\nu \neq 0$ then $\Gamma_{\underline{\gamma}^\nu} \subset \Gamma_{\underline{\gamma}}$ this map is given by the inclusion $\mathcal{M}^{\Gamma_{\underline{\gamma}^\nu}} \hookrightarrow \mathcal{M}^{\Gamma_{\underline{\gamma}}}$. But if $\nu = 0$ the first member of $\{\hat{e}, \gamma_1, \ldots, \gamma_\nu, \ldots, \gamma_q\}$ is γ_1 and hence $\underline{\gamma}^0 = \{\hat{\gamma}_1^{-1}, e, \gamma_1^{-1}\gamma_2, \ldots, \gamma_1^{-1}\gamma_q\}$. Therefore the map $\mathcal{M}^{\Gamma_{\underline{\gamma}^0}} \rightarrow \mathcal{M}^{\Gamma_{\underline{\gamma}}}$ is given by the multiplication by γ_1, followed by the restriction, i.e. it is the composition

$$\tilde{\mathcal{M}}(\tilde{U}_{\tilde{x}_{i_1}} \cap \cdots \cap \gamma_1^{-1}\gamma_q \tilde{U}_{\tilde{x}_q}) \xrightarrow{\gamma_1} \tilde{\mathcal{M}}(\gamma_1 \tilde{U}_{\tilde{x}_{i_1}} \cap \cdots \cap \gamma_q \tilde{U}_{\tilde{x}_q}) \rightarrow \tilde{\mathcal{M}}(\tilde{U}_{\tilde{x}_0} \cap \cdots \cap \gamma_q \tilde{U}_{\tilde{x}_q})$$

I think using this explicit description of the Čech complex one should be able to write a code which computes the cohomology. Of course we left the problem to find an orbiconvex covering and our module $\mathcal{M}$ should not to weird. It should be a finitely generated module over $\mathbb{Z}$ or $\mathbb{Z}[\frac{1}{N}]$. We will discuss special cases where $\mathcal{M}$ is finitely generated over a finite filed $\mathbb{F}_p$ or over a finite ring $\mathbb{Z}/p^n$.

We may also look at this situation from a different point of view: if $\tilde{x} \in X$ is any point and $\Gamma_{\tilde{x}} \subset \Gamma$ its stabiliser, then we define the compactly supported induced Γ module

$$\text{c-Ind}_{\Gamma_{\tilde{x}}}^{\Gamma} \mathbb{Z} := \{f : \Gamma \to \mathbb{Z} \mid f \text{ has finite support and } f(a\gamma) = f(\gamma),\ \forall a \in \Gamma_x, \gamma \in \Gamma\}. \tag{2.11}$$

If $V_{\tilde{x}}$ is an open neighbourhood of $\tilde{x}$ which satisfies a) and b) (i.e. it is orbiconvex), then we have $\pi^{-1}(\pi(V_{\tilde{x}})) = \bigcup_{\gamma \in \Gamma/\Gamma_{\tilde{x}}} \gamma V_{\tilde{x}}$ and

$$\pi^*(\tilde{\mathcal{M}}) \left(\bigcup_{\gamma \in \Gamma/\Gamma_{\tilde{x}}} \gamma V_{\tilde{x}} \right) = \text{Hom}(\text{c-Ind}_{\Gamma_{\tilde{x}}}^{\Gamma} \mathbb{Z}, \mathcal{M}).$$

We have the covering

$$\tilde{\mathfrak{U}} = \{\tilde{U}_{\tilde{x}_i}\}_{\{\tilde{\xi},\gamma\}};\ \bigcup_{i,\gamma \in \Gamma/\Gamma_{\tilde{\xi}}} \gamma \tilde{U}_{\tilde{x}_i} = X$$

The Čech-complex $C^\bullet(\tilde{\mathfrak{U}}, \pi^*(\tilde{\mathcal{M}}))$ computes the cohomology groups $H^q(X, \pi^*(\tilde{\mathcal{M}}))$ which are trivial for $q > 0$. Our considerations above yield

$$C^\bullet(\tilde{\mathfrak{U}}, \pi^*(\tilde{\mathcal{M}})) = 0 \to \bigoplus_{i=1}^{r} \text{Hom}(\text{Ind}_{\Gamma_{x_i}}^{\Gamma} \mathbb{Z}, \mathcal{M}) \xrightarrow{d^1} \bigoplus_{i<j,\tilde{x}_{i,j}} \text{Hom}(\text{Ind}_{\Gamma_{\tilde{x}_{i,j}}}^{\Gamma} \mathbb{Z}, \mathcal{M}) \xrightarrow{d^2}$$

Now it is easy to see that the boundary maps are induced by maps between the induced modules

$$\cdots \xrightarrow{\delta^2} \bigoplus_{i<j,\tilde{x}_{i,j}} \text{Ind}_{\Gamma_{\tilde{x}_{i,j}}}^{\Gamma} \mathbb{Z} \xrightarrow{\delta^1} \bigoplus_{i=1}^{r} \text{Ind}_{\Gamma_{\tilde{x}_i}}^{\Gamma} \mathbb{Z} \to 0,$$

where for $f \in \bigoplus_{\tilde{\xi}} c - Ind_{\Gamma_{\tilde{x}_J}}^{\Gamma} \mathbb{Z}$, in degree ν and $\omega \in C^{\nu-1}(\tilde{\mathfrak{U}}, \pi^*(\tilde{\mathcal{M}}))$, the relation $\omega(\delta^\nu(f)) = d^{\nu-1}(\omega)(f)$ defines δ^ν. We get an augmented complex

$$P^\bullet := \to \bigoplus_{\tilde{x}_J} \text{Ind}_{\Gamma_{\tilde{x}_J}}^{\Gamma} \mathbb{Z} \to \cdots \to \bigoplus_{\tilde{x}_{i,j}} \text{Ind}_{\Gamma_{\tilde{x}_{i,j}}}^{\Gamma} \mathbb{Z} \to \bigoplus_{i=1}^{r} \text{Ind}_{\Gamma_{\tilde{x}_i}}^{\Gamma} \mathbb{Z} \to \mathbb{Z} \to 0 \tag{2.12}$$

and since $C^\bullet(\tilde{\mathfrak{U}}, \pi^*(\tilde{\mathcal{M}}))$ is acyclic of degree > 0, we get that $P^\bullet$ must be a resolution of the trivial module $\mathbb{Z}$, i.e. the complex is acyclic.

Let $N = \prod_i \#\Gamma_{\tilde{x}_i}$ and $R := \mathbb{Z}[\frac{1}{N}]$. Then the $R[\Gamma]$ module $\text{Ind}_{\Gamma_x}^{\Gamma} \otimes R$ is a direct summand in $R[\Gamma]$ and hence a projective $R[\Gamma]$ module. This implies of course that

$$P^\bullet \otimes R = \to \bigoplus_{\tilde{x}_J} \text{Ind}_{\Gamma_{\tilde{x}_J}}^{\Gamma} R \to \cdots \to \bigoplus_{\tilde{x}_{i,j}} \text{Ind}_{\Gamma_{\tilde{x}_{i,j}}}^{\Gamma} R \to \bigoplus_{i=1}^{r} \text{Ind}_{\Gamma_{\tilde{x}_i}}^{\Gamma} R \to R \to 0 \tag{2.13}$$

is indeed a projective resolution of the trivial Γ-module R. Therefore we know that

$$\begin{aligned} H^\bullet(\Gamma, \mathcal{M}_R) = H^\bullet(0 \to \mathrm{Hom}_\Gamma(\bigoplus_{i=1}^{r} \mathrm{Ind}_{\Gamma_{\tilde{x}_i}}^{\Gamma} R, \mathcal{M}_R)) \\ \to \bigoplus_{i<j, \tilde{x}_{i,j}} \mathrm{Hom}_\Gamma(\mathrm{Ind}_{\Gamma_{\tilde{x}_{i,j}}}^{\Gamma} R, \mathcal{M}_R) \to \end{aligned} \tag{2.14}$$

where now on the left hand side we have the group cohomology. If we do not tensor by R then the complex

$$0 \to \bigoplus_{i=1}^{r} \mathrm{Hom}_\Gamma(\mathrm{Ind}_{\Gamma_{x_i}}^{\Gamma} \mathbb{Z}, \mathcal{M}) \to \bigoplus_{i<j, i,j} \mathrm{Hom}_\Gamma(\mathrm{Ind}_{\Gamma_{\tilde{x}_{i,j}}}^{\Gamma} \mathbb{Z}, \mathcal{M}) \to \dots \tag{2.15}$$

is isomorphic to the Čech complex (2.1.3) and it computes the sheaf cohomology $H^\bullet(\Gamma\backslash X, \tilde{\mathcal{M}})$.

It follows from reduction theory that:

Theorem 2.1.1. *We can construct a finite orbiconvex covering* $\mathfrak{U} = \{U_{x_i}\}_{\dots, x_i, \dots}$ *of* $\Gamma\backslash X = \bigcup_{i \in E} \cdot\,\cdot$

Proof. This is rather clear. We start from the covering by the sets $X^P(c_{\pi'}, r(c_{\pi'}))$. The set of "almost stable" points $X^G(r) \subset X$ is relatively compact modulo Γ. For any point $\tilde{x} \in X$ we look at the minimum distance

$$d(\tilde{x}) := \min_{\gamma \in \Gamma \setminus \Gamma_{\tilde{x}}} d(\tilde{x}, \gamma\tilde{x}).$$

Since the action of Γ is properly discontinuous $d(\tilde{x}) > 0$. Let $D(\tilde{x}, d(\tilde{x})/2) := \{\tilde{y} \mid d(\tilde{y}, \tilde{x}) < d(\tilde{x})/2\}$ (the Dirichlet-ball around $\tilde{x}$), then $D(\tilde{x}, d(\tilde{x})/2)$ is an orbiconvex neighbourhood of $\tilde{x}$. We can find finitely many points $\tilde{x}_1, \dots, \tilde{x}_r$ such that

$$\bigcup_{i=1}^{r} \bigcup_{\gamma \in \Gamma} \gamma D(\tilde{x}_i, d(\tilde{x}_i)/2)) \supset X^G(r).$$

We have to find a covering for the $X^P(c_{\pi'}, r(\pi'))$. We recall the fibration (see (1.2.7))

$$r_P : X^P(c_{\pi'}, r(\underline{c}_{\pi'})) \to X^M(r(\underline{c}_{\pi'})) \times \prod_{\alpha \in \pi'} (0, c_\alpha].$$

We apply our previous argument and find a finite covering

$$\bigcup_{i=1}^{s} \bigcup_{\gamma \in \Gamma_M} \gamma D(\tilde{y}_i, d(\tilde{y}_i)/2)) \supset X^M(r(\underline{c}_{\pi'})).$$

We pick a point $\underline{c}_0 \in \prod_{\alpha \in \pi'} (0, c_\alpha]$, then the inverse image

$$r_P^{-1}(D(\tilde{y}_i, d(\tilde{y}_i)/2)) \times \underline{c}_0$$

is relatively compact and we can find an orbiconvex covering $\mathfrak{V} = \{V_{\tilde{x}_\mu}\}$ of this set. Then the products $V_{\tilde{x}_\mu} \times \prod_{\alpha \in \pi'} (0, c_\alpha]$ provide an orbiconvex covering of $X^P(c_{\pi'}, r(\underline{c}_{\pi'}))$. Of course these sets are not (relatively) compact anymore. □

I think that it is a very important problem to write algorithms which compute cohomology effectively. The main goal would be to collect experimental data which may suggest conjectures or give support for conjectures which come from different sources. We will come back to this aspect in the following chapter and also in the final chapter.

I do not claim that my proposal using orbiconvex coverings provides an acceptable solution to this problem. It solves the problem in principle but it is not clear how far it reaches in practice. We see that the fixed points create some problems if we want to write down the explicit complexes. But it is certainly no solution to avoid these problems by passing to a congruence subgroup. Then the number of members in the covering grow rapidly and the complexes become much bigger. For the groups $\mathrm{Sl}_n(\mathbb{Z})$ (for some small values of n and for some congruence subgroups) various authors have computed the cohomology (Ash, Stevens, Gunnells) using the Voronoi decomposition of the cone of positive-definite symmetric matrices.

Here I want to propose a slightly modified strategy to compute the cohomology. A first step would be to effectively find an optimal orbiconvex covering $\{U_{\tilde{x}_\nu}\}$ of the set $X^G(r)$ of almost stable points. The covering sets must not necessarily be Dirichlet balls. We could proceed and construct orbiconvex coverings $\{V^M_{\tilde{y}_\mu}\}$ of the different reductive quotients $X^{M_1}(r(\underline{c}_{\pi'}))$. .Then we may consider the inverse images $(p_{M_1\times n^{\pi'}})^{-1}(V^M_{\tilde{y}_\mu}\times\prod_{\alpha\in\pi'}(0,c_\alpha]) = \tilde{V}_{\tilde{y}^M_\mu}$. (1.90). This family of sets $\{\gamma U_{\tilde{x}_\nu},\dots,\gamma_1\tilde{V}^M_{\tilde{y}_\mu},\dots\}$ provide a covering of X by open sets, hence the images under the projection provide a covering

$$\mathfrak{W} = \{W_i\}_{i\in I} = \{U_{x_\nu},\dots,\tilde{V}_{y^M_\mu},\dots\}$$

of $\Gamma\backslash X$, here the index set I is the union of the $\tilde{x}_\nu\tilde{x}_\nu,\dots,y^M_{\tilde{y}_\mu}$.

Of course we have a problem: the sets $\tilde{V}^M_{y_\mu}$ are not acyclic anymore, so we can not use the Čech complex of this covering for the computation of the cohomology. But we know that

$$\tilde{V}^M_{y_\mu} \to V^M_{y_\mu} \times \prod_{\alpha\in\pi'}(0,c_\alpha]$$

is a fiber bundle with fiber $U(\mathbb{Z})\backslash U(\mathbb{R})$. Since the base $V^M_{y_\mu}\times\prod_{\alpha\in\pi'}(0,c_\alpha]$ is acyclic we know that

$$H^\bullet(\tilde{V}^M_{y_\mu}) \xrightarrow{\sim} H^\bullet(U(\mathbb{Z})\backslash U(\mathbb{R}),\tilde{\mathcal{M}}) \tag{2.16}$$

and we have a good understanding of the cohomology on the right. If for instance we tensor by the rationals, the Theorem of Kostant (see 6.2.3) gives us a complete description of the cohomology $H^\bullet(U(\mathbb{Z})\backslash U(\mathbb{R}),\tilde{\mathcal{M}}\otimes\mathbb{Q})$.

For $\underline{i}\in I^{p+1}$ we put $W_{\underline{i}} = W_{i_0}\cap W_{i_1}\cap\dots\cap W_{i_p}$. Now we follow [48], 4.6.6, and for any $q\geq 0$ write the Čech complexe

$$C^\bullet(\mathfrak{W},\mathcal{H}^q):\dots\to\prod_{\underline{i}\in I^{p+1}} H^q(W_{\underline{i}})\to\prod_{\underline{i}\in I^{p+2}} H^q(W_{\underline{i}})\to\dots \tag{2.17}$$

and then we know that we get a spectral sequence

$$E_1^{p,q} = H^p(C^\bullet(\mathfrak{W},\mathcal{H}^q)) \implies H^{p+q}(\Gamma\backslash X,\tilde{\mathcal{M}}) \tag{2.18}$$

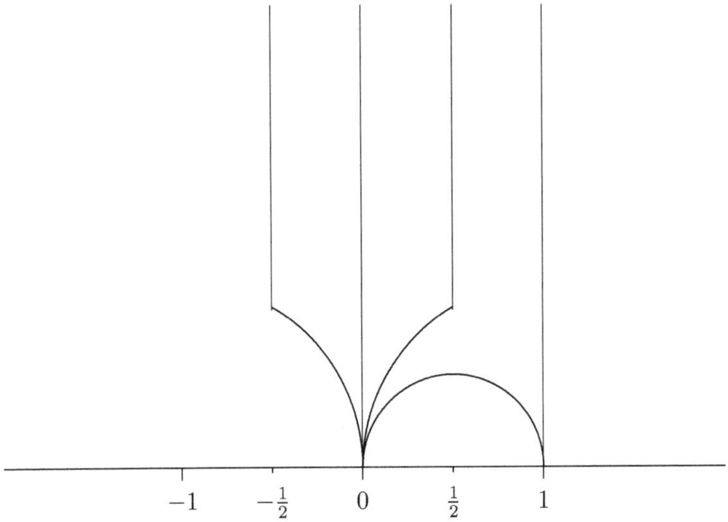

Figure 2.1: An orbifold cover

2.1.4 Special examples in low dimensions

We consider the group $\Gamma = \mathrm{Sl}_2(\mathbb{Z})$, and its action on the upper half plane $\mathbb{H}$. We want to investigate the cohomology groups $H^i(\Gamma\backslash\mathbb{H}, \tilde{\mathcal{M}})$ for any Γ-module $\mathcal{M}$. Let $\pi : \mathbb{H} \to \Gamma\backslash\mathbb{H}$ be the projection. We have the two special points i and ρ in $\mathbb{H}$. They are - up to conjugation by Γ - the only points which have a stabiliser not equal to $\{\pm\mathrm{Id}\}$. We construct two nice orbiconvex neighbourhoods of these two points. The stabilisers Γ_i, resp. Γ_ρ are cyclic and generated by the two elements

$$S = \begin{pmatrix} 0 & 1 \\ -1 & 0 \end{pmatrix}, R = \begin{pmatrix} 1 & -1 \\ 1 & 0 \end{pmatrix}$$

respectively.

We begin with i. We consider the strip $V_i = \{z | -1/2 < \Re(z) < 1/2\}$, the element S maps the two vertical boundary lines $\Re(z) = \pm\frac{1}{2}$ into geodesic circles starting from 0 and ending in ± 2. Then the intersection $\tilde{U}_i = V_i \cap S(V_i)$ is an orbiconvex neighbourhood of i.

Let us look at ρ. We consider the strip $V_\rho = \{z \mid -0 < \Re(z) < 1\}$ and now we define $\tilde{U}_\rho = V_\rho \cap R(V_\rho) \cap R^2(V_\rho)$. This is a nice orbiconvex neighbourhood of ρ.

Now it is clear that these two sets provide an orbiconvex covering of $\mathbb{H}$: if $U_i = \pi(\tilde{U}_i), U_\rho = \pi(\tilde{U}_\rho)$ then

$$\Gamma\backslash\mathbb{H} = U_i \cup U_\rho. \tag{2.19}$$

We have $\tilde{\mathcal{M}}(U_i) = \mathrm{sh}_\Gamma(\mathcal{M})(U_i) = \mathcal{M}^{\Gamma_i}, \tilde{\mathcal{M}}(U_\rho) = \mathcal{M}^{\Gamma_\rho}$ and hence the cohomology groups are given by the cohomology of the complex

$$0 \to \mathcal{M}^{\Gamma_i} \oplus \mathcal{M}^{\Gamma_\rho} \to \mathcal{M} \to 0. \tag{2.20}$$

This gives us $H^0(\Gamma\backslash\mathbb{H},\tilde{\mathcal{M}}) = \mathcal{M}^\Gamma = \mathcal{M}^{\Gamma_i}\cap\mathcal{M}^{\Gamma_\rho}$. Since this is true for any Γ-module -for instance for $\mathbb{Z}[\Gamma]$- we easily conclude that Γ is generated by Γ_i,Γ_ρ. Furthermore

$$H^1(\mathrm{Sl}_2(\mathbb{Z})\backslash\mathbb{H},\tilde{\mathcal{M}}) = \mathcal{M}/(\mathcal{M}^{(<S>}+\mathcal{M}^{<R>}), \tag{2.21}$$

The cohomology vanishes in higher degrees.

Exercise 1: *Let $\Gamma'\subset\Gamma=\mathrm{Sl}_2(\mathbb{Z})$ be a subgroup of finite index. Prove that we have (Shapiro s lemma)*

$$H^1(\Gamma'\backslash\mathbb{H},\mathbb{Z}) = H^1(\Gamma\backslash\mathbb{H},\widetilde{\mathrm{Ind}_{\Gamma'}^{\Gamma}\mathbb{Z}}).$$

These cohomology groups are free of rank

$$[\Gamma:\Gamma'] - n_i - n_\rho + 1,$$

where n_i (resp. n_ρ) is the number of orbits of Γ_i (resp. Γ_ρ) on $\Gamma'\backslash\Gamma$. If Γ' is torsion free then

$$\mathrm{rank}\, H^1(\Gamma\backslash\mathbb{H},\widetilde{\mathrm{Ind}_{\Gamma'}^{\Gamma}\mathbb{Z}}) = \frac{1}{6}[\Gamma:\Gamma']+1.$$

The Euler-characteristic of $\Gamma'\backslash\mathbb{H}$ is $-\frac{1}{6}[\Gamma:\Gamma']$.

Exercise 2: *Let $\mathcal{M}_n$ be the $\mathrm{Sl}_2(\mathbb{Z})$-module of homogenous polynomials in the two variables X,Y and coefficients in $\mathbb{Z}$ (see 1.2.1). We have the usual action of $\mathrm{Sl}_2(\mathbb{Z})$ on this module by*

$$\begin{pmatrix} a & b\\ c & d\end{pmatrix} P(X,Y) = P(aX+cY, bX+dY).$$

These modules define a sheaf $\tilde{\mathcal{M}}_n$ on $\Gamma\backslash\mathbb{H}$., We intend to compute the cohomology groups $H^\bullet(\Gamma\backslash\mathbb{H},\tilde{\mathcal{M}}_n)$.

Prove:

i) If n is odd, then $\tilde{\mathcal{M}}_n = 0$.
Hence we assume $n\geq 2$ and n even from now on.

ii) For $n>0$ we have $H^0(\Gamma\backslash\mathbb{H},\tilde{\mathcal{M}}_n)=0$.

iii) If we tensorize by $\mathbb{Q}$, then $H^1(\Gamma\backslash\mathbb{H},\tilde{\mathcal{M}}_n\otimes\mathbb{Q})$ is a vector space of rank equal to $n-1-2\left[\frac{n}{4}\right]-2\left[\frac{n}{6}\right]$.

Hint: *Diagonalise the action of Γ_i and Γ_ρ on $\mathcal{M}_n\otimes\overline{\mathbb{Q}}$ separately and look at the eigenspaces. To say it differently: over $\bar{\mathbb{Q}}$ we can conjugate the matrices $\begin{pmatrix} 0 & -1\\ 1 & 0\end{pmatrix}$, $\begin{pmatrix} 1 & 1\\ -1 & 0\end{pmatrix}$ into the diagonal maximal torus $\begin{pmatrix} t & 0\\ 0 & t^{-1}\end{pmatrix}$, and then look at the decomposition of $\mathcal{M}_n$ into eigenspaces.*

iv) Compute the torsion in $H^1(\Gamma, \mathcal{M}_n)$. *Investigate the torsion in* $H^1(\Gamma\backslash\mathbb{H}, \tilde{\mathcal{M}}_n)$. *Start from a prime* ℓ *and investigate the long exact sequence obtained from the sequence* $0 \to \mathcal{M}_n \xrightarrow{\times\ell} \mathcal{M}_n \to \mathcal{M}_n/\ell\mathcal{M}_n \to 0$. *If you get stuck wait until the proof of (3.3.3).*

v) Now we consider $\Gamma = \mathrm{Sl}_2(\mathbb{Z})$. *The two matrices* $S = \begin{pmatrix} 0 & -1 \\ 1 & 0 \end{pmatrix}$ *and* $R = \begin{pmatrix} -1 & 1 \\ -1 & 0 \end{pmatrix}$ *are generators of the stabilisers of* i *and* ρ *respectively.*

We take for our module $\mathcal{M}$ *the cyclic group* $\mathbb{Z}/12\mathbb{Z}$, *and consider the* E_2 *term of the spectral sequence*

$$E_2^{p,q} = H^p(\Gamma\backslash\mathbb{H}, R^q\mathrm{sh}_\Gamma(\mathbb{Z}/12\mathbb{Z})) \implies H^{p+q}(\Gamma, \mathbb{Z}/12\mathbb{Z}).$$

Show that $H^0(\Gamma\backslash\mathbb{H}, R^1\mathrm{sh}_\Gamma(\mathbb{Z}/12\mathbb{Z})) = \mathbb{Z}/12\mathbb{Z}$. *Show that the differential*

$$H^0(\Gamma\backslash\mathbb{H}, R^1\mathrm{sh}_\Gamma(\mathbb{Z}/12\mathbb{Z})) \to H^2(\Gamma\backslash\mathbb{H}, \mathrm{sh}_\Gamma(\mathbb{Z}/12\mathbb{Z}))$$

vanishes and conclude

$$H^1(\Gamma, \mathbb{Z}/12\mathbb{Z}) = \mathbb{Z}/12\mathbb{Z}.$$

The group $\Gamma = \mathrm{Gl}_2(\mathbb{Z}[i])$

A similar computation can be made to compute the cohomology in the case of $\tilde{\Gamma} = \mathrm{Gl}_2(\mathbb{Z}[i]))$. We have the three special points x_{12}, x_{13} and x_{23} (see (1.2.4)), and we choose closed sets A_{ij} containing these points which just leave out a small open strip containing the opposite face. If $\tilde{A}_{ij}$ is a connected component of the inverse image of A_{ij} in $\mathbb{H}_3$, then

$$A_{ij} = \Gamma_{ij}\backslash\tilde{A}_{ij}.$$

The intersections $A_{ij} \cap A_{i'j'} = A_\nu$ are closed sets. They are of the form

$$A_\nu = \Gamma_\nu\backslash\tilde{A}_\nu$$

where Γ_ν is the stabiliser of the arc joining x_{ij} and $x_{i'j'}$. The restrictions of our sheaves $\tilde{\mathcal{M}}$ to the A_{ij} and A_ν and to $A = A_{12} \cap A_{23} \cap A_{13}$ are acyclic, and hence we get a complex

$$0 \longrightarrow \tilde{\mathcal{M}} \longrightarrow \bigoplus_{(i,j)} \tilde{\mathcal{M}}_{A_{ij}} \longrightarrow \bigoplus \tilde{\mathcal{M}}_{A_\nu} \longrightarrow \tilde{\mathcal{M}}_A \longrightarrow 0$$

where the $\tilde{\mathcal{M}}_?$ are the restrictions of $\tilde{\mathcal{M}}$ to ? and then extended to the space again.

Hence we find that our cohomology groups are equal to the cohomology groups of the complex

$$0 \longrightarrow \bigoplus_{(i,j)} \mathcal{M}^{\Gamma_{ij}} \xrightarrow{d^1} \bigoplus_{\nu} \mathcal{M}^{\Gamma_\nu} \xrightarrow{d^2} \mathcal{M} \longrightarrow 0 \tag{2.22}$$

with boundary maps

$$d^1 : (m_{12}, m_{13}, m_{23}) \longmapsto (m_{12} - m_{13}, m_{23} - m_{12}, m_{13} - m_{23})$$
$$d^2 : (m_1, m_2, m_3) \longmapsto m_1 + m_2 + m_3.$$

If we take for instance $\tilde{\mathcal{M}} = \mathbb{Z}$ then we get $H^0(\tilde{\Gamma}\backslash\mathbb{H}_3, \mathbb{Z}) = \mathbb{Z}$ and $H^i(\tilde{\Gamma}\backslash\mathbb{H}_3, \mathbb{Z}) = 0$ for $i > 0$ as it should be.

We do not get a satisfying answer to our question. We consider the special case $\Gamma = \mathrm{Sl}_2(\mathbb{Z}[i])$ and the coefficient system $\mathcal{M}_{n_1} \otimes_{\mathbb{Z}[i]} \mathcal{M}_{n_2}$ where

$$\mathcal{M}_{n_1} := \{P(X,Y) = \sum_{\nu=0}^{n_1} a_\nu X^\nu Y^{n_1-\nu} \mid a_\nu \in \mathbb{Z}[i]\},$$
$$\mathcal{M}_{n_2} := \{Q(\bar{X},\bar{Y}) = \sum_{\nu=0}^{n_2} a_\nu \bar{X}^\nu \bar{Y}^{n_2-\nu} \mid a_\nu \in \mathbb{Z}[i]\},$$

and where a matrix $\gamma \in \Gamma$ acts by

$$\gamma(P(X,Y) \otimes Q(\bar{X},\bar{Y})) = P(aX+cY, bX+dY) \otimes Q(\bar{a}\bar{X}+\bar{c}\bar{Y}, \bar{b}\bar{X}+\bar{d}\bar{Y}). \tag{2.23}$$

If we now choose the module $\mathcal{M} = \mathcal{M}_{n_1} \otimes_{\mathbb{Z}[i]} \mathcal{M}_{n_2}$ for our module $\mathcal{M}$ in (2.22), then it is clear that for any given (not too large) values n_1, n_2 we can compute the cohomology explicitly. But we do not get any theoretical insight, for instance, we do not get a formula for the dimensions of the cohomology groups (tensorised by $\mathbb{Q}$). In his paper [97] H. Sengun presents some experimental experimental material, it seems to indicate that the cohomology groups on this case have a rather erratic behavior.

2.1.5 Homology, cohomology with compact support and Poincaré duality

Here we have to use the theory of compactifications. For any locally symmetric space we can embed $\Gamma\backslash X$ into its Borel-Serre compactification

$$i : \Gamma\backslash X \longrightarrow \Gamma\backslash\overline{X}_{BS},$$

this process was explained in detail for our low dimensional examples. Especially we gave an explicit description of a neighbourhood of a point $x \in \partial(\Gamma\backslash\overline{X}_{BS})$. If we have a sheaf $\tilde{\mathcal{M}}$ on $\Gamma\backslash X$, we can extend it to the compactification by using the functor i_*. We get a sheaf $i_*(\tilde{\mathcal{M}})$ on $\Gamma\backslash\overline{X}_{BS}$, and it is clear from the description of a neighbourhood of a point in the boundary that i_* is exact (this is not true for the Baily-Borel compactification).

Our construction $\mathcal{M} \to \tilde{\mathcal{M}}$ can be extended to the action of Γ on $\overline{X}_{BS}$ and clearly

$$i_*(\tilde{\mathcal{M}}) = \text{ result of the construction } \mathcal{M} \to \tilde{\mathcal{M}} \text{ on } \Gamma\backslash\overline{X}_{BS}.$$

Hence we get

$$H^\bullet(\Gamma\backslash X, \tilde{\mathcal{M}}) = H^\bullet(\Gamma\backslash\overline{X}_{BS}, i_*(\tilde{\mathcal{M}})).$$

But we have another construction of extending the sheaf $\tilde{\mathcal{M}}$ from $\Gamma\backslash X$ to $\Gamma\backslash\overline{X}_{BS}$. This is the so called extension by zero. We define the sheaf $i_!(\tilde{\mathcal{M}})$ on $\Gamma\backslash\overline{X}_{BS}$ by giving the stalks. For $x \in \Gamma\backslash\overline{X}_{BS}$ we put

$$i_!(\tilde{\mathcal{M}})_x = \begin{cases} \tilde{\mathcal{M}}_x & \text{if} \quad x \in \Gamma\backslash X \\ 0 & \text{if} \quad x \notin \Gamma\backslash X \end{cases}.$$

It is clear that $i_!$ is an exact functor sending sheaves on $\Gamma\backslash X$ to sheaves on $\Gamma\backslash\overline{X}_{BS}$, and we have for an arbitrary sheaf $\mathcal{F}$

$$H^0(\Gamma\backslash\overline{X}_{BS}, i_!\mathcal{F}) = H^0_c(\Gamma\backslash X, \mathcal{F})$$

where $H^0_c(\Gamma\backslash X, \mathcal{F})$ is the abelian group of sections with compact support.

Hence we define the cohomology with compact supports as

$$H^q_c(\Gamma\backslash X, \mathcal{F}) := H^q(\Gamma\backslash\overline{X}_{BS}, i_!\mathcal{F}).$$

If $\tilde{\mathcal{M}}$ is a sheaf on $\Gamma\backslash X$ which is obtained from a Γ-module $\mathcal{M}$, then it is quite clear that

$$H^0_c(\Gamma\backslash X, \tilde{\mathcal{M}}) = 0,$$

provided our quotient $\Gamma\backslash X$ is not compact.

The cohomology with compact supports is actually related to the homology of the group. The group homology groups $H_i(\Gamma, \mathcal{M})$ are defined in [48] 2.4. Assume that X is connected and that $\mathcal{M}$ a finitely generated R module, where R is a Dedekind ring. I want to indicate that we have a natural isomorphism

$$H_i(\Gamma, \mathcal{M}) \simeq H^{d-i}_c(\Gamma\backslash X, \tilde{\mathcal{M}})$$

provided the orders of the stabilisers $\#\Gamma_{\tilde{x}}$ are invertible in R.

We dualise the reasoning in (2.1.1). Our starting point is the fact that the projective Γ-modules have analogous vanishing properties as the induced modules.

Lemma 2.1.1. *Let us assume that Γ acts on the connected symmetric space X. If P is a projective Γ-module then*

$$H^i_c(\Gamma\backslash X, \tilde{P}) = \begin{cases} 0 & \text{if} \quad i \neq \dim X \\ \\ P_\Gamma & \text{if} \quad i = \dim X. \end{cases}$$

Let us believe this lemma. Then it is quite clear that we have

$$H_i(\Gamma, \mathcal{M}) \simeq H^{d-i}_c(\Gamma\backslash X, \tilde{\mathcal{M}}),$$

because both sides can be computed from a projective resolution.

2.1.6 The homology as singular homology

We recall the definition of the homology groups as homology groups of a chain complex $C_\bullet(\Gamma\backslash X, \underline{\mathcal{M}})$, this is the chain complex with coefficients in the "cosheaf"

$\underline{\mathcal{M}}$. I explain what I mean by that. In the case that $\mathcal{M}$ is a local system, i.e. we do not have fixed points, the complex is defined in [48] 2.8.6. But if we have fixed points the situation is slightly more complicated. The module $C_q(\Gamma\backslash X, \underline{\mathcal{M}})$ will be the free abelian group generated the q-simplices $\sigma \otimes \underline{m}$. Here $\sigma : \Delta^q \to \Gamma\backslash X$ is a continuous (or differentiable) map from the q simplex to $\Gamma\backslash X$, and where $\underline{m}$ is a "locally coconstant" function which attaches to any $v \in \Delta^q$ an element $\underline{m}(v)$ in the costalk $\underline{\mathcal{M}}_{\sigma(v)}$, Again this needs some explanation.

For any point $x \in \Gamma\backslash X$ we define the *costalk* $\underline{\mathcal{M}}_x$:, this will be the module of maps

$$\underline{m}_x : \pi_\Gamma^{-1}(x) \to \mathcal{M} \text{ which satisfy } \gamma^{-1}\underline{m}_x(\tilde{x})) = \underline{m}_x(\gamma\tilde{x})).$$

Of course such a $\underline{m}_x$ is determined by its value $\underline{m}_x(\tilde{x})$ at any point $\tilde{x} \in \pi_\Gamma^{-1}(x))$ and the this value is an element in the coinvariants $\mathcal{M}_{\Gamma_{\tilde{x}}}$. Hence we get an isomorphism $\underline{\mathcal{M}}_x \xrightarrow{j_{\tilde{x}}} \mathcal{M}_{\Gamma_{\tilde{x}}}$. Now we choose an open neighborhood U_x, we assume that x is central in this neighborhood. We choose a point $\tilde{x} \in \pi_\Gamma(x)$, then let $\tilde{U}_{\tilde{x}}$ be the connected component of $\pi_\Gamma^{-1}(V_x)$ which contains $\tilde{x}$. If now $y \in U_x$ and $\tilde{y} \in \pi_\Gamma^{-1}(y) \cap \tilde{U}_{\tilde{x}}$ then clearly $\Gamma_{\tilde{y}} \subset \Gamma_{\tilde{x}}$ and the identity on $\mathcal{M}$ induces map $\tilde{\mathcal{M}}_{\Gamma_{\tilde{y}}} \xrightarrow{r_{\tilde{y},\tilde{x}}} \tilde{\mathcal{M}}_{\Gamma_{\tilde{x}}}$. This map induces a corestriction map between the costalks

$$r_{y,x} \; : \; \underline{\mathcal{M}}_y \to \underline{\mathcal{M}}_x \tag{2.24}$$

If now $v_0 \in \Delta^q$ then we say that $\underline{m}$ is locally coconstant at v_0 if for all v in a suitably small neighborhood of v_0 we have

$$r_{\sigma(v),\sigma(v_0)}(\underline{m}(\sigma(v)) = \underline{m}(\sigma(v_0)$$

and $\underline{m}$ is coconstant if it is locally coconstant at all $v \in \Delta^q$.

Now we proceed as in in [48] 2.8.6. and define the the singular chain complex

$$\to C_q(\Gamma\backslash X, \underline{\mathcal{M}}) \to C_{q-1}(\Gamma\backslash X, \underline{\mathcal{M}}) \to \cdots \to C_0(\Gamma\backslash X, \underline{\mathcal{M}}) \to 0.$$

The homology groups are defined as

$$H_i(\Gamma\backslash X, \underline{\mathcal{M}}) = \ker(d_i)/\mathrm{Im}(d_{i+1}) \tag{2.25}$$

Now we can give a different description of the group of i-chains $C_i(\Gamma\backslash X, \underline{\mathcal{M}})$: an i-chain with values in the cosheaf $\underline{\mathcal{M}}$ is sum of terms of the form $\sigma \otimes f$ where $\sigma : \Delta^i \to \Gamma\backslash X$ is a continuous (differentiable) map from the i-dimensional simplex Δ^i to $\Gamma\backslash X$, and where f is a section in the cosheaf, i.e. $f_x \in \underline{\mathcal{M}}_{\sigma(x)}$ and where f_x varies continuously (this means: if $\sigma(y)$ specializes to $\sigma(x)$ then $r_{\sigma(y),\sigma(x)}(f_y) = f_x$).

Then $C_i(\Gamma\backslash X, \underline{\mathcal{M}})$ is the free abelian group generated by these i chains with values in $\underline{\mathcal{M}}$. The boundary maps d_i are defined in the usual way, and we get a slightly different description of the homology groups $H_i(\Gamma\backslash X, \underline{\mathcal{M}})$.

But we may choose for our module $\mathcal{M}$ simply the group ring. Then

$$(C_\bullet(X) \otimes \mathbb{Z}[\Gamma])_\Gamma \simeq C_\bullet(X),$$

and hence we have, since X is a cell, that

$$H_i(\Gamma\backslash X, \underline{\mathbb{Z}[\Gamma]}) = 0 \qquad \text{for} \qquad i > 0.$$

On the other hand we have

$$H_0(\Gamma\backslash X, \underline{\mathcal{M}}) = \mathcal{M}_\Gamma.$$

This follows directly from looking at the complex

$$(C_1(X) \otimes \mathcal{M})_\Gamma \longrightarrow (C_0(X) \otimes \mathcal{M})_\Gamma.$$

First of all we observe that 0-cycles

$$x_1 \otimes m - x_0 \otimes m$$

are boundaries since X is pathwise connected. On the other hand we have that

$$x_0 \otimes m - \gamma x_0 \otimes \gamma m \in C_0(X) \otimes \mathcal{M}$$

becomes zero if we go to the coinvariants, which implies the assertion.

If we have in addition that the orders of the stabilisers are invertible in R then it is clear that a short exact sequence of R-Γ-modules

$$0 \longrightarrow \mathcal{M}' \longrightarrow \mathcal{M} \longrightarrow \mathcal{M}'' \longrightarrow 0$$

leads to an exact sequence of complexes

$$0 \longrightarrow C_\bullet(\Gamma\backslash X, \underline{\mathcal{M}'}) \longrightarrow C_\bullet(\Gamma\backslash X, \underline{\mathcal{M}}) \longrightarrow C_\bullet(\Gamma\backslash X, \underline{\mathcal{M}''}) \longrightarrow 0,$$

and hence to a long exact (homology) sequence

$$\ldots \to H_i(\Gamma\backslash X, \underline{\mathcal{M}'}) \to H_i(\Gamma\backslash X, \underline{\mathcal{M}}) \to H_i(\Gamma\backslash X, \underline{\mathcal{M}''}) \to H_{i-1}(\Gamma\backslash X, \underline{\mathcal{M}'}) \to \ldots.$$

Now it is clear that

$$H_i(\Gamma, \mathcal{M}) \simeq H_i(\Gamma\backslash X, \underline{\mathcal{M}}) \simeq H_c^{d-i}(\Gamma\backslash X, \tilde{\mathcal{M}}).$$

2.2 The fundamental exact sequence

2.2.1 The fundamental exact sequence

In the notation of chapter (2.1.5) by construction we have the exact sequence

$$0 \to i_!(\tilde{\mathcal{M}}) \to i_*(\tilde{\mathcal{M}}) \to i_*(\tilde{\mathcal{M}})/i_!(\tilde{\mathcal{M}}) \to 0$$

of sheaves and clearly $i_*(\mathcal{M})/i_!(\mathcal{M})$ is simply the restriction of $i_*(\tilde{\mathcal{M}})$ to the boundary extended by zero to the entire space. This yields the *fundamental exact sequence*

$$\begin{aligned} &\to H^{q-1}(\partial(\Gamma\backslash X), \tilde{\mathcal{M}}) \to H_c^q(\Gamma\backslash X, \tilde{\mathcal{M}}) \\ &\qquad \to H^q(\Gamma\backslash \bar{X}, \tilde{\mathcal{M}}) \xrightarrow{r} H^q(\partial(\Gamma\backslash X), \tilde{\mathcal{M}}) \to \ldots \end{aligned}$$

We define the "inner cohomology"

$$H_!^q(\Gamma\backslash X,\tilde{\mathcal{M}}) := \operatorname{Im}\big(H_c^q(\Gamma\backslash X,\tilde{\mathcal{M}}) \to H^q(\Gamma\backslash \bar{X},\tilde{\mathcal{M}})\big) = \ker\big(H^q(\Gamma\backslash \bar{X},\tilde{\mathcal{M}}) \xrightarrow{r} H^q(\partial(\Gamma\backslash X),\tilde{\mathcal{M}})\big) \tag{2.26}$$

(This is a little bit misleading because these groups are not honest cohomology groups, as they are not the cohomology groups of a space with coefficients in a sheaf. An exact sequence of sheaves $0 \to \mathcal{M}' \to \mathcal{M} \to \mathcal{M}'' \to 0$ does not provide an exact sequence for these $H_!$ groups.)

In the special case that the underlying group $G/\mathbb{Q}$ is anisotropic, which implies that the quotient $\Gamma\backslash X$ is compact, the fundamental exact sequence becomes trivial and we have

$$H^\bullet(\Gamma\backslash X,\tilde{\mathcal{M}}) = H_c^\bullet(\Gamma\backslash X,\tilde{\mathcal{M}}) = H_!^\bullet(\Gamma\backslash X,\tilde{\mathcal{M}}).$$

Many authors prefer to consider the case of a compact quotient $\Gamma\backslash X$, but I think we lose some very interesting phenomena if we concentrate only on this case.

The cohomology of the boundary

We want to have a slightly different look at this sequence. We recall the covering (see (1.96), (1.97))

$$\Gamma\backslash X = \Gamma\backslash X(r) \cup \overset{\bullet}{\mathcal{N}}(\Gamma\backslash X) = \Gamma\backslash X(r) \cup \bigcup_{P:P \text{ proper}} \Gamma_P\backslash X^P(\underline{c}_{\pi'}, r(\underline{c}_{\pi'})) \tag{2.27}$$

where the union runs over Γ-conjugacy classes of proper parabolic subgroups over $\mathbb{Q}$ and $\overset{\bullet}{\mathcal{N}}(\Gamma\backslash X)$ is a punctured tubular neighbourhood of ∞, i.e. the boundary of the Borel-Serre compactification.

It is well-known (see for instance [48] , 4.5) that from a covering $\Gamma\backslash X = \bigcup_i V_i$ we get a Čech complex and a spectral sequence with E_1-term

$$E_1^{p,q} = \prod_{\underline{i}=\{i_0,i_1\dots,i_p\}} H^q(V_{\underline{i}},\tilde{\mathcal{M}}) \implies H^{p+q}(\Gamma\backslash X,\tilde{\mathcal{M}}), \tag{2.28}$$

where $V_{\underline{i}} = V_{i_0} \cap \dots \cap V_{i_p}$. The boundary in the Čech complex gives us the differential

$$d_1^{p,q}: \prod_{\underline{i}=\{i_0,i_1\dots,i_p\}} H^q(V_{\underline{i}},\tilde{\mathcal{M}}) \to \prod_{\underline{j}=\{j_0,j_1\dots,j_{p+1}\}} H^q(V_{\underline{j}},\tilde{\mathcal{M}}) \tag{2.29}$$

Here we work with the alternating Čech complex, where we also assume that we have an ordering on the set of simple positive roots. If such a $V_{\underline{i}}$ is non-empty then it of the form $\Gamma_Q\backslash X^Q(C(\underline{\tilde{c}}))$.

We return to the diagram (1.102) - on the left hand side we can divide by Γ_Q. We have the map which maps a Cartan involution on X to a Cartan involution on M. Then we get a diagram

$$\begin{array}{ccc} f^\dagger : X^Q(C(\underline{c},\tilde{\underline{r}})) & \to & X^M(r) \times C_{U_Q}(\underline{c},\tilde{\underline{r}}) \\ \downarrow p_Q & & \downarrow p_M \\ f : \Gamma_Q\backslash X^Q(C(\underline{c},\tilde{\underline{r}})) & \to & \Gamma_M\backslash X^M(r) \times C_{U_Q}(\underline{c},\tilde{\underline{r}})) \end{array} \tag{2.30}$$

where the bottom line is a fibration. To describe the fiber at a point $\tilde{x}$ we pick a point $x \in (p_M \circ f^\dagger)^{-1}(\tilde{x})$. Then $U_Q(\mathbb{R})$ acts simply transitively on the fiber $(f^\dagger)^{-1}(f^\dagger(x))$, hence $U_Q(\mathbb{R}) = (f^\dagger)^{-1}(f^\dagger(x))$. Then $p_Q : U_Q(\mathbb{R}) \to \Gamma_{U_Q}\backslash U_Q(\mathbb{R})$ yields the identification $i_x : \Gamma_{U_Q}\backslash U_Q(\mathbb{R}) \xrightarrow{\sim} f^{-1}(\tilde{x})$. If we replace x by $\gamma x = x_1$ with $\gamma \in \Gamma_Q$ then we get $i_{x_1} \circ \mathrm{Ad}(\gamma) = \mathrm{Ad}(\gamma) \circ i_x$, where $\mathrm{Ad}(\gamma)(u) = \gamma u \gamma^{-1}$ for $u \in U_Q(\mathbb{R})$ under this action of Γ_Q.

We have the spectral sequence

$$E_2^{p,q} = H^p(\Gamma_M\backslash X^M(r), R^q f_*(\tilde{\mathcal{M}})) \Rightarrow H^{p+q}(\Gamma_Q\backslash X^Q(C(\underline{c},\underline{r}), \tilde{\mathcal{M}})$$

and clearly $R^q f_*(\tilde{\mathcal{M}})$ is a locally constant sheaf. This sheaf is easy to determine. Under the above identification we get an isomorphism

$$i_x^\bullet : H^\bullet(\Gamma_{U_Q}\backslash U_Q(\mathbb{R}), \tilde{\mathcal{M}}) \xrightarrow{\sim} R^\bullet f_*(\tilde{\mathcal{M}})_{\tilde{x}}.$$

The adjoint action $\mathrm{Ad} : \Gamma_Q \to \mathrm{Aut}(\Gamma_{U_Q}\backslash U_Q(\mathbb{R}))$ induces an action of Γ_Q on the cohomology $H^\bullet(\Gamma_{U_Q}\backslash U_Q(\mathbb{R}), \tilde{\mathcal{M}})$. Since the functor cohomology is the derived functor of taking Γ_{U_Q} invariants, it follows that the restriction of Ad to Γ_{U_Q} acts trivially on $H^\bullet(\Gamma_{U_Q}\backslash U_Q(\mathbb{R}), \tilde{\mathcal{M}})$. Consequently $H^\bullet(\Gamma_{U_Q}\backslash U_Q(\mathbb{R}), \tilde{\mathcal{M}})$ is a Γ_M-module. We get

$$R^\bullet f_*(\tilde{\mathcal{M}}) \xrightarrow{\sim} H^\bullet(\widetilde{\Gamma_{U_Q}\backslash U_Q(\mathbb{R}), \tilde{\mathcal{M}}}),$$

and hence our spectral sequence becomes:

$$H^p(\Gamma_M\backslash X^M(r), H^\bullet(\widetilde{\Gamma_{U_Q}\backslash U_Q(\mathbb{R}), \tilde{\mathcal{M}}})) \Rightarrow H^{p+q}(\Gamma_Q\backslash X^Q(C(\tilde{\underline{c}})), \tilde{\mathcal{M}}). \tag{2.31}$$

We can take the composition $r_Q \circ f$. Then it is obvious that for any point $c_0 \in C_{U_Q}(\tilde{\underline{c}}))$ the restriction map

$$H^\bullet(X^Q(C(\tilde{\underline{c}})), \tilde{\mathcal{M}}) \to H^\bullet(X^Q((r_Q \circ f)^{-1}(c_0)), \tilde{\mathcal{M}}) \tag{2.32}$$

is an isomorphism. On the other hand it is clear that we may vary our parameter $\tilde{\underline{c}}$ in order to assume that the $C_{U_Q}(\tilde{\underline{c}})$ go to infinity. Then we may enlarge the parameter r without violating the assumptions in Proposition 1.2.4. Hence we get that the inclusion $\Gamma_Q\backslash X^Q(C(\tilde{\underline{c}})) \subset \Gamma_Q\backslash X^Q$ induces an isomorphism in cohomology

$$H^\bullet(\Gamma_Q\backslash X^Q(C(\tilde{\underline{c}})), \tilde{\mathcal{M}}) \xrightarrow{\sim} H^\bullet(\Gamma_Q\backslash X, \tilde{\mathcal{M}}) \tag{2.33}$$

We choose a total ordering on the set of Γ-conjugacy classes of parabolic subgroups, i.e. we enumerate them by a finite interval of integers $[1, N]$. We also enumerate the set of simple roots $\{\alpha_1, \ldots, \alpha_d\}$, where in our special case $\alpha_i = \alpha_{i,i+1}$. For any conjugacy class $[P]$ we define the type of P to be $t(P) = \pi^{U_P}$, the subset of unipotent simple roots and $d(P) = \#\pi^{U_P}$, the cardinality of this set. If

$P_{i_1}, \dots, P_{i_r}$ are maximal parabolics with $i_1 < i_2 \cdots < i_r$ and $P_{i_1} \cap \cdots \cap P_{i_r} = Q$ a parabolic subgroup, then we require that $t(P_{i_1}) < \cdots < t(P_{i_r})$.

The indexing set $\mathrm{Par}(\Gamma)$ of our covering is the set of Γ-conjugacy classes of parabolic subgroups over $\mathbb{Q}$. If we have a finite set $[P_{i_0}], [P_{i_1}], \dots, [P_{i_p}]$ of conjugacy classes, then we say $[Q] \in [P_{i_0}] \cap [P_{i_1}] \cap \cdots \cap [P_{i_p}]$ if we can find representatives $P'_{i_\nu} \in [P_{i_\nu}]$ and $Q' \in [Q]$ such that $Q' = P'_{i_0} \cap \cdots \cap P'_{i_p}$.

Hence we see that the $E_1^{\bullet,q}$ complex in our spectral sequence (2.29) is given by

$$\prod_i H^q(\Gamma_{P_i} \backslash X^{P_i}(C(\tilde{\underline{c}})), \tilde{\mathcal{M}}) \\ \to \prod_{i<j} \prod_{[R]\in[P_i]\cap[P_j]} H^q(\Gamma_R \backslash X^R(C(\tilde{\underline{c}})), \tilde{\mathcal{M}}) \to \tag{2.34}$$

which is obtained from our covering (1.97). Now we replace our covering by a simplicial space, i.e. we consider the diagram of maps between spaces

$$\mathfrak{Par} := \coprod_i \Gamma_{Q_i} \backslash X \underset{p_2}{\overset{p_1}{\leftleftarrows}} \coprod_{i<j} \coprod_{[R]\in[Q_i]\cap[Q_j]} \Gamma_R \backslash X \begin{smallmatrix}\longleftarrow\\ \longleftarrow\\ \longleftarrow\end{smallmatrix} \tag{2.35}$$

which yields a spectral sequence with $E_1^{\bullet,q}$ term

$$\prod_i H^q(\Gamma_{Q_i} \backslash X, \tilde{\mathcal{M}}) \xrightarrow{d^{(0)}} \prod_{i<j} \prod_{[R]\in[P_i]\cap[P_j]} H^q(\Gamma_R \backslash X^R, \tilde{\mathcal{M}}) \xrightarrow{d^{(1)}} \tag{2.36}$$

Our covering also yields a simplicial space which is a simplicial subspace of (2.35), wherefrom we get a map between (2.34) and (2.36), and this map is an isomorphism of complexes.

We replace $\mathfrak{Par}$ by another simplicial space

$$\mathfrak{Parmax} := \coprod_{[P]:d(P)=1} \Gamma_P \backslash X \underset{p_2}{\overset{p_1}{\leftleftarrows}} \coprod_{[Q]:d(Q)=2} \Gamma_Q \backslash X \begin{smallmatrix}\longleftarrow\\ \longleftarrow\\ \longleftarrow\end{smallmatrix} \tag{2.37}$$

We have an obvious projection $\Pi : \mathfrak{Par} \to \mathfrak{Parmax}$ which induces a homomorphism

$$\begin{array}{ccccc} \prod_i H^q(\Gamma_{Q_i} \backslash X, \tilde{\mathcal{M}}) & \xrightarrow{d^{(0)}} & \prod_{i<j} \prod_{[R]\in[P_i]\cap[P_j]} H^q(\Gamma_R \backslash X^R, \tilde{\mathcal{M}}) & \xrightarrow{d^{(1)}} \\ \uparrow & & \uparrow & \\ \prod_{[P]:d(P)=1} H^q(\Gamma_P \backslash X, \tilde{\mathcal{M}}) & \xrightarrow{d^{(0)}} & \prod_{[R]:d(R)=2} H^q(\Gamma_R \backslash X^R, \tilde{\mathcal{M}}) & \xrightarrow{d^{(1)}} \end{array}$$

and an easy argument in homological algebra shows that this induces an isomorphism in cohomology, or in other words, an isomorphism of the $E_2^{p,q}$ terms of the two spectral sequences.

We had the covering

$$\overset{\bullet}{\mathcal{N}}(\Gamma \backslash X) = \bigcup_{P:P \text{ proper}} \Gamma_P \backslash X^P(\underline{c}_{\pi'}, r(\underline{c}_{\pi'})) \tag{2.38}$$

which gives us the spectral sequence converging to $H^\bullet(\overset{\bullet}{\mathcal{N}}(\Gamma\backslash X), \tilde{\mathcal{M}})$ with

$$E_1^{p,q} = \bigoplus_{i_0<i_1<\cdots<i_p} \bigoplus_{[Q]\in[P_{i_0}]\cap[P_{i_1}]\cap\cdots\cap[P_{i_p}]} H^q(\Gamma_Q\backslash X^Q(\underline{c}_{\pi'}, r(\underline{c}_{\pi'})), \tilde{\mathcal{M}}). \quad (2.39)$$

Our covering of $\overset{\bullet}{\mathcal{N}}(\Gamma\backslash X)$ gives us a simplicial space $\mathfrak{Cov}(\overset{\bullet}{\mathcal{N}}(\Gamma\backslash X))$ and we have maps

$$\mathfrak{Cov}(\overset{\bullet}{\mathcal{N}}(\Gamma\backslash X)) \hookrightarrow \mathfrak{Par} \to \mathfrak{Parmax}. \quad (2.40)$$

We saw that the resulting maps induced an isomorphism in the $E_2^{p,q}$ terms of the spectral sequences. Hence we see that $\mathfrak{Parmax}$ yields a spectral sequence

$$E_1^{p,q} = \bigoplus_{[P]:d(P)=p+1} H^q(\Gamma_P\backslash X, \tilde{\mathcal{M}}) \Rightarrow H^{p+q}(\overset{\bullet}{\mathcal{N}}(\Gamma\backslash X), \tilde{\mathcal{M}}) \quad (2.41)$$

the differentials $d_1^{p,q} : E_1^{p,q} \to E_1^{p+1,q}$ are simply obtained from the restriction maps.

This is of course the "same" spectral sequence as the one above (2.36), and we may also restrict to the cohomology of the Borel-Serre boundary and get

$$E_1^{p,q} = \bigoplus_{[P]:d(P)=p+1} H^q(\partial_P(\Gamma\backslash \bar{X}), \tilde{\mathcal{M}}) \Rightarrow H^{p+q}(\partial((\Gamma\backslash \bar{X}, \tilde{\mathcal{M}})) \quad (2.42)$$

At this point we want to raise an interesting question:

Does this spectral sequence degenerate at $E_2^{p,q}$ level?

The author of this book is hoping that the answer to this question is no! And this is so for interesting reasons! We come back to this question when we discuss the Eisenstein cohomology in the last section of this book.

The complement of $\overset{\bullet}{\mathcal{N}}(\Gamma\backslash X)$ is a relatively compact open set $V \subset \Gamma\backslash X$ which contains the stable points. We define $\tilde{\mathcal{M}}_V^! = i_{V,!}(\tilde{\mathcal{M}})$ and we get an exact sequence

$$0 \to \tilde{\mathcal{M}}_V^! \to \tilde{\mathcal{M}} \to \tilde{\mathcal{M}}/\tilde{\mathcal{M}}_V^! \to 0 \quad (2.43)$$

and $\tilde{\mathcal{M}}/\tilde{\mathcal{M}}_V^!$ is obviously the extension of the restriction of $\tilde{\mathcal{M}}$ to $\overset{\bullet}{\mathcal{N}}(\Gamma\backslash X)$ by zero to $\Gamma\backslash X$. We claim (easy proof later) that

$$H_c^\bullet(\Gamma\backslash X, \tilde{\mathcal{M}}) = H^\bullet(\Gamma\backslash X, \tilde{\mathcal{M}}_V^!) \quad (2.44)$$

and this gives us again the fundamental exact sequence

$$\begin{aligned} H^{q-1}(\overset{\bullet}{\mathcal{N}}(\Gamma\backslash X), \tilde{\mathcal{M}}) &\to H^q(\Gamma\backslash X, \tilde{\mathcal{M}}_V^!) \\ &\to H^q(\Gamma\backslash X, \tilde{\mathcal{M}}) \to H^q(\overset{\bullet}{\mathcal{N}}(\Gamma\backslash X), \tilde{\mathcal{M}}) \to \end{aligned} \quad (2.45)$$

2.2.2 How to compute the cohomology groups $H_c^q(\Gamma\backslash X, \tilde{\mathcal{M}})$

We apply the sheaf-theoretic considerations in Section 4.8 of [48]. Again we cover $\Gamma\backslash X$ by orbiconvex open neighbourhoods U_{x_i}, and now we define

$$\tilde{\mathcal{M}}^!_{\underline{x}} = (i_{\underline{x}})_! i^*_{\underline{x}}(\tilde{\mathcal{M}}).$$

These sheaves have properties which are dual to those of the sheaves $\tilde{\mathcal{M}}_{\underline{x}}$. If $\underline{x} = (x_{i_1}, \dots, x_{i_s})$, and if we add another point $\underline{x}' = (x_{i_1}, \dots, x_{i_s}, x_{i_{s+1}})$, then we have th1e restriction $\tilde{\mathcal{M}}_{\underline{x}} \to \tilde{\mathcal{M}}_{\underline{x}'}$, which was used to construct the Čech resolution.

Now let $d = \dim(X)$. For the ! sheaves we get (see [48] , loc. cit.) a morphism $\tilde{\mathcal{M}}^!_{\underline{x}'} \to \tilde{\mathcal{M}}^!_{\underline{x}}$. For $\underline{x} = (x_1, \dots, x_s)$ we define the degree $d(\underline{x}) = d + 1 - s$. Then we construct the Čech-coresolution (see [48] , 4.8.3)

$$\to \prod_{\underline{x}:d(\underline{x})=q} \tilde{\mathcal{M}}^!_{\underline{x}} \to \cdots \to \prod_{(x_i,x_j)} \tilde{\mathcal{M}}^!_{x_i,x_j} \to \prod_{x_i} \tilde{\mathcal{M}}^!_{x_i} \to i_!(\tilde{\mathcal{M}}) \to 0.$$

Now the sets $U_{\underline{x}}$ are not necessarily connected, they are unions of convex connected component U_{z_μ} where z_μ is a central point, we choose $\tilde{z}_\mu \in \pi^{-1}(z_\mu)$. Then we have a dual statement to assertion (2.9):

Proposition 2.2.1. *If* $d = \dim(X)$ *then*

$$H^q(\Gamma\backslash X, \tilde{\mathcal{M}}^!_{\underline{x}}) = \begin{cases} \bigoplus_{\tilde{z}_\mu} \mathcal{M}_{\Gamma_{\tilde{z}_\mu}} & q = d \\ 0 & q \neq d \end{cases}$$

Hence the above complex of sheaves provides a complex of modules

$$\begin{aligned} C_!^\bullet(\Gamma\backslash X, \tilde{\mathcal{M}}) :=& \\ \to \prod_{\underline{x}:d(\underline{x})=q} H^d(\Gamma\backslash X, \tilde{\mathcal{M}}^!_{\underline{x}}) \to \cdots \to& \prod_{(x_i,x_j)} H^d(\Gamma\backslash X, \tilde{\mathcal{M}}^!_{x_i,x_j}) \\ &\to \prod_{x_i} H^d(\Gamma\backslash X, \tilde{\mathcal{M}}^!_{x_i}) \to 0. \end{aligned} \tag{2.46}$$

and then

$$H^q(\Gamma\backslash X, i_!(\tilde{\mathcal{M}})) = H_c^q(\Gamma\backslash X, \tilde{\mathcal{M}}) = H^q(C_!^\bullet(\mathfrak{U}, \tilde{\mathcal{M}})).$$

Now let us assume that our Γ module $\mathcal{M}$ is a finitely generated module over some commutative noetherian ring R with identity. Then clearly all our cohomology groups will be R-modules.

Then our general considerations above imply

Theorem (Raghunathan [91]) *Under our general assumptions all the cohomology groups* $H_c^q(\Gamma\backslash X, \tilde{\mathcal{M}}), H^q(\Gamma\backslash X, \tilde{\mathcal{M}}), H_!^q(\Gamma\backslash X, \tilde{\mathcal{M}}), H^q(\partial(\Gamma\backslash X), \tilde{\mathcal{M}})$ *are finitely generated* R *modules.*

2.2.3 Modified cohomology groups

Most of the time our module $\mathcal{M}$ will be a finitely generated free $\mathbb{Z}$-module. The theorem of Raghunathan asserts that the cohomology groups are also finitely generated $\mathbb{Z}$-modules. Sometimes we replace $\mathbb{Z}$ by the ring of integers $\mathcal{O}_F \subset F$ where $F/\mathbb{Q}$ is a finite extension. Then we will even invert some finite numbers of primes. Hence our coefficient modules will be finitely generated torsion free R-modules $\mathcal{M}_R$ where $\mathcal{O}_F \subset R$. In any case these rings R will be Dedekind rings.

Starting from the fundamental exact sequence we have introduced the modified cohomology groups $H^q_!(\cdot)$. There is a second process of modification: if $H^\bullet_?(\cdot)$ is any of our cohomology groups then we define

$$H^\bullet_{?,\,\mathrm{int}}(\cdot) := H^\bullet_?(\cdot)/\mathrm{Tors} = \mathrm{Im}\big(H^\bullet_?(\cdot) \to H^\bullet(\cdot) \otimes \mathbb{Q}\big). \tag{2.47}$$

We have to discuss a minor problem: these two processes of modification do not quite commute. This is due to the fact that the resulting sequence

$$\to H^{q-1}(\partial(\Gamma\backslash X), \tilde{\mathcal{M}}_R)_{\mathrm{int}} \to H^q_c(\Gamma\backslash X, \tilde{\mathcal{M}}_R)_{\mathrm{int}}$$
$$\xrightarrow{j} H^q(\Gamma\backslash \bar{X}, \tilde{\mathcal{M}}_R)_{\mathrm{int}} \xrightarrow{r} H^q(\partial(\Gamma\backslash X), \tilde{\mathcal{M}}_R)_{\mathrm{int}}$$

is not necessarily exact anymore. Clearly we have $H^q_!(\Gamma\backslash X, \tilde{\mathcal{M}}_R)_{\mathrm{int}} = \mathrm{Im}(j)$ and if we now define $H^q(\Gamma\backslash X, \tilde{\mathcal{M}}_R)_{\mathrm{int},!} := \ker(r)$ then we have

$$H^q_!(\Gamma\backslash X, \tilde{\mathcal{M}}_R)_{\mathrm{int}} \subset H^q(\Gamma\backslash X, \tilde{\mathcal{M}}_R)_{\mathrm{int},!},$$

but this inclusion may be proper. The following proposition is an elementary exercise in homological algebra.

Proposition 2.2.2. *The quotient $H^q(\Gamma\backslash X, \tilde{\mathcal{M}}_R)_{\mathrm{int},!}/H^q_!(\Gamma\backslash X, \tilde{\mathcal{M}}_R)_{\mathrm{int}}$ is finite and isomorphic to a sub quotient of $H^q(\partial(\Gamma\backslash X), \tilde{\mathcal{M}}_R)$.*

If we have a direct sum $V \supset \oplus_\nu V_\nu$ of submodules, then we say that this direct sum is a *decomposition up to isogeny* of V, if the quotient $V/\oplus_\nu V_\nu$ is a torsion module, and if for any ν the quotient V/V_ν is torsion free. Sometimes we also call this a *saturated decomposition* We can also say that $V \otimes \mathbb{Q} = \oplus V_\nu \otimes \mathbb{Q}$ and $V_\nu = V \cap V_\nu \otimes \mathbb{Q}$.

2.2.4 Poincaré duality in general

Let us assume that the symmetric space $X = G(\mathbb{R})/K_\infty$ is connected and let us also assume that Γ acts orientation preserving, which implies that any finite stabiliser Γ_x acts trivially on $\Lambda^d(T_{X,x})$. Then we know that $H^d_c(\Gamma\backslash X, \mathbb{Z}) \xrightarrow{\sim} \mathbb{Z}$, where the isomorphism depends on the choice of an orientation.

Assume that $\mathcal{M}$ is a finitely generated torsion free $\mathbb{Z}$-module with an action of Γ on it, and let $\mathcal{M}^\vee = \mathrm{Hom}(\mathcal{M}, \mathbb{Z})$ be the dual module. Then we have an obvious pairing between the two complexes $C^\bullet_!(\mathfrak{U}, \tilde{\mathcal{M}})$ and $C^\bullet(\mathfrak{U}, \tilde{\mathcal{M}}^\vee)$ (see (2.46), (2.10)) which are induced by the obvious pairing $H^d_c(U_{\underline{x}}, \tilde{\mathcal{M}}) \times H^0(U_{\underline{x}}, \tilde{\mathcal{M}}^\vee)$

and summation over the components. This pairing induces a pairing on the cohomology groups which are computed by these complexes (see [48], 4.8.4.)

$$H^q_c(\Gamma\backslash X,\tilde{\mathcal{M}})\otimes H^{d-q}(\Gamma\backslash X,\tilde{\mathcal{M}}^\vee)\stackrel{\langle\,,\,\rangle_{PD}}{\longrightarrow} H^d_c(\Gamma\backslash X,\tilde{\mathcal{M}}\otimes\tilde{\mathcal{M}}^\vee)\to H^d_c(\Gamma\backslash X,\mathbb{Z})=\mathbb{Z}.$$

Of course $\langle x,y\rangle_{PD}=0$ if one of the entries is a torsion element, and hence we get a pairing between the modified cohomology groups

$$H^q_c(\Gamma\backslash X,\tilde{\mathcal{M}})_{\text{ int}}\otimes H^{d-q}(\Gamma\backslash X,\tilde{\mathcal{M}}^\vee)_{\text{ int}}\stackrel{\langle\,,\,\rangle}{\longrightarrow}_{PD}\mathbb{Z}. \tag{2.48}$$

If Γ is torsion free, this is a non-degenerate pairing (see for instance [48], Thm. 4.8.4). If Γ has non-trivial torsion, the same is true if we replace $\mathbb{Z}$ by $R=\mathbb{Z}[1/N]$ where N is the product of the orders of the finite stabilisers.

If we analyse the fundamental exact sequence, we also find a non degenerate pairing

$$H^q_!(\Gamma\backslash X,\tilde{\mathcal{M}})_{\text{ int}}\otimes H^{d-q}(\Gamma\backslash X,\tilde{\mathcal{M}}^\vee)_{\text{ int},!}\stackrel{\langle\,,\,\rangle_{PD}}{\longrightarrow}\mathbb{Z} \tag{2.49}$$

Eventually we can pass to rational coefficient systems, and we get a non degenerate pairing

$$H^q_!(\Gamma\backslash X,\tilde{\mathcal{M}}_{\mathbb{Q}})\otimes H^{d-q}_!(\Gamma\backslash X,\tilde{\mathcal{M}}^\vee_{\mathbb{Q}})\stackrel{\langle\,,\,\rangle_{PD}}{\longrightarrow}\mathbb{Q} \tag{2.50}$$

If we look again at the fundamental exact sequence we see the two homomorphisms

$$\begin{aligned}&\delta: H^{q-1}(\partial(\Gamma\backslash X),\tilde{\mathcal{M}}_{\mathbb{Q}})\to H^q_c(\Gamma\backslash X,\tilde{\mathcal{M}}_{\mathbb{Q}})\\ &\text{and } r: H^{d-q}(\Gamma\backslash X,\tilde{\mathcal{M}}^\vee_{\mathbb{Q}})\to H^{d-q}(\partial(\Gamma\backslash X),\tilde{\mathcal{M}}^\vee_{\mathbb{Q}})\end{aligned}$$

The Borel-Serre boundary has the homotopy type of a $d-1$ dimensional manifold, hence we also have a non degenerate pairing

$$\langle\,,\,\rangle_\partial: H^{q-1}(\partial(\Gamma\backslash X),\tilde{\mathcal{M}}_{\mathbb{Q}})\times H^{d-q}(\partial(\Gamma\backslash X),\tilde{\mathcal{M}}^\vee_{\mathbb{Q}})\to\mathbb{Q}. \tag{2.51}$$

Now an easy computation shows that for $x\in H^{q-1}(\partial(\Gamma\backslash X),\tilde{\mathcal{M}}_{\mathbb{Q}}), y\in H^{d-q}(\Gamma\backslash X,\tilde{\mathcal{M}}^\vee_{\mathbb{Q}})$ we have the formula

$$\langle x,r(y)\rangle_\partial=<\delta(x),y>_!$$

and this has the following important consequence:

Proposition 2.2.3. *The spaces* $r(H^{q-1}(\Gamma\backslash X,\tilde{\mathcal{M}}_{\mathbb{Q}}))$ *and* $r(H^{d-q}(\Gamma\backslash X,\tilde{\mathcal{M}}^\vee_{\mathbb{Q}}))$ *are mutual orthogonal complements of each other, i.e.*

$$r(H^{d-q}(\Gamma\backslash X,\tilde{\mathcal{M}}^\vee_{\mathbb{Q}}))=\{y\mid\left\langle y,r(H^{q-1}(\Gamma\backslash X,\tilde{\mathcal{M}}_{\mathbb{Q}}))\right\rangle_\partial=0\}.$$

2.2.5 The case $\Gamma = \mathrm{Sl}_2(\mathbb{Z})$

We return to the case $\Gamma = \mathrm{Sl}_2(\mathbb{Z})$ (see section 2.1.4). We will see that for this seemingly very easy case we can formulate and prove some deep results, for instance we understand the denominators of Eisenstein classes (Theorem 3.74).

In the following $\mathcal{M}$ can be any Γ-module. We investigate the fundamental exact sequence for this special group. We computed already the cohomology groups $H^\bullet(\Gamma\backslash\mathbb{H}, \tilde{\mathcal{M}})$ in (2.1.4).

We compute the cohomology with compact supports. Of course we start again from our covering $\Gamma\backslash\mathbb{H} = U_i \cup U_\rho$. The cohomology with compact supports is the cohomology of the complex (see (2.46))

$$0 \to H^2(U_i \cap U_\rho, \tilde{\mathcal{M}}^!_{i,\rho}) \to H^2(U_i\ , \tilde{\mathcal{M}}^!_i) \oplus H^2(U_\rho, \tilde{\mathcal{M}}^!_\rho) \to 0.$$

We know $H^2(U_i \cap U_\rho, \tilde{\mathcal{M}}^!_{i,\rho}) = \mathcal{M}, H^2(U_i\ , \tilde{\mathcal{M}}^!_i) = \mathcal{M}_{\Gamma_i} = \mathcal{M}/(\mathrm{Id}-S)\mathcal{M}$ and $H^2(U_\rho\ , \tilde{\mathcal{M}}^!_\rho) = \mathcal{M}_{\Gamma_\rho} = \mathcal{M}/(\mathrm{Id}-R)\tilde{\mathcal{M}}$, and hence we get the complex

$$0 \to \mathcal{M} \to \mathcal{M}_{\Gamma_i} \oplus \mathcal{M}_{\Gamma_\rho} \to 0,$$

and from this we obtain

$$H^1(\Gamma\backslash\mathbb{H}, i_!(\mathcal{M})) = \ \ker\big(\mathcal{M} \to (\mathcal{M}/(\mathrm{Id}-S)\mathcal{M}) \oplus (\mathcal{M}/(\mathrm{Id}-R)\mathcal{M})\big),$$

and

$$H^0(\Gamma\backslash\mathbb{H}, i_!(\mathcal{M})) = 0, H^2(\Gamma\backslash\mathbb{H}, i_!(\mathcal{M})) = \mathcal{M}_\Gamma.$$

We look at the cohomology of the boundary $H^\bullet(\partial(\Gamma\backslash\mathbb{H}), \tilde{\mathcal{M}})$. We discussed the Borel-Serre compactification and saw that in this case we get this compactification if we add a circle at infinity to our picture of the quotient. But we may as well cut the cylinder at any level $c > 1$, i.e. we consider the level line $\mathbb{H}(c) = \{z = x + ic \mid x \in \mathbb{R}\}$ and divide this level line by the action of the translation group

$$\Gamma_U = \left\{\begin{pmatrix} 1 & n \\ 0 & 1 \end{pmatrix} \mid n \in \mathbb{Z}\right\} = \left\{\begin{pmatrix} \epsilon & n \\ 0 & \epsilon \end{pmatrix} \mid n \in \mathbb{Z}, \epsilon = \pm 1\right\} / \{\pm\mathrm{Id}\}.$$

But this quotient is homotopy equivalent to the cylinder

$$\Gamma_U\backslash\mathbb{H} \simeq \Gamma_U\backslash\mathbb{H}(c).$$

We apply our general consideration on cohomology of arithmetic groups to this situation and find

$$H^\bullet(\partial(\Gamma\backslash\mathbb{H}), \tilde{\mathcal{M}}) = H^\bullet(\Gamma_U\backslash\mathbb{H}, \mathrm{sh}_{\Gamma_U}(\mathcal{M})) = H^\bullet(\Gamma_U\backslash\mathbb{H}(c), \mathrm{sh}_{\Gamma_U}(\mathcal{M})).$$

This cohomology is easy to compute. The group Γ_U is generated by the element $T = \begin{pmatrix} 1 & 1 \\ 0 & 1 \end{pmatrix}$. It is rather clear that

$$H^0(\Gamma_U\backslash\mathbb{H}, \mathrm{sh}_{\Gamma_U}(\mathcal{M})) = \mathcal{M}^{\Gamma_U}, H^1(\Gamma_U\backslash\mathbb{H}, \mathrm{sh}_{\Gamma_U}(\mathcal{M})) = \mathcal{M}_{\Gamma_U} = \mathcal{M}/(\mathrm{Id}-T)\mathcal{M}.$$

Then our fundamental exact sequence becomes (see (2.21)):

$$\begin{aligned} 0 \to \mathcal{M}^{\Gamma} \to \mathcal{M}^{\Gamma_U} \to \ & \ker\big(\mathcal{M} \to (\mathcal{M}/(\mathrm{Id}-S)\mathcal{M}) \oplus (\mathcal{M}/(\mathrm{Id}-R)\mathcal{M})\big) \\ \xrightarrow{j} \mathcal{M}/(\mathcal{M}^{\Gamma_i} \oplus \mathcal{M}^{\Gamma_\rho}) & \xrightarrow{r} \mathcal{M}/(\mathrm{Id}-T)\mathcal{M} \to \mathcal{M}_{\Gamma} \to 0 \end{aligned} \tag{2.52}$$

Now it may come as a little surprise to the readers, that we can formulate a little exercise which is not entirely trivial.

Exercise: *Write down explicitly all the arrows in the above fundamental sequence.*

We give the answer without proof. I change notation slightly and work with the matrices

$$S = \begin{pmatrix} 0 & -1 \\ 1 & 0 \end{pmatrix}, R = \begin{pmatrix} 1 & -1 \\ 1 & 0 \end{pmatrix},$$

and we have the relation

$$-RS = T = \begin{pmatrix} 1 & 1 \\ 0 & 1 \end{pmatrix}.$$

Then $\Gamma_i = \langle S \rangle, \Gamma_\rho = \langle R \rangle$. The map

$$r : \mathcal{M}/(\mathcal{M}^{\langle S \rangle} \oplus \mathcal{M}^{\langle R \rangle}) \to \mathcal{M}/(\mathrm{Id}-T)\mathcal{M}$$

is given by

$$m \mapsto m - Sm.$$

We have to show that this map is well-defined: if $m \in \mathcal{M}^{\langle S \rangle}$ then $m \mapsto 0$. If $m \in \mathcal{M}^{\langle R \rangle}$ then

$$m - Sm = m - SR^{-1}m = m - T^{-1}m,$$

and this is zero in $\mathcal{M}/(\mathrm{Id}-T)\mathcal{M}$.

The map

$$j: \ \ker\big(\mathcal{M} \to (\mathcal{M}/(\mathrm{Id}-S)\mathcal{M}) \oplus (\mathcal{M}/(\mathrm{Id}-R)\mathcal{M})\big) \to \mathcal{M}/(\mathcal{M}^{\langle S \rangle} \oplus \mathcal{M}^{\langle R \rangle})$$

is a little bit delicate. We pick an element m in the kernel, hence we can write it as

$$m = m_1 - Sm_1 = m_2 - Rm_2$$

and send $m \mapsto m_1 - m_2$ (here we have to use the orientation). If we modify m_1, m_2 to $m_1' = m_1 + n_1, m_2' = m_2 + n_2$, then $m_1' - m_2'$ gives the same element in $\mathcal{M}/(\mathcal{M}^{\langle S \rangle} \oplus \mathcal{M}^{\langle R \rangle})$.

This answer can only be right if $m_1 - m_2$ goes to zero under the map r, i.e. we have to show that

$$m_1 - m_2 - S(m_1 - m_2) \in (\mathrm{Id}-T)\mathcal{M}.$$

We compute

$$m_1 - m_2 - S(m_1 - m_2) = m - m_2 + Sm_2 = m - m_2 + Rm_2 - Rm_2 + Sm_2 =$$
$$-Rm_2 + Sm_2 = -TSm_2 + Sm_2 \in (\mathrm{Id}-T)\mathcal{M}.$$

Finally we claim that the map

$$\mathcal{M}^{\langle T \rangle} \to \ker(\mathcal{M} \to (\mathcal{M}/(\mathrm{Id}-S)\mathcal{M} \oplus \mathcal{M}/(\mathrm{Id}-R)\mathcal{M}))$$

is given by $m \mapsto m - S^{-1}m = m - R^{-1}Tm = m - R^{-1}m$.

The complex conjugation

There is still another element of structure. The map $c : z \mapsto -\bar{z}$ induces an (differentiable) involution of $\mathbb{H}$. We put $S_1 = \begin{pmatrix} -1 & 0 \\ 0 & 1 \end{pmatrix}$, then $\gamma c z = c S_1 \gamma S_1^{-1} z$, and therefore c induces an involution on $\Gamma\backslash\mathbb{H}$. We get an isomorphism of cohomology groups

$$c^{(1)} : H^1(\Gamma\backslash\mathbb{H}, \tilde{\mathcal{M}}) \xrightarrow{\sim} H^1(\Gamma\backslash\mathbb{H}, c_*(\tilde{\mathcal{M}})). \tag{2.53}$$

The direct image sheaf $c_*(\tilde{\mathcal{M}})$ is by definition the sheaf attached to the Γ-module $\mathcal{M}^{[S_1]}$: this is the module which is equal to $\mathcal{M}$ as an abstract module, but the action is twisted by a conjugation by the above matrix S_1, i.e.

$$\gamma * m = S_1 \gamma S_1^{-1} m. \tag{2.54}$$

Now we assume that $\mathcal{M}$ is actually a $\mathrm{Gl}_2(\mathbb{Z})$-module. Then the map $m \to S_1 m$ provides an isomorphism $\mathcal{M}^{(S_1)} \xrightarrow{\sim} \mathcal{M}$ and hence we get an involution on the cohomology groups

$$\mathbf{c}^\bullet : H^\bullet(\Gamma\backslash\mathbb{H}, \tilde{\mathcal{M}}) \to H^\bullet(\Gamma\backslash\mathbb{H}, \tilde{\mathcal{M}}). \tag{2.55}$$

We have the explicit description of the cohomology groups $H^1(\Gamma\backslash\mathbb{H}, \tilde{\mathcal{M}})$ and we can compute this involution in terms of this description. The map $m \mapsto S_1 m$ induces an isomorphism

$$S_1^\bullet : \mathcal{M}/(\mathcal{M}^{\Gamma_i} + \mathcal{M}^{\Gamma_\rho}) \to \mathcal{M}/(\mathcal{M}^{\Gamma_i} + \mathcal{M}^{\Gamma_{\rho^2}}),$$

where both sides are equal to the cohomology $H^1(\Gamma\backslash\mathbb{H}, \tilde{\mathcal{M}})$. Hence $\mathbf{c}^\bullet$ is the map induced by $S_1^\bullet$.

The cohomology has a $+$ and a $-$ eigen-submodule under this involution

$$H^1(\Gamma\backslash\mathbb{H}, \tilde{\mathcal{M}}) \supset H^1(\Gamma\backslash\mathbb{H}, \tilde{\mathcal{M}})_+ \oplus H^1(\Gamma\backslash\mathbb{H}, \tilde{\mathcal{M}})_-,$$

and the sum of the two eigenmodules has finite index which is a power of 2.

Poincaré duality

We assume that our Γ-module $\mathcal{M}$ is a finitely generated and locally free module over R , where R is a Dedekind ring or a field. We assume $\frac{1}{2} \in R$. In section 2.1.5 we discuss Poincare duality in greater generality, whereas here we consider the pairing (see 2.49)

$$H^1_!(\Gamma\backslash\mathbb{H}, \tilde{\mathcal{M}})_{\text{ int}} \times H^1(\Gamma\backslash\mathbb{H}, \tilde{\mathcal{M}}^\vee)_{\text{ int},!} \to H^2_c(\Gamma\backslash\mathbb{H}, R) = R \tag{2.56}$$

It is clear that the involution $\mathbf{c}$ induces multiplication by -1 on $H^2_!(\Gamma\backslash\mathbb{H}, R)$. On the other hand we have the decompositions of the above cohomology groups into $\pm$ eigenmodules. The pairings of the $+,+$ parts and the $-,-$ yield zero and then we get pairings

$$H^1_!(\Gamma\backslash\mathbb{H}, \tilde{\mathcal{M}})_{\text{ int},+} \times H^1(\Gamma\backslash\mathbb{H}, \tilde{\mathcal{M}}^\vee)_{\text{ int},!,-} \to R$$
$$H^1(\Gamma\backslash\mathbb{H}, \tilde{\mathcal{M}})_{\text{ int},!,+} \times H^1_!(\Gamma\backslash\mathbb{H}, \tilde{\mathcal{M}}^\vee)_{\text{ int},-} \to R$$

. If we replace R by its quotient field then these two pairings are non degenerate.
If we have $\mathcal{M} = \mathcal{M}^\vee$ then we get

$$\text{rank}(H^1_!(\Gamma\backslash\mathbb{H}, \tilde{\mathcal{M}})_{\text{ int},+}) = \text{rank}(H^1_!(\Gamma\backslash\mathbb{H}, \tilde{\mathcal{M}})_{\text{ int},-}) \tag{2.57}$$

Final remark: The reader may get the impression that - at least in the case $\Gamma = \text{Sl}_2(\mathbb{Z})$ - it is easy to compute the cohomology, but the contrary is true. In the case $\Gamma = \text{Sl}_2(\mathbb{Z})/\pm\text{Id}$ we found formulae for the rank of the cohomology groups, which seems to give a satisfactory answer, but it is not. The point is that there is an additional structure. In the next section we will introduce the Hecke operators, where these Hecke operators form an algebra of endomorphisms of the cohomology groups.

It is a fundamental question (see further down) to understand the cohomology as a module under the action of this Hecke algebra.

It is not so easy to write down the effect of a Hecke operator on a module like $\mathcal{M}/(\mathcal{M}^{\Gamma_i} + \mathcal{M}^{\Gamma_\rho})$. We will discuss an explicit example in Section 3.3.

We mentioned already that the situation is even less satisfying if we consider the case $\Gamma = \text{Sl}_2(\mathbb{Z}[\mathbf{i}])$. In this case we considered the coefficient system $\mathcal{M} = \mathcal{M}_{n_1} \otimes \mathcal{M}_{n_2}$ (see (2.23)). Then it turns out that the complex (2.22) does not tell us very much about the cohomology.

We may for instance employ the complex (2.22) and the resulting complex for the cohomology with compact supports, then we can compute - for a given pair n_1, n_2 - the modules ($\mathbb{Q}[\mathbf{i}]$-vector spaces) in the exact sequence

$$0 \to H^1_!(\Gamma\backslash\mathbb{H}_3, \tilde{\mathcal{M}} \otimes \mathbb{Q}) \to H^1(\Gamma\backslash\mathbb{H}_3, \tilde{\mathcal{M}} \otimes \mathbb{Q}) \xrightarrow{r} H^1(\partial(\Gamma\backslash\mathbb{H}_3), \tilde{\mathcal{M}} \otimes \mathbb{Q}).$$

An easy computation (see later) shows that

$$H^1(\partial(\Gamma\backslash\mathbb{H}_3), \tilde{\mathcal{M}} \otimes \mathbb{Q}) = \mathbb{Q}[\mathbf{i}]e_{0,1} \oplus \mathbb{Q}[\mathbf{i}]e_{1,0}$$

where the $e_{i,j}$ are some naturally given generators. Another simple argument using Poincaré duality shows that the image under the restriction map r will be a one dimensional subspace

$$r(H^1(\Gamma\backslash\mathbb{H}_3, \tilde{\mathcal{M}} \otimes \mathbb{Q})) = \mathbb{Q}[\mathbf{i}](\mathcal{L}(n_1, n_2)e_{0,1} + \mathcal{L}(n_2, n_1)e_{1,0}).$$

This far we get with our topological methods.

In chapter 4 we will use tools from analysis and prove a vanishing theorem

$$H^\bullet_!(\Gamma\backslash\mathbb{H}_3, \tilde{\mathcal{M}} \otimes \mathbb{Q}) = 0 \text{ if } n_1 \neq n_2$$

(see 4.2). I do not see how such a result can not be obtained from studying the complex (2.22).

At this point we ask a natural question: can we compute the position of the one dimensional subspace $\text{Im}(r)$, the cohomology of the boundary, i.e. can we compute the point $(\mathcal{L}(n_1, n_2), \mathcal{L}(n_2, n_1)) \in \mathbb{P}^1(\mathbb{Z}[\mathbf{i}])$?

Here the answer is yes, but we have to use transcendental tools. The ratio $\frac{\mathcal{L}(n_1,n_2)}{\mathcal{L}(n_2,n_1)}$ (or its inverse) is the quotient of a Hecke L-function evaluated at two consecutive critical integer arguments (depending on n_1, n_2 - we come back to this later). So the complex can not give an elementary expression for it.

On the other hand our purely combinatorial considerations imply that the ratios $\frac{\mathcal{L}(n_1,n_2)}{\mathcal{L}(n_2,n_1)} \in \mathbb{Q}[\mathbf{i}]$ (or $=\infty$), which implies a non-trivial rationality relation between special values of the Hecke L-function.

The rationality relations are easy consequences of the results of Hurwitz on the special values of his Zeta-function.

The complex computes the $\mathbb{Z}[\mathbf{i}]$-modules $H^i_!(\Gamma\backslash\mathbb{H}_3, \tilde{\mathcal{M}}_\lambda)$ and therefore it also computes the torsion. Again it seems to be difficult to derive general theorems, except for the obvious finiteness assertions.

Chapter 3

Hecke Operators

3.1 The construction of Hecke operators

We mentioned already that the cohomology and homology groups of an arithmetic group have an additional structure: they are modules for the Hecke algebra. The following description of the Hecke algebra is somewhat provisional - we get a richer Hecke algebra if we work in the adelic context (see Chapter 6). The description here is more intuitive.

We start from the arithmetic group $\Gamma \subset G(\mathbb{Q})$ and an arbitrary Γ-module $\mathcal{M}$. The module $\mathcal{M}$ is also a module over a ring R which in the beginning may be simply $\mathbb{Z}$. More generally, R may be the ring of integers of an algebraic number field, where we also have inverted a finite number of primes. At this point it is better to have a notation for this action

$$\Gamma \times \mathcal{M} \to \mathcal{M}, (\gamma, m) \mapsto r(\gamma)(m)$$

where now $r : \Gamma \to \mathrm{Aut}_R(\mathcal{M})$. We abbreviate $r(\gamma)m = \gamma m$.

As before we have the symmetric space $X = G(\mathbb{R})/K_\infty$. If $\Gamma' \subset \Gamma$ is a subgroup of finite index, then we constructed maps

$$\pi^\bullet_{\Gamma',\Gamma} : H^\bullet(\Gamma\backslash X, \tilde{\mathcal{M}}) \longrightarrow H^\bullet(\Gamma'\backslash X, \tilde{\mathcal{M}})$$
$$\pi_{\Gamma',\Gamma,\bullet} : H^\bullet(\Gamma'\backslash X, \tilde{\mathcal{M}}) \longrightarrow H^\bullet(\Gamma\backslash X, \tilde{\mathcal{M}})$$

(see Section 2.1.2). We pick an element $\alpha \in G(\mathbb{Q})$. The group

$$\Gamma(\alpha^{-1}) = \alpha^{-1}\Gamma\alpha \cap \Gamma$$

is a subgroup of finite index in Γ, and the conjugation by α induces an isomorphism

$$\mathrm{inn}(\alpha) : \Gamma(\alpha^{-1}) \longrightarrow \Gamma(\alpha).$$

We get an isomorphism

$$j(\alpha) : \Gamma(\alpha^{-1})\backslash X \longrightarrow \Gamma(\alpha)\backslash X$$

which is induced by the map $x \mapsto \alpha x$ on the space X. This yields an isomorphism of cohomology groups

$$j(\alpha)^\bullet : H^\bullet(\Gamma(\alpha^{-1})\backslash X, \tilde{\mathcal{M}}) \longrightarrow H^\bullet(\Gamma(\alpha)\backslash X, j(\alpha)_*(\tilde{\mathcal{M}})).$$

G. Harder, *Cohomology of Arithmetic Groups*,
Springer Monographs in Mathematics, https://doi.org/10.1007/978-3-032-10378-9_3

We compute the sheaf $j(\alpha)_*(\tilde{\mathcal{M}})$. Given any point $x \in \Gamma(\alpha)\backslash X$ we have $j(\alpha)_*(\tilde{\mathcal{M}})_x = \tilde{\mathcal{M}}_{x'}$ where $j(\alpha)(x') = x$. In addition, we have the projection $\pi_{\Gamma(\alpha^{-1})} : X \to \Gamma(\alpha^{-1})\backslash X$, and the definition yields

$$\tilde{\mathcal{M}}_{x'} = \left\{ s : \pi^{-1}_{\Gamma(\alpha^{-1})}(x') \to \mathcal{M} \mid s(\gamma m) = \gamma s(m) \text{ for all } \gamma \in \Gamma(\alpha^{-1}) \right\}.$$

The map $z \mapsto \alpha z$ provides an identification $\pi^{-1}_{\Gamma(\alpha^{-1})}(x') \xrightarrow{\sim} \pi^{-1}_{\Gamma(\alpha)}(x)$, and in terms of this fibre we can describe the stalk at x as

$$j(\alpha)_*(\tilde{\mathcal{M}})_x = \left\{ s : \pi^{-1}_{\Gamma(\alpha)}(x) \to \mathcal{M} \mid s(\gamma v) = \alpha^{-1}\gamma\alpha s(v) \text{ for all } \gamma \in \Gamma(\alpha) \right\}.$$

Hence we see that we can use α to define a new $\Gamma(\alpha)$-module $\mathcal{M}^{(\alpha)}$: the underlying abelian group of $\mathcal{M}^{(\alpha)}$ is $\mathcal{M}$ but the operation of $\Gamma(\alpha)$ is given by

$$(\gamma, m) \longrightarrow (\alpha^{-1}\gamma\alpha)m =: \gamma *_\alpha m,$$

the sheaf $j(\alpha)_*(\tilde{\mathcal{M}})$ is equal to $\tilde{\mathcal{M}}^{(\alpha)}$. Now every element

$$u_\alpha \in \operatorname{Hom}_{\Gamma(\alpha)}(\mathcal{M}^{(\alpha)}, \mathcal{M})$$

defines a map $\tilde{u}_\alpha : j(\alpha)_*(\tilde{\mathcal{M}}) \to \tilde{\mathcal{M}}$, and we get a commuting diagram

$$\begin{array}{ccccc}
H^\bullet(\Gamma(\alpha^{-1})\backslash X, \tilde{\mathcal{M}}) & \xrightarrow{j(\alpha)_\bullet} & H^\bullet(\Gamma(\alpha)\backslash X, j(\alpha)_*(\tilde{\mathcal{M}})) & \xrightarrow{\tilde{u}_{\alpha\bullet}} & H^\bullet(\Gamma(\alpha)\backslash X, \tilde{\mathcal{M}}) \\
\uparrow \pi^\bullet & & & & \downarrow \pi_\bullet \\
H^\bullet(\Gamma\backslash X, \tilde{\mathcal{M}}) & & \xrightarrow{T(\alpha, u_\alpha)} & & H^\bullet(\Gamma\backslash X, \tilde{\mathcal{M}})
\end{array}, \tag{3.1}$$

and the operator $T(\alpha, u_\alpha)$ is the Hecke operator.

The Hecke operator depends on two data:

a) the element $\alpha \in G(\mathbb{Q})$,

b) the choice of $u_\alpha \in \operatorname{Hom}_{\Gamma(\alpha)}(\mathcal{M}^{(\alpha)}, \mathcal{M})$.

It is not difficult to show that the operator $T(\alpha, u_\alpha)$ only depends on the double coset $\Gamma\alpha\,\Gamma$, provided we adapt the choice of u_α. To be more precise, if

$$\alpha_1 = \gamma_1\alpha\gamma_2, \qquad \gamma_1, \gamma_2 \in \Gamma,$$

then we have an obvious bijection

$$\Phi_{\gamma_1,\gamma_2} : \operatorname{Hom}_{\Gamma(\alpha)}(\mathcal{M}^{(\alpha)}, \mathcal{M}) \longrightarrow \operatorname{Hom}_{\Gamma(\alpha_1)}(\mathcal{M}^{(\alpha_1)}, \mathcal{M})$$

which is given by

$$\Phi_{\gamma_1,\gamma_2}(u_\alpha) = u_{\alpha_1} = \gamma_1 u_\alpha \gamma_2.$$

The reader will verify without any difficulties that

$$T(\alpha, u_\alpha) = T(\alpha_1, u_{\alpha_1}).$$

(verify this for H^0 and then use some kind of resolution, see next section).

The choice of u_α may be delicate in some situations. There are cases where we also have a canonical choice of u_α. The first case is that our Γ-module $\mathcal{M}$ is of arithmetic origin. In this case $G(\mathbb{Q})$ acts upon $\mathcal{M}_\mathbb{Q} = \mathcal{M} \otimes \mathbb{Q}$. Then the natural choice of $u_{\alpha,\mathbb{Q}} : \mathcal{M}_\mathbb{Q}^{(\alpha)} \longrightarrow \mathcal{M}_\mathbb{Q}$ is given by $u_\alpha : m \mapsto \alpha m$. Hence we can speak of the Hecke-operator $T(\alpha) : H^\bullet(\Gamma\backslash X, \mathcal{M}_\mathbb{Q}) \to H^\bullet(\Gamma\backslash X, \mathcal{M}_\mathbb{Q})$.

But if we return to the R-module sheaf $\mathcal{M}$, this morphism $u_{\alpha,\mathbb{Q}}$ will not necessarily map the lattice $\mathcal{M}^{(\alpha)}$ into $\mathcal{M}$. Clearly we can find a rational number $d(\alpha) > 0$ for which

$$d(\alpha) \cdot u_{\alpha,\mathbb{Q}} : \mathcal{M}^{(\alpha)} \longrightarrow \mathcal{M} \text{ and } d(\alpha) \cdot u_{\alpha,\mathbb{Q}}(\mathcal{M}^{(\alpha)}) \not\subset b\mathcal{M}$$

for any integer $b > 1$ with $b\mathcal{M} \neq \mathcal{M}$. Then $u_\alpha = d(\alpha) \cdot u_{\alpha,\mathbb{Q}}$ is called the *normalised choice*, and $T(\alpha, u_\alpha)$ will be the *normalised Hecke operator.*

The canonical choice defines endomorphisms on the rational cohomology, i.e. the cohomology with coefficients in $\tilde{\mathcal{M}}_\mathbb{Q}$, whereas the normalised Hecke operators induce endomorphisms of the integral cohomology. The normalised choice and the canonical choice differ only by a scalar factor. This scalar factor is unique up to a unit in $R^\times$.

In the second case we assume that $G/\mathbb{Z}$ is a reductive group, $\Gamma_0 = G(\mathbb{Z})$. Let $\Gamma(N) \subset \Gamma_0$ be the full congruence subgroup mod N. Then $\Gamma_0/\Gamma(N) \subset G(\mathbb{Z}/N\mathbb{Z})$. Now we assume that $\mathcal{V}$ is a $G(\mathbb{Z}/N\mathbb{Z})$-module. In Section 1.2.1 we called this a module of congruence origin. Then we have some constraints on the choice of elements α. We introduce the semi-local ring $\mathbb{Z}_{(N)}$ where we invert all primes not dividing N. Now we pick our elements $\alpha \in G(\mathbb{Z}_{(N)})$. Since we have the homomorphism $\mathbb{Z}_{(N)} \to \mathbb{Z}/N\mathbb{Z}$ our module $\mathcal{V}$ is also a $G(\mathbb{Z}_{(N)})$-module. Therefore we can simply choose $u_\alpha := (m \mapsto \alpha m.)$

Consequently we can derive that we can construct many endomorphisms $T(\alpha, u_\alpha) : H^\bullet(\Gamma\backslash X, \tilde{\mathcal{V}}) \to H^\bullet(\Gamma\backslash X, \tilde{\mathcal{V}})$. These endomorphisms will generate an algebra

$$\mathcal{H}_{N,\tilde{\mathcal{V}}} \subset \mathrm{End}(H^\bullet(\Gamma\backslash X, \tilde{\mathcal{V}})). \tag{3.2}$$

This is now the *Hecke algebra.*

We can also define endomorphisms $T(\alpha, u_\alpha)$ on the cohomology with compact supports, on the inner cohomology and the cohomology of the boundary. Since the operators are compatible with all the arrows in the fundamental exact sequence we denote them by the same symbol.

The Hecke algebra also acts on the inner cohomology $H^q_!(\Gamma\backslash X, \tilde{\mathcal{M}})$. Of course we may tensorize our coefficient system with any number field $L \supset \mathbb{Q}$, then we write $\mathcal{M}_L = \mathcal{M} \otimes L$.

We state without proof the following fundamental theorem:

Theorem 3.1.1. *Let $\mathcal{M}$ be a module of arithmetic origin. For any extension $L/\mathbb{Q}$ the $\mathcal{H}_\Gamma \otimes L$-module $H^q_!(\Gamma\backslash X, \tilde{\mathcal{M}}_L)$ is semi-simple, i.e. a direct sum of irreducible $\mathcal{H}_\Gamma \otimes L$ modules.*

The proof of this theorem will be discussed in Chapter 6 (Section 6.2.4), and it requires some input from analysis. We give a brief sketch. We tensorize our coefficient system by $\mathbb{C}$, i.e. we consider $\mathcal{M}_L \otimes_L \mathbb{C} = \mathcal{M}_{\mathbb{C}}$. Let us assume that Γ is torsion-free. First of all start from the well-known fact that the cohomology $H^\bullet(\Gamma\backslash X, \tilde{\mathcal{M}}_{\mathbb{C}})$ can be computed from the de-Rham-complex (see 6.1.3)

$$H^\bullet(\Gamma\backslash X, \tilde{\mathcal{M}}_{\mathbb{C}}) = H^\bullet((\Omega^\bullet(X) \otimes \mathcal{M}_{\mathbb{C}})^\Gamma).$$

We will show in Chapter 6 that the inner cohomology is trivial unless the representation $\mathcal{M}_{\mathbb{C}}$ is c-autodual. If this is the case we can put some specific positive-definite hermitian form on $\mathcal{M}_{\mathbb{C}}$, and hence we define a hermitian scalar product between two $\tilde{\mathcal{M}}_{\mathbb{C}}$-valued p-forms

$$\langle \omega_1, \omega_2 \rangle = \int_{\Gamma\backslash X} \omega_1 \wedge *\omega_2,$$

provided one of the forms is compactly supported.

Now we can define a positive-definite scalar product on $H^p_!(\Gamma\backslash\mathbb{H}, \tilde{\mathcal{M}}_{n,\mathbb{C}})$. We apply Theorem 6.2.1, which tells us that we can find representatives ω_1^h, ω_2^h which are harmonic (they satisfy certain differential equations) and then

$$\langle [\omega_1], [\omega_2] \rangle := \int_{\Gamma\backslash X} \omega_1^h \wedge *\omega_2^h, \tag{3.3}$$

defines a positive-definite hermitian scalar product on $H^q_!(\Gamma\backslash X, \tilde{\mathcal{M}}_{\mathbb{C}})$. Finally we show that $\mathcal{H}_\Gamma$ is self-adjoint with respect to this scalar product (see 6.1), and then semi-simplicity follows from the standard argument.

For the groups $\Gamma \subset \mathrm{Sl}_2(\mathbb{Z})$ and the cohomology groups $H^1_!(\Gamma\backslash\mathbb{H}, \mathcal{M}_n \otimes \mathbb{C})$, these harmonic representatives are given by linear combinations of holomorphic and antiholomorphic cusp forms of weight $n+2$ (see 4.2). The scalar product on this space of modular forms is given by the Petersson scalar product (see section 4.2.1.)

3.1.1 Commuting relations

We want to say some words concerning the structure of the Hecke algebra.

To begin, we discuss the action of the Hecke-algebra on $H^0(\Gamma\backslash X, \tilde{\mathcal{M}})$. We do this since we defined the cohomology in terms of injective (or acyclic) resolutions, and therefore the general results concerning the structure of the Hecke algebra can be reduced to this special case.

If we have a Γ-module $\mathcal{M}$ and if we look at the diagram defining the Hecke operators, then we see that we get in degree 0

$$\begin{array}{ccccc} \mathcal{M}^{\Gamma(\alpha^{-1})} & \longrightarrow & (\mathcal{M}^{(\alpha)})^{\Gamma(\alpha)} \xrightarrow{u_\alpha} & & \mathcal{M}^{\Gamma(\alpha)} \\ \uparrow & & & & \downarrow \\ \mathcal{M}^{\Gamma} & & \xrightarrow{\quad T(\alpha,u_\alpha) \quad} & & \mathcal{M}^{\Gamma}. \end{array}$$

The first arrow on the top line is induced by the identity $\mathcal{M} \to \mathcal{M}^{(\alpha)} = \mathcal{M}$ and the second is given by a map $u_\alpha \in \mathrm{Hom}_{\mathbf{Ab}}(\mathcal{M}, \mathcal{M})$ which satisfies

$u_\alpha((\alpha^{-1}\gamma\alpha)m) = \gamma u_\alpha(m)$. Recalling the definition of the vertical arrow on the right, we find

$$T(\alpha, u_\alpha)(v) = \sum_{\gamma\in\Gamma/\Gamma(\alpha)} \gamma \cdot u_\alpha(v).$$

We are interested to get formulae for the product of Hecke operators, for instance, we would like to show that under certain assumptions on α, β and certain adjustments of u_α, u_β and $u_{\alpha\beta}$ we can show

$$T(\alpha, u_\alpha) \cdot T(\beta, u_\beta) = T(\beta, u_\beta) \cdot T(\alpha, u_\alpha) = T(\alpha\beta, u_{\alpha\beta}).$$

It is easy to see what the conditions are if we want such a formula to be true. We look at what happens in H^0. For $v \in \mathcal{M}^\Gamma$ we get

$$T(\alpha, u_\alpha) \cdot T(\beta, u_\beta)(v) = \sum_{\gamma\in\Gamma/\Gamma(\alpha)} \gamma u_\alpha(\sum_{\eta\in\Gamma/\Gamma(\beta)} \eta u_\beta(v)).$$

We assume that the following three conditions hold

(i) for each η we can find an $\eta' \in \Gamma$ such that

$$\eta' \circ u_\alpha = u_\alpha \circ \eta,$$

(ii) The elements $\gamma\eta'$ form a system of representatives for $\Gamma/\Gamma(\alpha\beta)$,

(iii) $u_\alpha u_\beta(v) = u_\beta u_\alpha(v) = u_{\alpha\beta}(v)$.

Then we get

$$T(\alpha, u_\alpha) \cdot T(\beta, u_\beta)(v) = \sum_{\gamma\in\Gamma/\Gamma(\alpha)} \sum_{\eta'\in\Gamma/\Gamma(\beta)} \gamma\eta' u_\alpha u_\beta(v) = \sum_{\xi\in\Gamma/\Gamma(\alpha\beta)} \xi u_{\alpha\beta}(v) =$$
$$T(\alpha\beta, u_{\alpha\beta})(v).$$

We want to explain in a special case that we may have relations like the one above.

Let S be a finite set of primes, and $|S|$ be the product of these primes. Then we define $\Gamma_S = G(\mathbb{Z}[\frac{1}{|S|}])$. We say that $\alpha \in G(\mathbb{Q})$ has support in S if $\alpha \in G(\mathbb{Z}[\frac{1}{|S|}])$.

We take the group $\Gamma = \mathrm{Sl}_d(\mathbb{Z})$, and we take two disjoint sets of primes S_1, S_2. For the group Γ one can prove the *strong approximation* theorem (see [72]) which asserts that for any natural number m the map

$$\mathrm{Sl}_d(\mathbb{Z}) \longrightarrow \mathrm{Sl}_d(\mathbb{Z}/m\mathbb{Z})$$

is surjective. (This special case is actually not so difficult. The theorem holds for many other arithmetic groups, for instance for simply connected Chevalley schemes over $\mathrm{Spec}(\mathbb{Z})$.)

We assume that our module $\mathcal{M}$ is an R-lattice in $\mathcal{M}_\mathbb{Q}$, where $\mathcal{M}_\mathbb{Q}$ is a rational $G(\mathbb{Q})$-module. Then we saw that we can write

$$u_\alpha = d(\alpha) \cdot \alpha$$

where $d(\alpha)$ will be a product of powers of the primes p dividing n_1. The same holds for β and n_2.

We consider the case

$$\alpha = \begin{pmatrix} a_1 & & & \\ & a_2 & & \\ & & \ddots & \\ & & & a_d \end{pmatrix} \in \Gamma_{S_1}, \beta = \begin{pmatrix} b_1 & & & \\ & b_2 & & \\ & & \ddots & \\ & & & b_d \end{pmatrix} \in \Gamma_{S_2},$$

where $a_d|a_{d-1}\cdots|a_1$ and $b_d|b_{d-1}|\cdots|b_1$. It is clear that we can find integers n_1 and n_2 which are only divisible by the primes in S_1 and S_2 respectively, so that

$$\Gamma(n_1) \subset \Gamma(\alpha^{-1}), \Gamma(n_2) \subset \Gamma(\beta^{-1}),$$

where the $\Gamma(n_i)$ are the full congruence subgroups $\mod n_i$. Since we have

$$\mathrm{Sl}_d(\mathbb{Z}/n_1n_2\mathbb{Z}) = \mathrm{Sl}_d(\mathbb{Z}/n_1\mathbb{Z}) \times \mathrm{Sl}_d(\mathbb{Z}/n_2\mathbb{Z})$$

we get

$$\Gamma/\Gamma(\alpha^{-1}\beta^{-1}) \xrightarrow{\sim} \Gamma/\Gamma(\alpha^{-1}) \times \Gamma/\Gamma(\beta^{-1}).$$

On the right hand side we can choose representatives γ for $\Gamma/\Gamma(\alpha^{-1})$ which satisfy $\gamma \equiv \mathrm{Id} \mod n_2$ and η for $\Gamma/\Gamma(\beta^{-1})$ which satisfy $\eta \equiv \mathrm{Id} \mod n_1$. Then the products $\gamma\eta$ will form a system of representatives for $\Gamma/\Gamma(\alpha^{-1}\beta^{-1})$. We can achieve $u_\alpha\eta = \alpha\eta\alpha^{-1}u_\alpha$ where $\alpha\eta\alpha^{-1} \in \eta\Gamma(\beta^{-1})$. We see that the conditions (i) and (ii) above are satisfied. We can put $u_{\alpha\beta} = u_\alpha u_\beta$.

Since we have $\alpha\beta = \beta\alpha$ and since clearly $d(\alpha)d(\beta) = d(\alpha\beta)$ we also get the commutation relation (iii).

So far we proved this relation only for the action on $H^0(\Gamma\backslash X, \tilde{\mathcal{M}})$. To prove it in arbitrary degrees we simply remember that we can compute the cohomology groups from the Čech complex of an orbiconvex covering $\mathfrak{U} = \{U_{x_i}\}_{\{x_i\}}$. We simply observe that we can apply the same reasoning to $\tilde{\mathcal{M}}(U_{\underline{x}_I})$ where $U_{\underline{x}_I} = \cap_{i\in I}U_{x_i}$ We replace the space X in our argument above by $\pi^{-1}(U_{\underline{x}_I})$ and it becomes clear that we get the commutation relation on the Čech complex and hence on the cohomology.

3.1.2 More relations between Hecke operators

We look at the algebra of Hecke operators in the special case that $G/\mathbb{Z} = \mathrm{Gl}_2/\mathbb{Z}$. We consider the action on $H^1(\Gamma\backslash\mathbb{H}, \tilde{\mathcal{M}})$, where $\Gamma = \mathrm{Sl}_2(\mathbb{Z})$, and we assume n even and $\mathcal{M} = \mathcal{M}_n[-\frac{n}{2}]$. This has the effect that the centre of $G/\mathbb{Z}$ acts trivially on $\mathcal{M}$, and this makes life simpler.

Hier ev. Übergang zu $\mathrm{PGL}_2(\mathbb{Z}) = \mathrm{Gl}_2(\mathbb{Z})/\pm 1$

We attach a Hecke operator to any coset $\Gamma\alpha\Gamma$ where $\alpha \in \mathrm{Gl}_2^+(\mathbb{Q})$ (i.e. $\det(\alpha) > 0$, as we want α to act on the upper half plane). Then α and $\lambda\alpha$ with $\lambda \in \mathbb{Q}^*$ define the same operator. Hence we may assume that the matrix entries of α are integers. The theorem of elementary divisors asserts that the double cosets

$$\Gamma \cdot M_n(\mathbb{Z})_{\det>0} \cdot \Gamma \subset \mathrm{Gl}_2^+(\mathbb{Q})$$

are represented by matrices of the form

$$\begin{pmatrix} a & 0 \\ 0 & b \end{pmatrix}$$

where $b \mid a$. But here we can divide by b, and we are left with the matrix

$$\alpha = \begin{pmatrix} a & 0 \\ 0 & 1 \end{pmatrix} \quad , \quad a \in \mathbb{N}.$$

We can attach a Hecke operator to this matrix provided we choose u_α. We see that α induces on the basis vectors of our module $\mathcal{M}$

$$X^\nu Y^{n-\nu} \longrightarrow a^{\nu-n/2} \cdot X^\nu Y^{n-\nu}.$$

Hence we see that we have the following natural choice for u_α

$$u_\alpha : P(X,Y) \longrightarrow a^{n/2}\alpha \cdot P(X,Y).$$

We get a family of endomorphisms

$$T\left(\begin{pmatrix} a & 0 \\ 0 & 1 \end{pmatrix}, u_{\begin{pmatrix} a & 0 \\ 0 & 1 \end{pmatrix}}\right) = T(a) \tag{3.4}$$

of the cohomology $H^i(\Gamma\backslash\mathbb{H}, \tilde{\mathcal{M}})$. We have seen already that we have $T_a T_b = T_{ab}$ if a, b are coprime.

Hence we have to investigate the local algebra $\mathcal{H}_p$ which is generated by the operators

$$T_{p^r} = T\left(\begin{pmatrix} p^r & 0 \\ 0 & 1 \end{pmatrix}, u_{\begin{pmatrix} p^r & 0 \\ 0 & 1 \end{pmatrix}}\right) \tag{3.5}$$

for the special case of the group $\Gamma = \mathrm{Sl}_2(\mathbb{Z})$ and the coefficient system $\mathcal{M} = \mathcal{M}_n[-\frac{n}{2}]$. To do this we compute the product

$$T_{p^r} \cdot T_p = T\left(\begin{pmatrix} p^r & 0 \\ 0 & 1 \end{pmatrix}, u_{\alpha_p^r}\right) \cdot T\left(\begin{pmatrix} p & 0 \\ 0 & 1 \end{pmatrix}, u_{\alpha_p}\right)$$

where the u_{α^r} are the natural choices.

Again we investigate first what happens in degree zero, i.e. on $H^0(\Gamma\backslash\mathbb{H}, \tilde{I})$ - here I can be any Γ-module. Let $\alpha = \begin{pmatrix} p & 0 \\ 0 & 1 \end{pmatrix}$, $\xi \in H^0(\Gamma\backslash X, \tilde{I})$ then

$$T(\alpha^r, u_{\alpha^r})T(\alpha, u_\alpha)(\xi) = (\sum_{\gamma\in\Gamma/\Gamma(\alpha^r)} \gamma u_{\alpha^r})(\sum_{\eta\in\Gamma/\Gamma(\alpha)} \eta u_\alpha)(\xi).$$

We have the classical system of representatives

$$\Gamma/\Gamma(\alpha^r) = \bigcup_{j \bmod p^r} \begin{pmatrix} 1 & j \\ 0 & 1 \end{pmatrix} \Gamma(\alpha^r) \cup \bigcup_{j' \bmod p^{r-1}} \begin{pmatrix} 1 & 0 \\ j'p & 1 \end{pmatrix} \begin{pmatrix} 0 & 1 \\ -1 & 0 \end{pmatrix} \Gamma(\alpha^r),$$

and our product of Hecke operators becomes

$$\left(\sum_{j \bmod p^r} \begin{pmatrix} 1 & j \\ 0 & 1 \end{pmatrix} + \sum_{j' \bmod p^{r-1}} \begin{pmatrix} 1 & 0 \\ j'p & 1 \end{pmatrix} \begin{pmatrix} 0 & 1 \\ -1 & 0 \end{pmatrix} \right) u_{\alpha^r}$$
$$\left(\sum_{j_1 \bmod p} \begin{pmatrix} 1 & j_1 \\ 0 & 1 \end{pmatrix} + \begin{pmatrix} 0 & 1 \\ -1 & 0 \end{pmatrix} \right) u_\alpha(\xi)$$
$$=$$
$$[\sum_{j \bmod p^r, j_1 \bmod p} \begin{pmatrix} 1 & j \\ 0 & 1 \end{pmatrix} u_{\alpha^r} \begin{pmatrix} 1 & j_1 \\ 0 & 1 \end{pmatrix} u_\alpha(\xi)$$
$$+ \sum_{j' \bmod p^{r-1}, j_1 \bmod p} \begin{pmatrix} 1 & 0 \\ j'p & 1 \end{pmatrix} \begin{pmatrix} 0 & 1 \\ -1 & 0 \end{pmatrix} u_{\alpha^r} \begin{pmatrix} 1 & j_1 \\ 0 & 1 \end{pmatrix} u_\alpha(\xi)]$$
$$+[\sum_{j \bmod p^r} \begin{pmatrix} 1 & j \\ 0 & 1 \end{pmatrix} u_{\alpha^r} \begin{pmatrix} 0 & 1 \\ -1 & 0 \end{pmatrix} u_\alpha(\xi)$$
$$+ \sum_{j' \bmod p^{r-1}} \begin{pmatrix} 1 & 0 \\ j'p & 1 \end{pmatrix} \begin{pmatrix} 0 & 1 \\ -1 & 0 \end{pmatrix} u_{\alpha^r} \begin{pmatrix} 0 & 1 \\ -1 & 0 \end{pmatrix} u_\alpha(\xi)]$$

where we group the summation into a sum of two subsummations.

We know that u_{α^ν} satisfy commutation rules

$$u_{\alpha^r} u_\alpha = u_{\alpha^{r+1}}$$
$$u_{\alpha^r} \begin{pmatrix} 1 & j_1 \\ 0 & 1 \end{pmatrix} = \begin{pmatrix} 1 & j_1 p^r \\ 0 & 1 \end{pmatrix} u_{\alpha^r} \tag{3.6}$$
$$u_{\alpha^r} \begin{pmatrix} 0 & 1 \\ -1 & 0 \end{pmatrix} u_\alpha \begin{pmatrix} 0 & 1 \\ -1 & 0 \end{pmatrix} = c_I(p) u_{\alpha^{r-1}}$$

where $c_I(p)$ is a non-zero integer. If we exploit the first two commutation relations then we get as the sum in the first $[\ldots]$

$$(\sum_{j \bmod p^r, j_1 \bmod p} \begin{pmatrix} 1 & j + p^r j_1 \\ 0 & 1 \end{pmatrix} +$$
$$\sum_{j' \bmod p^{r-1}, j_1 \bmod p} \left(\begin{pmatrix} 1 & 0 \\ (j' + p^{r-1} j_1)p & 1 \end{pmatrix} \begin{pmatrix} 0 & 1 \\ -1 & 0 \end{pmatrix} \right) u_{\alpha^{r+1}}(\xi) \tag{3.7}$$
$$= T(p^{r+1}, u_{\alpha^{r+1}})(\xi)$$

To compute the contribution of the second $\left[\ldots\right]$ we observe that $w = \begin{pmatrix} 0 & 1 \\ -1 & 0 \end{pmatrix} \in \Gamma$ and hence we have $w\xi = \xi$. Then the second commutation relation yields for the second $\left[\ldots\right]$

$$c_I(p)\left(\sum_{j \mod p^r} \begin{pmatrix} 1 & j \\ 0 & 1 \end{pmatrix} + \sum_{j' \mod p^{r-1}} \begin{pmatrix} 1 & 0 \\ j'p & 1 \end{pmatrix}\begin{pmatrix} 0 & 1 \\ -1 & 0 \end{pmatrix}\right) u_{\alpha^{r-1}}(\xi). \quad (3.8)$$

We observe that for $j \equiv 0 \mod p^{r-1}$ we get

$$\begin{pmatrix} 1 & j \\ 0 & 1 \end{pmatrix} u_{\alpha^{r-1}}(\xi) = u_{\alpha^{r-1}}\begin{pmatrix} 1 & \frac{j}{p^{r-1}} \\ 0 & 1 \end{pmatrix}(\xi) = u_{\alpha^{r-1}}(\xi)$$

and in case $r > 1$ for $j' \equiv 0 \mod p^{r-2}$:

$$\begin{aligned}\begin{pmatrix} 1 & 0 \\ j'p & 1 \end{pmatrix}\begin{pmatrix} 0 & 1 \\ -1 & 0 \end{pmatrix} u_{\alpha^{r-1}}(\xi) &= \begin{pmatrix} 0 & 1 \\ -1 & 0 \end{pmatrix} u_{\alpha^{r-1}}\begin{pmatrix} 1 & -\frac{pj'}{p^{r-1}} \\ 0 & 1 \end{pmatrix}(\xi) \\ &= \begin{pmatrix} 0 & 1 \\ -1 & 0 \end{pmatrix} u_{\alpha^{r-1}}(\xi).\end{aligned} \quad (3.9)$$

Here we used again (3.6) and $\xi \in H^0(\Gamma\backslash X, \tilde{I})$. In other words, in the summation (3.8) the first term only depends on $j \mod p^{r-1}$ and the second only on $j' \mod p^{r-2}$. For $r > 1$ this yields for the second term (3.8)

$$\begin{aligned}&pc_I(p)\left(\sum_{j \mod p^{r-1}} \begin{pmatrix} 1 & j \\ 0 & 1 \end{pmatrix} + \sum_{j' \mod p^{r-2}} \begin{pmatrix} 1 & 0 \\ j'p & 1 \end{pmatrix}\begin{pmatrix} 0 & 1 \\ -1 & 0 \end{pmatrix}\right) u_{\alpha^{r-1}}(\xi) \\ &= pc_I(p)T(p^{r-1})(\xi).\end{aligned}$$

If $r = 1$ the value for (3.8) is $c_I(p)(p+1)u_{\alpha^0}$ and hence we get the general formula

$$T_{p^r} \cdot T_p = T_{p^{r+1}} + (p + \epsilon(r))c_I(p)T_{p^{r-1}}$$

where $\epsilon(r) = 0$ if $r > 1$ and $\epsilon(r) = 1$ for $r = 1$.

This formula is valid for all values of $r \geq 0$ if we put $T_{p^{-1}} = 0$.

We want to know what this means for the action on $H^1(\Gamma\backslash\mathbb{H}, \mathcal{M})$. We apply this to the modules $I = \mathcal{M}$ and $I = \mathcal{M}(\pi^{-1}(U_{\underline{x}_J}))$ A simple calculation gives that the u_{α^r} satisfy the relations (3.6) with $c_{\mathcal{M}}(p) = p^n$. Hence we get for the action on $H^1(\Gamma\backslash\mathbb{H}, \mathcal{M})$

$$T_{p^r} \cdot T_p = T_{p^{r+1}} + p^{n+1}T_{p^{r-1}} + \epsilon(r)p^nT_{p^{r-1}} \quad (3.10)$$

where $\epsilon(r) = 0$ if $r > 1$ and $\epsilon(r) = 1$ for $r = 1$.

This has consequences for the structure of the action of the Hecke-algebra on $H^1(\Gamma\backslash\mathbb{H}, \tilde{\mathcal{M}})$. We are aware that we are discussing the unramified case, our group Γ is the group of $\mathbb{Z}$ valued point of the reductive group scheme $\mathrm{Gl}_2/\mathbb{Z}$.

Then we have seen that the Hecke algebra can be viewed as the restricted tensor product

$$\mathcal{H}_\Gamma = \otimes'_p \mathbb{Z}[T_p] = \mathbb{Z}[T_2] \otimes \mathbb{Z}[T_3] \otimes \dots \tag{3.11}$$

where restricted means that our tensors have the element $1 \in \mathbb{Z}$ at almost all places. This construction gives us a homomorphism to a finite dimensonal commutative algebra of endomorphisms

$$\mathcal{H}_\Gamma \to \mathcal{H}_{\Gamma,n} \subset \mathrm{End}(H^1(\Gamma\backslash\mathbb{H}, \tilde{\mathcal{M}}_n)). \tag{3.12}$$

This homomorphism has a huge kernel. This kernel contains the "local" kernels $J_{n,p} \in \mathbb{Z}[T_p]$ and these local kernels are the principal ideals $(P(T_p, X))$. We have still the complex conjugation on the cohomology, this complex conjugation decomposes the inner cohomoloogy into a + and a - eigenspace, and these two eigenspaces are isomorphic Hecke-modules. This implies the the characteristic polynomial of T_p on the inner cohomology is a square, i.e.

$$\det(T_p - \mathrm{Id}\cdot X)_{|H^1(\Gamma\backslash\mathbb{H},\tilde{\mathcal{M}}_n)} = P_!(T_p, X)^2.$$

Hence we get surjections

$$\begin{aligned} r_n &: \textstyle\bigotimes'_p \mathbb{Z}[T_p]/(P(T_p, X)) \to \mathcal{H}_{\Gamma,n} \subset \mathrm{End}(H^1(\Gamma\backslash\mathbb{H}, \tilde{\mathcal{M}}_n)) \\ r_{!,n} &: \textstyle\bigotimes'_p \mathbb{Z}[T_p]/(P_!(T_p, X)) \to \mathcal{H}_{\Gamma,n,!} \subset \mathrm{End}(H^1_!(\Gamma\backslash\mathbb{H}, \tilde{\mathcal{M}}_n)) \end{aligned} \tag{3.13}$$

At this point it seems to be appropriate to mention Maeda's conjecture: This conjecture says that the algebra $\mathcal{H}_{\Gamma,n,!}$ is a field. This field should be a maximal commutative subfield of $\mathrm{End}(H^1_!(\Gamma\backslash\mathbb{H}, \tilde{\mathcal{M}}_n)_\pm$, hence it is of degree $d(n) = \frac{1}{2}\dim(H^1_!(\Gamma\backslash\mathbb{H}, \tilde{\mathcal{M}}_n))$. Moreover people believe that the normal closure of this field has degree $d(n)!$ and hence its Galois group is $\mathfrak{S}_{d(n)}$, the full symmetric group in $d(n)$ letters. Furthermore we can ask whether or not already a single Hecke operator $r_{!,n}T_p$ generates the field $\mathcal{H}_{\Gamma,n,!}$.

This conjecture has been verified for $n \leq 12000$.

Interlude

We assume that a majority of the readers have seen Hecke operators in the context of modular forms and have also seen the recursion formulas for these Hecke operators. (see [102]), [62]). This is of course not accidental, In the following chapter 4 I will discuss the Eichler-Shimura isomorphism. This isomorphism provides an injection of the space $M_k(\Gamma)$ of modular forms of weight k into the cohomology, i.e.

$$M_k(\Gamma) \hookrightarrow H^1(\Gamma\backslash\mathbb{H}, \mathcal{M}_{k-2} \otimes \mathbb{C})$$

(see Thm. 4.2.1). This is an injection of Hecke modules.

The reader may notice that there is a slight discrepancy in the recursion formulae. This is due to the fact that for $r > 1$ the "classical" Hecke operators $T(p^r)$ and the Hecke operators T_{p^r} are not the same, but the T_p defined here correspond to the classical $T(p)$. ([102] and 4.86).

Here we want to stress that in this text so far - except in the introduction - there is no mentioning of modular forms, and this is intentional.

End Interlude

This can be generalised. We choose an integer $N > 1$ and we take as our arithmetic group the full congruence group $\Gamma = \Gamma(N)$. For any prime $p \nmid N$ the $T(\alpha, u_\alpha)$ with $\alpha \in \mathrm{Gl}_2^+(\mathbb{Z}[1/p])$ form a commutative subalgebra $\mathcal{H}_p$ which is generated by T_p. It has an identity element $e_p = T_{p^0}$. This is the *unramified Hecke algebra.*

For $p|N$ we can also consider the $T(\alpha, u_\alpha)$ with $\alpha \in \mathrm{Gl}_2^+(\mathbb{Z}[1/p])$. They will also generate a local algebra $\mathcal{H}_p$ of endomorphisms in all of our cohomology groups, but this algebra will not necessarily be commutative. But if we have two different primes p, p_1 then $\mathcal{H}_p$ and $\mathcal{H}_{p_1}$ commute

$$\mathcal{H}_\Gamma = \bigotimes_p{}' \mathcal{H}_p \tag{3.14}$$

where the superscript indicates that a tensor $h_f = \bigotimes_p h_p, h_p \in \mathcal{H}_p$ has a factor e_p for almost all p (in part II of this book we will give a better construction of the Hecke-algebra which uses the language of adeles). This algebra acts on all our cohomology groups. We recall that the action of $\mathcal{H}_\Gamma$ on the inner cohomology groups is semi-simple (see Thm. 3.1.1). This has important consequences, which we discuss after a brief recapitulation of the theory of semi-simple modules.

3.2 Some results on semi-simple $\mathcal{A}$-modules

We fix a field L and its algebraic closure $\bar{L}$, for simplicity we assume that the characteristic of L is zero, or that L is perfect. We consider an L-algebra $\mathcal{A}$, not necessarily commutative, but with identity. We need a few results and concepts from the theory of finite-dimensional vector spaces V/L with an action of $\mathcal{A}$, i.e. equipped with a homomorphism $\mathcal{A} \to \mathrm{End}_L(V)$.

Such an $\mathcal{A}$-module V is called *irreducible* if it does not contain an $\mathcal{A}$ invariant proper submodule $W \subset V$, i.e. $\{0\} \neq W \neq V$. It is called *absolutely irreducible* if the $\mathcal{A} \otimes \bar{L}$-module $V \otimes \bar{L}$ is irreducible. We say that V is *indecomposable* if it cannot be written as the direct sum of two non-zero submodules. An irreducible module is also indecomposable.

We say that the action of $\mathcal{A}$ on V is semi-simple, if the action of $\mathcal{A} \otimes \bar{L}$ on $V \otimes \bar{L}$ is semi-simple and this means that any $\mathcal{A}$-submodule $W \subset V \otimes \bar{L}$ has a complement, i.e. we can find an $\mathcal{A}$-submodule $W^\perp \subset V \otimes \bar{L}$ such that $V \otimes \bar{L} = W \oplus W^\perp$.

Then it is clear that we get a decomposition indexed by a finite set E

$$V \otimes \bar{L} = \bigoplus_{i \in E} W_i$$

where the W_i are (absolutely) irreducible submodules. In general this decomposition will not be unique. For any two W_i, W_j of these submodules we have (Schur's lemma)

$$\mathrm{Hom}_{\mathcal{A}}(W_i, W_j) \cong \begin{cases} \bar{L} & \text{if they are isomorphic as } \mathcal{A}\text{-modules} \\ 0 & \text{else} \end{cases}.$$

We decompose the indexing set $E = E_1 \cup E_2 \cup \cdots \cup E_k$ according to isomorphism types. For any E_ν we choose an $\mathcal{A}$-module $W_{[\nu]}$ of this given isomorphism type. Then by definition

$$\mathrm{Hom}_{\mathcal{A}}(W_{[\nu]}, W_j) \cong \begin{cases} \bar{L} & \text{if } j \in E_\nu \\ 0 & \text{else} \end{cases}.$$

Further, we define $H_{[\nu]} = \mathrm{Hom}_{\mathcal{A}}(W_{[\nu]}, V \otimes \bar{L})$, and we get an inclusion $H_{[\nu]} \otimes W_{[\nu]} \hookrightarrow V \otimes \bar{L}$. The image X_ν will be an $\mathcal{A}$-submodule, which is a direct sum of $\dim_{\bar{L}} H_{[\nu]}$ many copies of $W_{[\nu]}$ - it is the unique such submodule.

We get a direct sum decomposition

$$V \otimes \bar{L} = \bigoplus_{\nu} \bigoplus_{i \in E_\nu} W_i = \bigoplus_{\nu} X_\nu,$$

where this last decomposition is easily seen to be unique, and it is called the *isotypical decomposition.*

If V is a semi-simple $\mathcal{A}$-module then any submodule $W \subset V$ also has a complement (this is not entirely obvious because by definition only $W \otimes_L \bar{L}$ has a complement in $V \otimes_L \bar{L}$. But a small moment of meditation gives us that finding such a complement is the same as solving a system of linear equations over L. If this system has a solution over $\bar{L}$ it also has a solution over L.) Therefore we can also decompose the $\mathcal{A}$-module V into irreducibles. Again we can group the irreducibles according to isomorphism types and we get the *isotypical* decomposition

$$V = \bigoplus_{i \in E} U_i = \bigoplus_{\nu} \bigoplus_{i \in E_\nu} U_i = \bigoplus_{\nu} Y_\nu. \tag{3.15}$$

But of course a summand U_i may become reducible if we extend the scalars to $\bar{L}$ (see example below). Since it is clear that for any two $\mathcal{A}$-modules V_1, V_2 we have

$$\mathrm{Hom}_{\mathcal{A}}(V_1, V_2) \otimes \bar{L} = \mathrm{Hom}_{\mathcal{A} \otimes \bar{L}}(V_1 \otimes \bar{L}, V_2 \otimes \bar{L}),$$

we know that we get the isotypical decomposition of $V \otimes \bar{L}$ by taking the isotypical decomposition of the $Y_\nu \otimes \bar{L}$ and then taking the direct sum over ν.

Example: Let L_1/L be a finite extension of degree > 1, then we put $\mathcal{A} = L_1$ and $V = L_1$, where the action is given by multiplication. Clearly V is irreducible, but $V \otimes \bar{L}$ is not. If L_1/L is separable then the module is semi-simple, otherwise it is not.

We have a classical result:

Proposition 3.2.1. *Let V be a semi-simple $\mathcal{A}$-module. Then the following assertions are equivalent*

i) The $\mathcal{A}$-module V is absolutely irreducible.

ii) The image of $\mathcal{A}$ in the ring of endomorphisms is $\mathrm{End}(V)$.

iii) $\mathrm{End}_{\mathcal{A}}(V) = L$ *holds.*

This can be an exercise for an algebra class. Where do we need the assumption that V is semi-simple?

We return to our algebra $\mathcal{A}$ over L. Let V be an irreducible semi-simple $\mathcal{A}$-module, which is not necessarily absolutely irreducible. Let I_V be the two-sided ideal which annihilates V, i.e. the kernel of $\mathcal{A} \to \mathrm{End}_L(V)$. Let $\mathcal{C}_L$ be the centre of $\mathcal{A}/I_V$. This centre is a field, because any $c \in \mathcal{C}_L$ is either zero or an isomorphism, in other words V is a $\mathcal{C}_L$-vector space. The $\mathcal{C}_L$-algebra $\mathcal{A}/I_V$ is a central simple algebra. There is a central division algebra $\mathcal{D}/\mathcal{C}_L$ such that $\mathcal{A}/I_V \xrightarrow{\sim} M_r(\mathcal{D})$, the algebra of (r,r) matrices with coefficients in $\mathcal{D}$. This algebra has exactly one - up to isomorphism - non-zero irreducible module, which is the module of column vectors $\mathcal{D}^r$, where the algebra acts by multiplication from the left. Let us denote this module by $\mathcal{X}[\mathcal{A}/I_V]$.

Theorem 3.2.1. *The extension $\mathcal{C}_L/L$ is separable. Let L_1/L be a normal closure of $\mathcal{C}_L$. Then we have the isotypical decomposition*

$$V \otimes_L L_1 = \bigoplus_{\sigma:\mathcal{C}_L \to L_1} V \otimes_{\mathcal{C}_L,\sigma} L_1$$

The Galois group $\mathrm{Gal}(L_1/L)$ permutes the summands simply transitively, and the $\mathcal{A}/I_V \otimes_{\mathcal{C}_L,\sigma} L_1$-module $V \otimes_{\mathcal{C}_L,\sigma} L_1$ is isomorphic to the standard module $\mathcal{X}[\mathcal{A}/I_V \otimes_{\mathcal{C}_L,\sigma} L_1]$.

Proposition 3.2.2. *For any semi-simple $\mathcal{A}$-module V we can find a finite extension L_2/L such that the irreducible submodules in the decomposition into irreducibles are absolutely irreducible.*

Clearly, we have to take an extension which splits the various $\mathcal{D}$ appearing in connection with V.

If V is any $\mathcal{A}$-module - not necessarily semi-simple but finite dimensional over L - then there is a finite extension L_2/L and a filtration

$$\{0\} \subset V_1 \subset V_2 \subset \cdots \subset V_{r-1} \subset V \otimes_L L_2$$

such that the successive quotients V_i/V_{i-1} are absolutely irreducible. A very elementary argument shows that the set of isomorphism types occurring in this filtration does not depend on the filtration, and let us denote this set of isomorphism types by $\mathrm{Spec}_V(\mathcal{A} \otimes L_2)$.

We say that an $\mathcal{A}$-submodule $W \subset V$ is *complete in V* if the two sets $\mathrm{Spec}_W(\mathcal{A} \otimes L_2)$ and $\mathrm{Spec}_{V/W}(\mathcal{A} \otimes L_2)$ are disjoint. We have the following simple proposition:

Proposition 3.2.3. *a) If V is a semi-simple $\mathcal{A}$-module and if $W \subset V$ is complete in V then we have a canonical splitting $V = W \oplus W'$.*

b) If V is not necessarily semi-simple but if $\mathcal{A}$ is commutative instead, then any $W \subset V$ which is complete in V also has a canonical complement W', i.e. $V = W \oplus W'$.

Proof. For the second assertion we observe that an absolutely irreducible $\mathcal{A}$-module U is simply one-dimensional over L_2 and given by a homomorphism $\pi : \mathcal{A} \to L_2$, i.e. it is an eigenspace for $\mathcal{A}$. □

Let us call such a decomposition an *isotypical decomposition* into complete summands.

Let us now assume that we have two algebras $\mathcal{A}, \mathcal{B}$ acting on V. Let us furthermore assume that these two operations commute, i.e. for $A \in \mathcal{A}, B \in \mathcal{B}, v \in V$ we have $A(Bv) = B(Av)$. This structure is the same as having an $\mathcal{A} \otimes_L \mathcal{B}$ structure on V. Let us assume that $\mathcal{A}$ acts semi-simply on V and let us assume that the irreducible $\mathcal{A}$-submodules of V are absolutely irreducible. Then it is clear that the isotypical summands $Y_\nu = \bigoplus W_i$ are invariant under the action of $\mathcal{B}$. Now we pick an index i_0 - then the evaluation maps give us a homomorpism

$$W_{i_0} \otimes \mathrm{Hom}_{\mathcal{A}}(W_{i_0}, Y_\nu) \to Y_\nu.$$

Under our assumptions this is an isomorphism. Then we see that we get

$$V = \bigoplus_\nu W_{i_\nu} \otimes \mathrm{Hom}_{\mathcal{A}}(W_{i_0}, Y_\nu)$$

where i_ν is any element in E_ν and where $\mathcal{A}$ acts upon the first factor and $\mathcal{B}$ acts upon the second factor via the action of $\mathcal{B}$ on Y_ν.

We apply this to our Hecke algebra $\mathcal{H}_\Gamma = \bigotimes_p' \mathcal{H}_p$, (here the $\otimes'$ means that we take finite sums of tensors $\cdots \otimes h_p \otimes \ldots$ where almost all h_p are the identity) its action on $H^1(\Gamma\backslash\mathbb{H}, \tilde{\mathcal{M}})$. We anticipate the theorem that this action is semi-simple. Hence we can find a finite normal extension $F/\mathbb{Q}$ such that we get an isotypical decomposition

$$H^1(\Gamma\backslash\mathbb{H}, \tilde{\mathcal{M}} \otimes F) = \bigoplus_{\pi_f} H^1(\Gamma\backslash\mathbb{H}, \tilde{\mathcal{M}} \otimes F)(\pi_f), \tag{3.16}$$

here π_f is the isomorphism type of an absolutely irreducible $\mathcal{H}_\Gamma$-module over F. We can realise this module by a vector space H_{π_f}/F with an absolutely irreducible action of $\mathcal{H}_\Gamma$ on it. Then $H_{\pi_f} = \bigotimes' H_{\pi_p}$ where H_{π_p} is an absolutely irreducible $\mathcal{H}_p$-module. For almost all primes, H_{π_p} is one dimensional and π_p is simply a homomorphism

$$\pi_p : \mathcal{H}_p \to F \text{ which is determined by its value } \pi_f(T_p) \in F. \tag{3.17}$$

In this book we mostly consider the case where $\mathcal{H}_\Gamma$ is commutative field then we get a homomorphism $\pi_f = \otimes_p' \mathcal{H}_\Gamma \to F$. The Galois group $\mathrm{Gal}(F/\mathbb{Q})$ acts on $H^1(\Gamma\backslash\mathbb{H}, \tilde{\mathcal{M}} \otimes F)$, and hence it permutes the π_f which occur in the decomposition. Then for any π_f, the Hecke module

$$\bigoplus_{\sigma \in \mathrm{Gal}(F/\mathbb{Q})} H^1(\Gamma\backslash\mathbb{H}, \tilde{\mathcal{M}} \otimes F)({}^\sigma\pi_f), \tag{3.18}$$

is invariant under the action of the Galois group, hence defined over $\mathbb{Q}$. Therefore we get an isotypical decomposition

$$H^1(\Gamma\backslash\mathbb{H}, \tilde{\mathcal{M}}) = \bigoplus_{\Pi_f} H^1(\Gamma\backslash\mathbb{H}, \tilde{\mathcal{M}})(\Pi_f).$$

3.2.1 Explicit formulas for the Hecke operators, a general strategy

In the following section we discuss the Hecke operators, and for numerical experiments it is useful to have an explicit procedure to compute them in a given case. The main obstruction to get such an explicit procedure is to find an explicit way to compute the arrow $j^\bullet(\alpha)$ in the top line of the diagram (3.1) (we change the notation $j(\alpha)$ used there to $m(\alpha)$).

Let us assume that we have computed the cohomology groups on both sides by means of orbiconvex coverings $\mathfrak{V} : \cup_{i\in I} V_{y_i} = \Gamma(\alpha^{-1})\backslash X$ and $\mathfrak{U} : \cup_{j\in J} U_{y_j} = \Gamma(\alpha)\backslash X$.

The map $m(\alpha)$ is an isomorphism between spaces and hence $m(\alpha)(\mathfrak{V})$ is an acyclic covering of $\Gamma(\alpha)\backslash X$. This induces an identification

$$C^\bullet(\mathfrak{V}, \tilde{\mathcal{M}}) = C^\bullet(m(\alpha)(\mathfrak{V}), \tilde{\mathcal{M}}^{(\alpha)}),$$

and the complex on the right hand side computes $H^\bullet(\Gamma(\alpha)\backslash X, \tilde{\mathcal{M}}^{(\alpha)})$. But this cohomology is also computable from the complex $C^\bullet(\mathfrak{U}, \tilde{\mathcal{M}}^{(\alpha)})$. We take the disjoint union of the two indexing sets $I \cup J$ and look at the covering $m_\alpha(\mathfrak{V}) \cup \mathfrak{U}$. To be precise: we consider the disjoint union $\tilde{I} = I \cup J$ and define a covering $\mathfrak{W}$ indexed by $\tilde{I}$. If $i \in \tilde{I}$ then $W_i = m(\alpha)(V_{y_i})$ and if $i \in J$ then we put $W_i = U_{x_i}$. We get a diagram of Čech complexes

$$\begin{array}{ccccc}
\to & \bigoplus_{\underline{i}\in I^q} \tilde{\mathcal{M}}^{(\alpha)}(W_{\underline{i}}) & \to & \bigoplus_{\underline{i}\in I^{q+1}} \tilde{\mathcal{M}}^{(\alpha)}(W_{\underline{i}}) \to \\
& \uparrow & & \uparrow \\
\to & \bigoplus_{\underline{i}\in \tilde{I}^q} \tilde{\mathcal{M}}^{(\alpha)}(W_{\underline{i}}) & \to & \bigoplus_{\underline{i}\in \tilde{I}^{q+1}} \tilde{\mathcal{M}}^{(\alpha)}(W_{\underline{i}}) \to \\
& \downarrow & & \downarrow \\
\to & \bigoplus_{\underline{i}\in J^q} \tilde{\mathcal{M}}^{(\alpha)}(W_{\underline{i}}) & \to & \bigoplus_{\underline{i}\in J^{q+1}} \tilde{\mathcal{M}}^{(\alpha)}(W_{\underline{i}}) \to
\end{array} \tag{3.19}$$

The sets $I^\bullet, J^\bullet$ are subsets of $\tilde{I}^\bullet$ and the up-and-down-arrows are the resulting projection maps. We know that these up-and-down-arrows induce isomorphisms in cohomology.

Hence we can start from a cohomology class $\xi \in H^q(\Gamma(\alpha)\backslash X, \tilde{\mathcal{M}}^{(\alpha)})$, and we represent it by a cocycle

$$c_\xi \in \bigoplus_{\underline{i}\in I^q} \tilde{\mathcal{M}}^{(\alpha)}(W_{\underline{i}}).$$

Then we can find a cocycle $\tilde{c}_\xi \in \bigoplus_{\underline{i}\in \tilde{I}^q} \tilde{\mathcal{M}}^{(\alpha)}(W_{\underline{i}})$ which maps to c_ξ under the up arrow. To get this cocycle we have to do the following: our cocycle c_ξ is an array with components $c_\xi(\underline{i})$ for $\underline{i} \in I^q$. We have $d_q(c_\xi) = 0$. To get $\tilde{c}_\xi$ we have to give the values $\tilde{c}_\xi(\underline{i})$ for all $\underline{i} \in \tilde{I}^q \setminus I^q$. We must have

$$d_q \tilde{c}_\xi = 0.$$

This yields a system of linear equations for the remaining entries. We know that this system of equations has a solution - this is then our $\tilde{c}_\xi$ - and this solution is unique up to a boundary $d_{q-1}(\xi')$. Then we apply the down arrow to $\tilde{c}_\xi$ and get a cocycle $c_\xi^\dagger$, which represents the same class ξ but this class is now represented by a cocycle with respect to the covering $\mathfrak{U}$. We apply the map $\tilde{u}^\alpha : \tilde{\mathcal{M}}^{(\alpha)} \to \tilde{\mathcal{M}}$ to this cocycle and then we get a cocycle which represents the image of our class ξ under T_α.

The adjunction formula

Let $\mathcal{M}^\vee$ be the dual module to $\mathcal{M}$. We will define Hecke operators.

$$T_c(\alpha, u_{u_{\alpha^\vee}}) : H_c^\bullet(\Gamma\backslash X, \tilde{\mathcal{M}}^\vee) \to H_c^\bullet(\Gamma\backslash X, \tilde{\mathcal{M}}^\vee)$$

such that for $x \in H^q(\Gamma\backslash X, \tilde{\mathcal{M}}), y \in H_c^{d-q}(\Gamma\backslash X, \tilde{\mathcal{M}}^\vee)$ we have

$$\langle T(\alpha, u_\alpha)x, y\rangle_{PD} = \langle x, T_c(\alpha^{-1}, u_{\alpha^\vee})y\rangle_{PD}. \tag{3.20}$$

We proceed as in Section 3.1. If $\Gamma' \subset \Gamma$ is of finite index we have again the two maps

$$\pi_c^\bullet : H_c^\bullet(\Gamma'\backslash X, \tilde{\mathcal{M}}) \to H_c^\bullet(\Gamma\backslash X, \tilde{\mathcal{M}}) \text{ and } \pi_{\bullet,c} : H_c^\bullet(\Gamma\backslash X, \tilde{\mathcal{M}}) \to H_c^\bullet(\Gamma'\backslash X, \tilde{\mathcal{M}}).$$

Then we can define

$$\begin{array}{ccccc}
H_c^\bullet(\Gamma(\alpha^{-1})\backslash X, \tilde{\mathcal{M}}^\vee) & \overset{j(\alpha^{-1})_\bullet}{\longleftarrow} & H_c^\bullet(\Gamma(\alpha)\backslash X, j(\alpha^{-1})^*(\tilde{\mathcal{M}}^\vee)) & \overset{\tilde{u}_\alpha^\bullet}{\longleftarrow} & H_c^\bullet(\Gamma(\alpha)\backslash X, \mathcal{M}^\vee) \\
\downarrow \pi_\bullet & & & & \uparrow \pi^\bullet \\
H_c^\bullet(\Gamma\backslash X, \tilde{\mathcal{M}}^\vee) & & \xleftarrow{\quad T_c(\alpha^{-1}, u_{\alpha^{-1}}) \quad} & & H_c^\bullet(\Gamma\backslash X, \tilde{\mathcal{M}}^\vee)
\end{array},$$

Now it is easy to see that $\pi^\bullet$ and $\pi_c^\bullet$ and as well $\pi_\bullet$ and $\pi_{\bullet,c}$ are adjoint to each other with respect to Poincaré duality, and then it becomes clear that (3.20) holds.

We apply this to the special case that $\Gamma = \mathrm{Gl}_2(\mathbb{Z})$ and $\mathcal{M} = \mathcal{M}_n[-\frac{n}{2}]$. If enlarge $\mathbb{Z}$ to the ring R where all primes $p < n$ are inverted and take the coefficient module $\mathcal{M} \otimes R$ then this coefficient module becomes self dual. Our adjunction formula tells us that the adjoint operator to T_p is the operator $T_{p^{-1}}, u_{p^{-1}})$ but this equal to T_p. This tells us that the Hecke modules $H_!^1(\Gamma\backslash X, \mathcal{M} \otimes R)_-$ and $H_!^1(\Gamma\backslash X, \mathcal{M} \otimes R)_+$ are canonically isomorphic.

In the following section we discuss the explicit computation of a Hecke operator in a very specific situation. We start from our computation in Section (2.1.4) and write down some $H^\bullet(\Gamma\backslash X, \tilde{\mathcal{M}})$ explicitly. On these modules we give explicit procedures to compute a Hecke operator. We get some supply of data and we look for some interesting laws, or we try to verify some conjectures (see (3.74)).

3.3 Hecke operators for Gl_2

For the rest of this chapter we discuss a very specific case. The algebraic group scheme will be $\mathrm{Gl}_2/\mathbb{Z}$. The symmetric space will be

$$X = \mathrm{Gl}_2(\mathbb{R})/K_\infty \text{ where } K_\infty = \mathrm{SO}(2) \times \left\{\begin{pmatrix} t & 0 \\ 0 & t \end{pmatrix} \mid t \in \mathbb{R}^\times, t > 0\right\}.$$

Then the space X is the union of an upper and a lower half plane. We choose $\tilde{\Gamma} = \mathrm{Gl}_2(\mathbb{Z})$, then

$$\tilde{\Gamma}\backslash G_\infty/K_\infty = \Gamma\backslash\mathbb{H},$$

where $\Gamma = \mathrm{Sl}_2(\mathbb{Z})$ and $\mathbb{H}$ is the upper half-plane.

Earlier we defined the Γ-modules $\mathcal{M}_n[m]$ (see 1.2.1), and in the following we put $\mathcal{M} = \mathcal{M}_n[0]$.

We refer to Chapter 2, section 2.1.4. We have the two open sets $\tilde{U}_{\mathbf{i}}$, resp. $\tilde{U}_\rho \subset \mathbb{H}$, and they are fixed under

$$S = \begin{pmatrix} 0 & 1 \\ -1 & 0 \end{pmatrix} \text{ and } R = \begin{pmatrix} 1 & -1 \\ 1 & 0 \end{pmatrix},$$

respectively. We will also use the elements

$$T_+ = \begin{pmatrix} 1 & 1 \\ 0 & 1 \end{pmatrix}, \; S_1^+ = T_- S T_-^{-1} = \begin{pmatrix} -1 & 1 \\ -2 & 1 \end{pmatrix} \in \Gamma_0^+(2)$$

$$T_- = \begin{pmatrix} 1 & 0 \\ 1 & 1 \end{pmatrix}, \; S_1^- = T_+ S T_+^{-1} = \begin{pmatrix} -1 & 2 \\ -1 & 1 \end{pmatrix} \in \Gamma_0^-(2)$$

The elements S_1^+ and S_1^- are elements of order four, i.e. $(S_1^+)^2 = (S_1^-)^2 = -\mathrm{Id}$, the corresponding fixed points are $\frac{\mathbf{i}+1}{2}$ and $\mathbf{i}+1$ respectively. Hence S_1^- fixes the sets $\tilde{U}_{\frac{\mathbf{i}+1}{2}}$ and $\tilde{U}_{\mathbf{i}+1}$. This is the only occurrence of a non-trivial stabiliser.

3.3.1 The boundary cohomology

It is easier to compute the action of the Hecke operator T_p on the cohomology of the boundary, i.e. to compute the endomorphism

$$T_p : H^1(\partial(\Gamma\backslash\mathbb{H}), \tilde{\mathcal{M}}) \to H^1(\partial(\Gamma\backslash\mathbb{H}), \tilde{\mathcal{M}}).$$

We know (see 2.52) that $H^1(\partial(\Gamma\backslash\mathbb{H}), \tilde{\mathcal{M}}) \cong \mathcal{M}/(1-T_+)\mathcal{M}$, and we collect some easy facts concerning this module. For $n \geq k \geq 0$ we define the submodules

$$\mathcal{M}^{(k)} = \mathbb{Z}\, X^k Y^{n-k} \oplus \mathbb{Z}\, X^{k+1} Y^{n-k-1} \oplus \cdots \oplus \mathbb{Z}\, X^n.$$

For $k = 0$ (resp. $k = n$) we have $\mathcal{M}^{(0)} = \mathcal{M}$ (resp. $\mathcal{M}^{(n)} = \mathbb{Z}X^n$). These modules are invariant under the action of T_+. We have $(1-T_+)\mathcal{M}^{(k)} \subset \mathcal{M}^{(k+1)}$ and $\mathcal{M}^{(0)}/\mathcal{M}^{(1)} \xrightarrow{\sim} \mathbb{Z}$. The map $(1-T_+)$ induces a map

$$\partial_k : \mathcal{M}^{(k)}/\mathcal{M}^{(k+1)} \to \mathcal{M}^{(k+1)}/\mathcal{M}^{(k+2)}$$

which is given by multiplication with $n-k$. Hence it is clear that

$$\mathcal{M}/(1-T_+)\mathcal{M} = \mathbb{Z}[Y^n] \oplus \mathcal{M}^{(1)}/(1-T_+)\mathcal{M},$$

and the second summand is a finite module. The filtration of $\mathcal{M}$ by the $\mathcal{M}^{(k)}$ induces a filtration on $H^1(\partial(\Gamma\backslash\mathbb{H}), \tilde{\mathcal{M}})$, and we set

$$H^1(\partial(\Gamma\backslash\mathbb{H}), \tilde{\mathcal{M}})^{(k)} := \mathrm{Im}(H^1(\partial(\Gamma\backslash\mathbb{H}), \tilde{\mathcal{M}}^{(k)}) \to H^1(\partial(\Gamma\backslash\mathbb{H}), \tilde{\mathcal{M}})). \qquad (3.21)$$

Then:

Proposition 3.3.1. *For $k > 0$ the quotient*

$$H^1(\partial(\Gamma\backslash\mathbb{H}),\tilde{\mathcal{M}})^{(k)}/H^1(\partial(\Gamma\backslash\mathbb{H}),\tilde{\mathcal{M}})^{(k+1)} \xrightarrow{\sim} \mathbb{Z}/(n-k+1)\mathbb{Z}.$$

The Hecke operator T_p acts on $H^1(\partial(\Gamma\backslash\mathbb{H}),\tilde{\mathcal{M}})^{(k)}/H^1(\partial(\Gamma\backslash\mathbb{H}),\tilde{\mathcal{M}})^{(k+1)}$ by multiplication with $p^k + p^{n-k+1}$. Especially we have

$$T_p[Y^n] = (p^{n+1}+1)[Y^n]$$

Proof. We introduce the polynomials

$$\epsilon_k(X,Y) := X^n \tfrac{Y}{X}(\tfrac{Y}{X}-1)\dots(\tfrac{Y}{X}-k+1) = X^n \textstyle\prod_{\nu=0}^{k-1}(\tfrac{Y}{X}-\nu) = k!X^n\dbinom{\frac{Y}{X}}{k}$$

$$= X^{n-k}Y(Y-X)\dots(Y-(k-1)X) = X^{n-k}Y^k + \dots + (-1)^k k! X^n.$$

Obviously these $\epsilon_k(X,Y)$ form a basis of $\mathcal{M}$. Pascal's rule for binomial coefficients says $\binom{\frac{Y}{X}+1}{k} = \binom{\frac{Y}{X}}{k} + \binom{\frac{Y}{X}}{k-1}$, and this yields

$$T_+\epsilon_k(X,Y) = \epsilon_k(X,X+Y) = \epsilon_k(X,Y) + k\epsilon_{k-1}(X,Y).$$

From this we get

$$\mathcal{M}/(1-T_+)\mathcal{M} = \mathbb{Z}\epsilon_n(X,Y) \oplus \bigoplus_{k=n-1}^{0} (\mathbb{Z}/(k+1)\mathbb{Z})\epsilon_k(X,Y), \tag{3.22}$$

this is the first assertion.

We pick a prime p and investigate the action of T_p on $H^1(\partial(\Gamma\backslash\mathbb{H}),\tilde{\mathcal{M}})$. We recall the definition of the Hecke operator: we start from the matrix $\alpha = \begin{pmatrix} p & 0 \\ 0 & 1 \end{pmatrix}$ and consider the diagram (3.1) adapted to our situation

$$\begin{array}{ccccc}
H^1(\partial(\Gamma(\alpha^{-1})\backslash\mathbb{H}),\tilde{\mathcal{M}}) & \xrightarrow{j(\alpha)^{(1)}} & H^1(\partial(\Gamma(\alpha)\backslash\mathbb{H}), j(\alpha)_*(\tilde{\mathcal{M}})) & \xrightarrow{\tilde{u}_\alpha^{(1)}} & H^1(\partial(\Gamma(\alpha)\backslash\mathbb{H}),\mathcal{M}) \\
\uparrow \pi^{(1)} & & & & \downarrow \pi_{(1)} \\
H^1(\partial(\Gamma\backslash\mathbb{H}),\tilde{\mathcal{M}}) & & \xrightarrow{\quad T(\alpha,u_\alpha) \quad} & & H^1(\partial(\Gamma\backslash\mathbb{H}),\tilde{\mathcal{M}})
\end{array}$$

The group $\Gamma(\alpha^{-1}) = \left\{\begin{pmatrix} a & b \\ c & d\end{pmatrix} \mid c \equiv 0 \mod p\right\}$ acts on $\P^1(\mathbb{Q})$ and has two orbits which can be represented by ∞ and 0. The stabilisers of these two cusps are $\Gamma_\infty = \{\pm T_+^\nu\}$ and $\Gamma_0 = \{\pm T_-^{p\nu}\}$, respectively. Hence we get

$$H^1(\partial(\Gamma(\alpha^{-1})\backslash\mathbb{H}),\tilde{\mathcal{M}}) = \mathcal{M}/(\mathrm{Id}-T_+)\mathcal{M} \oplus \mathcal{M}/(\mathrm{Id}-T_-^p)\mathcal{M}. \tag{3.23}$$

We identify $H^1(\partial(\Gamma\backslash\mathbb{H}),\tilde{\mathcal{M}}) = \mathcal{M}/(\mathrm{Id}-T_+)\mathcal{M} \xrightarrow{w_0} \mathcal{M}/(\mathrm{Id}-T_-)\mathcal{M}$ where the last arrow is induced by the map $m \mapsto w_0 m$ with $w_0 = \begin{pmatrix} 0 & 1 \\ -1 & 0 \end{pmatrix}$. Then

$$\pi^{(1)}(m) = \left(m, \sum_{j=0}^{p-1}\begin{pmatrix} 1 & 0 \\ j & 1\end{pmatrix} w_0 m\right). \tag{3.24}$$

Therefore the composition $u_\alpha^{(1)} \circ j(\alpha)^{(1)}$

$$\mathcal{M}/(\mathrm{Id}-T_+)\mathcal{M} \oplus \mathcal{M}/(\mathrm{Id}-T_-^p)\mathcal{M} \to \mathcal{M}/(\mathrm{Id}-T_+^p)\mathcal{M} \oplus \mathcal{M}/(\mathrm{Id}-T_-)\mathcal{M}$$

is given by the assignment $u_\alpha^{(1)} \circ j(\alpha)^{(1)}(m_\infty, m_0) \mapsto (\alpha m_\infty, \alpha m_0)$, and the assignment $\pi_{(1)}((n_\infty, n_0)) = n_\infty + w_0 n_0$. This yields

$$T_p(m) = \alpha m + w_0 \alpha w_0^{-1} \sum_{j=0}^{p-1} \begin{pmatrix} 1 & j \\ 0 & 1 \end{pmatrix} m.$$

On $\mathcal{M}^{(k)}/\mathcal{M}^{(k+1)}$ the element $\begin{pmatrix} 1 & j \\ 0 & 1 \end{pmatrix}$ acts as identity, α is multiplication by p^k and $w_0 \alpha w_0^{-1}$ is multiplication p^{n-k}.

We observe that the quotient $H^1(\Gamma\backslash\mathbb{H}, \mathcal{M})_{\mathrm{int},!}/H_!^1(\Gamma\backslash\mathbb{H}, \mathcal{M})_{\mathrm{int}}$ may become non-trivial and somewhat interesting (see (2.47)). We have to consider the exact sequence

$$0 \to H_!^1(\Gamma\backslash\mathbb{H}, \tilde{\mathcal{M}}) \to H^1(\Gamma\backslash\mathbb{H}, \tilde{\mathcal{M}}) \xrightarrow{r} H^1(\partial(\Gamma\backslash\mathbb{H}), \tilde{\mathcal{M}}). \tag{3.25}$$

Our cohomology groups may have some torsion $\mathcal{T}_1 \subset H^1(\Gamma\backslash\mathbb{H}, \tilde{\mathcal{M}})$, resp. $\mathcal{T}_2 \subset H^1(\partial(\Gamma\backslash\mathbb{H}), \tilde{\mathcal{M}})$ and the map r maps the torsion $\mathcal{T}_1$ to a submodule $r(\mathcal{T}_1) \subset \mathcal{T}_2$. But it will happen that $r(r^{-1}(\mathcal{T}_2))$ is strictly larger than $r(\mathcal{T}_1)$ - this means that some non-torsion elements are mapped to torsion elements under r. By definition $H^1(\Gamma\backslash\mathbb{H}, \tilde{\mathcal{M}})_{\mathrm{int},!} = r^{-1}(\mathcal{T}_2)/\mathcal{T}_1$ and therefore

$$H^1(\Gamma\backslash\mathbb{H}, \tilde{\mathcal{M}})_{\mathrm{int},!}/H_!^1(\Gamma\backslash\mathbb{H}, \tilde{\mathcal{M}})_{\mathrm{int}} = r(r^{-1}(\mathcal{T}_2)/\mathcal{T}_1). \tag{3.26}$$

This has been investigated extensively by Taiwang Deng in [34].

We return to the cohomology $H^1(\Gamma\backslash X, \mathcal{M})$. We obtain a covering $\Gamma\backslash\mathbb{H} = \pi_1(\tilde{U}_{\mathbf{i}}) \cup \pi_1(\tilde{U}_\rho) = U_{\mathbf{i}} \cap U_\rho$ where $\pi_1 : \mathbb{H} \to \Gamma\backslash\mathbb{H}$ is the projection. From this covering we get the Čech complex

$$\begin{array}{ccccc} 0 \to & \tilde{\mathcal{M}}(U_{\mathbf{i}}) \oplus \tilde{\mathcal{M}}(U_\rho) & \to & \tilde{\mathcal{M}}(U_{\mathbf{i}} \cap U_\rho) & \to 0 \\ & \downarrow ev_{\tilde{U}_{\mathbf{i}}} \oplus ev_{\tilde{U}_\rho} & & \downarrow ev_{\tilde{U}_{\mathbf{i}} \cap \tilde{U}_\rho} & , \\ & \mathcal{M}^{<S>} \oplus \mathcal{M}^{<R>} & \to & \mathcal{M} & \to 0 \end{array} \tag{3.27}$$

and this gives us our formula for the first cohomology

$$H^1(\Gamma\backslash\mathbb{H}, \tilde{\mathcal{M}}) = \mathcal{M}/(\mathcal{M}^{\langle S\rangle} \oplus \mathcal{M}^{\langle R\rangle}). \tag{3.28}$$

We want to discuss the Hecke operator T_2. To do this we pass to the subgroups

$$\left.\begin{array}{l} \Gamma_0^+(2) = \left\{ \begin{pmatrix} a & b \\ c & d \end{pmatrix} \mid c \equiv 0 \mod 2 \right\} \\ \Gamma_0^-(2) = \left\{ \begin{pmatrix} a & b \\ c & d \end{pmatrix} \mid b \equiv 0 \mod 2 \right\} \end{array}\right\}. \tag{3.29}$$

We form the two quotients. We have the projection maps $\pi_2^{\pm} : \mathbb{H} \to \Gamma_0^{\pm}(2)\backslash\mathbb{H}$. We have an isomorphism between the spaces

$$\Gamma_0^{+}(2)\backslash\mathbb{H} \xrightarrow{\alpha_2} \Gamma_0^{-}(2)\backslash\mathbb{H}$$

which is induced by the map $\alpha : z \mapsto \begin{pmatrix} 2 & 0 \\ 0 & 1 \end{pmatrix} z = 2z$. This map induces an isomorphism

$$\alpha_{2_\bullet} : H^1(\Gamma_0^{+}(2)\backslash\mathbb{H}, \tilde{\mathcal{M}}) \xrightarrow{\sim} H^1(\Gamma_0^{-}(2)\backslash\mathbb{H}, \tilde{\mathcal{M}}^{(\alpha)}). \tag{3.30}$$

We also have the map between sheaves $u_2 : m \mapsto \begin{pmatrix} 2 & 0 \\ 0 & 1 \end{pmatrix} m$ and the composition with this map induces a homomorphism in cohomology

$$H^1(\Gamma_0^{+}(2)\backslash\mathbb{H}, \tilde{\mathcal{M}}) \xrightarrow{u_{2_\bullet} \circ \alpha_{2_\bullet}} H^1(\Gamma_0^{-}(2)\backslash\mathbb{H}, \tilde{\mathcal{M}}). \tag{3.31}$$

This is the homomorphism we need for the computation of the Hecke operator; it is easy to define but it may be difficult in practice to compute it.

Each of the spaces $\Gamma_0^{+}(2)\backslash\mathbb{H}, \Gamma_0^{-}(2)\backslash\mathbb{H}$ has two cusps which can be represented by the points $\infty, 0 \in \mathbb{P}^1(\mathbb{Q})$. The stabilisers of these two cusps in $\Gamma_0^{+}(2)$(resp. $\Gamma_0^{-}(2)$) are

$$\langle T_+ \rangle \times \{\pm \mathrm{Id}\} \text{ and } \langle T_-^2 \rangle \times \{\pm \mathrm{Id}\} \subset \Gamma_0^{+}(2)$$

resp.

$$\langle T_+^2 \rangle \times \{\pm \mathrm{Id}\} \text{ and } \langle T_- \rangle \times \{\pm \mathrm{Id}\} \subset \Gamma_0^{-}(2);$$

the factor $\{\pm \mathrm{Id}\}$ can be ignored. Then we get

$$H^1(\partial(\Gamma_0^{+}(2)\backslash\mathbb{H}), \tilde{\mathcal{M}}) \xrightarrow{\sim} \mathcal{M}/(\mathrm{Id} - T_+)\mathcal{M} \oplus \mathcal{M}/(\mathrm{Id} - T_-^2)\mathcal{M}$$

$$H^1(\partial(\Gamma_0^{-}(2)\backslash\mathbb{H}), \tilde{\mathcal{M}}) \xrightarrow{\sim} \mathcal{M}/(\mathrm{Id} - T_+^2)\mathcal{M} \oplus \mathcal{M}/(\mathrm{Id} - T_-)\mathcal{M}.$$

But now it is obvious that α maps the cusp ∞ to ∞ and 0 to 0 and then it is also clear that for the boundary cohomology the map

$$\alpha_2^\bullet : \mathcal{M}/(\mathrm{Id} - T_+)\mathcal{M} \oplus \mathcal{M}/(\mathrm{Id} - T_-^2)\mathcal{M} \to \mathcal{M}/(\mathrm{Id} - T_+^2)\mathcal{M} \oplus \mathcal{M}/(\mathrm{Id} - T_-)\mathcal{M}$$

is simply the map which is induced by $u_2 : \mathcal{M} \to \mathcal{M}$. If we ignore torsion then the individual summands are infinite cyclic.

Our module $\mathcal{M}$ is the module of homogenous polynomials of degree n in 2 variables X, Y with integer coefficients. Then the classes $[Y^n]$ (resp.$[X^n]$) of the polynomials Y^n (resp. X^n) are generators of $(\mathcal{M}/(\mathrm{Id} - T_+^\nu)\mathcal{M})/\mathrm{tors}$ (resp. $(\mathcal{M}/(\mathrm{Id} - T_-^\nu)\mathcal{M})/\mathrm{tors}$) where $\nu = 1$ or 2. Then we get for the homomorphism $\alpha_2^\bullet$

$$\alpha_2^\bullet : [Y^n] \mapsto [Y^n], \alpha_2^\bullet : [X^n] \mapsto 2^n[X^n]. \tag{3.32}$$

3.3.2 The explicit description of the cohomology

We give the explicit description of the cohomology $H^1(\Gamma_0^+(2)\backslash\mathbb{H},\tilde{\mathcal{M}})$. We introduce the projections

$$\mathbb{H} \xrightarrow{\pi_2^+} \Gamma_0^+(2)\backslash\mathbb{H};\ \mathbb{H} \xrightarrow{\pi_2^-} \Gamma_0^-(2)\backslash\mathbb{H},$$

and get the covering $\mathfrak{U}_2$

$$\Gamma_0^+(2)\backslash\mathbb{H} = \pi_2^+(\tilde{U}_{\mathbf{i}}) \cup \pi_2^+(T_-\tilde{U}_{\mathbf{i}}) \cup \pi_2^+(\tilde{U}_\rho) = \pi_2^+(\tilde{U}_{\mathbf{i}}) \cup \pi_2^+(\tilde{U}_{\frac{\mathbf{i}+1}{2}}) \cup \pi_2^+(\tilde{U}_\rho)$$

where we put $T_-\tilde{U}_{\mathbf{i}} = \tilde{U}_{\frac{\mathbf{i}+1}{2}}$. Our set $\{x_\nu\}$ of indexing points is $\mathbf{i}, \frac{\mathbf{i}+1}{2}, \rho$, we set $U_{x_i}^+ := \pi_2^+(\tilde{U}_{x_i})$. Note $T_- \notin \Gamma_0^+(2), T_+ \in \Gamma_0^+(2)$.

Again the cohomology is computed by the complex

$$0 \to \tilde{\mathcal{M}}(U_{\mathbf{i}}^+) \oplus \tilde{\mathcal{M}}(T_-\tilde{U}_{\mathbf{i}}^+) \oplus \tilde{\mathcal{M}}(U_\rho^+) \to \tilde{\mathcal{M}}(U_{\mathbf{i}}^+ \cap U_\rho^+) \oplus \tilde{\mathcal{M}}(T_-\tilde{U}_{\mathbf{i}}^+ \cap U_\rho^+) \to 0.$$

We have to identify the terms as submodules of some $\bigoplus \mathcal{M}$ and write down the boundary map explicitly. We have

$$\begin{array}{ccc} \tilde{\mathcal{M}}(U_{\mathbf{i}}^+) \oplus \tilde{\mathcal{M}}(U_{\frac{\mathbf{i}+1}{2}}^+) \oplus \tilde{\mathcal{M}}(U_\rho^+) & \xrightarrow{d_0} & \tilde{\mathcal{M}}(U_i^+ \cap U_\rho^+) \oplus \tilde{\mathcal{M}}(U_{\frac{\mathbf{i}+1}{2}}^+ \cap U_\rho^+) \\ \downarrow ev_{\tilde{U}_{\mathbf{i}}} \oplus ev_{T_-\tilde{U}_{\mathbf{i}}} \oplus ev_{\tilde{U}_\rho} & & \downarrow ev_{\tilde{U}_{\mathbf{i}}\cap\tilde{U}_\rho} \oplus ev_{\tilde{U}_{\mathbf{i}}\cap T_+^{-1}\tilde{U}_\rho} \oplus ev_{T_-\tilde{U}_{\mathbf{i}}\cap\tilde{U}_\rho} \\ \mathcal{M} \oplus \mathcal{M}^{\langle S_1^+\rangle} \oplus \mathcal{M} & \xrightarrow{\bar{d}_0} & \mathcal{M} \oplus \mathcal{M} \oplus \mathcal{M} \end{array} \tag{3.33}$$

where the vertical arrows are isomorphisms. The boundary map $\bar{d}_0$ in the bottom row is given by

$$(m_1, m_2, m_3) \mapsto (m_1 - m_3, m_1 - T_+^{-1}m_3, m_2 - m_3) = (x, y, z).$$

We may look at the (quasi-isomorphic) subcomplex where $x = z = 0$ and $m_1 = m_2 = m_3$. Then we obtain the complex

$$0 \to \mathcal{M}^{\langle S_1^+\rangle} \to \mathcal{M} \to 0;\ m_2 \mapsto m_2 - T_+^{-1}m_2,$$

which provides an isomorphism

$$H^1(\Gamma_0^+(2)\backslash\mathbb{H},\tilde{\mathcal{M}}) \xrightarrow{\sim} \mathcal{M}/(\mathrm{Id} - T_+^{-1})\mathcal{M}^{\langle S_1^+\rangle}. \tag{3.34}$$

A simple computation shows that the cohomology class represented by the class (x, y, z) is equal to the class represented by $(0, y - x + T_+^{-1}z - z, 0)$, thus we write

$$[(x, y, z)] = [(0, y - x + T_+^{-1}z - z, 0)] \tag{3.35}$$

3.3.3 The map to the boundary cohomology

We have the restriction map for the cohomology of the boundary

$$\begin{array}{ccc} H^1(\Gamma_0^+(2)\backslash\mathbb{H},\tilde{\mathcal{M}}) & \xrightarrow{\sim} & \mathcal{M}/(\mathrm{Id}-T_+^{-1})\mathcal{M}^{\langle S_1^+\rangle} \\ \downarrow & & r^+\oplus r^-\downarrow \\ H^1(\partial(\Gamma_0^+(2)\backslash\mathbb{H}),\tilde{\mathcal{M}}) & \xrightarrow{\sim} & \mathcal{M}/(\mathrm{Id}-T_+)\mathcal{M}\oplus\mathcal{M}/(\mathrm{Id}-T_-^2)\mathcal{M} \end{array} . \quad (3.36)$$

We give a formula for the second vertical arrow. We represent a class $[m]$ by an element $m\in\mathcal{M}$ and send m to its class in each of the two summands, respectively. This is well-defined, for r^+ it is obvious, while for r^- we observe that if $m=x-T_+^{-1}x$ and $S_1^+x=x$ then $m=x-T_+^{-1}S_1^+x=x-T_-^2x$.

Restriction and corestriction

Now we have to give explicit formulas for the two maps π^*,π_* in the big diagram (3.1). Here we should change notation: the map π will now be denoted by:

$$\varpi_2^+:\Gamma_0^+(2)\backslash\mathbb{H}\to\Gamma\backslash\mathbb{H}. \quad (3.37)$$

We have the two complexes which compute the cohomology $H^1(\Gamma_0^+(2)\backslash\mathbb{H},\tilde{\mathcal{M}})$ and $H^1(\Gamma\backslash\mathbb{H},\tilde{\mathcal{M}})$, and we have defined arrows between them. We realized these two complexes explicitly in (3.33) resp. (3.27) and we have

$$\begin{array}{ccc} \tilde{\mathcal{M}}(U_i^+)\oplus\tilde{\mathcal{M}}(U_{\frac{i+1}{2}}^+)\oplus\tilde{\mathcal{M}}(U_\rho^+) & \xrightarrow{d_0} & \tilde{\mathcal{M}}(U_i^+\cap U_\rho^+)\oplus\tilde{\mathcal{M}}(U_{\frac{i+1}{2}}^+\cap U_\rho^+) \\ (\varpi_2^+)^{(0)}\uparrow\downarrow(\varpi_2^+)_{(0)} & & (\varpi_2^+)^{(1)}\uparrow\downarrow(\varpi_2^+)_{(1)} \\ \tilde{\mathcal{M}}(U_i)\oplus\tilde{\mathcal{M}}(U_\rho) & \xrightarrow{d_0} & \tilde{\mathcal{M}}(U_i\cap U_\rho) \end{array} \quad (3.38)$$

. In terms of our explicit realization in diagram (3.33) this gives

$$\begin{array}{ccc} \mathcal{M}\oplus\mathcal{M}^{\langle S_1\rangle}\oplus\mathcal{M} & \xrightarrow{d_0} & (\mathcal{M}\oplus\mathcal{M})\oplus\mathcal{M} \\ (\varpi_2^+)^{(0)}\uparrow\downarrow(\varpi_2^+)_{(0)} & & (\varpi_2^+)^{(1)}\uparrow\downarrow(\varpi_2^+)_{(1)} \\ \mathcal{M}^{\langle S\rangle}\oplus\mathcal{M}^{\langle R\rangle} & \xrightarrow{d_0} & \mathcal{M} \end{array} \quad (3.39)$$

Looking at the definitions we find

$$\begin{array}{c} (\varpi_2^+)^{(0)}:(m_1,m_2)\mapsto(m_1,T_-m_1,m_2) \\ (\varpi_2^+)_{(0)}:(m_1,m_2,m_3)\mapsto(m_1+Sm_1+T_-^{-1}m_2,(1+R+R^2)m_3) \end{array} \quad (3.40)$$

and we check easily that the composition $(\varpi_2^+)_{(0)}\circ(\varpi_2^+)^{(0)}$ is the multiplication by 3, as it should be, since this is the index of $\Gamma_0(2)^+$ in Γ.

For the two arrows in degree one we find

$$\begin{aligned}(\varpi_2^+)^{(1)} &: m \mapsto (m, Sm, T_- m) \\ (\varpi_2^+)_{(1)} &: (m_1, m_2, m_3) \mapsto (m_1 + Sm_2 + T_-^{-1} m_3)\end{aligned} \tag{3.41}$$

We apply equation (3.35) and we see that $(\varpi_2^+)^{(1)}(m)$ is represented by

$$[(\varpi_2^+)^{(1)}(m)] = [0, Sm + T_+^{-1}T_- m - m - T_- m, 0]. \tag{3.42}$$

We do the same calculation for $\Gamma_0^-(2)$. As before, we start from a covering

$$\Gamma_0^-(2)\backslash\mathbb{H} = \pi_2^-(\tilde{U}_\mathrm{i}) \cup \pi_2^-(T_+\tilde{U}_\mathrm{i}) \cup \pi_2^-(\tilde{U}_\rho) = \pi_2^-(\tilde{U}_\mathrm{i}) \cup \pi_2^-(\tilde{U}_{i+1}) \cup \pi_2^-(\tilde{U}_\rho)$$

and as before we put $U_{y_\nu}^- = \pi_2^-(\tilde{U}_{y_\nu})$. In this case $\tilde{U}_{i+1} = T_+\tilde{U}_i$ is fixed by $S_1^- = \begin{pmatrix} -1 & 2 \\ -1 & 1 \end{pmatrix} \in \Gamma_0^-(2)$ and we get a diagram for theČech complex

$$\begin{array}{ccc} \tilde{\mathcal{M}}(U_\mathrm{i}^-) \oplus \tilde{\mathcal{M}}(U_{\mathrm{i}+1}^-) \oplus \tilde{\mathcal{M}}(U_\rho^-) & \xrightarrow{d_0} & \tilde{\mathcal{M}}(U_\mathrm{i}^- \cap U_\rho^-) \oplus \tilde{\mathcal{M}}(U_{\mathrm{i}+1}^- \cap U_\rho^-) \\ ev_{\tilde{U}_\mathrm{i}} \oplus ev_{\tilde{U}_{\mathrm{i}+1}} \downarrow \oplus ev_{\tilde{U}_\rho} & & ev_{\tilde{U}_\mathrm{i}\cap\tilde{U}_\rho} \oplus ev_{\tilde{U}_\mathrm{i}\cap T_-^{-1}\tilde{U}_\rho} \downarrow \oplus ev_{\tilde{U}_{\mathrm{i}+1}\cap\tilde{U}_\rho} \\ \mathcal{M} \oplus \mathcal{M}^{\langle S_1^- \rangle} \oplus \mathcal{M} & \xrightarrow{\bar{d}_0} & \mathcal{M} \oplus \mathcal{M} \oplus \mathcal{M}. \end{array} \tag{3.43}$$

Again we can modify this complex and get

$$H^1(\Gamma_0^-(2)\backslash\mathbb{H}, \tilde{\mathcal{M}}) \xrightarrow{\sim} \mathcal{M}/(\mathrm{Id} - T_-^{-1})\mathcal{M}^{<S_1^->}. \tag{3.44}$$

We compute the arrows $(\varpi_2^-)^*, (\varpi_2^-)_*$ in degree one

$$\begin{aligned}(\varpi_2^-)^{(1)} &: m \mapsto (m, Sm, T_+ m), \\ (\varpi_2^-)_{(1)} &: (m_1, m_2, m_3) \mapsto (m_1 + Sm_2 + T_+^{-1} m_3).\end{aligned} \tag{3.45}$$

The computation of $\alpha_{2\bullet}$

We recall our isomorphism α between the spaces and the resulting isomorphism (3.30). The identity map of the module $\mathcal{M}$ and the isomorphism α on the space identifies the two complexes

$$\tilde{\mathcal{M}}(U_\mathrm{i}^+) \oplus \tilde{\mathcal{M}}(U_{\frac{\mathrm{i}+1}{2}}^+) \oplus \tilde{\mathcal{M}}(U_\rho^+) \xrightarrow{d_0} \tilde{\mathcal{M}}(U_\mathrm{i}^+ \cap U_\rho^+) \oplus \tilde{\mathcal{M}}(U_{\frac{\mathrm{i}+1}{2}}^+ \cap U_\rho^+) \text{ resp.}$$

$$\tilde{\mathcal{M}}^{(\alpha)}(\alpha(U_\mathrm{i}^+)) \oplus \tilde{\mathcal{M}}^{(\alpha)}(\alpha(U_{\frac{\mathrm{i}+1}{2}}^+)) \oplus \tilde{\mathcal{M}}^{(\alpha)}(\alpha(U_\rho^+)) \xrightarrow{d_0}$$

$$\tilde{\mathcal{M}}^{(\alpha)}(\alpha(U_\mathrm{i}^+ \cap U_\rho^+)) \oplus \tilde{\mathcal{M}}^{(\alpha)}(\alpha(U_{\frac{\mathrm{i}+1}{2}}^+ \cap U_\rho^+))$$

and if we consider their explicit realization then this identification is given by the equality of $\mathbb{Z}$-modules $\mathcal{M} = \mathcal{M}^{(\alpha)}$. This equality of complexes expresses the identification (3.30). We can compute the cohomology $H^1(\Gamma_0^-(2)\backslash\mathbb{H}, \tilde{\mathcal{M}}^{(\alpha)})$ from any of the two coverings

$$\Gamma_0^-(2)\backslash\mathbb{H} = \alpha(U_{\mathbf{i}}^+) \cup \alpha(U_{\frac{\mathbf{i}+1}{2}}^+) \cup \alpha(U_\rho^+) = U_{x_1} \cup U_{x_2} \cup U_{x_3}$$

and

$$\Gamma_0^-(2)\backslash\mathbb{H} = U_{\mathbf{i}}^- \cup U_{\mathbf{i}+1}^- \cup U_\rho^- = U_{x_4} \cup U_{x_5} \cup U_{x_6}.$$

We have to pick a class $\xi \in H^1(\Gamma_0^-(2)\backslash\mathbb{H}, \tilde{\mathcal{M}}^{(\alpha)})$ and represent it by a cocycle

$$c_\xi \in \bigoplus_{1\le i<j\le 3} \tilde{\mathcal{M}}^{(\alpha)}(U_{x_i} \cap U_{x_j})$$

(the cocycle condition is empty since $U_{x_1} \cap U_{x_2} \cap U_{x_3} = \emptyset$).

Then we have to produce a cocycle

$$c_\xi^{(\alpha)} \in \bigoplus_{4\le i<j\le 6} \tilde{\mathcal{M}}^{(\alpha)}(U_{x_i} \cap U_{x_j})$$

which represents the same class.

To get this cocycle we write down the three complexes

$$\begin{array}{ccc}
\bigoplus_{1\le i<j\le 3} \tilde{\mathcal{M}}^{(\alpha)}(U_{x_i} \cap U_{x_j}) & \to & 0 \\
\uparrow & & \uparrow \\
\bigoplus_{1\le i<j\le 6} \tilde{\mathcal{M}}^{(\alpha)}(U_{x_i} \cap U_{x_j}) & \to & \bigoplus_{1\le i<j<k\le 6} \tilde{\mathcal{M}}^{(\alpha)}(U_{x_i} \cap U_{x_j} \cap U_{x_k}) \\
\downarrow & & \downarrow \\
\bigoplus_{4\le i<j\le 6} \tilde{\mathcal{M}}^{(\alpha)}(U_{x_i} \cap U_{x_j}) & \to & 0
\end{array} \tag{3.46}$$

For our cocycle c_ξ we find a cocycle $c_\xi^\dagger$ in the complex in the middle which maps to the class of c_ξ under the upwards arrow and this cocycle is unique up to a coboundary. Then we project it down by the downwards arrow, i.e. we only take its $4 \le i < j \le 6$ components, and this is our cocycle $c_\xi^{(\alpha)}$.

We apply the principles from Section 2.1.3 and we write down these complexes explicitly. For any pair $\underline{i} = (i, j), i < j$ of indices we have to compute the set $\mathcal{F}_{\underline{i}}$. We drew some pictures and from these pictures we get (modulo errors) the following list (of lists):

$$\begin{array}{cccc}
\mathcal{F}_{1,2} = \emptyset & \mathcal{F}_{1,3} = \{\mathrm{Id}, T_+^{-2}\} & \mathcal{F}_{1,4} = \{\mathrm{Id}\} & \mathcal{F}_{1,5} = \{\mathrm{Id}, T_+^{-2}\} \\
\mathcal{F}_{1,6} := \{\mathrm{Id}, T_-^{-1}\} & \mathcal{F}_{2,3} = \{\mathrm{Id}\} & \mathcal{F}_{2,4} = \{\mathrm{Id}, T_-\} & \mathcal{F}_{2,5} = \{\mathrm{Id}\} \\
\mathcal{F}_{2,6} = \{\mathrm{Id}\} & \mathcal{F}_{3,4} = \{\mathrm{Id}, T_+^2\} & \mathcal{F}_{3,5} = \{\mathrm{Id}\} & \mathcal{F}_{3,6} = \{\mathrm{Id}, S_1^-\} \\
\mathcal{F}_{4,5} = \emptyset & \mathcal{F}_{4,6} = \{\mathrm{Id}, T_-^{-1}\} & \mathcal{F}_{5,6} = \{\mathrm{Id}\} &
\end{array}.$$

Now we have to follow the rules in the first section and we can write down an explicit version of the diagram (3.46). We refer to Section 2.1.3 and get

$$\begin{array}{ccc}
\bigoplus_{1\le i<j\le 3}\bigoplus_{\gamma\in\mathcal{F}_{i,j}}(\mathcal{M}^{(\alpha)})^{\Gamma_{i,j,\gamma}} & \to & 0\\
\uparrow & & \uparrow\\
\bigoplus_{1\le i<j\le 6}\bigoplus_{\gamma\in\mathcal{F}_{i,j}}(\mathcal{M}^{(\alpha)})^{\Gamma_{i,j,\gamma}} & \to & \bigoplus_{1\le i<j<k\le 6}\bigoplus_{\gamma\in\mathcal{F}_{i,j,k}}(\mathcal{M}^{(\alpha)})^{\Gamma_{i,j,k,\gamma}}\\
\downarrow & & \downarrow\\
\bigoplus_{4\le i<j\le 6}\bigoplus_{\gamma\in\mathcal{F}_{i,j}}(\mathcal{M}^{(\alpha)})^{\Gamma_{i,j,\gamma}} & \to & 0
\end{array} \tag{3.47}$$

Here we have to interpret this diagram. The module $\mathcal{M}^{(\alpha)}$ is equal to $\mathcal{M}$ as an abstract module, but an element $\gamma\in\Gamma_0^-(2)$ acts by the twisted action (see ChapII, 2.2)

$$m\mapsto \gamma *_\alpha m=\alpha^{-1}\gamma\alpha * m$$

here the $*$ denotes the original action. Hence we have to take the invariants $(\mathcal{M}^{(\alpha)})^{\Gamma_{i,j,\gamma}}$ with respect to this twisted action. In our special situation, this has very little effect, since almost all the $\Gamma_{i,j,\gamma}$ are trivial, except for the case of the intersection $\alpha(\tilde{U}_{\frac{\mathbf{i}+1}{2}})\cap\tilde{U}_{\mathbf{i}}$, in which case $\Gamma_{i,j,\gamma}=\langle S_1^-\rangle$. Hence

$$(\mathcal{M}^{(\alpha)})^{\langle S_1^-\rangle}=\mathcal{M}^{\langle S_1^+\rangle}.$$

Each of the complexes in (3.47) compute the cohomology group $H^1(\Gamma_0^-(2)\backslash\mathbb{H},\tilde{\mathcal{M}})$, and the diagram gives us a formula for the isomorphism in (3.30). To get $u_\alpha^\bullet$ in $\mathcal{M}^{(\alpha)}$ we apply the multiplication $m_2: m\mapsto \alpha m$ to the complex in the middle and the bottom. Then the cocycle c_ξ^α is now an element in $\bigoplus\tilde{\mathcal{M}}^{(\alpha)}$, and αc_ξ^α represents the cohomology class $u_\alpha^\bullet(\xi)\in H^1(\Gamma_0^-(2)\backslash\mathbb{H},\tilde{\mathcal{M}})$.

Now it is clear how we can compute the Hecke operator

$$T_2=T_{\begin{pmatrix}2&0\\0&1\end{pmatrix}}:\mathcal{M}/(\mathcal{M}^{\langle S\rangle}\oplus\mathcal{M}^{\langle R\rangle})\to\mathcal{M}/(\mathcal{M}^{\langle S\rangle}\oplus\mathcal{M}^{\langle R\rangle}).$$

We pick a representative $m\in\mathcal{M}$ of the cohomology class. We apply $(\varpi_2^+)^{(1)}$ in the diagram (3.39) to it and this gives the element $(Sm,m,T_-m)=c_\xi$. We apply the above process to compute $c_\xi^{(\alpha)}$. Then $\alpha c_\xi^{(\alpha)}=(m_1,m_2,m_3)$ is an element in $\tilde{\mathcal{M}}(U_{\mathbf{i}}^-\cap U_\rho^-)\oplus\tilde{\mathcal{M}}(U_{\mathbf{i}+1}^-\cap U_\rho^-)$, and this module is identified with $\mathcal{M}\oplus\mathcal{M}\oplus\mathcal{M}$ by the vertical arrow in (3.43). To this element we apply the trace

$$(\varpi_2^-)_{(1)}(m_1,m_2,m_3)=m_1+m_2+T_+^{-1}m_3,$$

and the latter element in $\mathcal{M}$ represents the class $T_2([m])$.

We have written a computer program which for a given $\mathcal{M}=\mathcal{M}_n$, i.e. for a given even positive integer n, computes the module $H^1(\Gamma\backslash\mathbb{H},\tilde{\mathcal{M}})$, and the endomorphism T_2 on it.

Looking at our data, we discovered the following (surprising?) fact: considering the isomorphism in equation (3.30), we have the explicit description of the cohomology in (3.34)

$$H^1(\Gamma_0^+(2)\backslash\mathbb{H},\tilde{\mathcal{M}}) \xrightarrow{\sim} \mathcal{M}/(\mathrm{Id}-T_+^{-1})\mathcal{M}^{\langle S_1^+\rangle},$$

and

$$H^1(\Gamma_0^-(2)\backslash\mathbb{H},\tilde{\mathcal{M}}^{(\alpha)}) \xrightarrow{\sim} \mathcal{M}^{(\alpha)}/(\mathrm{Id}-T_-^{-1})(\mathcal{M}^{(\alpha)})^{\langle S_1^-\rangle}.$$

We know that we may represent any cohomology class by a cocycle

$$c_\xi=(0,c_\xi,0)\in\begin{cases}\mathcal{M}^{(\alpha)}(\pi_2^-(\alpha(U_{\mathrm{i}})\cap\alpha(U_\rho)))\\ \oplus\\ \mathcal{M}^{(\alpha)}(\pi_2^-(\alpha(U_{\mathrm{i}})\cap\alpha(T_+^{-1}U_\rho)))\\ \oplus\\ \mathcal{M}^{(\alpha)}(\pi_2^-(\alpha(U_{\frac{\mathrm{i}+1}{2}})\cap\alpha(T_+^{-1}U_\rho)))\end{cases},$$

so it is non-zero only in the middle component, thus it is simply an element in $\mathcal{M}$. If we now look at our data, we see that $c_\xi^{(\alpha)}$ is also non-zero only in the middle, hence

$$c_\xi^{(\alpha)}=(0,c'_\xi,0)\in 0\oplus\mathcal{M}^{(\alpha)}(\pi_2^-(U_{\mathrm{i}}\cap T_-^{-1}U_\rho))\oplus 0$$

hence it is also in $\mathcal{M}^{(\alpha)}$, and we even get

$$c'_\xi=c_\xi.$$

Hence we see that the homomorphism in equation (3.30) is simply induced by

$$X^\nu Y^{n-\nu}\mapsto 2^\nu X^\nu Y^{n-\nu}.$$

We get an explicit formula for the Hecke operator T_2 : we pick an element $m\in\mathcal{M}$ representing the class $[m]$. We send it by $(\varpi_2^+)^{(1)}$ to $H^1(\Gamma_0^+(2)\backslash\mathbb{H},\tilde{\mathcal{M}})$, i.e.

$$(\varpi_2^+)^{(1)}: m\mapsto (m,Sm,T_-m) \tag{3.48}$$

We modify it so that the first and third entries become zero (see 3.35)

$$[(m,Sm,T_-m)]=[(0,Sm-m+T_+^{-1}T_-m-T_-m,0)]. \tag{3.49}$$

To the entry in the middle we apply $M_2=\begin{pmatrix}2&0\\0&1\end{pmatrix}$ and then apply $(\varpi_2^-)_{(1)}$

then

$$T_2([m])=[S\cdot M_2(Sm-m+T_+^{-1}T_-m-T_-m)] \tag{3.50}$$

Together with H. Gangl we wrote a code in Mathematica which provided explicit matrices for T_2 for small values $n\leq 150$. In the meantime we learned that there are more effective algorithms available which are using modular symbols. But I think that our algorithm is more systematic, and it might work in higher dimensions too. In any case it is very desirable to have an algorithm which works for $\Gamma=\mathrm{Sp}_2(\mathbb{Z})$. In this case the modular symbol method seems to fail.

3.3.4 The first interesting example

We give an explicit formula for the cohomology in the case of $\mathcal{M} = \mathcal{M}_{10}$. Here we use the algorithm we wrote with Herbert Gangl, for safety reasons we might invert the primes 2 and 3, but I think the program runs well if we do not do that.

We define the submodule

$$\mathcal{M}^{\mathrm{tr}} = \bigoplus_{\nu=0}^{5} \mathbb{Z} Y^{10-\nu} X^{\nu}$$

and we have the truncation operator $\mathrm{trunc} : \mathcal{M} \to \mathcal{M}^{\mathrm{tr}}$

$$\mathrm{trunc} : Y^{10-\nu} X^{\nu} \mapsto \begin{cases} Y^{10-\nu} X^{\nu} & \text{if } \nu \leq 5, \\ (-1)^{\nu+1} Y^{\nu} X^{10-\nu} & \text{else}, \end{cases},$$

it identifies the quotient module $\mathcal{M}/\mathcal{M}^{\langle S \rangle}$ with $\mathcal{M}^{\mathrm{tr}}$. To get the cohomology, we have to divide by the relations coming from $\mathcal{M}^{\langle R \rangle}$, i.e. we have to divide by the submodule $\mathrm{trunc}(\mathcal{M}^{\langle R \rangle})$. The module of these relations is generated by

$$\begin{gathered} \mathrm{R}_1 = 10Y^9 X + 20Y^7 X^3 + Y^5 X^5 \\ \mathrm{R}_2 = 9Y^8 X^2 - 36Y^7 X^3 + 14Y^6 X^4 - 45Y^5 X^5 \\ \mathrm{R}_3 = 8Y^7 X^3 + 10Y^5 X^5 \end{gathered}$$

and then

$$H^1(\Gamma \backslash \mathbb{H}, \tilde{\mathcal{M}}) = \bigoplus_{\nu=0}^{5} \mathbb{Z} Y^{10-\nu} X^{\nu} / \langle \mathrm{R}_1, \mathrm{R}_2, \mathrm{R}_3 \rangle.$$

We simplify the notation and set $e_\nu = Y^\nu X^{n-\nu}$. Using R_1 we can eliminate $e_5 = -10e_9 - 20e_7$ and then

$$H^1(\Gamma \backslash \mathbb{H}, \tilde{\mathcal{M}}) = \bigoplus_{\nu=10}^{\nu=6} \mathbb{Z} e_\nu / \langle -450e_9 + 9e_8 + 864e_7 + 14e_6, -100e_9 - 192e_7 \rangle, \tag{3.51}$$

and introduce a new basis $\{f_{10}, f_9, f_8, f_7, f_6, f_5\}$ of the $\mathbb{Z}$-module $\mathcal{M}^{\mathrm{tr}}$:

$$\begin{gathered} f_{10} = e_{10}; f_8 = -2e_8 - 3e_6; f_6 = 9e_8 + 14e_6 \\ f_9 = -12e_9 - 23e_7; f_7 = 25e_9 + 48e_7; f_5 = 10e_9 + 20e_7 + e_5 \end{gathered} \tag{3.52}$$

and hence in the quotient we get $\bar{f}_5 = 0$, $4\bar{f}_7 = 0$ and $18\bar{f}_7 = \bar{f}_6$, and therefore

$$H^1(\Gamma \backslash \mathbb{H}, \tilde{\mathcal{M}}) = \mathbb{Z}\bar{f}_{10} \oplus \mathbb{Z}\bar{f}_9 \oplus \mathbb{Z}\bar{f}_8 \oplus \mathbb{Z}/4\bar{f}_7 \tag{3.53}$$

We defined the action of complex conjugation (see (2.55)) on $H^1(\Gamma \backslash \mathbb{H}, \tilde{\mathcal{M}})$, and we leave it as an exercise to the reader to show that

$$\mathbf{c}^{(1)}(\bar{f}_{10}) = -\bar{f}_{10}, \mathbf{c}^{(1)}(\bar{f}_9) = -\bar{f}_9, \mathbf{c}^{(1)}(\bar{f}_8) = \bar{f}_8 \tag{3.54}$$

We can apply the above procedure to compute the action of T_2 on cohomology. It turns out to be reasonable to compute the matrix for T_2 with respect to the basis $\bar{f}_{10}, \bar{f}_8, \bar{f}_9, \bar{f}_7$, i.e. we reorder the basis elements. Our program with Gangl yields

$$T_2 = \begin{pmatrix} 2049 & 0 & -68040 & 0 \\ 0 & -24 & 0 & 0 \\ 0 & 0 & -24 & 0 \\ 0 & 0 & 0 & 2 \end{pmatrix}. \tag{3.55}$$

Hence we see that T_2 is non-trivial on the torsion subgroup. If we divide by the torsion, then the matrix reduces to a (3,3)-matrix and this matrix gives us the endomorphism on the "integral" cohomology (see 2.2.3), which is defined in general by

$$H^\bullet_{\text{int}}(\Gamma\backslash X, \tilde{\mathcal{M}}) = H^\bullet(\Gamma\backslash X, \tilde{\mathcal{M}})/\text{tors} \subset H^\bullet(\Gamma\backslash X, \tilde{\mathcal{M}}_{\mathbb{Q}}). \tag{3.56}$$

Here we should be careful: the functor $H^\bullet \to H^\bullet_{\text{int}}$ is not exact. In our case we get (perhaps up to a little piece of 2-torsion) exact sequences of Hecke modules

$$\begin{array}{ccccccc} 0 & \to & \mathbb{Z}\bar{f}_9 \oplus \mathbb{Z}\bar{f}_8 & \to & \mathbb{Z}\bar{f}_{10} \oplus \mathbb{Z}\bar{f}_9 \oplus \mathbb{Z}\bar{f}_8 & \xrightarrow{r} & \mathbb{Z}\bar{f}_{10} & \to & 0 \\ & & \| & & \| & & \| & & \\ 0 & \to & H^1_{\text{int},!}(\Gamma\backslash\mathbb{H}, \tilde{\mathcal{M}}) & \to & H^1_{\text{int}}(\Gamma\backslash\mathbb{H}, \tilde{\mathcal{M}}) & \xrightarrow{r} & H^1_{\text{int}}(\partial(\Gamma\backslash\mathbb{H}), \tilde{\mathcal{M}}) & \to & 0 \end{array} \tag{3.57}$$

where $T_2(\bar{f}_{10}) = (2^{11}+1)\bar{f}_{10}$. If we tensor by $\mathbb{Q}$ then we can find a unique element - *the Eisenstein class* - $f^\dagger_{10} \in H^1_{\text{int}}(\Gamma\backslash\mathbb{H}, \tilde{\mathcal{M}}) \otimes \mathbb{Q}$ which maps to $\bar{f}_{10}$ and which satisfies $T_2(f^\dagger_{10}) = (2^{11}+1)f^\dagger_{10}$. This class is not necessarily integral. In our case, an easy computation shows that $f^\dagger_{10} \notin H^1_{\text{int}}(\Gamma\backslash\mathbb{H}, \tilde{\mathcal{M}})$. But $691 f^\dagger_{10} \in H^1_{\text{int}}(\Gamma\backslash\mathbb{H}, \tilde{\mathcal{M}})$. and this means that 691 is the denominator of $f^\dagger_{10}$, i.e. 691 is the *denominator of the Eisenstein class.*

Hence we see that

$$H^1_{\text{int}}(\Gamma\backslash\mathbb{H}, \tilde{\mathcal{M}}) \supset H^1_{\text{int},!}(\Gamma\backslash\mathbb{H}, \tilde{\mathcal{M}}) \oplus \mathbb{Z}(691 f^\dagger_{10}), \tag{3.58}$$

the quotient of these modules is isomorphic to $\mathbb{Z}/691\mathbb{Z}$.

The exact sequence $\mathcal{X}_{10}$ in (3.57) is an exact sequence of modules for the Hecke algebra $\mathcal{H} \supset \mathbb{Z}[T_2]$, and hence it yields an element

$$[\mathcal{X}_{10}] \in \text{Ext}^1_{\mathcal{H}}(\mathbb{Z}\bar{f}_{10}, H^1_{\text{int},!}(\Gamma\backslash\mathbb{H}, \tilde{\mathcal{M}})), \tag{3.59}$$

and an easy calculation shows that this Ext^1 group is cyclic of order 691 and that it is generated by $\mathcal{X}_{10}$.

We look at the action of the full Hecke algebra $\mathcal{H}$ on these cohomology groups. It turns out that for any prime p the Hecke operator T_p acts by the eigenvalue $p^{11}+1$ on $\bar{f}_{10}$ (see Proposition 3.3.1). Simple argument using Poincaré duality and the self adjointness of the Hecke operators shows that

T_p acts by multiplication by a scalar $\tau(p)$ on the inner cohomology.

Then we can conclude

For all primes p we have

$$\tau(p) \equiv p^{11} + 1 \mod 691. \tag{3.60}$$

Here we see a first instance of the Manin-Drinfeld principle. We have the free module $H^1_{\mathrm{int}}(\Gamma\backslash\mathbb{H}, \tilde{\mathcal{M}})$ and the set of commuting opertors T_p. Any of these T_p has an eigenvector in $H^1_{\mathrm{int}}(\Gamma\backslash\mathbb{H}, \tilde{\mathcal{M}} \otimes \mathbb{Q})$ with eigenvalue $p^{n+1}+1$. We say that T_p satisfies Manin-Drinfeld if it occurs with multiplicity one, or in other words it does not occur as an eigenvalue in $H^1_{\mathrm{int},!}(\Gamma\backslash\mathbb{H}, \tilde{\mathcal{M}} \otimes \mathbb{Q})$, or still in another words the endomorphism

$$T_p - (p^{n+1}+1)\mathrm{Id} : H^1_{\mathrm{int},!}(\Gamma\backslash\mathbb{H}, \tilde{\mathcal{M}}) \to H^1_{\mathrm{int},!}(\Gamma\backslash\mathbb{H}, \tilde{\mathcal{M}})$$

is injective. Hence it is clear that we get a saturated decomposition (see 3.58) which is determined by T_p and which is respected by any other Hecke operator T_q. This means also that any other Hecke operator t_q which satisfies the Manin-Drinfeld principle yields the same saturated decomposition.

This has the important consequence:

A single Hecke operator T_p which satisfies Manin-Drinfeld determines the denominator of the Eisenstein class.

Interlude: Ramanujan's $\Delta(z)$

We want to stress that the previous considerations are of purely algebraic and combinatorial nature, no analysis is involved. In the next chapter, we will use analytic methods - especially the results from the theory modular forms - to obtain some further insight into the structure of the cohomology groups. In our special case here it comes down to the following.

In his paper [92], Ramanujan introduced the function

$$\Delta(z) = e^{2\pi i z} \prod_{n=1}^{\infty} (1 - e^{2\pi i n z})^{24}$$

which is a function on the upper half plane $\mathbb{H} = \{z | \Im(z) > 0\}$ and it satisfies

$$\Delta(\frac{az+b}{cz+d}) = (cz+d)^{12}\Delta(z)$$

for all $\begin{pmatrix} a & b \\ c & d \end{pmatrix} \in \mathrm{Sl}_2(\mathbb{Z})$. This means that it is a modular form of weight 12. Since it goes to zero when $z = iy \to \infty$ it is even a modular cusp form. It is the unique (up to a non-zero scalar) cusp form of weight 12 for $\mathrm{Sl}_2(\mathbb{Z})$ (see [102]).

We can expand

$$\Delta(z) = e^{2\pi i z} - 24 e^{4\pi i z} + 252 e^{6\pi i z} + \cdots + a_n e^{2n\pi i z} + \ldots$$

The coefficients satisfy (conjectured by Ramanujan, proved by Mordell [85]) the following recursions

$$\begin{aligned} a_{n_1 n_2} &= a_{n_1} a_{n_2} \text{ if } n_1, n_2 \text{ are coprime;} \\ a_{p^r} &= a_p a_{p^{r-1}} + p^{11} a_{p^{r-2}} \text{ if } p \text{ is a prime and } r \geq 2. \end{aligned} \tag{3.61}$$

Hence we know the values a_n if we know the values a_p for primes p. On the other hand, it follows from the Eichler-Shimura isomorphism (see 4.2) that $a_p = \tau(p)$. Ramanujan defined his function $n \to \tau(n)$ as the expansion coefficients of $\Delta(z)$, and he proved the congruence 3.60 with his definition of $n \to \tau(n)$.

Ramanujan also made the famous conjecture saying that for all primes p, we have the inequality

$$\tau(p) \leq 2\, p^{\frac{11}{2}}.$$

This inequality implies of course that for all primes p (and especially for $p = 2$) $\tau(p) \neq p^{11} + 1$, and this implies that any Hecke operator T_p provides a canonical splitting into eigenspaces $H^1(\Gamma\backslash\mathbb{H}, \tilde{\mathcal{M}} \otimes \mathbb{Q}) = H^1_!(\Gamma\backslash\mathbb{H}, \tilde{\mathcal{M}} \otimes \mathbb{Q}) \oplus \mathbb{Q} f^{\dagger}_{10}$. This is the simplest instance where the Manin-Drinfeld principle works.

Other congruences

We still assume $n = 10$. It is easy to check that $H^1(\Gamma\backslash\mathbb{H}, \tilde{\mathcal{M}})$ and $H^2_c(\Gamma\backslash\mathbb{H}, \tilde{\mathcal{M}})$ do not have 5 or 7 torsion. Therefore we have (see Proposition 3.3.1)

$$\mathbb{Z}/10\mathbb{Z}\epsilon_9(X,Y) \oplus \mathbb{Z}/5\mathbb{Z}\epsilon_4(X,Y) \oplus \mathbb{Z}/7\mathbb{Z}\epsilon_6(X,Y) \subset r(r^{-1}(\mathcal{T}_2))/\mathcal{T}_1$$

and this implies the well-known congruences

$$\tau(p) \equiv p^{10} + p \equiv p^6 + p^5 \pmod 5; \; \tau(p) \equiv p^7 + p^5 \pmod 7 \tag{3.62}$$

(c.f. [109], [34]). These congruences are called congruences of *local origin*, whereas the congruence mod 691 is a congruence of *global origin*.

End of interlude

Reduction mod 691

Of course our program also runs if we reduce mod 691, and actually it runs much faster. Since there is at most 2-torsion, we get an exact sequence of Hecke-modules

$$\begin{aligned} 0 \to H^1_{\text{int},!}(\Gamma\backslash\mathbb{H}, \tilde{\mathcal{M}} \otimes \mathbb{F}_{691}) &\to H^1_{\text{int}}(\Gamma\backslash\mathbb{H}, \tilde{\mathcal{M}} \otimes \mathbb{F}_{691}) \\ \xrightarrow{r} H^1_{\text{int}}(\partial(\Gamma\backslash\mathbb{H}), \tilde{\mathcal{M}} \otimes \mathbb{F}_{691}) &\to 0. \end{aligned} \tag{3.63}$$

The matrix giving the Hecke operator mod 691 becomes

$$T_2 = \begin{pmatrix} 667 & 0 & 369 \\ 0 & 667 & 0 \\ 0 & 0 & 667 \end{pmatrix}. \tag{3.64}$$

Now we see that our computation mod 691 yields the extension class $[\mathcal{X}_{10} \otimes \mathbb{F}_{691}]$, as an element of order 691, or in other words, the sequence (3.63) does not split under the action of T_2. Therefore we get from the computation mod 691 that 691 divides the order of $[\mathcal{X}_{10}]$, and hence divides the order of the denominator of the Eisenstein class.

Of course we may also consider the other Hecke operators T_p acting on $H^1_{\text{int}}(\Gamma\backslash\mathbb{H}, \tilde{\mathcal{M}})$, then the corresponding matrix would be

$$T_p = \begin{pmatrix} p^{11}+1 & 0 & t^{(p)} \\ 0 & \tau(p) & 0 \\ 0 & 0 & \tau(p) \end{pmatrix}. \tag{3.65}$$

But there is no reason that for the other Hecke operators T_p the sequence (3.63) is always non-split. Here we have a little proposition:

Proposition 3.3.2. *If the sequence (3.63) splits for the action of T_p then*

$$p^{11} + 1 - \tau(p) \equiv 0 \mod 691^2.$$

Proof. The sequence mod 691 splits for T_p if and only if $t^{(p)} \equiv 0 \mod 691$. But we have seen that the equation $x(p^{11} + 1 - \tau(p)) = t^{(p)}$ has no solution in the local ring $\mathbb{Z}_{(691)}$, and this implies the above congruence. □

For the curious reader, we mention that this happens for $p = 3559$, and for the first ten thousand primes this happens 13 times. The number 13 is roughly equal to 10000/691. If we believe the the value of $t^{(p)} \mod 691$ is randomly distributed this is what should be expected.

At the end of Chapter 5 we presume the result of Deligne which says that we have an action of the Galois group on $H^1_{\text{int}}(\Gamma\backslash\mathbb{H}, \tilde{\mathcal{M}} \otimes \mathbb{F}_{691})$. We will see that the structure of this cohomology as a module for the Hecke algebra has interesting consequences for this action (see Theorem 5.2.3).

3.3.5 The general case

Now we describe the general case $\mathcal{M} = \mathcal{M}_n$ where $n > 0$ is an even integer. We define $\mathcal{M}^{\text{tr}}$ as above, if $n/2$ is even, then we leave out the summand $X^{n/2}Y^{n/2}$, and we get

$$\mathcal{M}^{\text{tr}} = \mathcal{M}/\mathcal{M}^{\langle S \rangle}.$$

This gives us, for the cohomology and the restriction to the boundary cohomology

$$\begin{array}{ccc} H^1(\Gamma\backslash\mathbb{H}, \tilde{\mathcal{M}}) & \xrightarrow{\sim} & \mathcal{M}^{\text{tr}}/\text{Rel} \\ \downarrow & & \downarrow \\ H^1(\partial(\Gamma\backslash\mathbb{H}), \tilde{\mathcal{M}}) & \xrightarrow{\sim} & \mathcal{M}/(\text{Id} - T)\mathcal{M} \end{array}. \tag{3.66}$$

We have the basis

$$e_n = \text{trunc}(Y^n), e_{n-1} = \text{trunc}(Y^{n-1}X), \ldots, \begin{cases} Y^{n/2}X^{n/2} & n/2 \text{ odd} \\ 0 & \text{else} \end{cases}$$

for $\mathcal{M}^{\mathrm{tr}}$. Let us put $n_2 = n/2$ or $n/2+1$. Then the algorithm *Smithnormalform* provides a second basis $f_n = e_n, f_{n-1}, \dots, f_{n_2}$ such that the module of relations becomes

$$d_n f_n = 0, d_{n-1} f_{n-1} = 0, \dots, d_t f_t = 0, \dots, d_{n_2} f_{n_2} = 0$$

where $d_{n_2} | d_{n_2+1} | \cdots | d_n$. We have $d_n = d_{n-1} = \cdots = d_{n-2s} = 0$ where $2s+1 = \dim H^1(\Gamma\backslash\mathbb{H}, \tilde{\mathcal{M}}) \otimes \mathbb{Q}$ and $d_{n-2s-1} \neq 0$.

Now our computer program with H.Gangl, which for a given n, gives us an explicit matrix for T_2, of the form

$$T_2(f_i) = \sum_{j=n}^{n_2} t^{(2)}_{i,j} f_j \tag{3.67}$$

where we have (the enumeration of the rows and columns is downwards from n to n_2)

$$\begin{array}{c} t^{(2)}_{\nu,n} = 0 \text{ for } \nu < n \text{ and } t^{(2)}_{i,j} \in \mathrm{Hom}(\mathbb{Z}/(d_i), \mathbb{Z}/(d_j)) \\ \text{and } t^{(2)}_{i,j} = 0 \text{ for } i \geq n-2s, j < n-2s \end{array}.$$

If we divide by the torsion, we get for the restriction map to the boundary cohomology

$$H^1(\Gamma\backslash\mathbb{H}, \tilde{\mathcal{M}})_{\mathrm{int}} = \bigoplus_{\nu=n}^{n-2s} \mathbb{Z} f_\nu \xrightarrow{r} H^1(\partial(\Gamma\backslash\mathbb{H}), \tilde{\mathcal{M}})_{\mathrm{int}} = \mathbb{Z} Y^n, \tag{3.68}$$

where $f_n \mapsto Y^n$ and $T_2(Y^n) = (2^{n+1}+1)Y^n$. Now we will find that the endomorphism $T_2 - (2^{n+1}+1)\mathrm{Id}$ of $H^1_!(\Gamma\backslash\mathbb{H}, \tilde{\mathcal{M}})_{\mathrm{int}}$ is injective.

We mentioned the exercise (see (3.54)) to determine the action of $\mathbf{c}^{(1)}$ - in the general case we find $\mathbf{c}^{(1)}(f_n) = -f_n$ and for $\nu < n$ we find $\mathbf{c}^{(1)}(f_\nu) = (-1)^\nu (f_\nu)$. This has the consequence that $t^{(2)}_{n,j} = 0$ for $j \equiv 0 \mod 2$. It turns out that it may be wise to reorder the basis and take as a basis the list $\{f_n, f_{n-2}, f_{n-1}, f_{n-4}, \dots\}$. With respect to this basis the matrix for T_2 looks like this

$$\begin{pmatrix} 2^{n+1}+1 & 0 & t^{(2)}_{n,n-2} & 0 & t^{(2)}_{n,n-4} & \dots \\ 0 & t^{(2)}_{n-1,n-1} & 0 & t^{(2)}_{n-1,n-3} & \dots & \\ 0 & 0 & t^{(2)}_{n-2,n-2} & 0 & t^{(2)}_{n-2,n-4} & \dots \\ & & \vdots & & & \end{pmatrix} \tag{3.69}$$

Comment: We can verify this of course for any given n experimentally. But this assertion follows from the *Manin-Drinfeld principle.* This principle exploits the fact that we have estimates for the eigenvalues of T_2 (or more generally for T_p on $H^1_!(\Gamma\backslash\mathbb{H}, \tilde{\mathcal{M}} \otimes \mathbb{C})$). These estimates say that for all primes p we always have the Ramanujan inequality

$$|\pi_f(T_p)| \leq 2p^{\frac{n+1}{2}}$$

(this is a very deep theorem which has been proven by Deligne).

This implies that $2^{n+1}+1$ can not be an eigenvalue of T_2 on $H^1_!(\Gamma\backslash\mathbb{H},\tilde{\mathcal{M}}\otimes\mathbb{C})$ and this proves the injectivity. Hence we can find a vector

$$\mathrm{Eis}_n = f_n + \sum_{\nu=n-1}^{\nu=n-2s} x_\nu f_\nu,\ x_\nu \in \mathbb{Q} \tag{3.70}$$

which is an eigenvector for T_2, i.e.

$$T_2(\mathrm{Eis}_n) = (2^{n+1}+1)\mathrm{Eis}_n. \tag{3.71}$$

The least common multiple $\Delta(n)$ of the denominators of the x_ν is the denominator of the Eisenstein class, the smallest positive integer for which

$$\Delta(n)\mathrm{Eis}_n \in H^1(\Gamma\backslash\mathbb{H},\tilde{\mathcal{M}})_{\mathrm{int}}. \tag{3.72}$$

This denominator is of great interest, and our computer program allows us to compute it for any given not too large n. We simply have to compute the x_ν. We know that $T_2(f_n) = (2^{n+1}+1)f_n + \sum_{\mu=n-1}^{\mu=n-2s} t^{(2)}_{n,\mu} f_\mu$, and then the x_ν are the unique solution of

$$\sum_{\nu=n-1}^{\nu=n-2s} ((2^{n+1}+1)\delta_{\nu,\mu} - t^{(2)}_{\nu,\mu})x_\nu = 0;\quad \mu = n-1,\dots,n-2s. \tag{3.73}$$

We computed the x_ν, for some small values of n (roughly $n \leq 150$) and we found that for these values

$$\Delta(n) = \mathrm{numerator}(\zeta(-1-n)). \tag{3.74}$$

Here of course $\zeta(s)$ is the Riemann ζ-function, it is well-known that for any even positive integer n the value $\zeta(-1-n)$ is a rational number, hence it makes sense to speak of the numerator. A prime number p is called *irregular* if it divides the numerator of such a value of the Riemann ζ-function. The most famous irregular prime is $p = 691$ we have

$$\zeta(-11) = \frac{691}{32760}.$$

Actually (3.74) is a theorem, we will give a proof in Chapter 5 (Theorem 5.1.2).

The reader might argue, why do you make such efforts to find out some experimental evidence for something you know to be true?

There are several reasons for doing this, but the main motivation is the following. Our Theorem 5.1.2 is hopefully a special case of a much more general ensemble of assertions. The problem to determine (estimate) denominators of Eisenstein classes is ubiquitous in the cohomology of arithmetic groups. And we have many cases where we have conjectures relating these denominators to special values of L-functions. In our case above, these are the values of the numerator $\zeta(-1-n)$. Some further examples are discussed in [51]. But in many of these cases, the methods to prove theorems like Theorem 5.1.2 fail.

On the other hand we explained in Section 3.2.1 that in any given case we can compute the denominator - in principle. Therefore it is of interest to develop algorithms which compute the cohomology and the action of Hecke operators explicitly in given cases and verify or falsify these conjectures. A general strategy for such an algorithm has been outlined in Section 3.2.1, H. Gangl and I wrote a toy model program in the above case. See also [54].

We are aware that these algorithms may become very slow for more general reductive groups, and it is very likely that we need clever new ideas to achieve these tasks. Finally I want to say that in many cases, the resulting congruences have been checked for certain finite sets of primes

3.3.6 Localisation at a prime ℓ

We will see later that we should not consider the denominator of the Eisenstein class as a number, but rather as an ideal. Hence we are only interested in the decomposition into prime ideals, i.e. for a prime ℓ we want to know the exact power of ℓ which divides $\Delta(n)$. To achieve this we replace in the considerations above the coefficient system $\tilde{\mathcal{M}}$ by $\tilde{\mathcal{M}}_{(\ell)} := \tilde{\mathcal{M}} \otimes \mathbb{Z}_{(\ell)}$, here $\mathbb{Z}_{(\ell)} \subset \mathbb{Q}$ is the local ring at ℓ. Then our cohomology modules will be finitely generated $\mathbb{Z}_{(\ell)}$-modules $H^\bullet(\Gamma\backslash\mathbb{H}, \tilde{\mathcal{M}}_{(\ell)})$.

ℓ-ordinary endomorphisms

In this subsection we fix a prime ℓ, and we consider finitely generated modules M over the local ring $\mathbb{Z}_{(\ell)} \subset \mathbb{Q}$. We consider such a module together with an endomorphism $\Phi : M \to M$.

We say that Φ is *ordinary* if $\phi(M) = M$.

If M is any finitely generated $\mathbb{Z}_{(\ell)}$ torsion module then it is clear that $M = M_{\text{ord}} \oplus M_{\text{nilpt}}$ where M_{ord} is ordinary and M_{nilpt} is the maximal submodule for which $\bigcap^k(\Phi^k M_{\text{nilpt}}) = 0$. If M is torsion free then it is clear that M is ordinary if the constant coefficient of the characteristic polynomial $\det(\Phi - T\text{Id})$ is not divisible by ℓ. Hence we see that in this case M contains a unique maximal Φ submodule $M_{\text{ord}} = M_{\text{ord}}(\Phi)$ which is ordinary. Hence it is clear that we can always define the functor $M \to M_{\text{ord}}$ and this is an exact functor.

This has some nice consequences for our considerations above. Since the functor $X \to X_{\text{ int}}$ is not exact, the surjectivity in (3.68) is problematic, because $H^2_c(\Gamma\backslash\mathbb{H}, \tilde{\mathcal{M}}) \neq 0$. But if we localise our fundamental exact sequence

$$H^1_c(\Gamma\backslash\mathbb{H}, \tilde{\mathcal{M}}_{(\ell)}) \to H^1(\Gamma\backslash\mathbb{H}, \tilde{\mathcal{M}}_{(\ell)}) \to H^1(\partial(\Gamma\backslash\mathbb{H}), \tilde{\mathcal{M}}_{(\ell)}) \to H^2_c(\Gamma\backslash\mathbb{H}, \tilde{\mathcal{M}}_{(\ell)}),$$

and choose for Φ the Hecke operator T_ℓ, then it follows from our computations in Section 3.3.1 that T_ℓ acts nilpotently on $H^2_c(\Gamma\backslash\mathbb{H}, \tilde{\mathcal{M}}_{(\ell)})$. Therefore $H^2_{c,\text{ord}}(\Gamma\backslash\mathbb{H}, \tilde{\mathcal{M}}_{(\ell)}) = 0$. We get the exact sequence

$$H^1_{c,\text{ord}}(\Gamma\backslash\mathbb{H}, \tilde{\mathcal{M}}_{(\ell)}) \to H^1_{\text{ord}}(\Gamma\backslash\mathbb{H}, \tilde{\mathcal{M}}_{(\ell)}) \to H^1_{\text{ord}}(\partial(\Gamma\backslash\mathbb{H}), \tilde{\mathcal{M}}_{(\ell)}) \to 0.$$

It follows from our earlier computations (Prop. 3.3.1) that $H^1_{\mathrm{ord}}(\partial(\Gamma\backslash\mathbb{H}),\tilde{\mathcal{M}}_{(\ell)}) = \mathbb{Z}_{(\ell)}[Y^n]$. Then we get for all Hecke operators $T_p[Y^n] = (p^{n+1}+1)[Y^n]$, we denote this Hecke-module by $\mathbb{Z}_{(\ell)}[n]$.

Now we can replace the sequence (3.68) by the above sequence if we want to study the power $\ell^{\delta_\ell(n)} = \Delta_\ell(n)$ in $\Delta(n)$.

The $\mathrm{Gl}_2/\mathbb{Z}$-module $\mathcal{M}_n$ contains the submodule

$$\mathcal{M}_n^\flat = \left\{\sum a_\nu \binom{n}{\nu} X^\nu Y^{n-\nu} \mid a_\nu \in \mathbb{Z}\right\}$$

(see 4.1.1). This is actually the smallest submodule of $\mathcal{M}_n$ which contains X^n. Then we consider the cohomology $H^\bullet(\Gamma\backslash\mathbb{H},\tilde{\mathcal{M}}_n^\flat)$ and again we can ask for the denominator of the Eisenstein class. Here the method of localising at ℓ provides a simple answer. We consider the exact sequence of coefficients

$$0 \to \tilde{\mathcal{M}}_n^\flat \otimes \mathbb{Z}_{(\ell)} \to \tilde{\mathcal{M}}_n \otimes \mathbb{Z}_{(\ell)} \to \tilde{\mathcal{M}}_n/\tilde{\mathcal{M}}_n^\flat \otimes \mathbb{Z}_{(\ell)} \to 0.$$

Now it follows easily from the definition, that the Hecke operator T_ℓ acts nilpotently on the cohomology modules $H^\bullet(\Gamma\backslash\mathbb{H},\tilde{\mathcal{M}}_n/\tilde{\mathcal{M}}_n^\flat \otimes \mathbb{Z}_{(\ell)})$. Hence we see that

$$H^1_{?,\mathrm{ord}}(\Gamma\backslash\mathbb{H},\tilde{\mathcal{M}}_n^\flat \otimes \mathbb{Z}_{(\ell)}) \xrightarrow{\sim} H^1_{?,\mathrm{ord}}(\Gamma\backslash\mathbb{H},\tilde{\mathcal{M}}_n \otimes \mathbb{Z}_{(\ell)})$$

is an isomorphism. This implies that the denominator of the Eisenstein class does not depend on the choice of the coefficient system.

At this point it seems to be appropriate to use some homological algebra. We consider the exact sequence, say $\mathcal{X}_n$, of modules for the Hecke algebra $\mathcal{H}$

$$0 \to H^1_{\mathrm{ord},!}(\Gamma\backslash\mathbb{H},\tilde{\mathcal{M}}_n \otimes \mathbb{Z}_{(\ell)}) \to H^1_{\mathrm{ord}}(\Gamma\backslash\mathbb{H},\tilde{\mathcal{M}}_n \otimes \mathbb{Z}_{(\ell)}) \to \mathbb{Z}_{(\ell)}[n] \quad \to 0,$$

where $\mathbb{Z}[n]$ is the Hecke-module $\mathbb{Z}$ with the action $T_p x = (p^{n+1}+1)x$ for all primes p.

We consider the sequence $\mathrm{Hom}_{\mathcal{H}}(\mathbb{Z}_{(\ell)}[n],\mathcal{X}_n)$, which is not exact anymore. This sequence yields a long exact sequence, and we are interested in the boundary map

$$\begin{aligned}&\to \mathrm{Hom}_{\mathcal{H}}(\mathbb{Z}_{(\ell)}[n], H^1_{\mathrm{ord}}(\Gamma\backslash\mathbb{H},\tilde{\mathcal{M}}_n \otimes \mathbb{Z}_{(\ell)})) \to \mathrm{Hom}_{\mathcal{H}}(\mathbb{Z}_{(\ell)}[n],\mathbb{Z}_{(\ell)}[n])\\ &\xrightarrow{\delta} \mathrm{Ext}^1_{\mathcal{H}}(\mathbb{Z}_{(\ell)}[n], H^1_{\mathrm{ord},!}(\Gamma\backslash\mathbb{H},\tilde{\mathcal{M}}_n \otimes \mathbb{Z}_{(\ell)})) \to\end{aligned} \tag{3.75}$$

The boundary map δ maps the identity element $\mathbf{1} \in \mathrm{Hom}_{\mathcal{H}}(\mathbb{Z}_{(\ell)}[n],\mathbb{Z}_{(\ell)}[n])$ to an element of order $\Delta_\ell(n)$, in other words δ maps $\mathrm{Hom}_{\mathcal{H}}(\mathbb{Z}_{(\ell)}[n],\mathbb{Z}_{(\ell)}[n])$ to a cyclic subgroup of Ext^1 of order $\Delta_\ell(n)$.

We introduce the Eisenstein ideal $\mathcal{IE} \subset \mathcal{H}$, the ideal which is generated by the elements $(p^{n+1}+1)\mathrm{Id}-T_p$, where p runs through all primes. It is not difficult to see that

$$\mathrm{Ext}^1_{\mathcal{H}}(\mathbb{Z}_{(\ell)}[n], H^1_{\mathrm{ord},!}(\Gamma\backslash\mathbb{H},\tilde{\mathcal{M}}_n \otimes \mathbb{Z}_{(\ell)})) =$$

$$H^1_{\mathrm{ord},!}(\Gamma\backslash\mathbb{H},\tilde{\mathcal{M}}_n \otimes \mathbb{Z}_{(\ell)})/\mathcal{IE}H^1_{\mathrm{ord},!}(\Gamma\backslash\mathbb{H},\tilde{\mathcal{M}}_n \otimes \mathbb{Z}_{(\ell)}).$$

Now we choose a prime p and look at the subalgebra $\mathbb{Z}[T_p] \subset \mathcal{H}$ which is generated by the Hecke operator T_p. We consider the exact sequence (3.75) but we change the subscript $\mathcal{H}$ to $\mathbb{Z}[T_p]$. As before the map

$$\mathrm{Hom}_{\mathbb{Z}[T_p]}(\mathbb{Z}_{(\ell)}[n], \mathbb{Z}_{(\ell)}[n]) \xrightarrow{\delta} \mathrm{Ext}^1_{\mathbb{Z}[T_p]}(\mathbb{Z}_{(\ell)}[n], H^1_{\mathrm{ord},!}(\Gamma\backslash\mathbb{H}, \tilde{\mathcal{M}}_n \otimes \mathbb{Z}_{(\ell)})) \quad (3.76)$$

maps the identity element $\mathbf{1}$ to an element of order $\Delta_\ell(n)$. But now it follows from the definition of δ that

$$\delta(\mathbf{1}) = \mathbf{t}_n^{(p)} = \{\dots, t_{n,\nu}^{(p)}, \dots\}_\nu \mod ((p^{n+1}+1) - T_p)H^1_{\mathrm{ord},!}(\Gamma\backslash\mathbb{H}, \tilde{\mathcal{M}}_n \otimes \mathbb{Z}_{(\ell)})).$$

Hence we see that we simply have to determine the order of the element $\mathbf{t}_n^{(p)}$ in the group $\mathrm{Ext}^1_{\mathbb{Z}[T_p]}(\mathbb{Z}_{(\ell)}[n], H^1_{\mathrm{ord},!}(\Gamma\backslash\mathbb{H}, \tilde{\mathcal{M}}_n \otimes \mathbb{Z}_{(\ell)}))$ for just one prime p.

3.3.7 Computing mod ℓ

Of course the coefficients $t_{\nu,\mu}^{(p)}$ will become very large if n or p becomes larger, hence we can verify (3.74) only in a very small range of degrees n. But if we only want to verify that $\ell|\ \Delta_\ell(n)$, then it is sufficient to compute the coefficients $t_{\nu,\mu}^{(p)}$ modulo ℓ and to check whether $\mathbf{t}_n^{(p)}$ represents a non-zero class in $\mathrm{Ext}^1_{\mathbb{Z}[T_p]}(\mathbb{F}_\ell[n], H^1_{\mathrm{ord},!}(\Gamma\backslash\mathbb{H}, \tilde{\mathcal{M}}_n \otimes \mathbb{F}_\ell))$. Hence we see: we have $\ell|\Delta_\ell(n)$ if for some choice of p the equation

$$\sum_{\nu=n-1}^{\nu=n-2s} ((p^{n+1}+1)\delta_{\nu,\mu} - t_{\nu,\mu}^{(p)})x_\nu \equiv t_{n,\mu}^{(p)} \mod \ell \quad (3.77)$$

has no solution. But now the coefficients are elements in $\mathbb{F}_\ell$, and this reduces the computational complexity considerably.

We have to be careful, it may and will happen that there is a solution $t_{n,\mu}^{(p)}$ mod ℓ (for some values of p), but still $\ell|\Delta_\ell(n)$.

Higher powers of ℓ

This reasoning can also be applied if we look at higher powers of ℓ dividing a numerator of a $\zeta(-1-n)$. Let us assume that $\ell^{\delta_\ell(n)}||\mathrm{numerator}(\zeta(-1-n))$. We have to show that $\ell^{\delta_\ell(n)}$ divides the lcm of the denominators of the x_ν in equation (3.73). If we assume that $\mathbf{t}_n^{(p)}$ is not zero modulo ℓ, then this follows if we show that the equation

$$\sum_{\nu=n-1}^{\nu=n-2s} ((p^{n+1}+1)\delta_{\nu,\mu} - t_{\nu,\mu}^{(p)})x_\nu \equiv \ell^{\delta_\ell(n)-1} t_{n,\mu}^{(p)} \mod \ell^{\delta_\ell(n)} \quad (3.78)$$

has no solution in $\mathbb{Z}/(\ell^{\delta_\ell(n)})$. Then the class

$$[\mathcal{X}_n \otimes \mathbb{Z}/\ell^{\delta_\ell(n)}\mathbb{Z}] \in \mathrm{Ext}^1_{\mathcal{H}}((\mathbb{Z}/\ell^{\delta_\ell(n)}\mathbb{Z})(-1-n), H^1_{\mathrm{int},!}(\Gamma\backslash\mathbb{H}, \tilde{\mathcal{M}} \otimes (\mathbb{Z}/\ell^{\delta_\ell(n)}\mathbb{Z})))$$

has exact order $\ell^{\delta_\ell(n)}$.

If ℓ is an irregular prime, then there is always an even positive integer n_0 such that $n_0 < \ell - 1$ such that $\ell|\zeta(-1-n_0)$. One does not know any pair (ℓ, n)

with $n_0 < \ell - 1$ such that we even have $\ell^2 | \zeta(-1-n_0)$. But if we drop the assumption $n < \ell - 1$ then we may find arbitrarily high powers of ℓ dividing $\zeta(-1-n)$ (see also Section 3.3.10). We have some examples

$$\zeta(-31) \equiv 0 \mod 37;\ \zeta(-283) \equiv 0 \mod 37^2;$$
$$\zeta(-37579) \equiv 0 \mod 37^3;\ \zeta(-1072543) \equiv 0 \mod 37^4;\dots$$

$$\zeta(-43) \equiv 0 \mod 59;\ \zeta(-913) \equiv 0 \mod 59^2$$

$$\zeta(-23) \equiv 0 \mod 103\ ;\zeta(-227) \equiv 0 \mod 103^2$$

We verified (3.74) in the cases $(37, 282), (59, 912), (103, 226)$ using our program with Gangl. The case $(59, 912)$ used roughly 18 hours, our algorithm becomes very slow if n becomes large.

3.3.8 The denominator and the congruences

For the following we assume that (3.74) is correct. We discuss the denominator of the Eisenstein class in this special case. In [Talk-Lille] this is discussed in a more abstract way, so here we treat basically the simplest example of 4.3 in [Talk-Lille]. Remember that in this section $\tilde{\mathcal{M}} = \tilde{\mathcal{M}}_n^\flat$, or $\tilde{\mathcal{M}} = \tilde{\mathcal{M}}_n$ for some even positive integer n.

The fundamental exact sequence provides the short exact sequence

$$0 \to H^1_{\mathrm{int},!}(\Gamma\backslash\mathbb{H}, \tilde{\mathcal{M}}) \to H^1_{\mathrm{int}}(\Gamma\backslash\mathbb{H}, \tilde{\mathcal{M}}) \xrightarrow{r} H^1_{\mathrm{int}}(\partial(\Gamma\backslash\mathbb{H}), \tilde{\mathcal{M}}) \to 0. \tag{3.79}$$

It is clear that the restriction map r is surjective because it is surjective if we localise at primes. We have $H^1_{\mathrm{int}}(\partial(\Gamma\backslash\mathbb{H}), \tilde{\mathcal{M}}) = \mathbb{Z}e_n$ and $T_2(e_n) = (2^{n+1}+1)e_n$. We get a saturated decomposition into Hecke modules

$$H^1_{\mathrm{int},!}(\Gamma\backslash\mathbb{H}, \tilde{\mathcal{M}}) \oplus \mathbb{Z}\tilde{e}_n \subset H^1_{\mathrm{int}}(\Gamma\backslash\mathbb{H}, \tilde{\mathcal{M}}), \tag{3.80}$$

where $T_2\tilde{e}_n = (2^{n+1}+1)\tilde{e}_n$, $r(\tilde{e}_n) = \Delta(n)e_n$, and

$$H^1_{\mathrm{int}}(\Gamma\backslash\mathbb{H}, \tilde{\mathcal{M}})/(H^1_{\mathrm{int},!}(\Gamma\backslash\mathbb{H}, \tilde{\mathcal{M}}) \oplus \mathbb{Z}\tilde{e}_n) = \mathbb{Z}/\Delta(n)\mathbb{Z}. \tag{3.81}$$

If $e_n^\dagger \in H^1_{\mathrm{int}}(\Gamma\backslash\mathbb{H}, \tilde{\mathcal{M}})$ maps to e_n then we can write

$$e_n^\dagger = \frac{y' + \tilde{e}_n}{\Delta(n)}. \tag{3.82}$$

The element y' is unique, up to multiples of $\Delta(n)$ in $H^1_{\mathrm{int},!}(\Gamma\backslash\mathbb{H}, \tilde{\mathcal{M}})$.

Hence

Theorem 3.3.1. *The Hecke module $H^1_{\mathrm{int},!}(\Gamma\backslash\mathbb{H}, \tilde{\mathcal{M}}) \otimes \mathbb{Z}/\Delta(n)\mathbb{Z}$ contains a cyclic submodule $\mathbb{Z}/\Delta(n)\mathbb{Z}(-1-n)$, on which for all primes p the Hecke operator T_p acts by the eigenvalue $p^{n+1}+1 \mod \Delta(n)$*

Proof. The submodule is simply the cyclic submodule generated by y'. □

We discuss some consequences of this theorem. We anticipate some results from the following chapter and from Chapter 5. These results can be formulated in terms of the concepts and the language we used up to here, but the proofs require tools from analysis.

We mentioned already the theorem that the cohomology $H^1_!(\Gamma\backslash\mathbb{H},\tilde{\mathcal{M}}\otimes\mathbb{Q})$ is semi-simple as a module for the Hecke-algebra (Thm. 3.1.1). This theorem implies that we can find a finite normal field extension $F/\mathbb{Q}$ such that we have an isotypical decomposition (see 3.16)

$$H^1_{\mathrm{int},!}(\Gamma\backslash\mathbb{H},\tilde{\mathcal{M}})\otimes F=\bigoplus_{\pi_f}H^1_{\mathrm{int},!}(\Gamma\backslash\mathbb{H},\tilde{\mathcal{M}}\otimes F)(\pi_f). \tag{3.83}$$

Here π_f runs over a finite set of homomorphisms $\pi_f:\mathcal{H}\to\mathcal{O}_F$. We also have the action of complex conjugation on the cohomology (see (2.55)). The complex conjugation commutes with the action of the Hecke algebra, and under this action each eigenspace decomposes into a $+$ and a $-$ eigenspace. In Chapter 4 we will prove the famous multiplicity one theorem which says that the spaces $H^1_{\mathrm{int},!}(\Gamma\backslash\mathbb{H},\tilde{\mathcal{M}}\otimes F)(\pi_f)_\pm$ are one dimensional. Let us denote the set of $\pi_f:\mathcal{H}\to\mathcal{O}_F$ which occur with positive multiplicity (then 2) in the above decomposition by $\mathrm{Coh}_!(n)$.

We know that

$$H^1(\Gamma\backslash\mathbb{H},\tilde{\mathcal{M}})\otimes F=H^1_!(\Gamma\backslash\mathbb{H},\tilde{\mathcal{M}})\otimes F\oplus Fe_n,$$

where $T_pe_n=(p^{n+1}+1)e_n$. Let $\pi_f^{\mathrm{Eis}}:\mathcal{H}\to\mathbb{Z}$ be the homomorphism $\pi_f^{\mathrm{Eis}}:T_p\to p^{n+1}+1$, then $\mathrm{Coh}(n)=\mathrm{Coh}_!(n)\cup\{\pi_f^{\mathrm{Eis}}\}$.

We make a list $\{\pi_{1,f},\dots,\pi_{r,f}\}$ of the elements in $\mathrm{Coh}_!(n)$. This decomposition induces a Jordan-Hölder filtration on the integral cohomology

$$\begin{aligned}(0)\subset\mathcal{JH}^{(1)}H^1_{\mathrm{int},!}(\Gamma\backslash\mathbb{H},\tilde{\mathcal{M}}_{\mathcal{O}_F})\subset\mathcal{JH}^{(2)}H^1_{\mathrm{int},!}(\Gamma\backslash\mathbb{H},\tilde{\mathcal{M}}_{\mathcal{O}_F})\\ \subset\cdots\subset\mathcal{JH}^{(r)}H^1_{\mathrm{int},!}(\Gamma\backslash\mathbb{H},\tilde{\mathcal{M}}_{\mathcal{O}_F}).\end{aligned} \tag{3.84}$$

Here the first step $\mathcal{JH}^{(1)}H^1_{\mathrm{int},!}(\Gamma\backslash\mathbb{H},\tilde{\mathcal{M}}_{\mathcal{O}_F})=H^1(\Gamma\backslash\mathbb{H},\tilde{\mathcal{M}}\otimes\mathcal{O}_F)(\pi_{1,f})$, where the subquotients are locally free $\mathcal{O}_F$ modules of rank 2. After tensoring with F, they become isomorphic to the corresponding $\pi_{j,f}$ eigenspace.

We choose a prime ℓ which divides $\Delta(n)$, let $\ell^{\delta_\ell(n)}||\Delta(n)$. Let $\mathfrak{l}$ be a prime in $\mathcal{O}_F$ which lies above ℓ. If e_ℓ is the ramification index, then we have

$$\{0\}\subset\mathcal{O}_F/\mathfrak{l}^{e_\ell\delta_\ell(n)}(-1-n)\subset H^1_{\mathrm{int},!}(\Gamma\backslash\mathbb{H},\tilde{\mathcal{M}}_{\mathcal{O}_F})\otimes\mathcal{O}_F/\mathfrak{l}^{e_\ell\delta_\ell(n)} \tag{3.85}$$

The above Jordan-Hölder filtration induces a Jordan-Hölder filtration on the cohomology $\mod \mathfrak{l}^{e_\ell\delta_\ell(n)}$, we have

$$\{0\}\subset\mathcal{JH}^{(1)}H^1_{\mathrm{int},!}(\Gamma\backslash\mathbb{H},\tilde{\mathcal{M}}_{\mathcal{O}_F})\otimes\mathcal{O}_F/\mathfrak{l}^{e_p\delta_\ell(n)}\subset\mathcal{JH}^{(2)}\cdots \tag{3.86}$$

where the successive subquotients $\overline{\mathcal{JH}^{(\nu)}}H^1_{\mathrm{int},!}(\Gamma\backslash\mathbb{H},\tilde{\mathcal{M}}_{\mathcal{O}_F})$ are free $\mathcal{O}_F/\mathfrak{l}^{e_p\delta_\ell(n)}$ modules of rank 2.

Theorem 3.3.2. *We can find $\pi_{f,1}, \pi_{f,2} \dots, \pi_{f,r}$ and numbers $f_1 > 0, f_2 > 0, \dots, f_r > 0$ in the above filtration such that $\sum f_i = e_\ell \delta_\ell(n)$, and we have the congruence*

$$\pi_{f,i}(T_p) \equiv p^{n+1} + 1 \mod \mathfrak{l}^{f_i} \tag{3.87}$$

for all primes p.

Proof. We look at the map from our cyclic submodule into the top Jordan-Hölder quotient

$$\mathcal{O}_F/\mathfrak{l}^{e_\ell \delta_\ell(n)}(-1-n) \to \overline{\mathcal{JH}^{(r)}} H^1_{\mathrm{int},!}(\Gamma\backslash\mathbb{H}, \tilde{\mathcal{M}}_{\mathcal{O}_F}).$$

This map has a kernel $\mathfrak{l}^{f_r}/\mathfrak{l}^{e_\ell \delta_\ell(n)}$ and the image in the Jordan-Hölder quotient is the cyclic submodule $\mathcal{O}_F/\mathfrak{l}^{f_r}$. The Hecke operator T_p acts on the Jordan-Hölder quotient by multiplication by $\pi_{f,r}(T_p)$, and on the cyclic submodule by multiplication by $p^{n+1}+1$. Hence we get $\pi_{f,r}(T_p) \equiv p^{n+1}+1 \mod \mathfrak{l}^{f_r}$. Now we get an embedding $\mathfrak{l}^{f_r}/\mathfrak{l}^{e_\ell \delta_\ell(n)} \hookrightarrow \mathcal{JH}^{(r-1)} H^1_{\mathrm{int},!}(\Gamma\backslash\mathbb{H}, \tilde{\mathcal{M}}_{\mathcal{O}_F}) \otimes \mathcal{O}_F/\mathfrak{l}^{e_\ell \delta_\ell(n)}$ and we apply the same reasoning to this embedding. This process stops if the embedded cyclic submodule becomes trivial. This proves the claim. □

But now we have to be aware that the Jordan-Hölder filtration does not split. If we define

$$H^1_{\mathrm{int},!}(\Gamma\backslash\mathbb{H}, \tilde{\mathcal{M}}_{\mathcal{O}_F})(\pi_f) = H^1_{\mathrm{int},!}(\Gamma\backslash\mathbb{H}, \tilde{\mathcal{M}}_{\mathcal{O}_F}) \cap H^1(\Gamma\backslash\mathbb{H}, \tilde{\mathcal{M}} \otimes F)(\pi_f)$$

then we get a saturated decomposition (decomposition up to isogeny)

$$H^1_{\mathrm{int},!}(\Gamma\backslash\mathbb{H}, \tilde{\mathcal{M}}_{\mathcal{O}_F}) \supset \bigoplus_{\pi_f \in \mathrm{Coh}_!(n)} H^1_{\mathrm{int},!}(\Gamma\backslash\mathbb{H}, \tilde{\mathcal{M}}_{\mathcal{O}_F})(\pi_f).$$

□

Here we encounter another interesting problem:

What can we say about the structure of the quotient if we divide the left hand side by the right hand side?

We can formulate some more or less plausible assertions which we can verify experimentally, but which are very difficult to prove. We definitely have to use methods which go far beyond the very elementary tools we used so far. For instance we verified experimentally in (3.74) for a certain range of values of n but our proof in Chapter 5 requires some analysis.

In the following we choose a prime p, the role of the two primes p and ℓ will be exchanged. In a first step we consider the cohomology mod p we are mainly interested in the ordinary part. We start from the exact sequence of Γ-modules

$$0 \to \mathcal{M}_n \xrightarrow{\times p} \mathcal{M}_n \to \mathcal{M}_n \otimes \mathbb{F}_p \to 0. \tag{3.88}$$

Here we want to assume that $p \neq 2,3$, then we get the resulting exact sequence of sheaves, and hence a long exact sequence of cohomology groups

$$\begin{aligned}0 \to (\mathcal{M}_n^\Gamma)_{\text{ord}} \xrightarrow{\times p} (\mathcal{M}_n^\Gamma)_{\text{ord}} \to (\mathcal{M}_n \otimes \mathbb{F}_p)_{\text{ord}} \to H^1_{\text{ord}}(\Gamma\backslash\mathbb{H}, \tilde{\mathcal{M}}_n) \\ \xrightarrow{\times p} H^1_{\text{ord}}(\Gamma\backslash\mathbb{H}, \tilde{\mathcal{M}}_n) \to H^1_{\text{ord}}(\Gamma\backslash\mathbb{H}, \tilde{\mathcal{M}}_n \otimes \mathbb{F}_p) \to 0,\end{aligned}$$

and we can break this sequence into pieces

$$0 \to (\mathcal{M}_n^\Gamma)_{\text{ord}} \xrightarrow{\times p} (\mathcal{M}_n^\Gamma)_{\text{ord}} \to (\mathcal{M}_n \otimes \mathbb{F}_p)^\Gamma_{\text{ord}} \to H^1_{\text{ord}}(\Gamma\backslash\mathbb{H}, \tilde{\mathcal{M}}_n)[p] \to 0,$$

and

$$\begin{aligned}0 \to H^1_{\text{ord}}(\Gamma\backslash\mathbb{H}, \tilde{\mathcal{M}}_n)[p] \to H^1_{\text{ord}}(\Gamma\backslash\mathbb{H}, \tilde{\mathcal{M}}_n) \\ \xrightarrow{\times p} H^1_{\text{ord}}(\Gamma\backslash\mathbb{H}, \mathcal{M}_n) \to H^1_{\text{ord}}(\Gamma\backslash\mathbb{H}, \tilde{\mathcal{M}}_n \otimes \mathbb{F}_p) \to 0\end{aligned}$$

where of course $\ldots [p]$ means kernel of the multiplication by p, and the far most 0 on the right is the vanishing of H^2.

We analyse these two sequences and get:

Theorem 3.3.3. *The cohomology $H^1_{\text{ord}}(\Gamma\backslash\mathbb{H}, \mathcal{M}_n)$ is p-torsion free unless we have $n > 0$ and $n \equiv 0 \mod p(p-1)$. The cohomology groups $H^1_{c,\text{ord}}(\Gamma\backslash\mathbb{H}, \mathcal{M}_n)$ are always torsion free and $H^2_{c,\text{ord}}(\Gamma\backslash\mathbb{H}, \mathcal{M}_n) = 0$ for $n > 0$.*

Proof. We consider the polynomial ring in two variables $\mathbb{F}_p[X,Y]$. On this ring we have the action of $\mathrm{Sl}_2(\mathbb{Z})$. It is an old theorem of L.E. Dickson that the ring of invariants is generated by the two polynomials

$$\begin{aligned}f_1 &= X^pY - XY^p \\ \text{and } f_2 &= \frac{X^{p^2-1} - Y^{p^2-1}}{X^{p-1} - Y^{p-1}} = X^{(p-1)p} + X^{(p-1)(p-1)}Y^{p-1} + \cdots\end{aligned}$$

Now every element in $(\mathcal{M}_n \otimes \mathbb{F}_p)^\Gamma_{\text{ord}}$ is a sum of monomials $f_1^a f_2^b$ where we have $a(p+1) + bp(p-1) = n$. We see that

$$u_{\begin{pmatrix} p & 0 \\ 0 & 1 \end{pmatrix}} = u_\alpha : \mathcal{M}_n^{(\alpha)} \to \mathcal{M}_n$$

multiplies f_1 with a multiple of p, and hence we see that all the monomials with $a > 0$ are multiplied by a multiple of p. This means that $(\mathcal{M}_n \otimes \mathbb{F}_p)^\Gamma_{\text{ord}} \neq 0$ if and only if $n = bp(p-1)$. If $n = 0$ the map $\mathcal{M}_n^\Gamma = \mathbb{Z}_p \to (\mathcal{M}_n \otimes \mathbb{F}_p)^\Gamma$ is surjective. If $n > 0$ we have $\mathcal{M}_n^\Gamma = 0$ and hence the theorem.

For the assertions concerning the compactly supported cohomology we have to recall that $H^2_c(\Gamma\backslash\mathbb{H}, \mathcal{M}_n) = (\mathcal{M}_n)_\Gamma = \mathcal{M}_n/I_\Gamma\mathcal{M}_n$ [book vol I, Section 2 and 4.8.5]. We check easily that $X^n, Y^n \in I_\Gamma\mathcal{M}_n$ and the assertion is clear. ☐

We now briefly discuss some interesting questions concerning the cohomology groups $H^1_{\text{ord}}(\Gamma\backslash\mathbb{H}, \mathcal{M}_n \otimes \mathbb{F}_p)$. We assume that we are not in the exceptional case of $n \equiv 0 \mod p(p-1)$, hence we get that

$$H^1_{\text{ord}}(\Gamma\backslash\mathbb{H}, \mathcal{M}_n \otimes \mathbb{F}_p) = H^1_{\text{ord}}(\Gamma\backslash\mathbb{H}, \mathcal{M}_n) \otimes \mathbb{F}_p.$$

We can find a finite extension $\mathbb{F}_{p^r}/\mathbb{F}_p$ such that

$$H^1_{\mathrm{ord}}(\Gamma\backslash\mathbb{H},\tilde{\mathcal{M}}_n\otimes\mathbb{F}_{p^r})=\bigoplus_{\bar\pi_f}H^1_{\mathrm{ord}}(\Gamma\backslash\mathbb{H},\tilde{\mathcal{M}}_n\otimes\mathbb{F}_{p^r})\{\bar\pi_f\},\tag{3.89}$$

where $\bar\pi_f:\mathcal{H}\to\mathbb{F}_{p^r}$ is a homomorphism and

$$H^1_{\mathrm{ord}}(\Gamma\backslash\mathbb{H},\tilde{\mathcal{M}}_n\otimes\mathbb{F}_{p^r})\{\bar\pi_f\}=\\ \{x\in H^1_{\mathrm{ord}}(\Gamma\backslash\mathbb{H},\tilde{\mathcal{M}}_n\otimes\mathbb{F}_{p^r})\mid(T_\ell-\bar\pi_f(T_\ell))^Nx=0,N\text{ large enough}\}\tag{3.90}$$

is a generalised eigenspace. Such a generalised eigenspace has a *socle*, which is the space

$$H^1_{\mathrm{ord}}(\Gamma\backslash\mathbb{H},\tilde{\mathcal{M}}_n\otimes\mathbb{F}_{p^r})(\bar\pi_f)=\{x\in H^1_{\mathrm{ord}}(\Gamma\backslash\mathbb{H},\tilde{\mathcal{M}}_n\otimes\mathbb{F}_{p^r})\mid(T_\ell-\bar\pi_f(T_\ell))x=0\}.$$

(Note the difference between ( ) and { }.) We know that the inner cohomology $H^1_!(\Gamma\backslash\mathbb{H},\tilde{\mathcal{M}}\otimes\mathbb{Q})$ is a semi-simple Hecke module, but we cannot expect anymore that the inner cohomology $H^1_{\mathrm{ord},!}(\Gamma\backslash\mathbb{H},\tilde{\mathcal{M}}_n\otimes\mathbb{F}_{p^r})$ is semi-simple. For the restriction of the decomposition (3.89) to the inner cohomology, semi-simplicity means that the generalised eigenspaces are always equal to their socle.

We choose a prime $\mathfrak{p}\subset\mathcal{O}_F$ above p and assume $\mathcal{O}_F/\mathfrak{p}=\mathbb{F}_{p^r}$. Let $F_\mathfrak{p}$ be the completion of F at $\mathfrak{p}$, let $\mathcal{O}_\mathfrak{p}$ be its ring of integers. We consider the reduction maps

$$\begin{array}{ccc}H^1_{!,\mathrm{ord}}(\Gamma\backslash\mathbb{H},\tilde{\mathcal{M}}_n\otimes\mathcal{O}_\mathfrak{p}) & \to H^1_{!,\mathrm{ord}}(\Gamma\backslash\mathbb{H},\tilde{\mathcal{M}}_n\otimes\mathbb{F}_{p^r})\\ \downarrow & \downarrow\\ H^1_{\mathrm{ord}}(\Gamma\backslash\mathbb{H},\tilde{\mathcal{M}}_n\otimes\mathcal{O}_\mathfrak{p}) & \to H^1_{\mathrm{ord}}(\Gamma\backslash\mathbb{H},\tilde{\mathcal{M}}_n\otimes\mathbb{F}_{p^r}).\end{array}\tag{3.91}$$

Under this map an eigenspace maps into the socle, i.e.

$$r_{\pi_f}:H^1_{\mathrm{ord}}(\Gamma\backslash\mathbb{H},\tilde{\mathcal{M}}_n\otimes\mathcal{O}_\mathfrak{p})(\pi_f)\to H^1_{\mathrm{ord}}(\Gamma\backslash\mathbb{H},\tilde{\mathcal{M}}_n\otimes\mathbb{F}_{p^r})(\bar\pi_f).$$

The image $r_{\pi_f}(H^1_{\mathrm{ord},!}(\Gamma\backslash\mathbb{H},\tilde{\mathcal{M}}_n\otimes\mathcal{O}_\mathfrak{p})(\pi_f))$ is an $\mathbb{F}_{p^r}$-vector space of dimension two.

We are interested in the fibres of the surjective map

$$R_\mathfrak{p}:\mathrm{Coh}_!(n)\to\mathrm{Cohmodp}_!(n)\ ;\ \pi_f\mapsto\bar\pi_f.$$

For a subset $\Sigma\subset R_\mathfrak{p}^{-1}\bar\pi_f$ we define

$$H^1_{!,\mathrm{ord}}(\Gamma\backslash\mathbb{H},\mathcal{M}_n\otimes\mathcal{O}_\mathfrak{p})\{\Sigma\}=\\ (\textstyle\bigoplus_{\pi_f\in\Sigma}H^1_{!,\mathrm{ord}}(\Gamma\backslash\mathbb{H},\tilde{\mathcal{M}}_n\otimes F)(\pi_f))\cap H^1_{!,\mathrm{ord}}(\Gamma\backslash\mathbb{H},\tilde{\mathcal{M}}_n\otimes\mathcal{O}_\mathfrak{p}).\tag{3.92}$$

Given Σ we put $\Sigma'=R_\mathfrak{p}^{-1}(\bar\pi_f)\backslash\Sigma$, and we say that Σ (or Σ') is *closed* if we get a direct sum decomposition

$$H^1_{\mathrm{ord}}(\Gamma\backslash\mathbb{H},\tilde{\mathcal{M}}_n\otimes\mathcal{O}_\mathfrak{p})\{\Sigma\cup\Sigma'\}=\\ H^1_{\mathrm{ord}}(\Gamma\backslash\mathbb{H},\tilde{\mathcal{M}}_n\otimes\mathcal{O}_\mathfrak{p})\{\Sigma\}\oplus H^1_{\mathrm{ord}}(\Gamma\backslash\mathbb{H},\tilde{\mathcal{M}}_n\otimes\mathcal{O}_\mathfrak{p})\{\Sigma'\}$$

and consequently we say that the fibre $R_{\mathfrak{p}}^{-1}(\bar{\pi}_f)$ (or simply $\bar{\pi}_f$) is *connected* if $\emptyset$ and the fibre itself are the only closed subsets. If $\pi_f, \pi_f' \in R_p^{-1}(\bar{\pi}_f)$ then we say that they are *inner congruent*. We have an easy proposition

Proposition 3.3.3. *If $R_{\mathfrak{p}}^{-1}(\bar{\pi}_f)$ is not connected, then the dimension of the socle $H_!^1(\Gamma\backslash\mathbb{H}, \tilde{\mathcal{M}}_n \otimes \mathbb{F}_{p^r})(\bar{\pi}_f)$ over $\mathbb{F}_{p^r}$ is ≥ 4*

Proof. This is rather clear. If we have a non-trivial direct sum decomposition as above then we also get a direct sum decomposition

$$H^1_{!,\mathrm{ord}}(\Gamma\backslash\mathbb{H}, \tilde{\mathcal{M}}_n)\{\Sigma \cup \Sigma'\} \otimes \mathbb{F}_{p^r} = H^1_{!,\mathrm{ord}}(\Gamma\backslash\mathbb{H}, \tilde{\mathcal{M}}_n)\{\Sigma\} \otimes \mathbb{F}_{p^r} \\ \oplus \\ H^1_{!,\mathrm{ord}}(\Gamma\backslash\mathbb{H}, \tilde{\mathcal{M}}_n)\{\Sigma'\} \otimes \mathbb{F}_{p^r}.$$

Now any $\pi_f \in \Sigma$ (resp. $\pi_f' \in \Sigma'$) provides a two dimensional $\mathbb{F}_{p^r}$ vector space $r_{\pi_f}(H^1_{!,\mathrm{ord}}(\Gamma\backslash\mathbb{H}, \tilde{\mathcal{M}}_n \otimes \mathcal{O}_F)(\pi_f))$ (resp. $r_{\pi_f'}(H^1_{!,\mathrm{ord}}(\Gamma\backslash\mathbb{H}, \tilde{\mathcal{M}}_n \otimes \mathcal{O}_F)(\pi_f'))$. These two vector spaces lie in the socle and in two different summands. □

We say that $\bar{\pi}_f$ occurs *with weak multiplicity one* if the dimension of the socle $\dim H_!^1(\Gamma\backslash\mathbb{H}, \tilde{\mathcal{M}}_n \otimes \mathbb{F}_{p^r})(\bar{\pi}_f) = 2$. This socle is the direct sum of the $\pm$ eigenspaces under complex conjugation. Our above theorem implies then that the fibre of $\bar{\pi}_f$ must be connected.

The (plain) multiplicity of $\bar{\pi}_f$ is just the number of elements in the fibre $R_{\mathfrak{p}}^{-1}(\bar{\pi}_f)$, this is the number $m(\bar{\pi}_f) = \frac{1}{2}\dim H_!(\Gamma\backslash\mathbb{H}, \tilde{\mathcal{M}} \otimes \mathbb{F}_{p^r})\{\bar{\pi}_f\}$. One might expect that a $\bar{\pi}_f$ occurs with weak multiplicity one happens "very frequently". But it seems to be very difficult to say something substantial in this direction. We will come back to this issue (see discussion of the Wieferich dilemma).

3.3.9 L-values, weak multiplicity one and connectedness

It is not very difficult to produce examples of $\bar{\pi}_f \in \mathrm{Coh}_!(n)$ where we have $\dim H^1_{\mathrm{ord},!}(\Gamma\backslash\mathbb{H}, \tilde{\mathcal{M}}_n \otimes \mathbb{F}_{p^r})\{\bar{\pi}_f\} > 2$. If we take $n = 22$, then we know that $H_!^1(\Gamma\backslash\mathbb{H}, \tilde{\mathcal{M}}_{22} \otimes \mathbb{Q})$ is of dimension four, and the $\pm$ eigenspaces are of dimension 2. We can compute T_2 and find for the smallest field that decomposes the cohomology (see (3.83)) $F = \mathbb{Q}(\sqrt{144169})$, this was of course known to Hecke. In this case $\mathrm{Coh}_!(22)$ consists of 2 elements, which are conjugate under the Galois group $\mathrm{Gal}(F/\mathbb{Q})$.

If π_f is one of the elements in $\mathrm{Coh}_!(22)$, then our program yields $\pi_f(T_2) = -12(-45 + \sqrt{144169})$ (of course this value can be looked up in any table for modular forms), hence we see that for any prime $p > 3$ the image of the ring $\mathbb{Z}_{(p)}[T_2] = \mathcal{O}_F \otimes \mathbb{Z}_{(p)}$. This implies that we do not have inner congruences except for $p = 144169$.

But now our program with Gangl provides an explicit matrix for $T_2 \mod 144169$ and this matrix has only one eigenvalue $\mod 144169$.

This matrix for T_2 is not a diagonal matrix, and this implies that the fibre is connected, hence π_f occurs with weak multiplicity one.

Essentially the same happens if we look at the values $n = 22, 26, 28, 30, 32, 36$ for which the degree of the splitting field F is 2. It will happen that the discriminant is not a prime, so we get inner congruences modulo several primes.

In the previous section, we investigated questions regarding the structure of the integral cohomology as a module for the Hecke algebra. We discussed the denominator of the Eisenstein classes and our experimental data suggested an answer in terms of special values of L-functions (see 3.74 and also (5.1.2)).

We briefly mention some results about the questions concerning inner congruences which we contemplated in the previous section. To state these results we have to anticipate the notion of L-functions attached to a $\pi_f \in \mathrm{Coh}_!(n)$, and we anticipate the theorems on special values. We still localise at a prime p, and we only look at the ordinary part. We apply the results from Section 5.1.2 .

Assume we have two elements $\pi_f, \pi'_f \in R_{\mathfrak{p}}^{-1}(\bar{\pi}_f)$. We say that these two elements are *linked* if they lie in the same connected component. I refer to Theorem 5.1.1.

Theorem 3.3.4. *The Hecke modules $\pi_f, \pi'_f \in R_{\mathfrak{p}}^{-1}(\bar{\pi}_f)$ are linked if and only if*

$$\frac{Z(\nu)}{\Omega(\epsilon(\nu) \times \pi_f)}\Lambda^{\mathrm{coh}}(\pi, n+1-\nu) \equiv \frac{Z(\nu)}{\Omega(\epsilon(\nu) \times \pi'_f)}\Lambda^{\mathrm{coh}}(\pi', n+1-\nu) \mod \mathfrak{p}$$

for all $\nu = 0, 1, \dots, n$ and where $Z(\nu) = 1$ for $\nu \neq 0, n$ and $Z(0) = Z(n) =$ $\mathrm{numerator}(\zeta(-1-n))$ and $\epsilon(\nu) = \pm 1$ depending on the parity of ν.

Note: the factor $Z(\nu)$ is needed to make the expressions integers.
This is a slightly strengthened version of a theorem of Vatsal [112].

We have a second theorem which is due to Hida. We pick a $\pi_f \in R_{\mathfrak{p}}^{-1}(\bar{\pi}_f)$ and we say that π_f is *isolated* if $\{\pi_f\}$ is open. Then Hida's theorem says (c.f. [64])

Theorem 3.3.5. *The Hecke module $\pi_f \in R_{\mathfrak{p}}^{-1}(\bar{\pi}_f)$ is isolated if and only if*

$$\frac{\Lambda^{\mathrm{coh}}(\pi_f, \mathrm{Sym}^2, 1)}{\Omega(+ \times \pi_f)\Omega(- \times \pi_f)} \notin \mathfrak{p}.$$

Higher Eisenstein congruences

We are especially interested in the case of the *Eisenstein homomorphism* $\bar{\pi}_f^{\mathrm{Eis}}$: $T_\ell \to \ell^{n+1} + 1 \mod p$. It certainly occurs in the cohomology $\mod p$ and it follows from Theorems 3.3.1 and 5.1.2 that $\bar{\pi}_f^{\mathrm{Eis}}$ occurs in $\mathrm{Cohmodp}_!(n)$ if $p|\zeta(-1-n)$.

Here it is very tempting to ask whether or not $\bar{\pi}_f^{\mathrm{Eis}}$ always occurs with weak multiplicity one in the inner cohomology.

This question can be checked experimentally. For any prime ℓ, we look at the operator T_ℓ on $H^1_!(\Gamma\backslash\mathbb{H}, \tilde{\mathcal{M}}_n \otimes \mathbb{F}_p)$. We compute the characteristic polynomial

$$P_\ell(X) = \det(X\mathrm{Id} - ((\ell^{n+1}+1)\mathrm{Id} - T_\ell) \mid H^1_{\mathrm{ord},!}(\Gamma\backslash\mathbb{H}, \tilde{\mathcal{M}}_n) \otimes \mathbb{F}_p) = Q_\ell(X)^2.$$

The characteristic polynomial is a square because the $\pm$ eigenspaces are isomorphic as Hecke modules. Our computer program with Gangl gives us an explicit expression for $Q_2(X)$ for a large number of pairs (n,p). We are interested in pairs (n,p) with $p \mid \zeta(-1-n)$. Then we find

$$Q_2(X) = a_1(n,p)X + a_2(n,p)X^2 \dots \mod p$$

Then we found $a_1(n,p) \not\equiv 0 \mod p$ for $n \le 200$. This means that in these cases $\bar{\pi}_f^{\text{Eis}}$ occurs with multiplicity one. This is in no way surprising, as we expect that $p|\zeta(-1-n)$ and $p|a_1(n,p)$ will be a very rare event.

Almost at the same time, when I checked this question Tobias Berger drew my attention to the paper [6]. Here the authors consider the same problem in a slightly different context. They show that for $p < 10^5$ we always have $p \nmid a_1(n,p)$ with the one exception $p = 547, n = 484$.

Assume we have such a pair (p,n)-nobody knows whether there is one besides $(541, 484)$ - and we know in addition that $p \nmid a_2(n,p)$. Then the space $H^1_{\text{ord},!}(\Gamma\backslash\mathbb{H}, \tilde{\mathcal{M}}_n \otimes \mathbb{F}_p)\{\bar{\pi}_f^{\text{Eis}}\}$ is 4-dimensional, $T_2^{\text{Eis}} = T_2 - (2^{n+1}+1)\text{Id}$ acts nilpotently on it and its $(-)$ eigenspace under complex conjugation contains a 1-dimensional subspace annihilated by all Hecke operators $T_\ell^{\text{Eis}} = T_\ell - (\ell^{n+1}+1)\text{Id}$. In the case that T_2^{Eis} is not acting by 0, we see easily that all Hecke operators $T_\ell^{\text{Eis}} = T_\ell - (\ell^{n+1}+1)\text{Id}$ are acting nilpotently on this 4-dimensional space. In this case (i.e. if all T_ℓ^{Eis} are acting nilpotently) we can ask the next question:

Is T_ℓ^{Eis} the zero operator, for all choices of ℓ? Then this means that $\bar{\pi}_f^{\text{Eis}}$ has weak multiplicity 2.

If we restrict T_ℓ^{Eis} to one of the two-dimensional $\pm$ subspaces, then T_ℓ^{Eis} is a nilpotent endomorphism, i.e. a nilpotent 2×2 matrix with entries in $\mathbb{F}_p$. If I am not mistaken, then there are exactly p^2 such matrices and the zero matrix is just one of them. So one might argue that the probability for $T_\ell^{\text{Eis}} = 0$ is roughly $\frac{1}{p^2}$. So with a high probability the answer to the above question is No. Since the probability that $p|a_1(n,p)$ is also very small it may be safe to conjecture that $\bar{\pi}_f^{\text{Eis}}$ always occurs with weak multiplicity one.

Of course I checked the case $(484, 547)$ for the operator T_2 and indeed in this case $\bar{\pi}_f^{\text{Eis}}$ occurs with weak multiplicity one.

We briefly return to Theorem 3.3.2, I want to say something about higher congruences. Assume we have a pair (n,p) with $p^\delta|\zeta(-1-n)$ and in addition $p|a_1(n,p)$. We also assume $p \nmid a_2(n,p)$, then we expect that $\bar{\pi}_f^{\text{Eis}}$ occurs with weak multiplicity one. Let $F_\mathfrak{p}/\mathbb{Q}_p$ be a smallest extension such that we get a decomposition

$$H^1_!(\Gamma\backslash\mathbb{H}, \tilde{\mathcal{M}}_n\otimes F_\mathfrak{p})\{\pi_1,\pi_2\} = H^1_!(\Gamma\backslash\mathbb{H}, \tilde{\mathcal{M}}_n\otimes F_\mathfrak{p})(\pi_{f,1}) \oplus H^1_!(\Gamma\backslash\mathbb{H}, \tilde{\mathcal{M}}_n\otimes F_\mathfrak{p})(\pi_{f,2}).$$

Let $\mathcal{O}_\mathfrak{p} \subset F_\mathfrak{p}$ its ring of integers and $\mathfrak{p}$ its maximal prime ideal. We know $(p) = \mathfrak{p}^{e_\mathfrak{p}}$. This extension is either trivial or a quadratic extension of $\mathbb{Q}_p$. Let us assume that $F_\mathfrak{p} = \mathbb{Q}_p[\sqrt{\beta(n)\cdot p}]$ with $\beta(n) \in \mathbb{Z}_p$. The above decomposition induces a 2-step filtration (see 3.84)

$$\{0\} \subset H^1(\Gamma\backslash\mathbb{H}, \tilde{\mathcal{M}}_n \otimes \mathcal{O}_\mathfrak{p}/p^\delta)(\pi_{f,1}) \subset H^1_!(\Gamma\backslash\mathbb{H}, \tilde{\mathcal{M}}_n \otimes \mathcal{O}_\mathfrak{p}/p^\delta)\{\pi_1,\pi_2\},$$

we denote the quotient of the second step by the first

$$\mathcal{F}^1 H^1_!(\Gamma\backslash\mathbb{H}, \tilde{\mathcal{M}}_n\otimes\mathcal{O}_p/p^\delta) := H^1_!(\Gamma\backslash\mathbb{H}, \tilde{\mathcal{M}}_n\otimes\mathcal{O}_p/p^\delta)/H^1(\Gamma\backslash\mathbb{H}, \tilde{\mathcal{M}}_n\otimes\mathcal{O}_\mathfrak{p}/p^\delta)(\pi_{f,1}).$$

We have the inclusion $j : \mathcal{O}_\mathfrak{p}/p^\delta(-n-1) \hookrightarrow H^1_!(\Gamma\backslash\mathbb{H}, \tilde{\mathcal{M}}_n \otimes \mathcal{O}_p/p^\delta)\{\pi_1,\pi_2\}$, we compose this map with the projection to the quotient $\mathcal{F}^1 H^1_!(\Gamma\backslash\mathbb{H}, \tilde{\mathcal{M}}_n \otimes \mathcal{O}_p/p^\delta)$ and then this composition has a kernel $\mathfrak{p}^{f_1}\mathcal{O}_\mathfrak{p}/\mathfrak{p}^{e_\mathfrak{p}\delta}$. Hence we get an embedding

$$j_2 : \mathcal{O}_\mathfrak{p}/\mathfrak{p}^{e_\mathfrak{p}\delta}(-n-1)/\mathfrak{p}^{f_1}\mathcal{O}_\mathfrak{p}/\mathfrak{p}^{e_\mathfrak{p}\delta}(-n-1) \hookrightarrow \mathcal{F}^1 H^1_!(\Gamma\backslash\mathbb{H}, \tilde{\mathcal{M}}_n \otimes \mathcal{O}_p/p^\delta).$$

and as we have seen earlier this implies that we get a congruence

$$\pi_{f,2}(T_\ell) \equiv \ell^{n+1} + 1 \mod \mathfrak{p}^{e_\mathfrak{p}\delta - f_1}$$

But we also know that $\mathfrak{p}^{f_1}\mathcal{O}_\mathfrak{p}/\mathfrak{p}^{e_\mathfrak{p}\delta})(-n-1) \hookrightarrow H^1(\Gamma\backslash\mathbb{H}, \tilde{\mathcal{M}}_n \otimes \mathcal{O}_\mathfrak{p}/p^\delta)(\pi_{f,1})$, and hence we get a congruence

$$\pi_{f,1}(T_\ell) \equiv \ell^{n+1} + 1 \mod \mathfrak{p}^{f_1}$$

Now let $r > 0$ be any integer such that we have the congruence $\pi_{f,2}(T_\ell) \equiv \ell^{n+1} + 1 \mod \mathfrak{p}^r$ for all Hecke operators T_ℓ. Our first assumption is that we may take $r = 1$. Now we consider T_ℓ^{Eis} acting on $H^1_!(\Gamma\backslash\backslash\mathbb{H}, \tilde{\mathcal{M}}_n \otimes \mathcal{O}_\mathfrak{p}/\mathfrak{p}^r)\{\pi_1,\pi_2\}$. It respects the filtration and it is zero on the bottom step of the filtration and hence it yields a homomorphism

$$\partial(T_\ell^{\mathrm{Eis}}) : \mathcal{F}^1 H^1_!(\Gamma\backslash\mathbb{H}, \tilde{\mathcal{M}}_n \otimes \mathcal{O}_\mathfrak{p}/\mathfrak{p}^r)\{\pi_1,\pi_2\} \to H^1(\Gamma\backslash\mathbb{H}, \tilde{\mathcal{M}}_n \otimes \mathcal{O}_\mathfrak{p}/\mathfrak{p}^r))(\pi_{f,1}).$$

Then our assumption weak multiplicity equal one implies that for $r = 1$ this homomorphism is non zero. This implies that for any such r the image of $\partial(T_\ell^{\mathrm{Eis}})$ contains a cyclic direct summand $\mathcal{O}_\mathfrak{p}/\mathfrak{p}^r$. Since $\partial(T_\ell^{\mathrm{Eis}})$ is a Hecke-module homomorphism we conclude

$$\pi_{f,1}(T_\ell) \equiv \pi_{f,2}(T_\ell) \equiv \ell^{n+1} + 1 \mod \mathfrak{p}^r.$$

Since we can exchange $\pi_{f,1}$ and $\pi_{f,2}$ we see that we may take $r = \max(f_1, e_\mathfrak{p} \cdot \delta - f_1)$ and hence we can take $r = [\frac{e_\mathfrak{p}\delta}{2}]$.

At this point the extension $F_\mathfrak{p}$ plays a role. If it is a quadratic extension, then the two homomorphisms $\pi_{f,1}$ and $\pi_{f,2}$ are conjugate under the action of the Galois group and hence the highest power $\mathfrak{p}^{r_1}, \mathfrak{p}^{r_2}$ for which we the above congruence must be the same. We do not need weak multiplicity one.

But if $F_\mathfrak{p} = \mathbb{Q}_p$ then we badly need weak multiplicity one. We can not exclude the possibility that $\pi_{f,1}$ satisfies the congruence modulo $\mathfrak{p}^\delta$ and there is no congruence for $\pi_{f,2}$.

3.3.10 p-adic interpolation

The p-adic ζ-function

Let p be an irregular prime, i.e. $p \mid \zeta(-1-n_0)$, here we assume $0 < n_0 < p-1$. We consider $\zeta(-1-n) = \zeta(-1-n_0-\alpha(p-1))$ as a function in the variable $\alpha \in \mathbb{N}$, and we want to find values $n = -1 - n_0 - \alpha(p-1)$ such that $\zeta(-1-n)$ is divisible by higher powers of p. We know that there exist a p-adic ζ-function (see [74], [69], [114]) and this function has an expansion

$$\begin{aligned}\zeta(-1-n) &= \zeta(-1-n_0-\alpha(p-1)) \\ &\equiv \zeta(-1-n_0) + a(n_0,1)\alpha p + a(n_0,2)\alpha^2 p^2 + \cdots\end{aligned} \tag{3.93}$$

where the coefficients $a(n_0,\nu) \in \mathbb{Z}_p$ are only defined $\mod p$. If $p \not| a(n_0,1)$ then we can apply Newton's method, and we find a converging sequence $\alpha_1, \alpha_2, \ldots$ such that

$$\alpha_\nu \equiv \alpha_{\nu+1} \mod p^\nu \text{ and } \zeta(-1-n_0-\alpha_\nu(p-1)) \equiv 0 \mod p^{\nu+1} \tag{3.94}$$

The sequence converges to a zero α_∞ of the p-adic ζ-function.

It is not always possible to raise the power of p which divides the term $\zeta(-1-n_0-\alpha(p-1))$. If for instance $p^2 \not| \zeta(-1-n_0)$ and in addition $p|a(n_0,1)$, then we never find a higher power of p dividing some $\zeta(-1-n_0-\alpha(p-1))$. Again, a naïve probabilistic argument suggests that this is an extremely rare event, but the argument also suggests that such a prime exists (see the section on the Wieferich Dilemma 3.3.11).

Again we write $n = n_0 + (p-1)\alpha$ where we assume $0 < n_0 < p-1$, and we want to study how the cohomology $H^1_{\mathrm{ord}}(\Gamma\backslash\mathbb{H}, \tilde{\mathcal{M}}_n \otimes \mathbb{Z}_p)$ varies with n. We know for instance that the denominator of the Eisenstein class may become larger. We have seen already that

$$H^1_{\mathrm{ord}}(\Gamma\backslash\mathbb{H}, \tilde{\mathcal{M}}_n) \otimes \mathbb{Z}/p^r\mathbb{Z} \xrightarrow{\sim} H^1_{\mathrm{ord}}(\Gamma\backslash\mathbb{H}, \tilde{\mathcal{M}}_n \otimes \mathbb{Z}/p^r). \tag{3.95}$$

Now we have the following theorem which follows from the work of Hida ([65],[66]), see also ([50])

Theorem 3.3.6. *If $n = n_0 + (p-1)\alpha$, $n' = n_0 + (p-1)\alpha'$ and $\alpha \equiv \alpha' \mod p^{r-1}$, (i.e. $n \equiv n' \mod (p-1)p^{r-1}$), then we have a canonical Hecke invariant isomorphism*

$$\Phi(n,n')_r : H^1_{\mathrm{ord}}(\Gamma\backslash\mathbb{H}, \tilde{\mathcal{M}}_n \otimes \mathbb{Z}/p^r) \xrightarrow{\sim} H^1_{\mathrm{ord}}(\Gamma\backslash\mathbb{H}, \tilde{\mathcal{M}}_{n'} \otimes \mathbb{Z}/p^r). \tag{3.96}$$

This system of isomorphisms is consistent with change of the parameter α, α' and r.

Proof. This is an immediate consequence of Theorem 2.1 in [50]. □

We find a finite extension $F_\mathfrak{p}/\mathbb{Q}_p$ such that we have a decomposition into eigenspaces

$$H^1_{\mathrm{ord}}(\Gamma\backslash\mathbb{H}, \mathcal{M}_n \otimes F_\mathfrak{p}) = \bigoplus_{\pi_f} H^1_{!,\mathrm{ord}}(\Gamma\backslash\mathbb{H}, \mathcal{M}_n \otimes F_\mathfrak{p})(\pi_f) \oplus F_\mathfrak{p} e_n \tag{3.97}$$

where the first summation on the right hand side goes over those $\pi_f \in \mathrm{Coh}_!(n)$ for which $\pi_f(T_p)$ is a unit in $\mathcal{O}_\mathfrak{p}$, the ring of integers in F. Let us denote this set by $\mathrm{Coh}^{(n)}_{!,\mathrm{ord}}$. Then the summation goes over the set $\mathrm{Coh}^{(n)}_{\mathrm{ord}} = \mathrm{Coh}^{(n)}_{!,\mathrm{ord}} \cup \{\pi_f^{\mathrm{Eis}}\}$. Intersecting this decomposition with $H^1_{\mathrm{ord}}(\Gamma\backslash\mathbb{H}, \mathcal{M}_n \otimes \mathcal{O}_\mathfrak{p})$ gives us a submodule of finite index

$$H^1_{\mathrm{ord}}(\Gamma\backslash\mathbb{H}, \mathcal{M}_n \otimes \mathcal{O}_\mathfrak{p}) \supset \bigoplus_{\pi_f} H^1_{!,\mathrm{ord}}(\Gamma\backslash\mathbb{H}, \mathcal{M}_n \otimes \mathcal{O}_\mathfrak{p})(\pi_f) \oplus \mathcal{O}_\mathfrak{p} e_n, \tag{3.98}$$

and this also gives us a Jordan-Hölder filtration as in (3.84).

For $\bar{\pi}_f \in \mathrm{Cohmodp}_!(n)$ we define

$$H^1_{!,\mathrm{ord}}(\Gamma\backslash\mathbb{H}, \mathcal{M}_n \otimes \mathcal{O}_\mathfrak{p})\{\bar{\pi}_f\} := H^1_{!,\mathrm{ord}}(\Gamma\backslash\mathbb{H}, \mathcal{M}_n \otimes \mathcal{O}_\mathfrak{p})\{R_\mathfrak{p}^{-1}\{\bar{\pi}_f)\}$$

and then we get a direct sum decomposition

$$H^1_{!,\mathrm{ord}}(\Gamma\backslash\mathbb{H},\mathcal{M}_n\otimes\mathcal{O}_\mathfrak{p})=\bigoplus_{\bar\pi_f}H^1_{!,\mathrm{ord}}(\Gamma\backslash\mathbb{H},\mathcal{M}_n\otimes\mathcal{O}_\mathfrak{p})\{\bar\pi_f\},$$

and for any $\bar\pi_f$ we get a decomposition up to isogeny

$$H^1_{!,\mathrm{ord}}(\Gamma\backslash\mathbb{H},\mathcal{M}_n\otimes\mathcal{O}_\mathfrak{p})\{\bar\pi_f\}\supset\bigoplus_{\pi_f\in R_\mathfrak{p}^{-1}(\bar\pi_f)}H^1_{!,\mathrm{ord}}(\Gamma\backslash\mathbb{H},\mathcal{M}_n\otimes\mathcal{O}_\mathfrak{p})(\pi_f)\qquad(3.99)$$

We are mainly interested in the Hecke module $H^1_{!,\mathrm{ord}}(\Gamma\backslash\mathbb{H},\mathcal{M}_n\otimes\mathcal{O}_\mathfrak{p})\{\bar\pi_f^{\mathrm{Eis}}\}$, how this varies as α varies. Our theorem above implies that the Hecke module $H^1_{!,\mathrm{ord}}(\Gamma\backslash\mathbb{H},\mathcal{M}_n\otimes\mathcal{O}_\mathfrak{p}/\mathfrak{p})$ does not depend on α.

Let us now assume that $\bar\pi_f^{\mathrm{Eis}}$ has multiplicity one. Then we know that the decomposition (3.99) applied to $\bar\pi_f^{\mathrm{Eis}}$ has only one summand $\pi_{!,f}^{\mathrm{Eis}}$. We still assume that $n=n_0+(p-1)\alpha$,

Theorem 3.3.7. *If $p^{\delta_p(n)}\mid\zeta(-1-n)$ we have the congruence*

$$\pi_{!,f}^{Eis}(T_\ell)\equiv\ell^{n+1}+1\mod p^{\delta_p(n)}\ \forall\text{ primes }\ell.$$

Finally we get $\pi_f(T_\ell)\in\mathbb{Z}_p$ for all primes ℓ and hence we may take $\mathcal{O}_\mathfrak{p}=\mathbb{Z}_p$. We can find a basis f_0,f_1,f_2 of $H^1_{\mathrm{ord}}(\Gamma\backslash\mathbb{H},\tilde{\mathcal{M}}_n\otimes\mathbb{Z}_p)\{\bar\pi_f^{\mathrm{Eis}}\}$ where

a) f_1,f_2 form a basis of $H^1_{!,\mathrm{ord}}(\Gamma\backslash\mathbb{H},\tilde{\mathcal{M}}_n\otimes\mathbb{Z}_p)$ and f_0 maps to a generator of $H^1(\partial(\Gamma\backslash\mathbb{H}),\tilde{\mathcal{M}}_n\otimes\mathbb{Z}_p)$.

b) The complex conjugation c acts by $c(f_i)=(-1)^{i+1}f_i$.

Finally
c) the matrix T_ℓ^{ord} with respect to this basis satisfies

$$T_\ell^{\mathrm{ord}}\equiv\begin{pmatrix}\ell^{n+1}+1&0&t^{(\ell)}\\0&\ell^{n+1}+1&0\\0&0&\ell^{n+1}+1\end{pmatrix}\mod p^{\delta_\ell(n)}$$

Proof. Clear. □

Now we assume that $p\nmid a(1,n_0)$ and we choose a sequence $\alpha_0=0,\alpha_1,\dots,\alpha_\nu$ as in (3.94), then we get Hecke-module maps

$$\begin{aligned}H^1_{!,\mathrm{ord}}(\Gamma\backslash\mathbb{H},\tilde{\mathcal{M}}_{n_{\nu+1}}\otimes\mathbb{Z}_p/p^{\nu+1})&\to H^1_{!,\mathrm{ord}}(\Gamma\backslash\mathbb{H},\tilde{\mathcal{M}}_{n_{\nu+1}}\otimes\mathbb{Z}_p/p^{\nu})\\&\xrightarrow{\sim}H^1_{!,\mathrm{ord}}(\Gamma\backslash\mathbb{H},\tilde{\mathcal{M}}_{n_\nu}\otimes\mathbb{Z}_p/p^{\nu}).\end{aligned}$$

The sequence n_ν converges to a p-adic integer n_∞, and we can form the projective limit and define

$$H^1_{\mathrm{ord}}(\Gamma\backslash\mathbb{H},\tilde{\mathcal{M}}_{n_\infty})=\varprojlim H^1_{\mathrm{ord}}(\Gamma\backslash\mathbb{H},\tilde{\mathcal{M}}_{n_\nu}\otimes\mathbb{Z}/p^\nu\mathbb{Z}).\qquad(3.100)$$

Under our assumptions $H^1_{\mathrm{ord},!}(\Gamma\backslash\mathbb{H},\tilde{\mathcal{M}}_{n_\infty})\{\bar{\pi}_f^{\mathrm{Eis}}\}$ is a free $\mathbb{Z}_p$-module of rank 3. The Hecke operators T_ℓ^{ord} acts on $H^1_{\mathrm{ord}}(\Gamma\backslash\mathbb{H},\tilde{\mathcal{M}}_{n_\nu}\otimes\mathbb{Z}/p^\nu\mathbb{Z})$ by a matrix of the shape as in Theorem 3.3.7, and the eigenvalues on the diagonal are

$$\ell^{n_\nu+1}+1=\ell^{n_0+(p-1)\alpha_\nu}+1 \mod p^\nu.$$

For $\ell\neq p$ we write

$$\ell^{p-1}=1+p\delta_p(\ell),\ \delta_p(\ell)\in\mathbb{N}$$

and then

$$\ell^{n_0+(p-1)\alpha_\nu}=\ell^{n_0}(1+\delta_p(\ell)p)^{\alpha_\nu}=l^{n_0}(1+\alpha_\nu p\delta_p(\ell)+\binom{\alpha_\nu}{2}p^2\delta_p(\ell)^2+\cdots.$$

We see that we can define $\ell^{n_0+\alpha(p-1)}$ for any $\alpha\in\mathbb{Z}_p$, and then we clearly have $\lim_{\nu\to\infty}\ell^{n_\nu}=\ell^{n_\infty}$. Hence we see that T_ℓ^{ord} acts on the Eisenstein part of $H^1_{\mathrm{ord}}(\Gamma\backslash\mathbb{H},\tilde{\mathcal{M}}_{n_\infty})$ by the matrix

$$T_\ell^{\mathrm{ord}}\equiv\begin{pmatrix}\ell^{n_\infty+1}+1 & 0 & t^{(\ell)}\\ 0 & \ell^{n_\infty+1}+1 & 0\\ 0 & 0 & \ell^{n_\infty+1}+1\end{pmatrix}$$

where we have $t^{(\ell)}\neq 0$ (this follows from earlier arguments).

If we drop the assumption that $\bar{\pi}_f^{\mathrm{Eis}}$ has multiplicity one, then the situation becomes definitely much more complicated. We believe that this is a rare event, as we have seen that for $p<10^5$ this happens only once. But on the other hand our naïve probabilistic argument suggests that it should happen again. Then the same probabilistic argument suggests that $m(\bar{\pi}_f^{\mathrm{Eis}})>2$ never happens.

Therefore we make the assumption that $m(\bar{\pi}_f^{\mathrm{Eis}})=2$. We look at the decomposition (3.99) applied to $\bar{\pi}_f=\bar{\pi}_f^{\mathrm{Eis}}$. We consider the characteristic polynomial

$$\begin{aligned}\det\Big(T_\ell^{\mathrm{Eis}}-X\mathrm{Id}-(l^{n+1}+1)\mathrm{Id}|H^1_!(\Gamma\backslash\mathbb{H},\tilde{\mathcal{M}}\otimes\mathbb{Z}_p)\{\bar{\pi}_f^{\mathrm{Eis}}\}\Big)&\\ =(b_0(n,p)+b_1(n,p)X+X^2)^2.&\end{aligned}$$

Our assumption $m(\bar{\pi}_f^{\mathrm{Eis}})=2$ implies that $b_1(n,p),b_0(n,p)\equiv 0 \mod p$. Let us write

$$P^{\mathrm{Eis}}(T_\ell,n,X):=b_0(n,p)+b_1(n,p)X+X^2.$$

If we now find an ℓ such that $p^2\nmid b_0(n,p)$, then the zeroes of $P^{\mathrm{Eis}}(T_\ell,n,X)$ are

$$\lambda(T_\ell,n)=-\frac{b_1(n,p)}{2}\pm\sqrt{\frac{b_1(n,p)^2}{4}-b_0(n,p)}$$

and hence it is clear that we need a ramified quadratic extension $F_{\mathfrak{p}}=\mathbb{Q}_p[\sqrt{\beta(n)p}]$ (where $\beta(n)$ is a unit in $\mathbb{Z}_p^\times$) if we want a decomposition into eigenspaces (3.97). If $\mathcal{O}_{\mathfrak{p}}$ is the ring of integers in $F_{\mathfrak{p}}$, then we get a decomposition up to isogeny into two conjugate eigenspaces

$$\begin{aligned}H^1_!(\Gamma\backslash\mathbb{H},\tilde{\mathcal{M}}_n\otimes\mathcal{O}_{\mathfrak{p}})\{\bar{\pi}_f^{\mathrm{Eis}}\}\supset H^1_!(\Gamma\backslash\mathbb{H},\tilde{\mathcal{M}}_n\otimes\mathcal{O}_{\mathfrak{p}})(\pi_f)&\\ \oplus H^1_!(\Gamma\backslash\mathbb{H},\tilde{\mathcal{M}}\otimes\mathcal{O}_{\mathfrak{p}})({}^\sigma\pi_f),&\end{aligned}\tag{3.101}$$

where σ is the non-trivial element in the Galois group of $F_{\mathfrak{p}}/\mathbb{Q}_p$.

Under our assumptions, the quotient of the left hand side by the right hand side is isomorphic to $H^1_!(\Gamma\backslash\mathbb{H}, \tilde{\mathcal{M}}\otimes\mathcal{O}_{\mathfrak{p}}/\mathfrak{p})(\bar{\pi}_f^{\mathrm{Eis}})$. For all ℓ' we have the congruence $\pi_f(T_{\ell'}) \equiv {}^\sigma\pi_f(T_{\ell'}) \mod \mathfrak{p}$. We recall that $n = n_0 + \alpha(p-1)$. It is clear that the condition $p^2 \nmid b_0(n,p)$ only depends on $\alpha \mod p$. We should expect that for a fixed α the "probability" that $p^2| \, b_0(n,p)$ is $1/p$. But there may be a value of α for which this divisibility holds.

We drop the condition that $p^2 \nmid b_0(n,p)$. We return to our sequence $\alpha_1, \dots$ (see (3.94)). Then it is easy to see that for $\nu \geq 1$ we have the decomposition up to isogeny (3.101)

$$H^1_!(\Gamma\backslash\mathbb{H}, \tilde{\mathcal{M}}_{n_\nu} \otimes \mathcal{O}_{\mathfrak{p}})\{\bar{\pi}_f^{\mathrm{Eis}}\} \supset H^1_!(\Gamma\backslash\mathbb{H}, \tilde{\mathcal{M}}_{n_\nu} \otimes \mathcal{O}_{\mathfrak{p}})(\pi_{f,1}^{(\nu)}) \\ \oplus H^1_!(\Gamma\backslash\mathbb{H}, \tilde{\mathcal{M}}_{n_\nu} \otimes \mathcal{O}_{\mathfrak{p}})(\pi_{f,2}^{(\nu)}).$$

We have the inclusion $\mathbb{Z}_p/(p^{\nu+1})(-n_\nu - 1) \hookrightarrow H^1_!(\Gamma\backslash\mathbb{H}, \tilde{\mathcal{M}}_{n_\nu} \otimes \mathbb{Z}/(p^{\nu+1}))$ and tensoring by $\mathcal{O}_{\mathfrak{p}}/\mathfrak{p}^{e_{\mathfrak{p}}(n_\nu+1)}$ yields the inclusion

$$j_{\mathfrak{p},\nu} : \mathcal{O}_{\mathfrak{p}}/\mathfrak{p}^{e_{\mathfrak{p}}(\nu+1)}(-n_\nu - 1) \hookrightarrow H^1_!(\Gamma\backslash\mathbb{H}, \tilde{\mathcal{M}}_{n_\nu} \otimes \mathcal{O}_{\mathfrak{p}}/(\mathfrak{p}^{e_{\mathfrak{p}}(\nu+1)}))$$

We have

$$H^1_!(\Gamma\backslash\mathbb{H}, \tilde{\mathcal{M}}_n \otimes \mathcal{O}_{\mathfrak{p}}/(\mathfrak{p}^{e_{\mathfrak{p}}(\nu+1)}))(\pi_{f,1}^\nu) \subset H^1_!(\Gamma\backslash\mathbb{H}, \tilde{\mathcal{M}}_n \otimes \mathcal{O}_{\mathfrak{p}}/(\mathfrak{p}^{e_{\mathfrak{p}}(\nu+1)}))\{\bar{\pi}_f^{\mathrm{Eis}}\}$$

this is the first step in the Jordan-Hölder filtration, and the quotient by this first step is isomorphic to $H^1_!(\Gamma\backslash\mathbb{H}, \tilde{\mathcal{M}}_n \otimes \mathcal{O}_{\mathfrak{p}}/(\mathfrak{p}^{e_{\mathfrak{p}}(\nu+1)}))(\pi_{f,2}^\nu)$. We repeat the argument in the proof of Theorem 3.3.2, and we conclude that we have congruences

$$\pi_{f,1}^\nu(T_\ell) \equiv \ell^{n_\nu+1} + 1 \mod \mathfrak{p}^{(\nu+1)}; \; \pi_{f,1}^\nu(T_\ell) \equiv \ell^{n_\nu+1} + 1 \mod \mathfrak{p}^{(\nu+1)}.$$

At this place we do not know anything about our field $F_{\mathfrak{p}}$ we need weak multiplicity one. Again we can pass to the limit $H^1(\Gamma\backslash\mathbb{H}, \tilde{\mathcal{M}}_{n_\infty})\{\bar{\pi}_f^{\mathrm{Eis}}\}$, this limit has a three step Jordan-Hölder filtration, the top quotient is $\mathbb{Z}_p(-n_\infty - 1) \otimes \mathcal{O}_{\mathfrak{p}}$ - which is the cohomology of the boundary. For the eigenvalues on the middle step and the bottom we also get the same limit. Hence if we choose our basis as before, i.e. we take the action of the complex conjugation into account, then we get for the matrix of the Hecke operator restricted to the Eisenstein part being:

$$T_\ell^\infty = \begin{pmatrix} \ell^{n_\infty+1}+1 & 0 & t^{(\ell)} & 0 & s^{(\ell)} \\ 0 & \ell^{n_\infty+1}+1 & 0 & u^{(\ell)} & 0 \\ 0 & 0 & \ell^{n_\infty+1}+1 & 0 & v^{(\ell)} \\ 0 & 0 & 0 & \ell^{n_\infty+1}+1 & 0 \\ 0 & 0 & 0 & 0 & \ell^{n_\infty+1}+1 \end{pmatrix} \tag{3.102}$$

where the non-zero entries are units for a suitable choice of ℓ.

In the special case $(484, 547)$ the number $\alpha_1 = 100$, we have $\zeta(-485 - 100 * 546) = \zeta(-55085) \equiv 0 \mod 547^2$. Earlier we checked $547^2 \nmid b_0(484, 547)$ for the Hecke operator T_2. But can we ever check $547^2 \nmid b_0(55084, 547)$? The matrices become too big.

But there is a way out. I think it is possible to prove that we have an expansion

$$b_0(484 + \alpha \cdot 546, 547) \equiv b_0(484, 547) + \alpha \cdot 547 \cdot b_0'(484, 457) \mod 547^2.$$

(This should follow from the general results which we announced in [50] and which we still hope to prove in a paper with J. Mahnkopf.).

Using our program with Gangl for T_2 and with the help of A.Weisse we computed $T_2 \mod p^2$ for the cases $\alpha = 0, 1, 2$. The program still works in reasonable time in these cases. We found (of course everything $\mod 547^2$)

$$\begin{aligned} b_0(484) &= 547 \cdot 10, \\ b_0(484 + 546, 547) &= 547 \cdot 174 = 547(10 + 164) \\ b_0(484 + 2 \cdot 546, 547) &= 547 \cdot 338 = 547(10 + 2 \cdot 164). \end{aligned}$$

Hence we expect $b_0'(484, 547) = 164 \mod 547$ and assuming the above linearity we get $b_0(55084, 547) = 547(10 + 100 \cdot 164) = 547 \cdot 16410 = 547^2 \cdot 30$.

When I saw this the first time I was surprised, but it seems to be clear that the considerations above imply $547^2|\zeta(-1-484-100\cdot 546)$ implies $547^2|b_0(484+100 \cdot 546)$

But we still do not know whether the field extension is trivial or not. To decide this we have to compute $P^{\mathrm{Eis}}(T_2, 484 + a \cdot 546, X) \mod 547^3$ for some small values of a. We computed the zeroes

$$\lambda(T_2, a) = \alpha \cdot 457 \pm \sqrt{\beta \cdot 547} \mod 547^3$$

of the quadratic factor $P^{\mathrm{Eis}}(T_2, 484 + a \cdot 546, X)$ for $a = 0, 1, 2$. We found

$$\begin{aligned} \lambda(T_2, 0) &= 268381 \cdot 547 \pm \sqrt{537 \cdot 547} \mod 547^3 \\ \lambda(T_2, 1) &= 189064 \cdot 547 \pm \sqrt{251993 \cdot 547} \mod 547^3 \\ \lambda(T_2, 2) &= 13475 \cdot 547 \pm \sqrt{169232 \cdot 547} \mod 547^3 \end{aligned}$$

and we see that these roots lie in a ramified quadratic extension of $\mathbb{Z}/(547^3)$. From the roots we get the coefficients of $P^{\mathrm{Eis}}(T_2, 484 + a * 546, X)$

$$\begin{aligned} b_0(484, 547) &= (268381 \cdot 547)^2 - 537 \cdot 547 = 37406595 \mod 547^3 \\ b_0(484 + 546, 547) &= (189064 \cdot 547)^2 - 251993 \cdot 547 = 135636855 \mod 547^3 \\ b_0(484 + 2 \cdot 546, 547) &= (13475 \cdot 547)^2 - 169232 \cdot 547 = 91742840 \mod 547^3 \\ b_1(484, 547) &= -2 \cdot 268381 \cdot 547 \mod 547^3 \\ b_1(484 + 546, 547) &= -2 \cdot 189064 \cdot 547 \mod 547^3 \end{aligned}$$

We need the value $\lambda(T_2, 100)$ Now we hope that in the still to be written paper with J. Mahnkopf we will show that we have an expansion

$$b_0(484 + \alpha 546, 547) = b_0(484, 547) + b_0'(484)\alpha 547 + b_0''(484)\alpha^2 547^2 \mod 547^3$$

where $b_0'(484), b_0''(484)$ are numbers mod 547^3, which we can compute from the three values above. We get

$$b_0'(484) = 82143154\ ;\ b_0''(484) = 81833424$$

We do the same for $b_1(484 + \alpha \cdot 546)$ and get

$$b_1(484 + \alpha \cdot 546) = b_1(484) + 158634 \cdot 547 \cdot \alpha \mod 547^2.$$

Now the discriminant is

$$\Delta(\alpha, 547) = -4b_0(484 + \alpha \cdot 546, 547) + b_1(484 + \alpha \cdot 546, 457)^2 \mod 547^3,$$

and if we believe in the interpolation formula we get for the discriminant

$$4 \cdot \Delta(100, 547) = 5^3 \cdot 547^2 \mod 547^3.$$

Hence we see that

$$\lambda(T_2, 100) = 97240737 \pm \frac{5}{2}\sqrt{5} \mod 547^3.$$

Now we check easily that 5 is not a square mod 547, and hence we see the roots now lie in the unique unramified quadratic extension $\mathbb{Z}/(547^3)[\sqrt{5}]$. Now we put again $\mathcal{O}_\mathfrak{p} = \mathbb{Z}_{547}[\sqrt{5}]$ and let $p = 547$ then $\mathfrak{p} = (p)$.

We consider the sequence $\alpha_0, \alpha_1, \dots, \alpha_\nu, \dots$ (see (3.94)). As before we get for any $\nu \geq 1$ (a decomposition up to isogeny (3.99))

$$H^1_!(\Gamma\backslash\mathbb{H}, \tilde{\mathcal{M}}_{n_\nu} \otimes \mathcal{O}_\mathfrak{p})\{\pi_f^{\mathrm{Eis}}\} \supset$$

$$H^1_!(\Gamma\backslash\mathbb{H}, \tilde{\mathcal{M}}_{n_\nu} \otimes \mathcal{O}_\mathfrak{p})(\pi_f^{(\nu)}) \oplus H^1_!(\Gamma\backslash\mathbb{H}, \tilde{\mathcal{M}}_{n_\nu} \otimes \mathcal{O}_\mathfrak{p}))({}^\sigma\pi_f^{(\nu)}).$$

We argue as before that our embedding

$$\mathbb{Z}/p^{\nu+1}(-1 - n_\nu) \hookrightarrow H^1_!(\Gamma\backslash\mathbb{H}, \tilde{\mathcal{M}}_{n_\nu} \otimes \mathbb{Z}/(p^{\nu+1}))$$

provides the Galois invariant embedding

$$\mathcal{O}_\mathfrak{p}/(p^{\nu+1})(-1 - n_\nu) \hookrightarrow H^1_!(\Gamma\backslash\mathbb{H}, \tilde{\mathcal{M}}_{n_\nu} \otimes \mathcal{O}_\mathfrak{p}/(p^{\nu+1})).$$

There is a smallest number $f_\nu \leq \nu$ such that the submodule $p^{f_\nu}\mathcal{O}_\mathfrak{p}/p^{\nu+1}(-1 - n_\nu)$ embeds into the submodule $H^1_!(\Gamma\backslash\mathbb{H}, \tilde{\mathcal{M}}_{n_1} \otimes \mathcal{O}_\mathfrak{p}/(p^{\nu+1}))(\pi_f^{(\nu)})$. But then we get an injection

$$\mathcal{O}_\mathfrak{p}/(p^{\nu+1})(-1 - n_\nu)/p^{f_\nu}(\mathcal{O}_\mathfrak{p}/p^{\nu+1}(-1 - n_\nu)) \hookrightarrow$$

$$H^1_!(\Gamma\backslash\mathbb{H}, \tilde{\mathcal{M}}_{n_\nu} \otimes \mathcal{O}_\mathfrak{p}/(p^{\nu+1})/H^1_!(\Gamma\backslash\mathbb{H}, \tilde{\mathcal{M}}_{n_\nu} \otimes \mathcal{O}_\mathfrak{p})(\pi_f^{(\nu)})) \otimes \mathcal{O}_\mathfrak{p}/(p^{\nu+1}).$$

The target module in the bottom line is isomorphic to the module

$$H^1_!(\Gamma\backslash\mathbb{H}, \tilde{\mathcal{M}}_{n_1} \otimes \mathcal{O}_\mathfrak{p})({}^\sigma\pi_f^{(\nu)}) \otimes \mathcal{O}_\mathfrak{p}/(p^\nu)$$

and hence we get congruences

$$\pi_f^{(\nu)}(T_\ell) \equiv \ell^{n_\nu+1} + 1 \mod p^{f_\nu},\ {}^\sigma\pi_f^{(\nu)}(T_\ell) \equiv \ell^{n_\nu+1} + 1 \mod p^{\nu+1-f_\nu}.$$

Since these congruences are invariant under the action of the Galois group, we get congruences

$$\pi_f^{(\nu)}(T_\ell) \equiv \ell^{n_\nu+1} + 1 \mod p^{[\frac{\nu+2}{2}]},\ {}^\sigma\pi_f^{(\nu)}(T_\ell) \equiv \ell^{n_\nu+1} + 1 \mod p^{[\frac{\nu+2}{2}]}$$

where $[x]$ denotes the usual Gauss bracket.

The Galois group

It is a fundamental fact that we have an action of the Galois group $\mathrm{Gal}(\bar{\mathbb{Q}}/\mathbb{Q})$ on the modules $H^1(\Gamma\backslash\mathbb{H}, \tilde{\mathcal{M}}_n \otimes \mathbb{Z}_p/p^\delta), H^1(\Gamma\backslash\mathbb{H}, \tilde{\mathcal{M}}_n \otimes \mathcal{O}_\mathfrak{p}), H^1(\Gamma\backslash\mathbb{H}, \tilde{\mathcal{M}}_{n_\infty})\{\bar{\pi}_f^{\mathrm{Eis}}\}$. This action commutes with the action of the Hecke algebra. Hence we get interesting representations of the Galois group, where these representations have been studied by many people (see for instance [83], [94],[109] or [59]).

We will explain this matter in a very cursory manner in Section 5.2.

3.3.11 The Wieferich dilemma

In 1909 the student Arthur Wieferich proved the following

If for a prime $p > 2$ the equation $x^p + y^p + z^p = 0$ has a solution in integers with $xyz \not\equiv 0 \mod p$, then $2^{p-1} - 1 \equiv 0 \mod p^2$.

Of course we know that $2^{p-1} - 1 \equiv 0 \mod p$ but checking a few of small primes suggests that the residue class $w(p) = \frac{2^{p-1}-1}{p} \mod p$ can be any number mod p. Hence we expect that it is a rare event that this residue class is zero, where the "probability" is $\frac{1}{p}$. Later these primes, for which $w(p) = 0$, are called Wieferich primes. At the present moment it seems that there are only two Wieferich primes $< 6.7 \times 10^5$.

But poor Arthur could not show that he had proved the first case of Fermat for infinitely many prime exponents p because he could not show that there are infinitely many non-Wieferich primes (and we still do not know it now).

This kind of phenomenon was not new and was well-known at the time when Wieferich proved his theorem (simply Google: The first case of Fermat's Theorem or "irregular primes“) but Wieferich's case is a striking example because it is so easy to state. That is the reason why we propose to call it the Wieferich dilemma.

In this book we encounter the Wieferich dilemma at several occasions. If $p|\zeta(-1-n)$ for some $n < p$ we raised the question whether $\bar{\pi}_f^{\mathrm{Eis}}$ occurs with multiplicity one. We saw that this is the case if $a_1(n,p) \not\equiv 0 \mod p$. Again we may argue that the probability that $p|a_1(n,p)$ is $\frac{1}{p}$ so we expect this to be a rare event that $\bar{\pi}_f^{\mathrm{Eis}}$ occurs with higher multiplicity. Actually we learned from [6] that there is only one exception for $p < 10^5$. On the other hand we believe that $\sum_{p \text{ irregular}} \frac{1}{p}$ is divergent so there is a certain chance that some larger such primes exist.

Assume that there is a prime p for which $\bar{\pi}_f^{\mathrm{Eis}}$ occurs with higher multiplicity. Then we may ask whether it occurs with weak multiplicity one, and again a probability argument shows that the probability that this is not so is $\frac{1}{p^2}$. This now suggests that it may always occur with weak multiplicity one.

Here I want to make a metamathematical statement.

It is very well conceivable that $\bar{\pi}_f^{\mathrm{Eis}}$ always occurs with weak multiplicity one, but we will never find a proof. But it is simply true because the probability that $\bar{\pi}_f^{\mathrm{Eis}}$ ever occurs with higher weak multiplicity is so small. It is simply "true" without a proof.

In this chapter, we discussed some questions concerning the structure of the cohomology of arithmetic groups as a module under the Hecke algebra. We executed computations and experiments to support and suggest certain hypotheses. But we only considered the unramified case. But there is much wider range where we can ask questions and make hypotheses.

We drop our assumption that we are in the totally unramified situation, this means that we can replace $\Gamma_0 = \mathrm{Sl}_2(\mathbb{Z})$ by a (normal) congruence subgroup $\Gamma \subset \Gamma_0$. We choose a free $\mathbb{Z}$-module of finite rank $\mathcal{V}$ with an action of Γ_0/Γ, i.e. we have an irreducible representation

$$\rho_{\mathcal{V}} : \Gamma_0/\Gamma \to \mathrm{Gl}(\mathcal{V}).$$

We assume that the matrix $-\mathrm{Id}$ acts by a scalar $\rho_{\mathcal{V}}(-\mathrm{Id}) = \pm\mathrm{Id}$. The Γ_0-modules $\mathcal{M}_n \otimes \mathcal{V}$ provide sheaves $\widetilde{\mathcal{M}_n \otimes \mathcal{V}}$, where we assume that $\rho_{\mathcal{V}}(-\mathrm{Id}) = (-1)^n\mathrm{Id}$. Again we study the cohomology groups, and especially we can study the fundamental exact sequence

$$\to H^1_c(\Gamma_0\backslash\mathbb{H}, \widetilde{\mathcal{M}_n \otimes \mathcal{V}}) \to H^1(\Gamma_0\backslash\mathbb{H}, \widetilde{\mathcal{M}_n \otimes \mathcal{V}}) \xrightarrow{r} H^1(\partial(\Gamma_0\backslash\mathbb{H}), \widetilde{\mathcal{M}_n \otimes \mathcal{V}})$$

On these cohomology groups we have the action the Hecke algebra $\mathcal{H}_\Gamma = \bigotimes' \mathcal{H}_p$. Here we have to be aware that for $p|N$ the action of the local factors $\mathcal{H}_p$ may not be commutative anymore. But this does not really matter, we can carry out the previous considerations. Especially we can discuss the structure of cohomology of the boundary as Hecke module and then we can ask for generalisations of Theorem 5.1.2.

Here the nature of the irreducible representation will play a role, we should start from an induced representation. We choose a positive integer N, and consider the congruence subgroup $\Gamma_0(N) = \left\{ \begin{pmatrix} a & b \\ c & d \end{pmatrix} \in \mathrm{Sl}_2(\mathbb{Z}) | c \equiv 0 \mod N \right\}$. Let $\Gamma_1(N) \subset \Gamma_0(N)$ be the subgroup where $a \equiv 1 \mod N$, then $\Gamma_0(N)/\Gamma_1(N) = (\mathbb{Z}/N\mathbb{Z})^\times$. We choose a character $\chi : \Gamma_0(N)/\Gamma_1(N) \to \mathbb{Z}[\zeta_{\phi(N)}]^\times$ and consider the induced representation

$$\mathcal{V}_\chi = \mathrm{Ind}_{\Gamma_0(N)}^{\Gamma_0} \chi = \{f : \Gamma_0 \to \mathbb{Z}[\zeta_{\phi(N)}] \mid f(\gamma x) = \chi(\gamma) f(x); \gamma \in \Gamma_0(N), x \in \Gamma_0\}.$$

Again we can study Eisenstein classes and their denominators. In this case the denominator should be related to the L values $L(\chi, -1-n)$. We should take for $N = p_0$ a small prime. The value $L(\chi, -1-n) \in \mathbb{Z}[\zeta_{p_0-1}]$. Then we expect that for a "large" primes p for which the ideal $\mathfrak{p} = (L(\chi, -1-n), p)$ is a prime ideal, this prime ideal divides the denominator of a certain Eisenstein class. It is certainly interesting to collect some data, which might allow us to formulate more precise hypotheses. For instance it is desirable to determine a very small set S_χ of exceptional primes which is determined by χ such that "large" prime can be replaced by p not in this exceptional set S_χ.

At this point it seems plausible to me, that the analytic methods (Eisenstein cohomology See Chapter 5) provide some precise estimates for the denominator. We may be able to find cases where the "large" primes are actually not so large.

This may be relevant if we investigate the multiplicity questions. Our cohomology groups are $\mathbb{Z}[\zeta_N]$-modules, let us assume that we have a prime ideal

$\mathfrak{p}|L(\chi,-1-n)$. We assume $\mathfrak{p} \not| p_0$ and let $\mathcal{O}_{\mathfrak{p}} = \mathbb{Z}[\zeta_{p_0-1}]_{\mathfrak{p}}$ be the completion at $\mathfrak{p}$. Then we can define the direct summand (see 3.90)

$$H^1(\Gamma_0\backslash\mathbb{H}, \mathcal{M}_n \widetilde{\otimes \mathcal{V}} \otimes \mathcal{O}_{\mathfrak{p}})\{\bar{\pi}^{\mathrm{Eis}}_{f,\chi}\} \subset H^1_{\mathrm{ord}}(\Gamma_0\backslash\mathbb{H}, \mathcal{M}_n \widetilde{\otimes \mathcal{V}} \otimes \mathcal{O}_{\mathfrak{p}})$$

this direct summand is a Hecke module. Again we should carry out some experiments in this more general context. For instance we can look for cases with weak multiplicity > 1. Our chances to find an example are a little better if we check cases where we have some "small large" primes. So we have some chances to falsify the conjecture about weak multiplicity one once we allow some ramification.

We will resume these considerations at the end of Chapter 5.

Chapter 4

Representation Theory, Eichler-Shimura Isomorphism

4.1 Harish-Chandra modules with cohomology

In Chapter 6 we will give a general discussion of the tools from representation theory and analysis which help us to understand the cohomology of arithmetic groups. In Chapter 6, Section 6.1.5, we will recall the results of Vogan-Zuckerman on the cohomology of Harish-Chandra modules.

Here we specialise these results to the specific cases $G = \mathrm{Gl}_2(\mathbb{R})$ (case A)) and $G = \mathrm{Gl}_2(\mathbb{C})$ (case B)). For the general definition of Harish-Chandra modules and for the definition of $(\mathfrak{g}, K_\infty)$-cohomology we refer to (6.1.2).

4.1.1 The finite rank highest weight modules

We consider the case A). In this case, our group $G/\mathbb{R}$ is the base extension of the reductive group scheme $\mathcal{G} = \mathrm{Gl}_2/\mathrm{Spec}(\mathbb{Z})$. In principle this is a pretentious language. At this point it simply means that for any commutative ring R with identity we can speak of $\mathcal{G}(R)$ - the group of R-valued points - and that $\mathcal{G}(R)$ depends functorially on R.

G. Harder, *Cohomology of Arithmetic Groups*,
Springer Monographs in Mathematics, https://doi.org/10.1007/978-3-032-10378-9_4

(Sometimes in the following we will replace Spec($\mathbb{Z}$) by $\mathbb{Z}$.) Then $\mathcal{G}^{(1)}/\mathbb{Z}$ is the kernel of the determinant map $\det : \mathcal{G}/\mathbb{Z} \to \mathbb{G}_m/\mathbb{Z}$. We have the standard maximal torus $\mathcal{T}/\mathbb{Z}$ and choose the Borel subgroup $\mathcal{B}/\mathbb{Z} \supset \mathcal{T}/\mathbb{Z}$ to be the group of upper triangular matrices. Let $X^*(\mathcal{T}) = X^*(T \times \mathbb{C})$ be the character module. This character module is $\mathbb{Z}e_1 \oplus \mathbb{Z}e_2$ where

$$e_i : \begin{pmatrix} t_1 & 0 \\ 0 & t_2 \end{pmatrix} \mapsto t_i.$$

Any character can be written as $\lambda = n\gamma + d\det$ where $\gamma = \frac{e_1 - e_2}{2}$ ($\notin X^*(\mathcal{T})$!), $\det = e_1 + e_2$ and where $n \in \mathbb{Z}, d \in \frac{1}{2}\mathbb{Z}$ and $n \equiv 2d \mod 2$. We say that λ is *dominant* if $n \geq 0$.

To any such character $\lambda = n\gamma + d\det$ we attach a highest weight module $\mathcal{M}_\lambda$. We consider the $\mathbb{Z}$-module of polynomials

$$\mathcal{M}_n = \{P(X,Y) \mid P(X,Y) = \sum_{\nu=0}^{n} a_\nu X^\nu Y^{n-\nu}, a_\nu \in \mathbb{Z}\}.$$

To a polynomial $P \in \mathcal{M}_n$ we attach the regular function (see 1.1.1)

$$f_P\left(\begin{pmatrix} x & y \\ u & v \end{pmatrix}\right) = P(u,v) \det\left(\begin{pmatrix} x & y \\ u & v \end{pmatrix}\right)^{-\frac{n}{2}+d}, \tag{4.1}$$

then

$$\begin{aligned} f_P\left(\begin{pmatrix} t_1 & w \\ 0 & t_2 \end{pmatrix}\begin{pmatrix} x & y \\ u & v \end{pmatrix}\right) &= t_2^n (t_1 t_2)^{-\frac{n}{2}+d} f_P\left(\begin{pmatrix} x & y \\ u & v \end{pmatrix}\right) \\ &= \lambda^-\left(\begin{pmatrix} t_1 & w \\ 0 & t_2 \end{pmatrix}\right) f_P\left(\begin{pmatrix} x & y \\ u & v \end{pmatrix}\right) \end{aligned} \tag{4.2}$$

where $\lambda^- = -n\gamma + d\det$. On this module of regular functions the group scheme $\mathcal{G}/\mathbb{Z}$ acts by right translations:

$$\rho_\lambda\left(\begin{pmatrix} a & b \\ c & d \end{pmatrix}\right)(f_P)\left(\begin{pmatrix} x & y \\ u & v \end{pmatrix}\right) = f_P\left(\begin{pmatrix} x & y \\ u & v \end{pmatrix}\begin{pmatrix} a & b \\ c & d \end{pmatrix}\right).$$

This is now the highest weight module $\mathcal{M}_\lambda$ for the group scheme $\mathcal{G}/\mathbb{Z}$. The highest weight vector is f_{X^n}, and clearly we have

$$\rho_\lambda\left(\begin{pmatrix} t_1 & w \\ 0 & t_2 \end{pmatrix}\right)(f_{X^n}) = \lambda\left(\begin{pmatrix} t_1 & w \\ 0 & t_2 \end{pmatrix}\right) f_{X^n} = \lambda\left(\begin{pmatrix} t_1 & 0 \\ 0 & t_2 \end{pmatrix}\right) f_{X^n}.$$

In the following we change the notation, instead of f_P we will simply write P.

Comment: when we say that $\mathcal{M}_\lambda$ is a module for the group scheme $\mathcal{G}/\mathbb{Z}$ we mean nothing more than that for any commutative ring R with identity, we have an action of $\mathcal{G}(R)$ on $\mathcal{M}_n \otimes R$, which is given by (4.1) and depends functorially on R. We can "evaluate" at $R = \mathbb{Z}$ and get the $\Gamma = \mathrm{Gl}_2(\mathbb{Z})$-module $\mathcal{M}_{\lambda,\mathbb{Z}}$. (actually we should not so much distinguish between the $\mathrm{Gl}_2(\mathbb{Z})$-module $\mathcal{M}_{\lambda,\mathbb{Z}}$ and $\mathcal{M}_\lambda$). Of course we have seen these $\mathrm{Gl}_2(\mathbb{Z})$-modules before, they are of course equal to the modules $\mathcal{M}_n[d - \frac{n}{2}]$ in Section 1.2.1.

The parameter d plays only a marginal role, sometimes we will consider $\mathcal{M}_n$ instead of $\mathcal{M}_\lambda$ and then we leave the choice of d open or in the case n that n is even we choose $d = -\frac{n}{2}$. This has the advantage that then $\mathcal{M}_n$ is a module for the projective linear group PGL_2.

Remark: There is a slightly more sophisticated interpretation of this module. We can form the flag manifold $\mathcal{B}\backslash\mathcal{G} = \mathbb{P}^1/\mathbb{Z}$, and the character λ yields a line bundle $\mathcal{L}_{\lambda^-}$. The group scheme $\mathcal{G}$ is acting on the pair $(\mathcal{B}\backslash\mathcal{G}, \mathcal{L}_{\lambda^-})$ and hence on $H^0(\mathcal{B}\backslash\mathcal{G}, \mathcal{L}_{\lambda^-})$ which is tautologically equal to $\mathcal{M}_\lambda$ (Borel-Weil theorem).

We can do essentially the same in the case B). In this case we start from an imaginary quadratic extension $F/\mathbb{Q}$ and let $\mathcal{O} = \mathcal{O}_F \subset F$ be its ring of integers. We form the group scheme $\mathcal{G}/\mathbb{Z} = R_{\mathcal{O}/\mathbb{Z}}(\mathrm{Gl}_2/\mathcal{O})$. Again $\mathcal{G}^{(1)}/\mathbb{Z}$ will be the kernel of $\det : \mathcal{G}/\mathbb{Z} \to \mathcal{Z}/\mathbb{Z} = R_{\mathcal{O}/\mathbb{Z}}(\mathbb{G}_m)$. Then $\mathcal{G}(\mathcal{O}) = \mathrm{Gl}_2(\mathcal{O}\otimes\mathcal{O}) \subset \mathrm{Gl}_2(\mathcal{O}) \times \mathrm{Gl}_2(\mathcal{O})$. The base change of the maximal torus $T/\mathbb{Q} \subset \mathcal{G}\times_{\mathbb{Z}}\mathbb{Q}$ is the product $T_1 \times T_2/F$, where the two factors are the standard maximal tori in the two factors Gl_2/F.

We get for the character module

$$X^*(T\times F) = X^*(T_1) \oplus X^*(T_2) = \{n_1\gamma_1 + d_1\mathrm{det}_1\} \oplus \{n_2\gamma_2 + d_2\mathrm{det}_2\} \quad (4.3)$$

where we have to observe the parity conditions $n_1 \equiv 2d_1 \mod 2$, and $n_2 \equiv 2d_2 \mod 2$.

Then the same procedure as in case A) provides a free $\mathcal{O}$-module $\mathcal{M}_\lambda$ with an action of $\mathcal{G}(\mathbb{Z})$ on it. To get this module and to see this action we embed the group $\mathcal{G}(\mathbb{Z}) = \mathrm{Gl}_2(\mathcal{O})$ into $\mathrm{Gl}_2(\mathcal{O})\times\mathrm{Gl}_2(\mathcal{O})$ by the map $g \mapsto (g,\bar{g})$ where $\bar{g}$ is of course the conjugate. If $\lambda = n_1\gamma_1 + d_1\det_1 + n_2\gamma_2 + d_2\mathrm{det}_2 = \lambda_1+\lambda_2$ then we have our two $\mathrm{Gl}_2(\mathcal{O})$-modules $\mathcal{M}_{\lambda_1,\mathcal{O}}, \mathcal{M}_{\lambda_2,\mathcal{O}}$, and this provides the $\mathrm{Gl}_2(\mathcal{O})\times\mathrm{Gl}_2(\mathcal{O})$-module $\mathcal{M}_{\lambda_1,\mathcal{O}}\otimes\mathcal{M}_{\lambda_2,\mathcal{O}}$. Our $\mathcal{M}_{\lambda,\mathcal{O}}$ is now simply the restriction of this tensor product module to $\mathcal{G}(\mathbb{Z})$. Sometimes we will also write our character as the sum of the semi-simple component and the central component, i.e.

$$\lambda = \lambda^{(1)} + \delta = (n_1\gamma_1 + n_2\gamma_2) + (d_1\mathrm{det}_1 + d_2\mathrm{det}_2).$$

The relevant term is the semi-simple component, the central component is not important at all, as it only serves to fulfill the parity condition. If we restrict the representation $\mathcal{M}_\lambda$ to $\mathcal{G}^{(1)}/\mathbb{Z}$, then the dependence on d disappears. In other words, representations with the same semi-simple highest weight component only differ by a twist, the role played by δ is marginal.

At this point we notice that the module $\mathcal{M}_{\lambda,\mathcal{O}}$ is only a module over $\mathcal{O}$. We may also say that $\mathcal{M}_{\lambda,\mathcal{O}}\otimes F$ is an absolutely irreducible highest weight module for the group $\mathcal{G}\otimes_{\mathcal{O}} F = \mathrm{Gl}_2\times\mathrm{Gl}_2/F$, this representation "is defined" over F. But in the special case that $\lambda_1 = \lambda_2$ we have an action of the Galois group $\mathrm{Gal}(F/\mathbb{Q})$: if $\mathbf{c}$ is the non-trivial element in this Galois group then

$$\begin{gathered}\mathbf{c}\left(\left(\sum_\nu a_\nu X^\nu Y^{n-\nu}\right)\otimes\left(\sum_\mu b_\mu \bar{X}^\mu \bar{Y}^{n-\mu}\right)\right) = \\ \left(\sum_\mu \mathbf{c}(b_\mu)X^\mu Y^{n-\mu}\right)\otimes\left(\sum_\nu \mathbf{c}(a_\nu)\bar{X}^\nu\bar{Y}^{n-\nu}\right),\end{gathered} \quad (4.4)$$

and for $g \in \mathcal{G}(\mathcal{O})$, $m \in \mathcal{M}_\lambda$ we have $\mathbf{c}(g)\mathbf{c}(m) = \mathbf{c}(gm)$, and therefore it is clear that the $\mathbb{Z}$-module $\mathcal{M}_\lambda{}^{\mathbf{c}}$ is a module for $\mathcal{G}/\mathbb{Z}$.

We return to $\mathrm{Gl}_2/\mathbb{Z}$. Given $\lambda = n\gamma + d\det$, we define the dual character as $\lambda^\vee = n\gamma - d\det$. For our finite dimensional modules we have

$$\mathcal{M}_\lambda^\vee \otimes \mathbb{Q} \xrightarrow{\sim} \mathcal{M}_{\lambda^\vee} \otimes \mathbb{Q}. \tag{4.5}$$

This isomorphism is induced by the Gl_2 invariant pairing

$$<\,,\,>: X^\nu Y^{n-\nu} \times X^{n-\mu} Y^\mu \mapsto \delta_{\nu,\mu} \binom{n}{\nu}^{-1}. \tag{4.6}$$

If we consider the modules over the integers this does not define a pairing over the integers. Thus we define submodule

$$\mathcal{M}_n^\flat = \left\{ P(X,Y) \mid P(X,Y) = \sum_{\nu=0}^{n} \binom{n}{\nu} a_\nu X^\nu Y^{n-\nu}, a_\nu \in \mathbb{Z} \right\}. \tag{4.7}$$

This is a submodule of $\mathcal{M}_n$ and the quotient $\mathcal{M}_n/\mathcal{M}_n^\flat$ is finite. It is also clear that this submodule is invariant under $\mathrm{Sl}_2/\mathbb{Z}$. We introduce some notation

$$e_\nu := X^\nu Y^{n-\nu} \text{ and } e_\nu^\flat := \binom{n}{\nu} X^{n-\nu} Y^\nu, \tag{4.8}$$

then the e_ν (resp. $e_\nu^\flat$) form a basis of $\mathcal{M}_n$ (resp. $\mathcal{M}_n^\flat$). These basis are dual basis of each other.

We always consider $\mathcal{M}_\lambda^\flat$ as a submodule of $\mathcal{M}_\lambda$. For the pairing we have

$$< e_\nu, e_\mu^\flat >= \delta_{\nu.\mu} \tag{4.9}$$

4.1.2 The principal series representations

We consider the two real algebraic groups $G = \mathrm{Gl}_2/\mathbb{R}$ (case A)) and $G = R_{\mathbb{C}/\mathbb{R}}(\mathrm{Gl}_2/\mathbb{C})$ (case B)). Let $T/\mathbb{R}$, (resp. $B/\mathbb{R}$) be the standard diagonal torus (resp. Borel subgroup of upper triangular matrices). Let us put $Z/\mathbb{R} = \mathbb{G}_m$ (resp. $R_{\mathbb{C}/\mathbb{R}}\mathbb{G}_m$). We have the determinant $\det : G/\mathbb{R} \to Z/\mathbb{R}$ and moreover $Z/\mathbb{R} = \mathrm{center}(G/\mathbb{R})$. If we restrict the determinant to the center then this becomes the map $z \mapsto z^2$. The kernel of the determinant is denoted by $G^{(1)}/\mathbb{R}$, of course $G^{(1)} = \mathrm{Sl}_2$, resp. $R_{\mathbb{C}/\mathbb{R}}(\mathrm{Sl}_2/\mathbb{C})$. Let us denote by $\mathfrak{g}, \mathfrak{g}^{(1)}, \mathfrak{t}, \mathfrak{b}, \mathfrak{z}$ the corresponding Lie-algebras.

The Cartan decompositions

In both cases we fix a maximal compact subgroup $K_\infty \subset G^{(1)}(\mathbb{R})$:

$$\begin{aligned} K_\infty = e(\phi) &= \left\{ \begin{pmatrix} \cos(\phi) & \sin(\phi) \\ -\sin(\phi) & \cos(\phi) \end{pmatrix} \mid \phi \in \mathbb{R} \right\} \\ K_\infty &= \left\{ \begin{pmatrix} \alpha & \beta \\ -\bar{\beta} & \bar{\alpha} \end{pmatrix} \mid \alpha\bar{\alpha} + \beta\bar{\beta} = 1 \right\} \end{aligned} \tag{4.10}$$

We define extensions $\tilde{K}_\infty = Z(\mathbb{R})^{(0)} K_\infty$, here of course $Z(\mathbb{R})^{(0)}$ is the connected component of the identity. In both cases the group K_∞ is the group of fixed points under the Cartan involution Θ_0, which is given by

$$\Theta_0 : g \mapsto {}^t g^{-1} \text{ resp. } g \mapsto {}^t \bar{g}^{-1} \text{ i.e. } \Theta_0\left(\begin{pmatrix} a & b \\ c & d \end{pmatrix}\right) = \begin{pmatrix} \bar{d} & -\bar{c} \\ -\bar{b} & \bar{a} \end{pmatrix}. \tag{4.11}$$

This involution induces an involution on $\mathfrak{g}^{(1)}$, and we can extend it to an involution acting on $\mathfrak{g} = \mathfrak{z} \oplus \mathfrak{g}^{(1)}$, we let it act trivially on $\mathfrak{z}$. Then the fixed point Lie algebra $\tilde{\mathfrak{k}} = \mathfrak{z} \oplus \mathfrak{k} \subset \mathfrak{z} \oplus \mathfrak{g}^{(1)}$ is the Lie-algebra of $\tilde{K}_\infty$.

Here are some arithmetic considerations, they may not be so relevant, but further down we make some choices of a basis in some of these algebras, and these choices can be justified by these arithmetic considerations.

We can write our group scheme $G/\mathbb{R}$ as a base extension of a group scheme $\mathcal{G}/\mathbb{Z}$, i.e. $G/\mathbb{R} = \mathcal{G} \times_\mathbb{Z} \mathbb{R}$. For this we simply take $\mathcal{G}/\mathbb{Z} = \mathrm{Gl}_2/\mathbb{Z}$ in case A). In case B) we take $\mathcal{G}/\mathbb{Z} = R_{\mathbb{Z}[\mathbf{i}]/\mathbb{Z}}(\mathrm{Gl}_2/\mathbb{Z}[\mathbf{i}])$. In case A) this gives a reductive group scheme over $\mathbb{Z}$, in case B) it is only a flat group scheme, but the base extension $\mathcal{G} \times_\mathbb{Z} \mathbb{Z}[1/2]$ is reductive (this group scheme over $\mathbb{Z}$ is not semi-simple since $\mathbb{Z}[\mathbf{i}]$ is ramified at the prime 2).

Now it is clear that Θ_0 is actually an automorphism of $\mathcal{G}^{(1)}/\mathbb{Z}$ and then it follows that the scheme of elements fixed by Θ_0 is again a group scheme $\mathcal{K}/\mathbb{Z}$. If we define $R = \mathbb{Z}[1/2]$ then $\mathcal{K} \times_\mathbb{Z} R$ is actually reductive (if we replace $\mathbb{Z}[\mathbf{i}]$ by the ring of integers of another imaginary quadratic extension, we have to modify R accordingly).

Consequently we see that all the above Lie-algebras are defined over R, hence they actually are free R-modules, and we denote them by $\mathfrak{g}_R$ and so on.

The Cartan Θ_0 involution induces an involution on the Lie algebras $\mathfrak{g}_R, \mathfrak{g}_R^{(1)}$. The module decomposes into a $+$ and a $-$ eigenspace

$$\mathfrak{g}_R = \tilde{\mathfrak{k}}_R \oplus \mathfrak{p}_R \text{ and } \mathfrak{g}_R^{(1)} = \mathfrak{k}_R \oplus \mathfrak{p}_R. \tag{4.12}$$

The $+$ eigenspaces $\tilde{\mathfrak{k}}_R, \mathfrak{k}_R$ are the Lie-algebras of $\tilde{\mathcal{K}}, \mathcal{K}$, both summands in the decompositions are $\tilde{\mathcal{K}}$-modules.

The Lie-algebra $\mathfrak{b}_R$ is not stable under Θ_0, and it is clear that the intersection

$$\mathfrak{b}_R \cap \Theta_0(\mathfrak{b}_R) = \mathfrak{t}_R,$$

where $\mathfrak{t}_R$ is the Lie-algebra of the standard maximal torus $\mathcal{T}/R \subset \mathcal{G}/R$. This torus is a product (up to isogeny) $\mathcal{T}/R = \mathcal{Z} \cdot \mathcal{T}^{(1)}/R$.

In case A) the torus $\mathcal{T}^{(1)}/R \xrightarrow{\sim} \mathbb{G}_m/R$, and the Cartan involution Θ_0 acts by $t \mapsto t^{-1}$. Therefore it acts by -1 on $\mathfrak{t}_R^{(1)}$. We write

$$\mathfrak{t}_R^{(1)} = R\begin{pmatrix} 1 & 0 \\ 0 & -1 \end{pmatrix} = RH. \tag{4.13}$$

The generator H is unique up to an element in $R^\times$, i.e. up to a sign and a power of 2.

In case B) the torus $\mathcal{T}^{(1)}/R$ is (up to isogeny) a product $\mathcal{T}_s^{(1)} \cdot \mathcal{T}_c^{(1)}/R$. The Cartan involution Θ_0 acts by $t \to t^{-1}$ on the split component $\mathcal{T}_s^{(1)}$ and by the identity on $\mathcal{T}_c^{(1)}$. The Lie-algebra decomposes accordingly into two summands of rank one:

$$\mathfrak{t}_R^{(1)} = R\begin{pmatrix}1 & 0\\ 0 & -1\end{pmatrix} \oplus R\begin{pmatrix}i & 0\\ 0 & -i\end{pmatrix} = RH \oplus RH_i.$$

In both cases the group scheme $\mathcal{K}$ acts on $\mathfrak{p}_R$ by the adjoint action, we can describe this action explicitly.

In case A) the group scheme $\mathcal{K}$ is the following group of matrices

$$\mathcal{K} = \left\{\alpha = \begin{pmatrix}a & b\\ -b & a\end{pmatrix} | a^2 + b^2 = 1\right\}.$$

This is a torus over R which splits over $R[i]$. We have

$$\mathfrak{p}_R = RH \oplus R\begin{pmatrix}0 & 1\\ 1 & 0\end{pmatrix} = RH \oplus RV,$$

and $\mathrm{Ad}(\alpha)(H) = (a^2 - b^2)H - 2abV$, $\mathrm{Ad}(\alpha)(V) = 2abH + (a^2 - b^2)V$. Since the torus splits over $\mathbb{Z}[\mathbf{i}]$, we can decompose $\mathfrak{p} \otimes R[i]$ into weight spaces. We introduce the basis elements

$$P_+ := H + V \otimes i,\ P_- := H - V \otimes i \in \mathfrak{p} \otimes R[i] \tag{4.14}$$

then

$$\mathrm{Ad}(\alpha)P_+ = (a + bi)^2 P_+,\ \mathrm{Ad}(\alpha)P_- = (a - bi)^2 P_-.$$

Hence we get - in case A) - the decomposition

$$\mathfrak{g}^{(1)}_{R[i]} = \mathfrak{k}_{R[i]} \oplus \mathfrak{p}_{R[i]} = R[i]\begin{pmatrix}0 & 1\\ -1 & 0\end{pmatrix} \oplus R[i]P_+ \oplus R[i]P_- = R[i]Y \oplus R[i]P_+ \oplus R[i]P_- \tag{4.15}$$

where the generators are unique up to an element in $R[i]^\times$.

In case B) the group scheme $\mathcal{K}/R$ is semi-simple. It contains $\mathcal{T}_c^{(1)}/R$ as maximal torus. The two $\mathcal{K}/R$ modules $\mathfrak{k}_R$ and $\mathfrak{p}_R$ are highest weight modules of rank 3. Since 2 is invertible in R, they are even isomorphic. Again we can decompose them into rank one weight spaces and give almost canonical generators for these weight spaces. The Lie algebra is:

$$\mathfrak{k}_R = RH_i \oplus R\begin{pmatrix}0 & 1\\ -1 & 0\end{pmatrix} \oplus R\begin{pmatrix}0 & i\\ i & 0\end{pmatrix} = RH_i \oplus RY \oplus RF_i. \tag{4.16}$$

We introduce the elements $P_{c+} = Y - F_i \otimes i$, $P_{c-} = Y + F_i \otimes i$, and then

$$\mathfrak{k}_R \otimes R[i] = R[i]H_i \oplus R[i]P_{c+} \oplus R[i]P_{c-}. \tag{4.17}$$

This is the decomposition into weight spaces under the action of $\mathcal{T}_c^{(1)}/R$, and the element $\alpha = \begin{pmatrix} x & 0 \\ 0 & \bar{x} \end{pmatrix}$ acts via the adjoint action

$$\mathrm{Ad}(\alpha)P_{c+} = x^2 P_{c+}\ ,\ \mathrm{Ad}(\alpha)H_i = H_i\ ,\ \mathrm{Ad}(\alpha)P_{c-} = x^{-2}P_{c-}.$$

Essentially the same can be done for $\mathfrak{p}_R \otimes R[i]$. We define

$$P_{\mathfrak{p},+} = V - \begin{pmatrix} 0 & i \\ -i & 0 \end{pmatrix} \otimes i,\ P_{\mathfrak{p},-} = V + \begin{pmatrix} 0 & i \\ -i & 0 \end{pmatrix} \otimes i.$$

Then we get the weight decomposition

$$\mathfrak{p}_R \otimes R[i] = R[i]P_{\mathfrak{p},+} \oplus R[i]H \oplus R[i]P_{\mathfrak{p},-}. \tag{4.18}$$

Rational characters versus continuous characters

Our aim is to construct certain irreducible (differentiable) representations of $G(\mathbb{R})$ together with their "algebraic skeleton", the associated *Harish-Chandra modules.*

For any torus $T/\mathbb{R}$ we consider the group $\mathrm{Hom}(T(\mathbb{R}), \mathbb{C}^\times)$ of continuous characters, and we write this group multiplicatively, i.e. $\chi_1 \cdot \chi_2(x) = \chi_1(x)\chi_2(x)$. We also have defined the group of (rational) characters $X^*(T \times_{\mathbb{R}} \mathbb{C}, \mathbb{G}_m)$ (see Chap. 1, 1.5), and we have the evaluation map

$$X^*(T \times_{\mathbb{R}} \mathbb{C}, \mathbb{G}_m) \xrightarrow{ev} \mathrm{Hom}(T(\mathbb{R}), \mathbb{C}^\times);\ ev : \gamma \mapsto \gamma_{\mathbb{R}} : x \mapsto |\gamma(x)|. \tag{4.19}$$

Sometimes we will also write $|\gamma_{\mathbb{R}}|$ for $\gamma_{\mathbb{R}}$. Since we wrote the group of (rational) characters additively we have

$$(\gamma_1 + \gamma_2)_{\mathbb{R}} = \gamma_{1\mathbb{R}} \cdot \gamma_{2\mathbb{R}}.$$

For $\gamma \otimes z \in X^*(T) \otimes \mathbb{C}$ we also consider the character $x \mapsto |\gamma(x)|^z$.

4.1.3 The induced representations

We start from a continuous homomorphism (a character) $\chi : T(\mathbb{R}) \to \mathbb{C}^\times$, of course, this can also be seen as a character $\chi : B(\mathbb{R}) \to \mathbb{C}^\times$. This allows us to define the induced module

$$I_B^G\chi := \{f : G(\mathbb{R}) \to \mathbb{C} | f \in \mathcal{C}_\infty(G(\mathbb{R})), f(bg) = \chi(b)f(g),\ \forall\, b \in B(\mathbb{R}), g \in G(\mathbb{R})\},$$

where we require that f should be smooth. Then this space of functions is a $G(\mathbb{R})$-module, where the group $G(\mathbb{R})$ acts by right translations: for $f \in I_B^G\chi$ and $g \in G(\mathbb{R})$ we put

$$R_g(f)(x) = f(xg).$$

If we modify our character χ by a character $\delta \circ \det$ where $\delta : Z(\mathbb{R}) \to \mathbb{C}^\times$, then the central character of the representation gets multiplied by δ^2.

We know that $G(\mathbb{R}) = B(\mathbb{R}) \cdot \tilde{K}_\infty$. This implies that a function $f \in I_B^G \chi$ is determined by its restriction to $\tilde{K}_\infty$. In other words we have an identification of vector spaces

$$I_B^G \chi = \{f : \tilde{K}_\infty \to \mathbb{C} \mid f(t_c k) = \chi(t_c) f(k), t_c \in \tilde{K}_\infty \cap B(\mathbb{R}), k \in \tilde{K}_\infty\}. \quad (4.20)$$

The center acts by the central character ω_χ, the restriction of χ to $Z(\mathbb{R})$.

We put $T_c = B(\mathbb{R}) \cap \tilde{K}_\infty$ and define χ_c to be the restriction of χ to T_c. Then the module on the right in the above equation can be written as $I_{T_c}^{\tilde{K}_\infty} \chi_c$. By its very definition, $I_{T_c}^{\tilde{K}_\infty} \chi_c$ is only a $\tilde{K}_\infty$-module. Inside $I_{T_c}^{\tilde{K}_\infty} \chi_c$ we have the submodule of vectors of finite type

$${}^\circ I_{T_c}^{\tilde{K}_\infty} \chi_c := \{f \in I_{T_c}^{\tilde{K}_\infty} \chi_c \mid R_k(f),\ k \in \tilde{K}_\infty, \text{ lies in a finite dimensional subspace}\}.$$

Here it suffices to consider only the translates $R_k(f)$ for $k \in K_\infty$, because $Z(\mathbb{R})^{(0)}$ acts by the character ω_χ. The famous Peter-Weyl theorem tells us that all irreducible representations (satisfying some continuity condition) are finite dimensional and occur with finite multiplicity in $I_{T_c}^{\tilde{K}_\infty} \chi_c$, and therefore we get

$${}^\circ I_{T_c}^{\tilde{K}_\infty} \chi_c = \bigoplus_{\vartheta \in \hat{K}_\infty} V_\vartheta^{m(\vartheta)} = \bigoplus_{\vartheta \in \hat{K}_\infty} {}^\circ I_{T_c}^{\tilde{K}_\infty} \chi_c[\vartheta], \quad (4.21)$$

where $\hat{K}_\infty$ is the set of isomorphism classes of irreducible representations of K_∞, where V_ϑ is an irreducible module of type ϑ, and where $m(\vartheta)$ is the multiplicity of ϑ in ${}^\circ I_{T_c}^{\tilde{K}_\infty} \chi_c$. Of course ${}^\circ I_{T_c}^{\tilde{K}_\infty} \chi_c$ is a submodule of $I_B^G \chi$, but this submodule is not invariant under the operation of $G(\mathbb{R})$, in other words, if $0 \neq f \in {}^\circ I_{T_c}^{\tilde{K}_\infty} \chi_c$ and $g \in G(\mathbb{R})$ a sufficiently general element, then $R_g(f) \notin {}^\circ I_{T_c}^{\tilde{K}_\infty} \chi_c$.

We can differentiate the action of $G(\mathbb{R})$ on $I_B^G \chi$. We have the well-known exponential map $\exp : \mathfrak{g} = \mathrm{Lie}(G/\mathbb{R}) \to G(\mathbb{R})$, and for $f \in I_B^G, X \in \mathfrak{g}$ we define

$$X f(g) = \lim_{t \to 0} \frac{f(g \exp(tX)) - f(g)}{t}, \quad (4.22)$$

and it is well-known and also easy to see, that this gives an action of the Lie-algebra on I_B^G, where we have $X_1(X_2 f) - X_2(X_1 f) = [X_1, X_2] f$. The Lie-algebra is a K_∞-module under the adjoint action and it is obvious that for $f \in {}^\circ I_{T_c}^{\tilde{K}_\infty} \chi_c[\vartheta]$, the element Xf lies in $\bigoplus_{\vartheta \in \hat{K}_\infty} {}^\circ I_{T_c}^{\tilde{K}_\infty} \chi_c[\vartheta']$ where ϑ' runs over the finitely many isomorphism types occurring in $V_\vartheta \otimes \mathfrak{g}$. Hence

Proposition 4.1.1. *The submodule ${}^\circ I_{T_c}^{\tilde{K}_\infty} \chi_c \subset I_B^G \chi$ is invariant under the action of $\mathfrak{g}$.*

The submodule ${}^\circ I_{T_c}^{\tilde{K}_\infty} \chi_c$ together with this action of $\mathfrak{g}$ will now be denoted by $\mathfrak{I}_B^G \chi$. We should think of this module as the algebraic skeleton of $I_B^G \chi$.

Such a module will be called a $(\mathfrak{g}, K_\infty)$-module or a Harish-Chandra module. This means that we have an action of the Lie-algebra $\mathfrak{g}$, an action of K_∞, and these two actions satisfy some obvious compatibility conditions.

We also observe that ${}^{\circ}I_{T_c}^{\tilde{K}_\infty}\chi_c$ is also invariant under right translation R_z for $z \in Z(\mathbb{R})$. Hence we can extend the action of K_∞ to the larger group $\tilde{K}_\infty = K_\infty \cdot Z(\mathbb{R})$. Then $\mathfrak{I}_B^G\chi$ becomes a $(\mathfrak{g}, \tilde{K}_\infty)$-module. Finally, observe that in the case A), the element

$$\mathbf{c} = \begin{pmatrix} -1 & 0 \\ 0 & 1 \end{pmatrix} \notin \tilde{K}_\infty, \tag{4.23}$$

and $R_{\mathbf{c}}$ induces an involution on $\mathfrak{I}_B^G$. We could also say that we can enlarge K_∞ (resp. $\tilde{K}_\infty$) to subgroups $K_\infty{}^*$ (resp. $\tilde{K}_\infty^*$) which contain $\mathbf{c}$ and contain K_∞(resp. $\tilde{K}_\infty$) as subgroups of index two. Then $\mathfrak{I}_B^G\chi$ also becomes a $(\mathfrak{g}, \tilde{K}_\infty^*)$-module.

These $(\mathfrak{g}, \tilde{K}_\infty)$-modules $\mathfrak{I}_B^G\chi$ are called the *principal series modules.* We have the following important theorem:

Theorem 4.1.1. *For any irreducible Harish-Chandra module $\mathcal{V}$ we can find a χ such that we have an embedding of $(\mathfrak{g}, \tilde{K}_\infty)$-modules*

$$i : \mathcal{V} \hookrightarrow \mathfrak{I}_B^G\chi.$$

This is actually a special case of a much more general theorem and applies *mutatis mutandis* to all reductive groups over $\mathbb{R}$. In the following we will see, that in our special cases we only have a very short list of submodules of the $\mathfrak{I}_B^G\chi$, and hence we get a complete list of irreducible Harish-Chandra modules.

We denote the restriction of χ to the central torus $Z = \left\{\begin{pmatrix} t & 0 \\ 0 & t \end{pmatrix}\right\}$ by ω_χ. Then $Z(\mathbb{R})$ acts on $\mathfrak{I}_B^G\chi$ by the *central character* ω_χ, i.e. $R_z(f) = \omega_\chi(z)f$. Once we fix the central character, then there is no difference between $(\mathfrak{g}, \tilde{K}_\infty)$ and $(\mathfrak{g}, K_\infty)$-modules. Hence we always assume that ω_χ is fixed.

Unitary induction

In the general theory of representations of real (or p-adic groups), people work with *unitary induction.* We introduce the special character

$$|\rho|_{\mathbb{R}} : \begin{pmatrix} t_1 & u \\ 0 & t_2 \end{pmatrix} \to \left|\frac{t_1}{t_2}\right|^{\frac{1}{2}}, \tag{4.24}$$

here the absolute value $|t|$ is the usual absolute value if we are in case A) and $|z| = z\bar{z}$ for $z \in \mathbb{C}$ if we are in case B).

Now we define the unitarily induced module (see 1.65).

$$\mathrm{Indunit}_B^G\chi := I_B^G\chi \cdot |\rho|_{\mathbb{R}}.$$

This concept of induction is of course equivalent to the previous one. It has certain advantages, as some statements have a more elegant formulation. But in this book we are also interested in the "arithmetic properties" of our modules, and here the naïve concept of induction has its own virtues.

The decomposition into K_∞-types

We look briefly at the K_∞-module ${}^\circ I_{T_c}^{\tilde{K}_\infty}\chi_c$. In case A) the group

$$K_\infty = \mathrm{SO}(2) = \left\{ \begin{pmatrix} \cos(\phi) & \sin(\phi) \\ -\sin(\phi) & \cos(\phi) \end{pmatrix} = e(\phi) \right\}, \tag{4.25}$$

and $T_c = {K_\infty}^T = T(\mathbb{R}) \cap K_\infty$ is cyclic of order two with generator $e(\pi)$. Then χ_c is given by an integer mod 2, i.e. $\chi_c(e(\pi)) = (-1)^m$. For any $n \equiv m \mod 2$ we define $\psi_n \in \mathfrak{I}_B^G\chi$ by

$$\psi_n(e(\phi))) = e^{in\phi} \tag{4.26}$$

and then decoKuA

$$\mathfrak{I}_B^G\chi = \bigoplus_{k\equiv m \mod 2} \mathbb{C}\psi_k. \tag{4.27}$$

The group $T(\mathbb{R}) = T_c \times (\mathbb{R}_{>0}^\times)^2$ and sometimes we write our character χ in the form

$$\chi(t) = \chi_c(t)|t_1|^{z_1}|t_2|^{z_2} = \chi_c(t)|\frac{t_1}{t_2}|^{\frac{z_1-z_2}{2}}|t_1t_2|^{\frac{z_1+z_2}{2}}, \tag{4.28}$$

where $z_1, z_2 \in \mathbb{C}$. We put $z = z_1 - z_2$ and $\zeta = z_1 + z_2$. The relevant variable is z. If we restrict this character to the split torus in $\mathrm{Sl}_2(\mathbb{R})$ it becomes

$$\chi : \begin{pmatrix} t & 0 \\ 0 & t^{-1} \end{pmatrix} \mapsto \chi_c(t)|\,t|^z$$

We leave it an exercise for the reader to show that

$$P_+(\psi_\nu) = (z+\nu)\psi_{\nu+2}\,; P_-(\psi_\nu) = (z-\nu)\psi_{\nu-2} \tag{4.29}$$

This implies that $\mathfrak{I}_B^G\chi$ is irreducible unless z is an integer. What happens in this case (the interesting case in this book) will be discussed later.

In the case B) the maximal compact subgroup is

$$U(2) \subset G(\mathbb{R}) = R_{\mathbb{C}/\mathbb{R}}(\mathrm{Gl}_2/\mathbb{C})(\mathbb{R}) \subset \mathrm{Gl}_2(\mathbb{C}) \times \mathrm{Gl}_2(\mathbb{C}),$$

which is the group of real points of the reductive group $U(2)/\mathbb{R}$. The intersection

$$T_c = {K_\infty}^T = T(\mathbb{R}) \cap K_\infty = \left\{ \begin{pmatrix} e^{i\phi_1} & 0 \\ 0 & e^{i\phi_2} \end{pmatrix} = e(\underline{\phi}) \right\}.$$

The base change $U(2) \times \mathbb{C} = \mathrm{Gl}_2/\mathbb{C}$ and $T_c \times \mathbb{C}$ becomes the standard maximal compact torus. The irreducible finite dimensional $U(2)$-modules are labelled by dominant highest weights $\lambda_c = n\gamma_c + d\det \in X^*(T_c \times \mathbb{C})$ (see section 4.1.1, here again $n \geq 0, n \in \mathbb{Z}, n \equiv 2d \mod 2$ and $\gamma_c(e(\underline{\phi})) = e^{i(\phi_1-\phi_2)/2}$).

We denote these modules by $\mathcal{M}_{\lambda_c}$; after base change to $\mathbb{C}$ they become the modules $\mathcal{M}_{\lambda,\mathbb{C}}$.

As a subgroup of $G(\mathbb{R}) \subset \mathrm{Gl}_2(\mathbb{C}) \times \mathrm{Gl}_2(\mathbb{C})$ our torus is

$$T_c = \left\{ \begin{pmatrix} e^{i\phi_1} & 0 \\ 0 & e^{i\phi_2} \end{pmatrix} \times \begin{pmatrix} e^{-i\phi_1} & 0 \\ 0 & e^{-i\phi_2} \end{pmatrix} \right\} \xrightarrow{\sim} \left\{ \begin{pmatrix} e^{i\phi_1} & 0 \\ 0 & e^{i\phi_2} \end{pmatrix} \right\},$$

and the restriction of χ to T_c is of the form

$$\chi_c(e(\underline{\phi})) = e^{ia\phi_1 + ib\phi_2} = e^{\frac{a-b}{2}(\phi_1-\phi_2)} e^{\frac{a+b}{2}(\phi_1+\phi_2)},$$

and this character is $(a-b)\gamma_c + \frac{a+b}{2}\det$. Then we know decoKuB

$$\circ I_{T_c}^{\tilde{K}_\infty} \chi_c = \mathfrak{I}_B^G \chi = \bigoplus_{\mu_c = k\gamma_c + \frac{a+b}{2}\det; k \equiv (a-b) \mod 2; k \geq |a-b|} \mathcal{M}_{\mu_c}. \tag{4.30}$$

4.1.4 Intertwining operators

Let $N(T)$ the normaliser of $T/\mathbb{R}$, the quotient $W = N(T)/T$ is a finite group scheme. Then in our case, the group $W(\mathbb{R})$ is cyclic of order 2 and is generated by

$$w_0 = \begin{pmatrix} 0 & 1 \\ -1 & 0 \end{pmatrix}.$$

In case A) we have $W(\mathbb{R}) = W(\mathbb{C})$, and in case B) we have

$$G \times_{\mathbb{R}} \mathbb{C} = (\mathrm{Gl}_2 \times \mathrm{Gl}_2)/\mathbb{C}\ ;\ T \times_{\mathbb{R}} \mathbb{C} = T_1 \times T_2\ ; W(\mathbb{C}) = \mathbb{Z}/2 \times \mathbb{Z}/2,$$

where the two factors are generated by $s_1 = (w_0, 1), s_2 = (1, w_0)$. The group $W(\mathbb{R})$ - the group of real points of the Weyl group - is the cyclic group of order two generated by (w_0, w_0). We also call this element w_0. The group $W(\mathbb{R})$ acts on $T(\mathbb{R})$ by conjugation, and hence it also acts on the group $\mathrm{Hom}(T(\mathbb{R}), \mathbb{C}^\times)$ of characters, and we denote this action by $\chi \mapsto \chi^w$. We write this group of characters multiplicatively and we define the twisted action

$$w \cdot \chi = (\chi|\rho_{\mathbb{R}}|)^w |\rho_{\mathbb{R}}|^{-1}.$$

We recall some well-known facts:

i) We have a non-degenerate $(\mathfrak{g}, K_\infty)$-invariant pairing

$$\mathfrak{I}_B^G \chi \times \mathfrak{I}_B^G \chi^{w_0} |\rho|_{\mathbb{R}}^2 \to \mathbb{C}(\omega_\chi \circ \det) \text{ given by } (f_1, f_2) \mapsto \int_{K_\infty} f_1(k) f_2(k) dk.$$

We define the dual $\mathfrak{I}_B^{G,\vee}\chi$ of a Harish-Chandra module as a submodule of $\mathrm{Hom}_{\mathbb{C}}(\mathfrak{I}_B^G\chi, \mathbb{C})$. It consists of those linear maps which vanish on almost all K_∞ types. It is clear that this is again a $(\mathfrak{g}, K_\infty)$-module. The above assertion can be reformulated as

ii) We have an isomorphism of $(\mathfrak{g}, K_\infty)$-modules

$$\mathfrak{I}_B^G \chi (\omega_\chi \circ \det)^{-1} \to \mathfrak{I}_B^{G,\vee} \chi^{w_0} |\rho|_{\mathbb{R}}^2. \tag{4.31}$$

For $f \in \mathfrak{I}_B^G\chi,\ g \in G(\mathbb{R})$ we consider the integral

$$T_\infty^{\mathrm{loc}}(f)(g) = \int_{U(\mathbb{R})} f(w_0 u g) du. \tag{4.32}$$

It is well-known and easy to check that these integrals converge absolutely and locally uniformly for $\Re(z) >> 0$ and provide an intertwining operator

$$T_\infty^{\mathrm{loc}}(\chi, z) : \mathfrak{I}_B^G\chi^{w_0}|\rho|_{\mathbb{R}}^z \to \mathfrak{I}_B^G\chi|\rho|_{\mathbb{R}}^2|\rho|_{\mathbb{R}}^{-z}. \tag{4.33}$$

It is also not hard to see that they extend to meromorphic functions on all of $\mathbb{C}^2$. To see this, we recall the decomposition into K_∞ types

$$\mathfrak{I}_B^G\chi_{\mathbb{R}}^{w_0}|\rho|_{\mathbb{R}}^z = \bigoplus_{\vartheta\in\hat{K}_\infty} {}^\circ I_{T_c}^{\tilde{K}_\infty}\chi_c[\vartheta] = \bigoplus_{\vartheta\in\hat{K}_\infty} \mathfrak{I}_B^G\chi_{\mathbb{R}}^{w_0}|\rho|_{\mathbb{R}}^z[\vartheta],$$

and our intertwining operator is a direct sum of linear maps between finite dimensional vector spaces

$$c(\chi, z, \vartheta) : \mathfrak{I}_B^G\chi^{w_0}|\rho|_{\mathbb{R}}^z[\vartheta] \to \mathfrak{I}_B^G\chi|\rho|_{\mathbb{R}}^{2-z}[\vartheta].$$

The finite dimensional vector spaces do not depend on z, and the $c(\chi|\rho|_{\mathbb{R}}^z, \vartheta)$ can be expressed in terms of values of the Γ-function. Especially they are meromorphic functions in the variable z sSee sl2neu.pdf). For any $z_0 \in \mathbb{C}$ where we have a pole we can find an integer $m \geq 0$ such that

$$(z - z_0)^m T_\infty^{\mathrm{loc}}(\chi^{w_0}, z)|_{z=z_0} : \mathfrak{I}_B^G\chi^{w_0} \to \mathfrak{I}_B^G\chi|\rho|_{\mathbb{R}}^2,$$

is a non-zero intertwining operator, and this is now our regularized operator $T_\infty^{\mathrm{loc,reg}}(\chi^{w_0})$.

iii) The regularized values define non-zero intertwining operators

$$T_\infty^{\mathrm{loc,reg}}(\chi^{w_0}, z) : \mathfrak{I}_B^G\chi \to \mathfrak{I}_B^G\chi^{w_0}|\rho|_{\mathbb{R}}^2. \tag{4.34}$$

These operators span the one-dimensional space of intertwining operators

$$\mathrm{Hom}_{(\mathfrak{g},K_\infty)}(\mathfrak{I}_B^G\chi, \mathfrak{I}_B^G w_0 \cdot \chi).$$

Of course we can translate this into the language of unitary induction, then the intertwining operators map

$$T^{\mathrm{loc,un}}(\chi) : \mathrm{Indunit}_B^G\chi \to \mathrm{Indunit}_B^G\chi^{w_0}.$$

This is definitely a more elegant formulation.

Finally we discuss the question of which of these representations are unitary. This means that we have to find a pairing

$$\psi : \mathfrak{I}_B^G\chi \times \mathfrak{I}_B^G\chi \to \mathbb{C} \tag{4.35}$$

which satisfies

a) It is linear in the first and conjugate linear in the second variable.

b) It is positive-definite, i.e. $\psi(f,f) > 0\ \forall f \neq 0 \in \mathfrak{I}_B^G\chi$.

c) It is invariant under the action of K_∞ and Lie-algebra invariant under the action of $\mathfrak{g}$, i.e. we have

For $f_1, f_2 \in \mathfrak{I}_B^G\chi$ and $X \in \mathfrak{g}$ we have $\psi(Xf_1, f_2) + \psi(f_1, Xf_2) = 0$.

We are also interested in quasi-unitary modules. This notion is perhaps best explained if instead of c) we require

d) There exists a continuous homomorphism (a character) $\eta : G(\mathbb{R}) \to \mathbb{R}^\times$ such that for all $X \in \mathfrak{g}$, $\psi(Xf_1, f_2)+\psi(f_1, Xf_2) = d\eta(X)\psi(f_1, f_2)$, $f_1, f_2 \in \mathfrak{I}_B^G\chi$.

It is clear that a non-zero pairing ψ which satisfies a) and c) is the same as a non-zero $(\mathfrak{g}, K_\infty)$-module linear map

$$i_\psi : \mathfrak{I}_B^G\chi \to \overline{(\mathfrak{I}_B^G\chi)^\vee}. \tag{4.36}$$

This means that i_ψ is a conjugate linear map from $\mathfrak{I}_B^G\chi$ to $(\mathfrak{I}_B^G\chi)^\vee$. The map i_ψ and the pairing ψ are related by the formula $\psi(v_1, v_2) = i_\psi(v_2)(v_1)$.

Of course we know that (see (4.31),4.28))

$$\overline{(\mathfrak{I}_B^G\chi)^\vee} \xrightarrow{\sim} \mathfrak{I}_B^G\overline{\chi^{w_0}|\rho|_{\mathbb{R}}^2(\omega_\chi \circ \det)^{-1}}. \tag{4.37}$$

Hence we find a i_ψ if

$$\chi = \overline{\chi^{w_0}|\rho|_{\mathbb{R}}^2\delta_\chi^{-1}} \text{ or } \chi^{w_0}|\rho|_{\mathbb{R}}^2 = \overline{\chi^{w_0}|\rho|_{\mathbb{R}}^2\delta_\chi^{-1}}. \tag{4.38}$$

We write our χ in the form (4.28). A necessary condition for the existence of a hermitian form is of course that all $|\omega_\chi(x)| = 1$ for $x \in Z(\mathbb{R})$. This means that $\Re(z_1 + z_2) = 0$, hence we write

$$z_1 = \sigma + i\tau_1, z_2 = -\sigma + i\tau_2. \tag{4.39}$$

Then the two conditions in (4.38) simply say

$$(\mathrm{un}_1) : \sigma = \frac{1}{2} \quad \text{or} \quad (\mathrm{un}_2) : \tau_1 = \tau_2 \text{ and } \chi_c = \chi_c^{w_0}. \tag{4.40}$$

In both cases we can write down a pairing which satisfies a) and c). We still have to check b). In the first case, i.e. $\sigma = \frac{1}{2}$, we simply take the map $i_\psi = \mathrm{Id}$ and then we get for $f_1, f_2 \in \mathfrak{I}_B^G\chi$ the formula

$$\psi(f_1, f_2) = \int_{K_\infty} f_1(k)\overline{f_2(k)}dk, \tag{4.41}$$

and this is clearly positive-definite. These are the representations of the *unitary principal series.*

In the second case we have to use the intertwining operator in (4.34) and write

$$\psi(f_1, f_2) = T_\infty^{\mathrm{loc,reg}}(f_2)(f_1). \tag{4.42}$$

Now it is not clear whether this pairing satisfies b). This will depend on the parameter σ. We can twist by a character $\eta : R_{\mathbb{C}/\mathbb{R}}\mathbb{G}_m(\mathbb{R}) \to \mathbb{C}^\times$ and achieve that $\chi_c = 1, \tau_1 = \tau_2 = 0$. We know that for $\sigma = \frac{1}{2}$, the intertwining operator T_∞^{loc} is regular at χ, and since in addition under these conditions $\mathfrak{I}_B^G\chi$ is irreducible we see that

$$T_\infty^{\text{loc}}(\chi) = \alpha \,\text{Id with } \alpha \in \mathbb{R}_{>0}^\times. \tag{4.43}$$

We see that the two pairings defined by the rule in case (un$_1$) and (un$_2$) differ by a positive real number. Hence the pairing defined in (4.42) is positive-definite if $\sigma = \frac{1}{2}$.

But now we can vary σ. It is well-known that $\mathfrak{I}_B^G\chi$ stays irreducible as long as $0 < \sigma < 1$ (see next section) and since $T_\infty^{\text{loc}}(\chi)(f)(f)$ varies continuously we see that (4.42) defines a positive-definite hermitian product on $\mathfrak{I}_B^G\chi$ as long as $0 < \sigma < 1$. This is the *supplementary series.* What happens if we leave this interval will be discussed in the next section.

4.1.5 Reducibility and representations with non-trivial cohomology

As usual we denote by $\rho \in X^*(T) \otimes \mathbb{Q}$ the half sum of positive roots we have $\rho = \gamma$ (resp. $\rho = \gamma_1 + \gamma_2 \in X^*(T) \otimes \mathbb{Q}$) in case A) (resp. B)).

For any character $\lambda \in X^*(T \times \mathbb{C})$ the character $\lambda_\mathbb{R}$ provides a homomorphism $B(\mathbb{R}) \to T(\mathbb{R})$, and hence we get the Harish-Chandra modules $\mathfrak{I}_B^G\lambda_\mathbb{R}$, which are of special interest for us because these are the only ones with non-trivial cohomology. It follows easily from (4.29 that $\mathfrak{I}_B^G\chi$ is always irreducible unless $\chi = \lambda_\mathbb{R}$ for some λ .

We return to the situation discussed in Section 4.1.1, especially we reintroduce the field $F/\mathbb{Q}$. Then we have $X^*(T \times F) = X^*(T \times \mathbb{C})$ and hence $\lambda \in X^*(T \times F)$. We assume that λ is dominant, i.e. $n \geq 0$ in case A) or $n_1, n_2 \geq 0$ in case B). Now we realise our modules $\mathcal{M}_\lambda$ as submodules in the algebra of regular functions on $\mathcal{G}/\mathbb{Z}$: if we look at the definition (see (4.2)) we see immediately that $\mathcal{M}_{\lambda,\mathbb{C}} \subset \mathfrak{I}_B^G\lambda_\mathbb{R}^{w_0}$, and hence we get an exact sequence of $(\mathfrak{g}, K_\infty)$ modules

$$0 \to \mathcal{M}_{\lambda,\mathbb{C}} \to \mathfrak{I}_B^G\lambda_\mathbb{R}^{w_0} \xrightarrow{r} \mathcal{D}_\lambda \to 0. \tag{4.44}$$

Hence we see that $\mathfrak{I}_B^G\lambda_\mathbb{R}^{w_0}$ is not irreducible. We may also argue that

$$\mathcal{M}_\lambda = \bigoplus_{\nu:\nu\equiv m \mod 2;|\nu|\leq n} \mathbb{C}\psi_\mu$$

is an invariant submodule because $P_+\psi_n = 0; P_-\psi_{-n} = 0$.

We can also look at the dual sequence. Here we recall that we wrote $\lambda = n\gamma + d\det$. Clearly $\mathcal{M}_{\lambda,\mathbb{C}}^\vee = \mathcal{M}_{\lambda-2d\det,\mathbb{C}}$, if we twist the dual sequence by $\det^{2d}$ then the dual sequence becomes

$$0 \to \mathcal{D}_\lambda^\vee \otimes \det_\mathbb{R}^{2d} \to (\mathfrak{I}_B^G\lambda_\mathbb{R}^{w_0})^\vee \otimes \det_\mathbb{R}^{2d} \to \mathcal{M}_{\lambda,\mathbb{C}} \to 0. \tag{4.45}$$

Equation (4.31) yields $(\mathfrak{I}_B^G\lambda_{\mathbb{R}}^{w_0})^\vee\otimes\det_{\mathbb{R}}^{2d}\xrightarrow{\sim}\mathfrak{I}_B^G\lambda_{\mathbb{R}}|\rho|_{\mathbb{R}}^2$, and our second sequence becomes

$$0\to\mathcal{D}_\lambda^\vee\otimes\det_{\mathbb{R}}^{2d}\to\mathfrak{I}_B^G\lambda_{\mathbb{R}}|\rho|_{\mathbb{R}}^2\to\mathcal{M}_{\lambda,\mathbb{C}}\to 0, \tag{4.46}$$

where we set $\mathcal{D}_{\lambda^\vee}:=\mathcal{D}_\lambda^\vee\otimes\det_{\mathbb{R}}^{2d}$. Again we can argue that $\mathcal{D}_\lambda^\vee\otimes\det_{\mathbb{R}}^{2d}$ is an invariant submodule because $P_+\psi_{-n-2}=0, P_-\psi_{n+2}=0$

Now we consider the two middle terms in the two exact sequences (4.44), (4.46) above. The equation (4.34) claims that we have two non-zero *regularized* intertwining operators

$$\begin{aligned}T_\infty^{\mathrm{loc,reg}}(\lambda_{\mathbb{R}}^{w_0}):\mathfrak{I}_B^G\lambda_{\mathbb{R}}^{w_0}\to\mathfrak{I}_B^G\lambda_{\mathbb{R}}|\rho|_{\mathbb{R}}^2\ ;\\ T_\infty^{\mathrm{loc,reg}}(\lambda_{\mathbb{R}}|\rho|_{\mathbb{R}}^2):\mathfrak{I}_B^G\lambda_{\mathbb{R}}|\rho|_{\mathbb{R}}^2\to\mathfrak{I}_B^G\lambda_{\mathbb{R}}^{w_0}.\end{aligned} \tag{4.47}$$

If we now look more carefully at our two regularized intertwining operators above then a simple computation yields (see sl2neu.pdf):

Proposition 4.1.2. *Recall that λ is dominant. The kernel of $T_\infty^{\mathrm{loc,reg}}(\lambda_{\mathbb{R}}^{w_0})$ is $\mathcal{M}_{\lambda,\mathbb{C}}$, and this operator induces an isomorphism*

$$\bar{T}(\lambda_R):\mathcal{D}_\lambda\xrightarrow{\sim}\mathcal{D}_\lambda^\vee\otimes\det_{\mathbb{R}}^{2d}.$$

The kernel of $T_\infty^{\mathrm{loc,reg}}(\lambda_{\mathbb{R}}|\rho|_{\mathbb{R}}^2)$ is $\mathcal{D}_\lambda^\vee\otimes\det_{\mathbb{R}}^{2d}$, and it induces an isomorphism of $\mathcal{M}_{\lambda,\mathbb{C}}$.

The module $\mathfrak{I}_B^G\chi$ is reducible if $T_\infty^{\mathrm{loc,reg}}(\chi)$ is not an isomorphism, and this happens if and only if $\chi=\lambda_{\mathbb{R}}^{w_0}$ or $\chi=\lambda_{\mathbb{R}}|\rho|_{\mathbb{R}}^2$ and λ is dominant. There is one exception to the converse of the above assertion, namely in the case A) and $\sigma=\frac{1}{2}$ and $\chi_c^{w_0}\neq\chi_c$.

Unitarity

For us it is of relevance to know whether we have a positive-definite hermitian form on the $(\mathfrak{g},K_\infty)$-modules $\mathcal{D}_\lambda$. To discuss this question we treat the cases A) and B) separately.

We look at the decomposition into K_∞-types (see (4.27)). In case A) it is clear that $\mathcal{M}_{\lambda,\mathbb{C}}$ is the direct sum of the K_∞ types $\mathbb{C}\psi_l$ with $|l|\le n$. Hence

$$\mathcal{D}_\lambda=\bigoplus_{k\le -n-2,k\equiv n(2)}\mathbb{C}\psi_k\oplus\bigoplus_{k\ge n+2,k\equiv n(2)}\mathbb{C}\psi_k=\mathcal{D}_\lambda^-\oplus\mathcal{D}_\lambda^+ \tag{4.48}$$

Proposition 4.1.3. *The representations $\mathcal{D}_\lambda^-,\mathcal{D}_\lambda^+$ are irreducible, these are the discrete series representations. The element* **c** *(see (4.23)) interchanges $\mathcal{D}_\lambda^-,\mathcal{D}_\lambda^+$, hence $\mathcal{D}_\lambda$ is an irreducible $(\mathfrak{g},\tilde{K}_\infty^*)$-module.*

The operator $\bar{T}(\lambda_R)$ induces a quasi-unitary structure on the $(\mathfrak{g},\tilde{K}_\infty)$-module $\mathcal{D}_\lambda$. The two sets of K_∞-types occurring in $\mathcal{M}_{\lambda,\mathbb{C}}$ and K_∞-types occuring in $\mathcal{D}_\lambda$ (resp.) are disjoint.

Proof. Remember that as a vector space $\mathcal{D}_\lambda^\vee\otimes\det_{\mathbb{R}}^{2d}=\mathcal{D}_\lambda^\vee$, only the way $\tilde{K}_\infty$ acts is twisted by $\det_{\mathbb{R}}^{2d}$. Then the form

$$h_\psi(f_1,\ f_2)=T_\infty^{\mathrm{loc,reg}}(\lambda_{\mathbb{R}}^{w_0})(\bar{f}_2)(f_1) \tag{4.49}$$

defines a quasi-invariant hermitian form. It is positive-definite (for more details see sl2neu.pdf). □

A similar argument works in case B). We restrict the $\mathrm{Gl}_2(\mathbb{C}) \times \mathrm{Gl}_2(\mathbb{C})$-module $\mathcal{M}_{\lambda,\mathbb{C}}$ to $U(2) \times U(2)$. Then it becomes the highest weight module $\mathcal{M}_{\lambda_c} = \mathcal{M}_{\lambda_{1,c}} \otimes \mathcal{M}_{\lambda_{2,c}}$ (see Section 4.1.1). Under the action of the group $U(2) \subset U(2) \times U(2)$ it decomposes into $U(2)$ types according to the Clebsch-Gordan formula

$$\mathcal{M}_{\lambda_c}|_{U(2)} = \bigoplus_{\mu_c = k\gamma_c + \frac{d_1+d_2}{2}\det;\ k\equiv(n_1-n_2) \mod 2;\ n_1+n_2\geq k\geq|n_1-n_2|} \mathcal{M}_{\mu_c}. \tag{4.50}$$

Hence we get

$$\mathcal{D}_{\lambda_c}|_{U(2)} = \bigoplus_{\mu_c = k\gamma_c + \frac{d_1+d_2}{2}\det;\ k\equiv(n_1-n_2) \mod 2;\ k\geq n_1+n_2+2} \mathcal{M}_{\mu_c}, \tag{4.51}$$

where again we have

Proposition 4.1.4. *The operator $T_\infty^{\mathrm{loc,reg}}(\lambda_\mathbb{R}^{w_0})$ induces an isomorphism*

$$\bar{T}(\lambda_R) : \mathcal{D}_\lambda \xrightarrow{\sim} \mathcal{D}_\lambda^\vee \otimes \det{}_\mathbb{R}^{2d}.$$

The $(\mathfrak{g}, K_\infty)$-modules are irreducible.

The operator $T_\infty^{\mathrm{loc,reg}}(\lambda_\mathbb{R}^{w_0})$ induces the structure of a quasi-unitary module on $\mathcal{D}_\lambda$ with $\lambda = n_1\gamma_1+n_2\gamma_2+\delta$ if and only if $n_1 = n_2$. This is the only case when we have a quasi-unitary structure on $\mathcal{D}_\lambda$. The two sets of K_∞-types occurring in $\mathcal{M}_{\lambda,\mathbb{C}}$ and in $\mathcal{D}_\lambda$ are disjoint.

The Weyl group W acts on the torus T by conjugation, hence on $X^*(T\times\mathbb{C})$ and we define the *twisted action* by

$$s\cdot\lambda = s(\lambda+\rho)-\rho. \tag{4.52}$$

Given a dominant λ we may consider the four characters $w\cdot\lambda, w \in W(\mathbb{C}) = W$ and the resulting induced modules $\mathfrak{I}_B^G w\cdot\lambda_\mathbb{R}$. We observe (using notation from Section 4.1.1)

$$\begin{aligned} s_1\cdot(n_1\gamma + d_1\det + n_2\bar\gamma + d_2\overline{\det}) &= (-n_1-2)\gamma + d_1\det + n_2\bar\gamma + d_2\overline{\det} \\ s_2\cdot(n_1\gamma + d_1\det + n_2\bar\gamma + d_2\overline{\det}) &= n_1\gamma + d_1\det + (-n_2-2)\bar\gamma + d_2\overline{\det} \end{aligned} \tag{4.53}$$

Looking closely we see that the K_∞-types occurring in $\mathfrak{I}_B^G s_1\cdot\lambda$ or $\mathfrak{I}_B^G s_2\cdot\lambda$ are exactly those which occur in $\mathcal{D}_\lambda$. This has a simple explanation, as we have

Proposition 4.1.5. *For a dominant character λ we have isomorphisms between the $(\mathfrak{g}, K_\infty)$-modules*

$$\mathcal{D}_\lambda \xrightarrow{\sim} \mathfrak{I}_B^G s_1\cdot\lambda,\ \mathcal{D}_\lambda \xrightarrow{\sim} \mathfrak{I}_B^G s_2\cdot\lambda. \tag{4.54}$$

The resulting isomorphism $\mathfrak{I}_B^G s_1\cdot\lambda_\mathbb{R} \xrightarrow{\sim} \mathfrak{I}_B^G s_2\cdot\lambda_\mathbb{R}$ is of course given by the map $T_\infty^{\mathrm{loc}}(s_1\cdot\lambda)$.

Interlude: Here we see a fundamental difference between the two cases A) and B). In the second case the infinite dimensional subquotients of the induced representations are again induced representations. In the case A) this is not so, the representations $\mathcal{D}_\lambda^\pm$ are not isomorphic to representations induced from the Borel subgroup.

These representation $\mathcal{D}_\lambda^\pm$ are called *discrete series* representations, and we want to briefly explain why. Let G be the group of real points of a reductive group over $\mathbb{R}$, for example our $G = G(\mathbb{R})$. Here we allow both cases. Let Z be the center of G, it can be written as $Z_0(\mathbb{R}) \cdot Z_c$ where Z_c is maximal compact and $Z_0 = Z_0(\mathbb{R}) = (\mathbb{R}^\times_{>0})^t$. Let $\omega^{(0)} : Z_0 \to \mathbb{R}^\times_{>0}$ be a character. Then we define the space

$$\mathcal{C}_\infty(G, \omega^{(0)}) := \{f \in \mathcal{C}_\infty(G) \mid f(zg) = \omega^{(0)}(z) f(g) \,; \forall z \in Z_0, g \in G\} \tag{4.55}$$

and we define the subspace

$$L^2_\infty(G, \omega^{(0)}) := \{f \in \mathcal{C}_\infty(G, \omega^{(0)}) \mid \int_{G/Z_0} f(g)\overline{f(g)}\delta(g)^2 dg < \infty\} \tag{4.56}$$

where of course dg is a Haar measure. As usual, $L^2(G, \omega^{(0)})$ will be the Hilbert space obtained by completion. This Hilbert space only depends in a very mild way on the choice of $\omega^{(0)}$. We can find a character $\delta : G \to \mathbb{R}^\times_{>0}$ such that $\omega^{(0)}\delta|_{Z_0} = 1$. Then $f \mapsto f\delta$ provides an isomorphism $L^2(G, \omega^{(0)}) \xrightarrow{\sim} L^2(G/Z_0)$.

We have an action of $G \times G$ on $L^2(G, \omega^{(0)})$ by left and right translations. Harish-Chandra investigated the question of how this "decomposes" into irreducible submodules. Let $\hat{G}_{\omega^{(0)}}$ be the set of isomorphism classes of irreducible unitary representations of G.

Harish-Chandra showed that there exists a positive measure μ on $\hat{G}_{\omega^{(0)}}$ and a measurable family H_ξ of irreducible unitary representations of G such that

$$L^2(G, \omega^{(0)}) = \int_{\hat{G}_{\omega^{(0)}}} H_\xi \otimes \overline{H_\xi}\; \mu(d\xi). \tag{4.57}$$

(If instead of a semi-simple Lie group we take a finite group G, then this is the fundamental theorem of Frobenius that the group ring $\mathbb{C}[G] = \oplus_\theta V_\theta \otimes V_\theta^\vee$ where V_θ are the irreducible representations.)

If we are in the case A), the sets consisting of just one point $\{\mathcal{D}_\lambda^\pm\}$ have strictly positive measure, i.e. $\mu(\{\mathcal{D}_\lambda^\pm\}) > 0$. This means that the irreducible unitary $G \times G$-modules $\mathcal{D}_\lambda^\pm \otimes \mathcal{D}_{\lambda^\vee}^\pm$ occur as direct summands (i.e. discretely in $L^2(G)$).

Such irreducible direct summands do not exist in the case B), as in this case for any $\xi \in \hat{G}$ we have $\mu(\{\xi\}) = 0$.

End Interlude

We return to the short exact sequences (4.44), (4.46). I claim that both sequences do not split as sequences of $(\mathfrak{g}, K_\infty)$-modules. Of course it follows from the above proposition that these sequences split canonically as sequence of K_∞ modules. But one sees easily that the complementary summand is not invariant under the action of $\mathfrak{g}$. This means that the sequence provides a non-trivial class in $\mathrm{Ext}^1_{(\mathfrak{g},K_\infty)}(\mathcal{D}_\lambda, \mathcal{M}_{\lambda,\mathbb{C}})$.

The general principles of homological algebra teach us that we can understand these extension groups in terms of relative Lie-algebra cohomology. Let $\mathfrak{k}$ resp. $\tilde{\mathfrak{k}}$ be the Lie-algebras of K_∞ resp. $\tilde{K}_\infty$ the group $\tilde{K}_\infty$ acts on $\mathfrak{g}, \tilde{\mathfrak{k}}$ via the adjoint action (see 1.1.4).

We start from a $(\mathfrak{g}, \tilde{K}_\infty)$-module $\mathfrak{I}_B^G\chi$ and a module $\mathcal{M}_{\lambda,\mathbb{C}}$. Our first goal is to compute the cohomology $H^\bullet(\mathfrak{g}, K_\infty, \mathfrak{I}_B^G\chi \otimes \mathcal{M}_{\lambda,\mathbb{C}})$ which is defined as the cohomology of the complex (see 6.1.2, (6.4))

$$H^\bullet(\mathfrak{g}, K_\infty, \mathfrak{I}_B^G\chi \otimes \mathcal{M}_{\lambda,\mathbb{C}}) := H^\bullet(\mathrm{Hom}_{\tilde{K}_\infty}(\Lambda^\bullet(\mathfrak{g}/\tilde{\mathfrak{k}}), \mathfrak{I}_B^G\chi \otimes \mathcal{M}_{\lambda,\mathbb{C}})). \quad (4.58)$$

Here we assume that $\chi : T(\mathbb{R}) \to \mathbb{C}^\times$ is any character, but we will see that there is only one χ for which we have non-trivial cohomology.

There is an obvious condition for the complex to be non-zero. The group $Z(\mathbb{R}) \subset \tilde{K}_\infty$ acts trivially on $\mathfrak{g}/\mathfrak{k}$, and hence we see that the complex is trivial unless we have $\omega_\chi^{-1} = \lambda_\mathbb{R}|_{Z(\mathbb{R})^{(0)}}$, we assume that this relation holds.

We will give a very explicit formula for these cohomology modules (Delorme's formula):

An element $\omega \in \mathrm{Hom}_{\tilde{K}_\infty}(\Lambda^n(\mathfrak{g}/\tilde{\mathfrak{k}}), \mathfrak{I}_B^G\chi \otimes \mathcal{M}_{\lambda,\mathbb{C}})$ attaches to any n-tuple $v_1, \ldots, v_n$ of elements in $\mathfrak{g}/\tilde{\mathfrak{k}}$ an element

$$\omega(v_1, \ldots, v_n) \in \mathfrak{I}_B^G\chi \otimes \mathcal{M}_{\lambda,\mathbb{C}} \quad (4.59)$$

such that $\omega(\mathrm{Ad}(k)v_1, \ldots, \mathrm{Ad}(k)v_n) = k\omega(v_1, \ldots, v_n)$ for all $k \in \tilde{K}_\infty$.

By construction

$$\omega(v_1, \ldots, v_n) = \sum_\nu f_\nu \otimes m_\nu \text{ where } f_\nu \in \mathfrak{I}_B^G\chi, m_\nu \in \mathcal{M}_{\lambda,\mathbb{C}},$$

and f_ν is a function in $\mathcal{C}_\infty$ which is determined by its restriction to $\tilde{K}_\infty$ (and this restriction is $\tilde{K}_\infty$ finite). We can evaluate this function at the identity $e_G \in G(\mathbb{R})$ and then

$$\omega(v_1, \ldots, v_n)(e_G) = \sum f_\nu(e) \otimes m_\nu \in \mathbb{C}\chi \otimes \mathcal{M}_{\lambda,\mathbb{C}}.$$

The $\tilde{K}_\infty$ invariance (4.59) implies that ω is determined by this evaluation at e_G. Let $\tilde{K}_\infty^T := T(\mathbb{R}) \cap \tilde{K}_\infty = Z(\mathbb{R}) \cdot T_c$. Then it is clear that $\omega^* : (v_1, \ldots, v_n) \mapsto \omega(v_1, \ldots, v_n)(e_G)$ is an element in $\omega^* \in \mathrm{Hom}_{\tilde{K}_\infty^T}(\Lambda^n(\mathfrak{g}/\tilde{\mathfrak{k}}), \mathbb{C}\chi \otimes \mathcal{M}_{\lambda,\mathbb{C}})$, the map $\omega \mapsto \omega^*$ yields an isomorphism of complexes

$$\mathrm{Hom}_{\tilde{K}_\infty}(\Lambda^\bullet(\mathfrak{g}/\tilde{\mathfrak{k}}), \mathfrak{I}_B^G\chi \otimes \mathcal{M}_{\lambda,\mathbb{C}}) \xrightarrow{\sim} \mathrm{Hom}_{\tilde{K}_\infty^T}(\Lambda^\bullet(\mathfrak{g}/\tilde{\mathfrak{k}}), \mathbb{C}\chi \otimes \mathcal{M}_{\lambda,\mathbb{C}}). \quad (4.60)$$

The Lie algebra $\mathfrak{g}$ can be written as a sum of $\mathfrak{c}$-invariant submodules

$$\mathfrak{g} = \mathfrak{b} + \tilde{\mathfrak{k}} = \mathfrak{t} + \mathfrak{u} + \tilde{\mathfrak{k}}. \quad (4.61)$$

In case B) this sum is not direct, and we have $\mathfrak{b} \cap \tilde{\mathfrak{k}} = \mathfrak{t} \cap \tilde{\mathfrak{k}} = \mathfrak{c}$. Hence we get the direct sum decomposition into $\tilde{K}_\infty^T$-invariant subspaces

$$\mathfrak{g}/\tilde{\mathfrak{k}} = \mathfrak{t}/\mathfrak{c} \oplus \mathfrak{u}. \quad (4.62)$$

We get an isomorphism of complexes (Frobenius reciprocity)

$$\begin{aligned}&\mathrm{Hom}_{\tilde{K}_\infty}(\Lambda^\bullet(\mathfrak{g}/\tilde{\mathfrak{k}}), \mathfrak{I}_B^G\chi \otimes \mathcal{M}_{\lambda,\mathbb{C}})\\ &\xrightarrow{\sim} \mathrm{Hom}_{\tilde{K}_\infty^T}(\Lambda^\bullet(\mathfrak{t}/\mathfrak{c}), \mathbb{C}\chi \otimes \mathrm{Hom}(\Lambda^\bullet(\mathfrak{u}), \mathcal{M}_{\lambda,\mathbb{C}}))\end{aligned} \tag{4.63}$$

where the complex on the left is isomorphic to the total complex of the double complex on the right. The next step is the computation of the cohomology of the complex $\mathrm{Hom}(\Lambda^\bullet(\mathfrak{u}), \mathcal{M}_{\lambda,\mathbb{C}})$.

Case A). We have $\mathfrak{u} = \mathbb{Q}E_+$ where $E_+ = \begin{pmatrix} 0 & 1 \\ 0 & 0 \end{pmatrix}$, and our module $\mathcal{M}_{\lambda,\mathbb{Q}}$ has a decomposition into weight spaces

$$\mathcal{M}_{\lambda,\mathbb{Q}} = \bigoplus_{\mu=0}^{n} \mathbb{Q}X^\mu Y^{n-\mu} = \bigoplus_{\mu=0}^{n} \mathbb{Q}e_\mu. \tag{4.64}$$

The torus $T^{(1)} = \left\{ \begin{pmatrix} t & 0 \\ 0 & t^{-1} \end{pmatrix} \right\}$ acts on $e_\mu = X^\mu Y^{n-\mu}$ by

$$\rho_\lambda\left(\begin{pmatrix} t & 0 \\ 0 & t^{-1} \end{pmatrix}\right) e_\mu = t^{2\mu-n} e_\mu. \tag{4.65}$$

If n is even and if we choose $d = -\frac{n}{2}$, then $\mathcal{M}_\lambda$ is a module for the group PGL_2, and

$$\rho_\lambda\left(\begin{pmatrix} t & 0 \\ 0 & 1 \end{pmatrix}\right) e_\mu = t^{\mu-n/2} e_\mu. \tag{4.66}$$

We also have the action of the Lie algebra on $\mathcal{M}_{\lambda,\mathbb{Q}}$, and by definition we get

$$d(\rho_\lambda)(E_+)e_\mu = E_+e_\mu = (n-\mu)e_{\mu+1}. \tag{4.67}$$

We write down our complex $\mathrm{Hom}(\Lambda^\bullet(\mathfrak{u}), \mathcal{M}_{\lambda,\mathbb{C}})$ very explicitly: Let $E_+^\vee \in \mathrm{Hom}(\mathfrak{u}, \mathbb{Q})$ be the element $E_+^\vee(E_+) = 1, E_+^\vee(H) = 0$. Then the complex becomes

$$0 \to \bigoplus_{\mu=0}^{n} \mathbb{Q}e_\mu \xrightarrow{d} \bigoplus_{\mu=0}^{n} \mathbb{Q}E_+^\vee \otimes e_\mu \to 0, \tag{4.68}$$

where $d(e_\mu) = (n-\mu)E_+^\vee \otimes e_{\mu+1}$. This gives us a decomposition of our complex into two subcomplexes

$$\mathrm{Hom}(\Lambda^\bullet(\mathfrak{u}), \mathcal{M}_{\lambda,\mathbb{C}}) = \mathbb{H}^\bullet(\mathfrak{u}, \mathcal{M}_{\lambda,\mathbb{Q}}) \oplus AC^\bullet, \tag{4.69}$$

where $AC^\bullet$ is acyclic (it has no cohomology) and

$$\mathbb{H}^\bullet(\mathfrak{u}, \mathcal{M}_{\lambda,\mathbb{Q}}) = \{0 \to \mathbb{Q}\, e_n \xrightarrow{d} \mathbb{Q}\, E_+^\vee \otimes e_0 \to 0\}, \tag{4.70}$$

where the differential $d = 0$. Hence we get

$$H^\bullet(\mathfrak{u}, \mathcal{M}_{\lambda,\mathbb{Q}}) = H^\bullet(\mathrm{Hom}(\Lambda^\bullet(\mathfrak{u}), \mathcal{M}_{\lambda,\mathbb{Q}})) = \mathbb{H}^\bullet(\mathfrak{u}, \mathcal{M}_{\lambda,\mathbb{Q}}). \tag{4.71}$$

We notice that the torus T acts on $H^\bullet(\mathfrak{u}, \mathcal{M}_{\lambda,\mathbb{Q}})$. (The Borel subgroup B acts on the complex $\mathrm{Hom}(\Lambda^\bullet(\mathfrak{u}), \mathcal{M}_{\lambda,\mathbb{Q}})$, but since the Lie algebra cohomology is the derived functor of taking invariants under U - elements annihilated by $\mathfrak{u}$ - it follows that this action is trivial on U.) Now it is clear that (4.63) yields

$$H^\bullet(\mathfrak{g}, K_\infty, \mathfrak{I}_B^G\chi \otimes \mathcal{M}_{\lambda,\mathbb{C}}) \xrightarrow{\sim} H^\bullet(\mathfrak{t}, K_\infty{}^T, \mathbb{C}\chi \otimes \mathbb{H}^\bullet(\mathfrak{u}, \mathcal{M}_{\lambda,\mathbb{Q}})). \tag{4.72}$$

Hence we see that T acts by the character λ on $\mathbb{Q}\, e_n = H^0(\mathfrak{u}, \mathcal{M}_{\lambda,\mathbb{Q}})$ and by the character $\lambda^- - \alpha = w_0 \cdot \lambda = \lambda^{w_0} - 2\rho$ on $\mathbb{Q}E_+^\vee \otimes e_{-n} = H^1(\mathfrak{u}, \mathcal{M}_{\lambda,\mathbb{Q}})$. Here we see the simplest example of the famous theorem of Kostant which will be discussed in Chap. 6 (see 6.2.3).

Then our cohomology groups $H^\bullet(\mathfrak{t}, K_\infty{}^T, \mathbb{C}\chi \otimes \mathbb{H}^\bullet(\mathfrak{u}, \mathcal{M}_{\lambda,\mathbb{Q}}))$ are given as the cohomology groups of the double complex with entries

$$\mathrm{Hom}_{K_\infty{}^T}(\Lambda^p(\mathfrak{t}/\mathfrak{t}\cap\mathfrak{k}), \mathbb{C}\chi \otimes \mathbb{H}^q(\mathfrak{u}, \mathcal{M}_{\lambda,\mathbb{Q}}))$$

where $p = 0, 1, q = 0, 1$ and where the differentials in direction q are zero. We have to compute the cohomology of the complexes

$$\begin{aligned} 0 \to\ & \mathrm{Hom}_{K_\infty{}^T}(\Lambda^0(\mathfrak{t}/\mathfrak{t}\cap\mathfrak{k}), \mathbb{C}\chi \otimes \mathbb{H}^q(\mathfrak{u}, \mathcal{M}_{\lambda,\mathbb{Q}})) \\ \xrightarrow{d}\ & \mathrm{Hom}_{K_\infty{}^T}(\Lambda^1(\mathfrak{t}/\mathfrak{t}\cap\mathfrak{k}), \mathbb{C}\chi \otimes \mathbb{H}^q(\mathfrak{u}, \mathcal{M}_{\lambda,\mathbb{Q}})) \to 0. \end{aligned}$$

In this complex we drop the subscript $K_\infty{}^T$. Then both spaces in the complex are one-dimensional, and the differential is - up to a non-zero factor - multiplication by $d\chi(H) + d(w\cdot\lambda)(H)$. Hence we have zero cohomology, unless we have $d\chi(H) + d(w\cdot\lambda)(H) = 0$. Hence we see (observe that $q = l(w)$)

$$H^\bullet(\mathfrak{t}, K_\infty{}^T, \mathbb{C}\chi \otimes \mathbb{H}^q(\mathfrak{u}, \mathcal{M}_{\lambda,\mathbb{Q}})) \neq 0 \implies \chi|T(\mathbb{R})^{(0)} = (w\cdot\lambda)_{\mathbb{R}}^{-1}|T(\mathbb{R})^{(0)}.$$

We reintroduce the subscript $K_\infty{}^T$. Since clearly $K_\infty{}^T \cdot T(\mathbb{R})^{(0)} = T(\mathbb{R})$, we see that we have non-trivial cohomology if and only if $\chi = (w\cdot\lambda)_{\mathbb{R}}^{-1}$. Putting everything together we see:

If we can find an element $w \in W(\mathbb{C})$ so that $\chi^{-1} = w\cdot\lambda_{\mathbb{R}}$ then

$$H^{\bullet+l(w)}(\mathfrak{g}, K_\infty, \mathfrak{I}_B^G\chi \otimes \mathcal{M}_{\lambda,\mathbb{Q}}) = \mathbb{H}^{l(w)}(\mathfrak{u}, \mathcal{M}_{\lambda,\mathbb{Q}})) \wedge \Lambda^\bullet(\mathfrak{t}/\mathfrak{t}\cap\mathfrak{k})^\vee. \tag{4.73}$$

If there is no such element w then the cohomology is zero.

Remark: For later purpose we discuss the complex which gives these cohomology groups. We assume that n is even and $d = -\frac{n}{2}$ hence we can work on the projective linear group. Then we have $\mathfrak{g}/\mathfrak{k} = \mathfrak{t} \oplus \mathfrak{u}$. On $\mathfrak{g}/\mathfrak{k}$ we have a positive definite K_∞ invariant metric which is given by $< H, H >_{\mathfrak{p}}= 1, < V, V >_{\mathfrak{p}}= 1$ and $< H, V >_{\mathfrak{p}}= 0$, we use this metric to identify $\mathfrak{g}/\mathfrak{k}$ to its dual and hence we also get an identification $\mathfrak{t} \oplus \mathfrak{u}$ with its dual. Then we see that elements in $\mathrm{Hom}(\mathfrak{t} \oplus \mathfrak{u}, \mathbb{C}\lambda_{\mathbb{R}}\rho_{\mathbb{R}}^2 \otimes \mathcal{M}_\lambda$ can be written as elements of the form $Z \otimes P(X, Y)$ with $\mathbb{Z} \in \mathfrak{t} \oplus \mathfrak{u}$ and P a homogeneous polynomial of degree n. The specific element $E_+ \otimes Y^n$ yields the generator in $H^1(\mathfrak{u}, \mathcal{M}_\lambda)$ an element $E_+ \otimes Q(X, Y)$ lies in the image of the differential d if the coefficient of $E_+ \otimes Y^n$ is zero.

We tensorize the sequence (4.44) with the dual $\mathcal{M}_{\lambda^\vee}$. We get an exact sequence of $(\mathfrak{g}, \tilde{K}_\infty)$-modules. We thus look at the resulting long exact sequence in cohomology. We know that $H^1(\mathfrak{g}, \tilde{K}_\infty, \mathcal{M}_\lambda \otimes M_{\lambda^\vee}) = 0$ and then we look at the piece

$$0 \to H^1(\mathfrak{g}, \tilde{K}_\infty, \mathfrak{I}_B^G \lambda^{w_0} \otimes \mathcal{M}_{\lambda^\vee}) \to H^1(\mathfrak{g}, \tilde{K}_\infty, \mathcal{D}_\lambda \otimes \mathcal{M}_{\lambda^\vee}) \to H^2(\mathfrak{g}, \tilde{K}_\infty, \mathcal{M}_\lambda \otimes \mathcal{M}_{\lambda^\vee}) \to 0.$$

We have seen, and we know, that the two extreme terms are equal to $\mathbb{C}$ and thus we easily get

$$H^1(\mathfrak{g}, \tilde{K}_\infty, \mathcal{D}_\lambda \otimes \mathcal{M}_{\lambda^\vee}) = \mathbb{C} \oplus \mathbb{C},$$

and this cohomology vanishes in all other degrees.

We have a slightly different look at this, we consider the dual situation. For simplicity we assume n even and $d = -\frac{n}{2}$, the central character is trivial. We start from the exact sequence

$$0 \to \mathcal{D}_{\lambda^\vee} \to \mathfrak{I}_B^G \lambda_{\mathbb{R}} \rho_{\mathbb{R}}^2 \to \mathcal{M}_\lambda \to 0.$$

We tensor it with $\mathcal{M}_\lambda$ and look at the resulting long exact sequence in cohomology

$$0 \to H^0(\mathfrak{g}, \tilde{K}_\infty, \mathcal{M}_\lambda \otimes \mathcal{M}_\lambda) \xrightarrow{\delta} H^1(\mathfrak{g}, \tilde{K}_\infty, \mathcal{D}_{\lambda^\vee} \otimes \mathcal{M}_\lambda) \to$$

$$\to H^1(\mathfrak{g}, \tilde{K}_\infty, \mathfrak{I}_B^G \lambda_{\mathbb{R}} \rho_{\mathbb{R}}^2 \otimes \mathcal{M}_\lambda) \to 0$$

We have still the action on complex conjugation (4.23) on the cohomology groups. This action decomposes $\mathrm{Hom}_{\tilde{K}_\infty}(\Lambda^1(\mathfrak{g}/\tilde{\mathfrak{k}}), \mathcal{D}_\lambda \otimes \mathcal{M}_\lambda)$ into a $+$ and a $-$ one dimensional eigenspaces

$$\mathrm{Hom}_{\tilde{K}_\infty}(\Lambda^1(\mathfrak{g}/\tilde{\mathfrak{k}}), \mathcal{D}_\lambda \otimes \mathcal{M}_\lambda) = H^1(\mathfrak{g}, \tilde{K}_\infty, \mathcal{D}_\lambda \otimes \mathcal{M}_\lambda) = \mathbb{C}\omega_\lambda^+ \oplus \mathbb{C}\omega_\lambda^-.$$

In Section 4.2.5 we give explicit expressions for the generating elements $\omega_\lambda^\pm$ in $\mathrm{Hom}_{\tilde{K}_\infty}(\Lambda^\bullet(\mathfrak{g}/\tilde{\mathfrak{k}}), \mathcal{D}_\lambda \otimes \mathcal{M}_{\lambda^\vee})$. It is rather clear that ω_λ^+ is in the image of the boundary map δ whereas $\mathbb{C}\omega_\lambda^-$ maps surjectively to $H^1(\mathfrak{g}, \tilde{K}_\infty, \mathfrak{I}_B^G \lambda_{\mathbb{R}} \rho_{\mathbb{R}}^2 \otimes \mathcal{M}_\lambda)$.

We briefly discuss the case B). Again we want that our group $G/\mathbb{R} = R_{\mathbb{C}/\mathbb{R}}(\mathrm{Gl}_2/\mathbb{C})$ is a base change from a group $G/\mathbb{Q}$ denoted by the same letter. We need an imaginary quadratic extension $F/\mathbb{Q}$ and put $G/\mathbb{Q} = R_{F/\mathbb{Q}}(\mathrm{Gl}_2/F)$. We choose a dominant weight $\lambda = \lambda_1 + \lambda_2 = n_1\gamma_1 + d_1 \det_1 + n_2\gamma_2 + d_2 \det_2$ and then $\mathcal{M}_{\lambda,F} = \mathcal{M}_{\lambda_1,F} \otimes \mathcal{M}_{\lambda_2,F}$ is an irreducible representation of $G \times_{\mathbb{Q}} F = \mathrm{Gl}_2 \times \mathrm{Gl}_2/F$. Now we have $\mathfrak{u} \otimes F = FE_+^1 \oplus FE_+^2$. Then basically the same computation yields that the cohomology $H^\bullet(\mathfrak{u}, \mathcal{M}_{\lambda,F})$ is equal to the complex

$$\begin{aligned}\mathbb{H}^\bullet(\mathfrak{u}, \mathcal{M}_{\lambda,F}) = (0 &\to Fe_{n_1}^{(1)} \otimes Fe_{n_2}^{(2)} \\ &\xrightarrow{d} FE_+^{1,\vee} \otimes e_0^{(1)} \otimes e_{n_2}^{(2)} \oplus Fe_{n_1}^{(1)} \otimes E_+^{2,\vee} \otimes e_0^{(2)} \\ &\xrightarrow{d} FE_+^{1,\vee} \otimes e_0^{(1)} \otimes E_+^{2,\vee} \otimes e_0^{(2)} \to 0),\end{aligned}$$

where all the differentials are zero. The torus T acts by the weights

$$\lambda \text{ in degree } 0,\ s_1 \cdot \lambda, s_2 \cdot \lambda \text{ in degree } 1,\ w_0 \cdot \lambda \text{ in degree } 2, \tag{4.74}$$

and we have a decomposition into one dimensional weight spaces

$$H^\bullet(\mathfrak{u}, \mathcal{M}_{\lambda,F}) = \bigoplus_{w\in W(\mathbb{C})} \mathbb{H}^\bullet(\mathfrak{u}, \mathcal{M}_{\lambda,F})(w\cdot\lambda)$$

We go back to (4.63) and get a homomorphism of complexes

$$\mathrm{Hom}_{\tilde{K}_\infty}(\Lambda^\bullet(\mathfrak{g}/\tilde{\mathfrak{k}}), \mathfrak{I}_B^G\chi \otimes \mathcal{M}_{\lambda,\mathbb{C}}) \to \mathrm{Hom}_{\tilde{K}_\infty^T}(\Lambda^\bullet(\mathfrak{t}/\mathfrak{t}\cap\tilde{\mathfrak{k}}), \mathbb{C}\chi \otimes \mathbb{H}^\bullet(\mathfrak{u}, \mathcal{M}_{\lambda,\mathbb{C}})),$$

which induces an isomorphism in cohomology, so that finally

$$H^\bullet(\mathfrak{g}, \tilde{K}_\infty, \mathfrak{I}_B^G\chi \otimes \mathcal{M}_{\lambda,\mathbb{C}}) \xrightarrow{\sim} H^\bullet(\mathrm{Hom}_{\tilde{K}_\infty^T}(\Lambda^\bullet(\mathfrak{t}/\mathfrak{t}\cap\tilde{\mathfrak{k}}), \mathbb{C}\chi \otimes H^\bullet(\mathfrak{u}, \mathcal{M}_{\lambda,\mathbb{C}})),$$

and combining this with the results above we get:

Theorem 4.1.2. *If we can find an element $w \in W(\mathbb{C})$ so that $\chi^{-1} = w\cdot\lambda_\mathbb{R}$ then*

$$H^{l(w)}(\mathfrak{u}, \mathcal{M}_{\lambda,\mathbb{C}})(w\cdot\lambda) \otimes \Lambda^\bullet(\mathfrak{t}/\mathfrak{t}\cap\tilde{\mathfrak{k}})^\vee \xrightarrow{\sim} H^\bullet(\mathfrak{g}, \tilde{K}_\infty, \mathfrak{I}_B^G\chi \otimes \mathcal{M}_{\lambda,\mathbb{C}}).$$

If there is no such w then the cohomology is zero.

Proof. Our torus $T(\mathbb{R}) = \tilde{K}_\infty^T \times \left\{\begin{pmatrix} t & 0 \\ 0 & t^{-1}\end{pmatrix}; t\in\mathbb{R}_{>0}^\times\right\} = \tilde{K}_\infty^T \times A$. Hence we see that $\dim \mathfrak{t}/\mathfrak{t}\cap\tilde{\mathfrak{k}} = 1$, and the element $H_0 = \begin{pmatrix} 1 & 0 \\ 0 & -1\end{pmatrix}$. Of course we must have that $\chi^{-1}w\cdot\lambda_\mathbb{R}|_\mathfrak{c}$ is the trivial character. The second factor A does act on $\mathbb{C}\chi$ by the character $\chi(t) = t^z$ and on $H^{l(w)}(\mathfrak{u}, \mathcal{M}_{\lambda,\mathbb{C}})(w\cdot\lambda)$ by $t\mapsto t^{m(w)}$. Differentiating we get the complex

$$0 \to H^{l(w)}(\mathfrak{u}, \mathcal{M}_{\lambda,\mathbb{C}})(w\cdot\lambda) \to \mathbb{C}\otimes H_0^\vee \otimes H^{l(w)}(\mathfrak{u}, \mathcal{M}_{\lambda,\mathbb{C}})(w\cdot\lambda) \to 0,$$

where the differential is multiplication by $m(w)+z$. Hence we see that the cohomology is trivial unless $m(w)+z=0$, but this means $\chi^{-1} = w\cdot\lambda_\mathbb{R}$. □

4.1.6 The cohomology of the modules $\mathcal{M}_{\lambda,\mathbb{C}}$, $\mathcal{D}_\lambda$, and the cohomology of unitary modules

Let $\mathrm{Irr}(G, K_\infty)$ be the set of isomorphism classes of irreducible $(\mathfrak{g}, K_\infty)$ Harish-Chandra modules. We are a little bit pedantic, if $\mathcal{V}$ is such an irreducible module, then its isomorphism class is $[\mathcal{V}]$. For any dominant λ we define the sets

$$\mathrm{Coh}(\lambda) := \{[\mathcal{V}] \in \mathrm{Irr}(G, K_\infty) \mid H^\bullet(\mathfrak{g}, K_\infty, \mathcal{V}\otimes\mathcal{M}_{\lambda,\mathbb{C}}) \neq 0\}.$$

We also define $\mathrm{Coh}_2(\lambda)$ to be the set of those $[\mathcal{V}]$ which are also unitary. This definition makes sense in greater generality (see (6.18)). In our special case, these sets are very small. Remember that we have a fixed central character ω.

At first we determine the finite dimensional elements in $\mathrm{Coh}(\lambda)$. Of course $\mathcal{M}_{\lambda,\mathbb{C}}$ itself is a Harish-Chandra module, and it follows from Wigner's lemma that $H^\bullet(\mathfrak{g}, K_\infty, \mathcal{M}_{\lambda,\mathbb{C}}) = 0$ unless $\lambda^{(1)} = 0$, i.e. $\mathcal{M}_{\lambda,\mathbb{C}}$ is one dimensional. Then it follows from Clebsch-Gordan that

Proposition 4.1.6. *In case A)*

$$H^0(\mathfrak{g},\tilde{K}_\infty,\mathcal{M}_{\lambda^\vee,\mathbb{C}}\otimes\mathcal{M}_{\lambda,\mathbb{C}})=H^2(\mathfrak{g},\tilde{K}_\infty,\mathcal{M}_{\lambda^\vee,\mathbb{C}}\otimes\mathcal{M}_{\lambda,\mathbb{C}})=\mathbb{C},$$

$$H^1(\mathfrak{g},\tilde{K}_\infty,\mathcal{M}_{\lambda^\vee,\mathbb{C}}\otimes\mathcal{M}_{\lambda,\mathbb{C}})=0.$$

In case B)

$$H^0(\mathfrak{g},\tilde{K}_\infty,\mathcal{M}_{\lambda^\vee,\mathbb{C}}\otimes\mathcal{M}_{\lambda,\mathbb{C}})=H^3(\mathfrak{g},\tilde{K}_\infty,\mathcal{M}_{\lambda^\vee,\mathbb{C}}\otimes\mathcal{M}_{\lambda,\mathbb{C}})=\mathbb{C},$$

$$H^1(\mathfrak{g},\tilde{K}_\infty,\mathcal{M}_{\lambda^\vee,\mathbb{C}}\otimes\mathcal{M}_{\lambda,\mathbb{C}})=H^2(\mathfrak{g},\tilde{K}_\infty,\mathcal{M}_{\lambda^\vee,\mathbb{C}}\otimes\mathcal{M}_{\lambda,\mathbb{C}})=0$$

Here we take notice of a point, which plays a role if it comes to questions concerning orientability. In case A) we can twist the $G(\mathbb{R})$-module $\mathcal{M}_{\lambda^\vee,\mathbb{C}}$ by the sign character $\eta: g\mapsto \operatorname{sgn}(\det(g))$, where it has the same central character. Obviously the twisted module $\mathcal{M}_{\lambda^\vee,\mathbb{C}}\otimes\eta$ provides the same $(\mathfrak{g},K_\infty)$-module. But this depends on the choice of K_∞, as if we replace K_∞ by the larger group $K_\infty{}^*$ (see Section 4.1.3) then the $(\mathfrak{g},K_\infty{}^*)$-modules $\mathcal{M}_{\lambda^\vee,\mathbb{C}}$ and $\mathcal{M}_{\lambda^\vee,\mathbb{C}}\otimes\eta$ are not isomorphic. If we replace in the above proposition K_∞ by $K_\infty{}^*$ and $\mathcal{M}_{\lambda^\vee,\mathbb{C}}$ by $\mathcal{M}_{\lambda^\vee,\mathbb{C}}\otimes\eta$, then the cohomology vanishes in all degrees.

Small remark: in general it is sapient to work with a connected K_∞ or $\tilde{K}_\infty$ and then keep track of the action of $K_\infty{}^*$ on $H^\bullet(\mathfrak{g},K_\infty,\mathcal{V}\otimes\mathcal{M}_{\lambda,\mathbb{C}})$.

Again we start from a dominant character λ. Then our considerations yield that in case A)

$$\operatorname{Coh}(\lambda^\vee)=\{\mathcal{M}_{\lambda,\mathbb{C}},\mathcal{D}_\lambda^+,\mathcal{D}_\lambda^-\},$$

and we even have $\mathcal{D}_\lambda^+,\mathcal{D}_\lambda^-\in\operatorname{Coh}_2(\lambda^\vee)$ and $\mathcal{M}_{\lambda,\mathbb{C}}\in\operatorname{Coh}_2(\lambda^\vee)$ if and only if $\lambda^{(1)}=0$.

For some reason we call $\{\mathcal{D}_\lambda^+,\mathcal{D}_\lambda^-\}=\operatorname{Coh}_{\mathrm{cusp}}(\lambda^\vee)$ and $\{\mathcal{M}_{\lambda,\mathbb{C}}\}=\operatorname{Coh}_{\mathrm{Eis}}(\lambda^\vee)$.

In case B) we take the tensor product of the exact sequence (4.44) by $\mathcal{M}_{\lambda^\vee,\mathbb{C}}$ and we get a long exact sequence of $(\mathfrak{g},K_\infty)$-cohomology modules (we insert the values for $H^\bullet(\mathfrak{g},K_\infty,\mathcal{M}_{\lambda,\mathbb{C}}\otimes\mathcal{M}_{\lambda^\vee,\mathbb{C}})$)

$$\begin{aligned}
0\to\mathbb{C}\to H^0(\mathfrak{g},K_\infty,\mathfrak{I}_B^G\lambda_{\mathbb{R}}^{w_0}\otimes\mathcal{M}_{\lambda^\vee,\mathbb{C}})&\xrightarrow{r^0}H^0(\mathfrak{g},K_\infty,\mathcal{D}_\lambda\otimes\mathcal{M}_{\lambda^\vee,\mathbb{C}})(=0)\\
\to 0\to H^1(\mathfrak{g},K_\infty,\mathfrak{I}_B^G\lambda_{\mathbb{R}}^{w_0}\otimes\mathcal{M}_{\lambda^\vee,\mathbb{C}})&\xrightarrow{r^1}H^1(\mathfrak{g},K_\infty,\mathcal{D}_\lambda\otimes\mathcal{M}_{\lambda^\vee,\mathbb{C}})\\
\to 0\to H^2(\mathfrak{g},K_\infty,\mathfrak{I}_B^G\lambda_{\mathbb{R}}^{w_0}\otimes\mathcal{M}_{\lambda^\vee,\mathbb{C}})&\xrightarrow{r^2}H^2(\mathfrak{g},K_\infty,\mathcal{D}_\lambda\otimes\mathcal{M}_{\lambda^\vee,\mathbb{C}})\\
\to\mathbb{C}\to H^3(\mathfrak{g},K_\infty,\mathfrak{I}_B^G\lambda_{\mathbb{R}}^{w_0}\otimes\mathcal{M}_{\lambda^\vee,\mathbb{C}})&\xrightarrow{r^3}0.
\end{aligned}$$

The homomorphisms r^1,r^2 are isomorphisms and all the $H^1,H^2=\mathbb{C}$. Hence we see that in this case

$$\operatorname{Coh}(\lambda^\vee)=\{\mathcal{M}_{\lambda,\mathbb{C}},\mathcal{D}_\lambda\}$$

and

$$\operatorname{Coh}_2(\lambda^\vee)=\begin{cases}\{\mathcal{M}_{\lambda,\mathbb{C}},\mathcal{D}_\lambda\} & \text{if } \lambda^{(1)}=0\\ \{\mathcal{D}_\lambda\} & \text{if } n_1=n_2>0\end{cases}$$

4.2 The Eichler-Shimura isomorphism

We want to apply these facts about representation theory to the study of cohomology groups $H^\bullet(\Gamma\backslash X, \tilde{\mathcal{M}}_\lambda \otimes \mathbb{C})$, where now Γ is a congruence subgroup of $\mathrm{Gl}_2(\mathbb{Z})$ or $\mathrm{Gl}_2(\mathcal{O})$. where $\mathcal{O}$ is the ring of integers in imaginary quadratic field.

We start again from a dominant weight $\lambda = n\gamma + d\det \in X^*(T \times \mathbb{C})$. Every $(\mathfrak{g}, \tilde{K}_\infty)$-invariant homomorphism $\Psi_\lambda : \mathfrak{I}_B^G w \cdot \lambda_\mathbb{R} \to \mathcal{C}_\infty(\Gamma\backslash G(\mathbb{R}))$ induces a homomorphism

$$\begin{aligned}\Psi_\lambda : &\operatorname{Hom}_{\tilde{K}_\infty}(\Lambda^\bullet(\mathfrak{g}/\tilde{\mathfrak{k}}), \mathfrak{I}_B^G w \cdot \lambda_\mathbb{R} \otimes \mathcal{M}_{\lambda^\vee,\mathbb{C}})\\ &\to \operatorname{Hom}_{\tilde{K}_\infty}(\Lambda^\bullet(\mathfrak{g}/\tilde{\mathfrak{k}}), \mathcal{C}_\infty(\Gamma\backslash G(\mathbb{R})) \otimes \mathcal{M}_{\lambda^\vee,\mathbb{C}}).\end{aligned}$$

We will show in Section 6.1.3 Proposition 6.1.1 that the complex on the right is isomorphic to the de-Rham complex:

$$\operatorname{Hom}_{\tilde{K}_\infty}(\Lambda^\bullet(\mathfrak{g}/\tilde{\mathfrak{k}}), \mathcal{C}_\infty(\Gamma\backslash G(\mathbb{R})) \otimes \mathcal{M}_{\lambda^\vee,\mathbb{C}}) \xrightarrow{\sim} \Omega^\bullet(\Gamma\backslash X, \tilde{\mathcal{M}}_{\lambda^\vee,\mathbb{C}}). \tag{4.75}$$

This de-Rham complex computes the cohomology, and hence we get a homomorphism

$$\Psi_\lambda^\bullet : H^\bullet(\mathfrak{g}, \tilde{K}_\infty, \mathfrak{I}_B^G w \cdot \lambda_\mathbb{R} \otimes \mathcal{M}_{\lambda^\vee,\mathbb{C}}) \to H^\bullet(\Gamma\backslash X, \tilde{\mathcal{M}}_{\lambda^\vee,\mathbb{C}}). \tag{4.76}$$

We denote by $\omega^{(0)}$ the restriction of the central character of $\mathfrak{I}_B^G w \cdot \lambda_\mathbb{R}$ to the subgroup Z_0, and we introduce the spaces

$$\begin{aligned}\mathcal{E}^{(\infty)}(\lambda, w, \Gamma) = \quad &\operatorname{Hom}_{(\mathfrak{g},K_\infty)}(\mathfrak{I}_B^G w \cdot \lambda_\mathbb{R},\ \mathcal{C}_\infty(\Gamma\backslash G(\mathbb{R}), \omega^{(0)}))\\ \cup \quad & \\ \mathcal{E}^{(2)}(\lambda, w, \Gamma) = \quad &\operatorname{Hom}_{(\mathfrak{g},K_\infty)}(\mathfrak{I}_B^G w \cdot \lambda_\mathbb{R},\ \mathcal{C}_\infty^{(2)}(\Gamma\backslash G(\mathbb{R}), \omega^{(0)}))\end{aligned} \tag{4.77}$$

where the superscript $^{(2)}$ means square integrable (see (6.14)). It is clear from the results in Chapter 6 that the spaces $\mathcal{E}^{(?)}$ are finite-dimensional.

We still have another subspace of $\mathcal{C}_\infty(\Gamma\backslash G(\mathbb{R}))$, namely the space of *cusp forms*

$$\mathcal{C}_\infty^{(\text{cusp})}(\Gamma\backslash G(\mathbb{R})) = \{f \in \mathcal{C}_\infty^{(2)}(\Gamma\backslash G(\mathbb{R})) \mid \int_{U(\mathbb{R})/\Gamma\cap U(\mathbb{R})} f(ug) = 0\},$$

where U runs over the set of unipotent radicals of Borel subgroups over $\mathbb{Q}$ and f should satisfy some mild growth condition. It is well-known that cusp forms are rapidly decreasing, and hence we have $\mathcal{C}_\infty^{(\text{cusp})}(\Gamma\backslash G(\mathbb{R})) \subset \mathcal{C}_\infty^{(2)}(\Gamma\backslash G(\mathbb{R}))$.

For $? \in \{\infty, 2, \text{cusp}\}$ in the superscript as above, we get maps in cohomology

$$\Phi^? : \mathcal{E}^{(?)}(\lambda, w, \Gamma) \otimes H^\bullet(\mathfrak{g}, \tilde{K}_\infty, \mathfrak{I}_B^G w \cdot \lambda_\mathbb{R} \otimes \mathcal{M}_{\lambda^\vee,\mathbb{C}}) \to H^\bullet(\Gamma\backslash X, \mathcal{M}_{\lambda^\vee,\mathbb{C}}).$$

Of course the module $\mathcal{E}^{(2)}(\lambda, w \cdot \lambda) = 0$ unless $\mathfrak{I}_B^G w \cdot \lambda_\mathbb{R}$ has a non-trivial quotient module which admits a positive-definite quasiunitary $(\mathfrak{g}, K_\infty)$-invariant metric. This means that $\mathcal{E}^{(2)}(\lambda, w \cdot \lambda) \neq 0$ implies that in case B) the coefficients satisfy

$$n_1 = n_2, \text{ i.e. } \lambda = n(\gamma_1 + \gamma_2) + d_1 \det + d_2 \det. \tag{4.78}$$

We will say that λ is unitary if this condition is fulfilled. Then the results in Section 4.1.5 yield that these irreducible quasiunitary quotient modules are $\mathcal{D}_\lambda^\pm$ in case A) and $\mathcal{D}_\lambda$ in case B).

Hence it is clear that a $\Psi_\lambda \in \mathcal{E}^{(2)}(\lambda, w_0 \cdot \lambda)$ must vanish on the finite dimensional submodule $\mathcal{M}_\lambda$ if $n > 0$ and hence under this condition we have

$$\mathcal{E}^{(2)}(\lambda, w_0 \cdot \lambda) = \mathrm{Hom}_{(\mathfrak{g}, K_\infty)}(\mathcal{D}_\lambda, \mathcal{C}_\infty^{(2)}(\Gamma\backslash G(\mathbb{R}), \omega^{(0)})).$$

v

We have the following fundamental theorem:

Theorem 4.2.1. *(Eichler-Shimura Isomorphism) For λ unitary, the map*

$$\Phi_\lambda^{(2)} : \mathcal{E}^{(2)}(\lambda, w_0, \Gamma) \otimes H^q(\mathfrak{g}, \tilde{K}_\infty, \mathcal{D}_\lambda \otimes \mathcal{M}_{\lambda^\vee,\mathbb{C}}) \to H_!^q(\Gamma\backslash X, \mathcal{M}_{\lambda^\vee,\mathbb{C}}) \quad (4.79)$$

is an isomorphism for $q = 1$ in case A) and for $q = 1, 2$ in case B).

This is well-known and proofs can be found everywhere in the literature. It is a special case of Theorem 6.2.1. But it says a little bit more, as it says that the image of $\Phi_\lambda^{(2)}$ lies in the inner cohomology and not only in $H^\bullet_{(2)}$. But this is very easy to see, if we apply our considerations from at the end of Section 6.2.3 to this special case.

We still have the special case $\lambda^{(1)} = 0$. In this case $\mathcal{M}_\lambda$ is one-dimensional and isomorphic to the one-dimensional subspace $\mathbb{C}[\lambda] \subset \mathcal{C}_\infty^{(2)}(\Gamma\backslash G(\mathbb{R}))$. Then the map

$$H^\bullet(\mathfrak{g}, \tilde{K}_\infty, \mathbb{C}[\lambda] \otimes \mathbb{C}[\lambda^\vee]) \to H^\bullet(\Gamma\backslash X, \mathcal{M}_{\lambda^\vee,\mathbb{C}})$$

is an isomorphism in degree zero, and zero in all other degrees.

For the case A) we want to relate this to the classical formulation. The group $\mathrm{Sl}_2(\mathbb{R})$ acts transitively on the upper half plane $\mathbb{H} = \mathrm{Sl}_2(\mathbb{R})/\mathrm{SO}(2)$. For $g = \begin{pmatrix} a & b \\ c & d \end{pmatrix}$ and $z \in \mathbb{H}$ we put $j(g, z) = cz + d$. To any

$$\Phi \in \mathrm{Hom}_{(\mathfrak{g}, K_\infty)}(\mathcal{D}_\lambda^+,\ \mathcal{C}_\infty^{(2)}(\Gamma\backslash G(\mathbb{R}), \omega^{(0)}))$$

we attach a function $f_{n+2}^\Phi : \mathbb{H} \to \mathbb{C}$: Write $z = gi$ with $g \in \mathrm{Sl}_2(\mathbb{R})$ and define

$$f_{n+2}^\Phi(z) = \Phi(\psi_{n+2})(g) j(g, i)^{n+2}. \quad (4.80)$$

An easy calculation shows that f_{n+2}^Φ is well-defined and holomorphic and for $\gamma = \begin{pmatrix} a & b \\ c & d \end{pmatrix} \in \mathrm{Sl}_2(\mathbb{Z})$ it satisfies

$$f_{n+2}^\Phi(\gamma z) = (cz + d)^{n+2} f_{n+2}^\Phi(z).$$

The condition that $\Phi(\psi_{n+2})(g)$ is square integrable implies that f_{n+2} is a holomorphic cusp form of weight $n + 2 = k$. It is a special case of the theorem of Gelfand-Graev that this provides an isomorphism

$$\mathrm{Hom}_{(\mathfrak{g},K_\infty)}(\mathcal{D}_\lambda^+, \mathcal{C}_\infty^{(2)}(\Gamma\backslash G(\mathbb{R}))) \xrightarrow{\sim} S_k(\Gamma), \tag{4.81}$$

where of course $S_k(\Gamma)$ is the space of holomorphic cusp forms for Γ.

We can do the same thing for $\mathcal{D}_\lambda^-$.Then we land in the spaces of antiholomorphic cusp forms. These two spaces are isomorphic under conjugation. Combining this with our results above gives the classical formulation of the Eichler-Shimura theorem.

We have a canonical isomorphism

$$S_k(\Gamma) \oplus \overline{S_k(\Gamma)} \xrightarrow{\sim} H_!^1(\Gamma\backslash\mathbb{H}, \tilde{\mathcal{M}}_{\lambda^\vee,\mathbb{C}}). \tag{4.82}$$

Of course the involution $\mathbf{c}^{(1)}$ interchanges the two summands. It is also clear that $\mathbf{c}^{(1)}$ is the extension of the involution.

There is an analogous formulation in case we have to work with Bianchi modular forms.

4.2.1 Petersson scalar product and semi-simplicity

Earlier on in Chapter 3 we stated a general theorem, Theorem 3.1.1, which in this case says that $H_!^1(\Gamma\backslash\mathbb{H}, \mathcal{M}_{\lambda^\vee,\mathbb{C}})$ is a semi-simple module for the Hecke algebra, and we also gave an outline of the proof. In this case the hermitian scalar product is obtained from the Petersson scalar product on $S_k(\Gamma)$. For two cusp forms $f, g \in S_k(\Gamma)$, this scalar product is given by

$$\langle f, g\rangle = \int_{\Gamma\backslash\mathbb{H}} f(z)\overline{g(z)}y^k i\frac{dz\wedge d\bar{z}}{y^2}.$$

For this metric the Hecke operators are self-adjoint, and thus $S_k(\Gamma)$ is semi-simple as a Hecke module.

We can decompose into eigenspaces

$$H_!^1(\Gamma\backslash\mathbb{H}, \tilde{\mathcal{M}}_{\lambda^\vee,F}) = \bigoplus_{\pi_f} H_!^1(\Gamma\backslash\mathbb{H}, \tilde{\mathcal{M}}_{\lambda^\vee,F})(\pi_f),$$

where $\pi_f : \mathcal{H} \to F$ is a homomorphism. In this case we know that each π_f which actually occurs, occurs with multiplicity 2 (it occurs with multiplicity one in $S_k(\Gamma)$ and $\overline{S_k(\Gamma)}$).

For any embedding $\iota : F \hookrightarrow \mathbb{C}$ we know the Ramanujan-Petersson conjecture, which says

$$\text{For all primes } p \text{ we have } |\iota(\pi_f(T_p))| \leq 2\, p^{\frac{n+1}{2}} \tag{4.83}$$

and again we can conclude that we get a canonical splitting of Hecke-modules

$$H^1(\Gamma\backslash\mathbb{H}, \tilde{\mathcal{M}}_{\lambda^\vee,F}) = H_!^1(\Gamma\backslash\mathbb{H}, \tilde{\mathcal{M}}_{\lambda^\vee,F}) \oplus F\mathrm{Eis}_n$$

where $T_p(\mathrm{Eis}_n) = (p^{n+1}+1)\mathrm{Eis}_n$. Note: the eigenvalue of T_p on Eis_n is different from the eigenvalues of T_p on $H_!^1(\Gamma\backslash\mathbb{H}, \tilde{\mathcal{M}}_{\lambda^\vee,F})$ (Manin-Drinfeld principle), and then a standard linear algebra argument gives us the splitting.

The cohomology of the boundary

In section (1.2.6) we discussed the Borel-Serre compactification in the special case $\Gamma = \mathrm{Sl}_2(\mathbb{Z})$ and very briefly also $\mathrm{Sl}_2(\mathbb{Z}[\mathbf{i}]$. Again we have the de-Rham isomorphism

$$H^1(\partial(\Gamma\backslash\mathbb{H}), \tilde{\mathcal{M}}_\lambda \otimes \mathbb{C}) \xrightarrow{\sim} H^1(\mathrm{Hom}_{K_\infty}(\Lambda^\bullet(\mathfrak{g}/\mathfrak{k}, \mathcal{C}_\infty(\Gamma_\infty\backslash G(\mathbb{R}) \otimes \mathcal{M}_\lambda)$$

We have the tautological embedding $\mathfrak{I}_B^G \lambda_\mathbb{R} \rho_\mathbb{R}^2 \hookrightarrow \mathcal{C}_\infty(\Gamma_\infty\backslash G(\mathbb{R}))$ and hence we get a homomorphism

$$H^1(\mathfrak{g}, K_\infty, \mathfrak{I}_B^G \lambda_\mathbb{R} \rho_\mathbb{R}^2 \otimes \mathcal{M}_\lambda) \to H^1(\partial(\Gamma\backslash\mathbb{H}), \tilde{\mathcal{M}}_\lambda \otimes \mathbb{C}) = \mathbb{C} \otimes (\mathcal{M}_\lambda)_{\Gamma_\infty} = \mathbb{C} \otimes Y^n$$

I think it is obvious that the class $E_+ \otimes Y^n \mapsto Y^n$.

4.2.2 Local Whittaker-models

How do we get such Ψ_λ? In our special situation we get them from Fourier expansions of Whittaker functions, and this will be explained next.

We recall some fundamental results from the representation theory of groups $\mathrm{Gl}_2(\mathbb{Q}_p)$. Let $F/\mathbb{Q}$ be a finite extension. An admissible representation of $\mathrm{Gl}_2(\mathbb{Q}_p)$ is an action of $\mathrm{Gl}_2(\mathbb{Q}_p)$ on an F-vector space V which fulfills the following two additional requirements

a) For any open subgroup $K_p \subset \mathrm{Gl}_2(\mathbb{Z}_p)$, the space of fixed vectors V^{K_p} is finite-dimensional.

b) For any $v \in V$ we find an open subgroup $K_p \subset \mathrm{Gl}_2(\mathbb{Z}_p)$ such that $v \in V^{K_p}$.

We say that V is a $\mathrm{Gl}_2(\mathbb{Q}_p)$-module and we denote the action of $\mathrm{Gl}_2(\mathbb{Q}_p)$ on V by $(g, v) \mapsto gv$. In addition we want to assume that our module has a central character, which means that the center $Z(\mathbb{Q}_p) = \mathbb{Q}_p^\times$ acts by a character $\omega_V : Z(\mathbb{Q}_p) \to F^\times$. Such a module is called irreducible if it does not contain a non-trivial invariant submodule.

Again we define a Hecke algebra. Given K_p, we consider the space of functions

$$\begin{aligned}\mathcal{H}_{K_p} = \{f : \mathrm{Gl}_2(\mathbb{Q}_p) \to F \mid f(zg) = \omega_V^{-1}(z) f(g); f(k_1 g k_2) = f(g), k_i \in K_p \\ f \text{ has compact support} \mod Z(\mathbb{Q}_p)\}.\end{aligned}$$

This gives an algebra by convolution, and this algebra acts on V^{K_p} by

$$f * v = \int_{\mathrm{Gl}_2(\mathbb{Q}_p)/Z(\mathbb{Q}_p)} f(x) x v dx.$$

We normalize the measure dx such that it gives volume one to $K_p/K_p \cap Z(\mathbb{Q}_p)$. If $K_p = \mathrm{Gl}_2(\mathbb{Z}_p)$ the Hecke algebra $\mathcal{H}_{K_p}$ is commutative and generated by the characteristic functions on the two double cosets (3.1.2)

$$T_p = \mathrm{Gl}_2(\mathbb{Z}_p) \begin{pmatrix} p & 0 \\ 0 & 1 \end{pmatrix} \mathrm{Gl}_2(\mathbb{Z}_p) \text{ and } C_p = \mathrm{Gl}_2(\mathbb{Z}_p) \begin{pmatrix} p & 0 \\ 0 & p \end{pmatrix}. \qquad (4.84)$$

We recall - and explain the meaning of - the fundamental fact that each isomorphism class of admissible irreducible modules has a unique Whittaker-model. We assume that $F \subset \mathbb{C}$, then we define the (additive) character

$$\begin{aligned} \tau_p : \mathbb{Q}_p \to \mathbb{Q}_p/\mathbb{Z}_p \cong \mathbb{Q}/\mathbb{Z} \to \mathbb{C}^\times; \\ a/p^m \mapsto e^{-\frac{2\pi i a}{p^m}}. \end{aligned} \tag{4.85}$$

It is clear that the kernel of τ_p is $\mathbb{Z}_p$. Since we have $U(\mathbb{Q}_p) = \mathbb{Q}_p$ we can view τ_p as a character $\tau_p : U(\mathbb{Q}_p) \to \mathbb{C}^\times$. We introduce the space

$$\mathcal{C}_{\tau_p}(\mathrm{Gl}_2(\mathbb{Q}_p)) = \{f : \mathrm{Gl}_2(\mathbb{Q}_p) \to \mathbb{C} \mid f(ug) = \tau_p(u)f(g)\},$$

where in addition we require that our f is invariant under a suitable open subgroup $K_f \subset \mathrm{Gl}_2(\mathbb{Z}_p)$. The group $\mathrm{Gl}_2(\mathbb{Q}_p)$ acts on this space by right translation. The action is not admissible but satisfies the above condition b).

Now we can state the theorem about existence and uniqueness of the Whittaker-model

Theorem 4.2.2. *For any infinite-dimensional, absolutely irreducible admissible* $\mathrm{Gl}_2(\mathbb{Q}_p)$*-module* V *we find a non-trivial (of course invariant under* $\mathrm{Gl}_2(\mathbb{Q}_p)$*) homomorphism*

$$\Psi : V \to \mathcal{C}_{\tau_p}(\mathrm{Gl}_2(\mathbb{Q}_p)),$$

and it is unique up to multiplication by a non-zero scalar.

Proof. We refer to the literature - [70], [40]. □

Spherical representations, their Whittaker-model and the Euler factor

An absolutely irreducible $\mathrm{Gl}_2(\mathbb{Q}_p)$-module is called *spherical* or unramified if for $K_p = \mathrm{Gl}_2(\mathbb{Z}_p)$ we have $V^{K_p} \neq \{0\}$. The space V^{K_p} is an absolutely irreducible module for $\mathcal{H}_{K_p}$, hence it is of rank one. Let h_0 be a generator. Our two operators act by scalars on V^{K_p}, and we write

$$T_p(h_0) = \pi_V(T_p)h_0 \text{ and } C_p(h_0) = \pi_V(C_p)h_0.$$

The module V is completely determined by these two eigenvalues, of course $\pi_V(C_p) = \omega_V(C_p)$.

We can formulate this a little bit differently. Let π_p be the isomorphism-type of our $\mathrm{Gl}_2(\mathbb{Q}_p)$-module V. Then our theorem above asserts that there is a unique $\mathrm{Gl}_2(\mathbb{Q}_p)$-module

$$\mathcal{W}(\pi_p) \subset \mathcal{C}_{\tau_p}(\mathrm{Gl}_2(\mathbb{Q}_p)),$$

with isomorphism-type equal to $\pi_p \times_F \mathbb{C}$. We call this module the *Whittaker realization* of π_p. If our isomorphism-type is unramified then the resulting homomorphism of $\mathcal{H}_p$ to F is also denoted by π_p.

We have the spherical vector $h^{(0)}_{\pi_p} \in \mathcal{W}(\pi_p)^{\mathrm{Gl}_2(\mathbb{Z}_p)}$ which is unique up to a scalar. Since $\mathrm{Gl}_2(\mathbb{Q}_p) = U(\mathbb{Q}_p)T(\mathbb{Q}_p)\mathrm{Gl}_2(\mathbb{Z}_p)$, this spherical vector is determined by its restriction to $T(\mathbb{Q}_p)$. We have a formula for this restriction. First of all we observe that

$$h^{(0)}_{\pi_p}\left(\begin{pmatrix} p^n & 0 \\ 0 & p^m \end{pmatrix}\right) = \pi_p(C_p^m)h^{(0)}_{\pi_p}\left(\begin{pmatrix} p^{n-m} & 0 \\ 0 & 1 \end{pmatrix}\right).$$

We claim that $h^{(0)}_{\pi_p}\left(\begin{pmatrix} p^n & 0 \\ 0 & 1 \end{pmatrix}\right) = 0$ if $n < 0$. To see this, we look at the equalities

$$\begin{aligned} h^{(0)}_{\pi_p}\left(\begin{pmatrix} 1 & u \\ 0 & 1 \end{pmatrix}\begin{pmatrix} p^n & 0 \\ 0 & 1 \end{pmatrix}\right) &= \tau_p(u)h^{(0)}_{\pi_p}\left(\begin{pmatrix} p^n & 0 \\ 0 & 1 \end{pmatrix}\right) \\ &= h^{(0)}_{\pi_p}\left(\begin{pmatrix} p^n & 0 \\ 0 & 1 \end{pmatrix}\begin{pmatrix} 1 & p^{-n}u \\ 0 & 1 \end{pmatrix}\right), \end{aligned}$$

and we can find an element $u \in \mathbb{Q}_p$ such that $p^{-n}u \in \mathbb{Z}_p$ and $\tau_p(u) \neq 1$. This implies the claim.

We exploit the eigenvalue equation $T_p(h^{(0)}_{\pi_p}) = \pi_p(T_p)h^{(0)}_{\pi_p}$, we write the double coset $K_p\begin{pmatrix} p & 0 \\ 0 & 1 \end{pmatrix}K_p$ as a union of right K_p cosets

$$K_p\begin{pmatrix} p & 0 \\ 0 & 1 \end{pmatrix}K_p = \bigcup_{x\in\mathbb{Z}/p\mathbb{Z}}\begin{pmatrix} 1 & x \\ 0 & 1 \end{pmatrix}\begin{pmatrix} p & 0 \\ 0 & 1 \end{pmatrix}K_p \bigcup \begin{pmatrix} 0 & 1 \\ -1 & 0 \end{pmatrix}\begin{pmatrix} p & 0 \\ 0 & 1 \end{pmatrix}K_p.$$

Clearly

$$h^{(0)}_{\pi_p}\left(\begin{pmatrix} p^n & 0 \\ 0 & 1 \end{pmatrix}\begin{pmatrix} 1 & x \\ 0 & 1 \end{pmatrix}\begin{pmatrix} p & 0 \\ 0 & 1 \end{pmatrix}\right) = h^{(0)}_{\pi_p}\left(\begin{pmatrix} p^{n+1} & 0 \\ 0 & 1 \end{pmatrix}\right)$$

$$h^{(0)}_{\pi_p}\left(\begin{pmatrix} p^n & 0 \\ 0 & 1 \end{pmatrix}\begin{pmatrix} 0 & 1 \\ -1 & 0 \end{pmatrix}\begin{pmatrix} p & 0 \\ 0 & 1 \end{pmatrix}\right) = \pi_p(C_p)h^{(0)}_{\pi_p}\left(\begin{pmatrix} p^{n-1} & 0 \\ 0 & 1 \end{pmatrix}\right),$$

and this implies the recursion formula

$$\begin{aligned} \pi_p(T_p)h^{(0)}_{\pi_p}\left(\begin{pmatrix} p^n & 0 \\ 0 & 1 \end{pmatrix}\right) &= \pi_p(C_p)h^{(0)}_{\pi_p}\left(\begin{pmatrix} p^{n-1} & 0 \\ 0 & 1 \end{pmatrix}\right) \\ &\quad + \begin{cases} ph^{(0)}_{\pi_p}\left(\begin{pmatrix} p^{n+1} & 0 \\ 0 & 1 \end{pmatrix}\right) & \text{if } n \geq 0 \\ 0 & \text{if } n < 0 \end{cases} \end{aligned} \tag{4.86}$$

We can normalize $h^{(0)}_{\pi_p}\left(\begin{pmatrix} 1 & 0 \\ 0 & 1 \end{pmatrix}\right) = 1$, and then the values for $n > 0$ follow from the recursion.

There is a more elegant way of writing this recursion. For our unramified π_p we define the local Euler factor

$$L(\pi_p, s) = \frac{1}{1 - \pi_p(T_p)p^{-s} + p\pi_p(C_p)p^{-2s}}. \tag{4.87}$$

We expand this into a power series in p^{-s}, and an elementary calculation shows that

$$L(\pi_p, s) = \sum_{n=0}^{\infty} h^{(0)}_{\pi_p}(\begin{pmatrix} p^n & 0 \\ 0 & 1 \end{pmatrix})p^n p^{-ns}. \tag{4.88}$$

Whittaker-models for Harish-Chandra modules

We also have a theory of Whittacker models for the irreducible Harish-Chandra modules studied in Section 4.1. The unipotent radical being $U(\mathbb{R}) = \mathbb{R}$ resp. $U(\mathbb{R}) = \mathbb{C}$. Again we fix characters $\tau_\infty : U(\mathbb{R}) \to \mathbb{C}^\times$, and we put

$$\tau_\infty(x) = \begin{cases} e^{2\pi i x} & \text{in case A)} \\ e^{2\pi i(x+\bar{x})} & \text{in case B)} \end{cases}, \tag{4.89}$$

and as in the p-adic case we define

$$\mathcal{C}_{\tau_\infty}(G(\mathbb{R})) = \{f : G(\mathbb{R}) \to \mathbb{C} \mid f(ug) = \tau_\infty(u)f(g), f \text{ is } \mathcal{C}_\infty\}.$$

Then again we have the following:

Theorem 4.2.3. *For any infinite-dimensional, absolutely irreducible admissible* $\mathrm{Gl}_2(\mathbb{R})$*-module* V *we find a non-trivial (of course invariant under* $\mathrm{Gl}_2(\mathbb{R})$*) homomorphism*

$$\Psi : V \to \mathcal{C}_{\tau_\infty}(G(\mathbb{R})). \tag{4.90}$$

This homomorphism is unique up to a scalar. The image of V *under the homomorphism* Ψ *will be denoted by* $\tilde{V}$.

Proof. Again we refer to the literature [40]. □

Hence we can say that for any isomorphism class π_∞ of irreducible infinite-dimensional Harish-Chandra modules there exists a unique Whittaker-model $\mathcal{W}(\pi_\infty) \subset \mathcal{C}_{\psi_\infty}(G(\mathbb{R}))$. In the book of Godement [40] we find explicit formulae for these Whittaker functions.

Actually, it is easy to write down such maps $\tilde{\Psi}_\pm$ resp. $\tilde{\Psi}$ explicitly for our induced modules. We start from a dominant weight given by $\lambda = n\gamma + \delta$ (resp. $n_1\gamma_1 + n_2\gamma_2 + \delta$) where $n \geq 0, n_1, n_2 \geq 0$. We define

$$\mathcal{F} : \mathfrak{I}_B^G \lambda_{\mathbb{R}} |\rho|^2_{\mathbb{R}} \to \mathcal{C}_{\tau_\infty}(G(\mathbb{R}))$$

by the integral

$$\mathcal{F}(f)(g) = \int_{U(\mathbb{R})} f(wug)\tau_\infty(-u)du.$$

There is no problem with convergence as long as $n > 0, n_1, n_2 > 0$. If one of these numbers is zero then there is a tiny difficulty to overcome, and we ignore it. In any case, we get an isomorphism

$$\mathcal{F} : \mathfrak{I}_B^G \lambda_{\mathbb{R}} |\rho|^2_{\mathbb{R}} \xrightarrow{\sim} \mathfrak{I}_B^{G,\dagger} \lambda_{\mathbb{R}} |\rho|^2_{\mathbb{R}},$$

we will write $\mathcal{F}(h) = h^\dagger$.

We consider the case A). Let n be even. We consider the induced module $\mathfrak{I}_B^G \lambda_\mathbb{R} \rho_\mathbb{R}^2 = \bigoplus_{\nu \equiv 0(2)} \mathbb{C}\phi_{\lambda,\nu}$ (see 4.27), and we have the exact sequence (see 4.46)

$$0 \to \mathcal{D}^+_{\lambda^\vee} \oplus \mathcal{D}^-_{\lambda^\vee} \to \mathfrak{I}_B^G \lambda_\mathbb{R} \rho_\mathbb{R}^2 \to \mathcal{M}_\lambda \to 0.$$

We have the Whittaker map

$$\mathcal{F} : \mathfrak{I}_B^G \lambda_\mathbb{R} \rho_\mathbb{R}^2 \to \mathcal{C}_\tau(G(\mathbb{R}))$$

which is defined by

$$\mathcal{F}(\phi_{\lambda,\nu})\left(\begin{pmatrix} t & 0 \\ 0 & 1 \end{pmatrix}\right) := \int_{-\infty}^{\infty} \phi_{\lambda,\nu}\left(w \begin{pmatrix} 1 & x \\ 0 & 1 \end{pmatrix} \begin{pmatrix} t & 0 \\ 0 & 1 \end{pmatrix}\right) e^{-2\pi i x} dx.$$

We change variables $x \to -x$ and then the Iwasawa decomposition gives

$$w \begin{pmatrix} 1 & -x \\ 0 & 1 \end{pmatrix} \begin{pmatrix} t & 0 \\ 0 & 1 \end{pmatrix} = \begin{pmatrix} 0 & 1 \\ -t & x \end{pmatrix} = \begin{pmatrix} \frac{t}{\sqrt{t^2+x^2}} & * \\ 0 & \sqrt{t^2+x^2} \end{pmatrix} \begin{pmatrix} \frac{x}{\sqrt{t^2+x^2}} & \frac{t}{\sqrt{t^2+x^2}} \\ \frac{-t}{\sqrt{t^2+x^2}} & \frac{x}{\sqrt{t^2+x^2}} \end{pmatrix},$$

and a straightforward computation gives us that we have to evaluate

$$\mathcal{F}(\phi_{\lambda,\nu})\left(\begin{pmatrix} t & 0 \\ 0 & 1 \end{pmatrix}\right) = t^{\frac{n}{2}+1} \int_{-\infty}^{\infty} \frac{e^{2\pi i x}}{(-x+ti)^{n/2+\nu/2+1}(-x-ti)^{n/2-\nu/2+1}} dx.$$

We apply the residue theorem and we consider the case $t > 0$. Let $R >> 0$ and consider the path from $-R$ to R on the real line followed by the half arc C_R in the upper half plane from from R back to $-R$. The integral along this path is the sum of the residues in the interior of this closed path. On the other hand it it is easy to see that for $R \to \infty$ the integral along C_R goes to zero, basically because $|e^{2\pi i z}|$ becomes very small if $\mathrm{Im}(z)$ becomes large. Our function has only one pole in the upper half plane, namely $x = ti$, and therefore

$$\begin{aligned} &\int_{-\infty}^{\infty} \phi_{\lambda,\nu}\left(w \begin{pmatrix} 1 & -x \\ 0 & 1 \end{pmatrix} \begin{pmatrix} t & 0 \\ 0 & 1 \end{pmatrix}\right) e^{2\pi i x} dx \\ &\qquad = (-2\pi i) t^{\frac{n}{2}+1} \mathrm{Res}_{x=ti} \frac{e^{2\pi i x}}{(-x+ti)^{n/2+\nu/2+1}(-x-ti)^{n/2-\nu/2+1}} \end{aligned}$$

If we put $z := x - ti$ then our integral becomes

$$(-2\pi i)(2i)^{-\frac{n}{2}+\frac{\nu}{2}-1} t^{\frac{\nu}{2}} e^{-2\pi t} \mathrm{Res}_{z=0} \frac{e^{2\pi i z}}{(1+\frac{z}{2ti})^{\frac{n}{2}-\frac{\nu}{2}+1} z^{\frac{n}{2}+\frac{\nu}{2}+1}} = P_{\lambda,\nu}(t) e^{-2\pi t},$$

where $P_{\lambda,\nu}(t)$ is a Laurent polynomial in $\mathbb{C}[t, t^{-1}]$.

If $t > 0$ then there is no pole for $\nu \leq -n-2$, which implies that $\mathcal{F}$ maps $\mathcal{D}^-_{\lambda^\vee}$ to zero. The group $\mathrm{Gl}_2(\mathbb{R})$ has the two connected components $\mathrm{Gl}_2(\mathbb{R})^+, \mathrm{Gl}_2(\mathbb{R})^-$ (corresponding to $\det > 0, \det < 0$ respectively). If we restrict the Whittaker functions to $\mathrm{Gl}_2(\mathbb{R})^+$ our map $\mathcal{F}$ induces an injection

$$\mathfrak{I}_B^G \lambda_\mathbb{R} \rho_\mathbb{R}^2 / \mathcal{D}^-_{\lambda^\vee} \hookrightarrow \mathcal{C}_{\tau_\infty}(\mathrm{Gl}_2(\mathbb{R})^+)$$

which is of course an intertwining operator. We let $\phi_{\lambda,n+2}$ be the lowest weight vector for the module $\mathcal{D}^+_{\lambda^\vee} \subset \mathfrak{I}_B^G \lambda_\mathbb{R} \rho_\mathbb{R}^2 / \mathcal{D}^+_{\lambda^\vee}$. Computing $\mathcal{F}(\phi_{\lambda,n+2})$ one gets that the nasty factor $(1+\frac{z}{2ti})^{n/2-\nu/2+1}$ is equal to one in this case, and hence we have, up to a non-zero constant

$$\mathcal{F}(\phi_{\lambda,n+2})\left(\begin{pmatrix} t & 0 \\ 0 & 1 \end{pmatrix}\right) = c_\lambda t^{\frac{n}{2}+1}e^{-2\pi t}.$$

If we now restrict to $t<0$ then basically the same computation shows that $\mathcal{F}$ sends $\mathcal{D}^-_{\lambda^\vee}$ to zero.

In the case A) we have the two discrete irreducible series representations $\mathcal{D}^+_{\lambda^\vee}, \mathcal{D}^-_{\lambda^\vee}$ attached to a dominant weight λ. We have their Whittaker-model

$$\mathcal{F}_\pm : \mathcal{D}^\pm_{\lambda^\vee} \hookrightarrow \mathcal{C}_{\tau_\infty}(\mathrm{Gl}_2(\mathbb{R})). \tag{4.91}$$

and we have

$$\mathcal{F}_+(\mathcal{D}^+_{\lambda^\vee}) = \mathcal{D}^{+,\dagger}_{\lambda^\vee} \text{ is supported on } \mathrm{Gl}_2(\mathbb{R})^+, \mathcal{D}^{-,\dagger}_{\lambda^\vee} \text{ is supported on } \mathrm{Gl}_2(\mathbb{R})^-.$$

Under the isomorphism $\tilde{\Psi}_\pm$ the elements $\psi_{\pm(n+2)}$ (see (4.26)) are mapped to functions $\psi^\dagger_{\pm(n+2)}$. We can normalize $\tilde{\Psi}_\pm$ such that

$$\psi^\dagger{}_{n+2}\left(\begin{pmatrix} t & 0 \\ 0 & 1 \end{pmatrix}\right) = \begin{cases} t^{\frac{n}{2}+1}e^{-2\pi t} & \text{if } t>0 \\ 0 & \text{else} \end{cases}; \tag{4.92}$$

then $\psi^\dagger{}_{-n-2}$ is given by the formula

$$\psi^\dagger{}_{-n-2}\left(\begin{pmatrix} t & 0 \\ 0 & 1 \end{pmatrix}\right) = \begin{cases} (-t)^{\frac{n}{2}+1}e^{2\pi t} & \text{if } t<0 \\ 0 & \text{else} \end{cases}. \tag{4.93}$$

We discuss the same issue for the group $\mathrm{Gl}_2(\mathbb{C})$ later in Section 4.2.5.

4.2.3 Global Whittaker-models, Fourier expansions and multiplicity one

We also have global Whittaker-models. To define them we recall some results from Tate's thesis [110]. We introduce the ring of adeles: For any finite subset Σ in the set of primes we put $\mathbb{A}_\Sigma = \prod_{p\in\Sigma}\mathbb{Q}_p \times \prod_{p\notin\Sigma}\mathbb{Z}_p$, if $\Sigma\subset\Sigma'$ then $\mathbb{A}_\Sigma\subset\mathbb{A}_{\Sigma'}$ and the ring of finite adeles is $\mathbb{A}_f := \bigcup_\Sigma \mathbb{A}_\Sigma$. On $\mathbb{A}_\Sigma$ we have the usual Tychonoff topology. Then $\mathbb{A}_\Sigma$ is a locally compact ring. A subset in $\mathbb{A}_f$ is open iff its intersection with each $\mathbb{A}_\Sigma$ is open in $\mathbb{A}_\Sigma$. Then $\mathbb{A}_f$ is a locally compact ring.

The ring of adeles is $\mathbb{A}=\mathbb{A}_\mathbb{Q}=\mathbb{Q}_\infty\times\mathbb{A}_f$, where of course $\mathbb{Q}_\infty=\mathbb{R}$. The field $\mathbb{Q}$ can be embedded into $\mathbb{A}_\mathbb{Q}$, we simply send $a\in\mathbb{Q}$ to the principal adele $(a,a,\dots)$. The ring of finite adeles contains the compact subring $\hat{\mathbb{Z}}=\mathbb{A}_\emptyset=\prod_p\mathbb{Z}_p$ of integral adeles.

We define a global character $\tau : U(\mathbb{A})/U(\mathbb{Q}) = \mathbb{A}/\mathbb{Q}\to\mathbb{C}^\times$ as the product

$$\tau(x_\infty,\dots,x_p,\dots) = \tau_\infty(x_\infty)\prod_p \tau_p(x_p) \tag{4.94}$$

where the local components τ_v are as above. We have to first check that τ is trivial on $U(\mathbb{Q})$ (see [110], "note the minus sign"). For any $a \in \mathbb{Q}$ we define $\tau^{[a]}(x) = \tau(ax)$, so $\tau = \tau^{[1]}$. In [110] it is shown that the map

$$\mathbb{Q} \to \mathrm{Hom}(\mathbb{A}/\mathbb{Q}, \mathbb{C}^\times);\ a \mapsto \tau^{[a]} \tag{4.95}$$

is an isomorphism between $\mathbb{Q}$ and the character group of $\mathbb{A}/\mathbb{Q}$. Hence we know that for any reasonable function $h : \mathbb{A}/\mathbb{Q} \to \mathbb{C}$ we have a Fourier expansion

$$h(\underline{u}) = \sum_{a\in\mathbb{Q}} \hat{h}(a)\tau(a\underline{u}) \tag{4.96}$$

where $\hat{h}(a) = \int_{\mathbb{A}/\mathbb{Q}} h(\underline{u})\tau(-a\underline{u})d\underline{u}$, and where $\mathrm{vol}_{d\underline{u}}(\mathbb{A}/\mathbb{Q}) = 1$. Then we put

$$\mathcal{C}_\tau(\mathrm{Gl}_2(\mathbb{R})\times\mathrm{Gl}_2(\mathbb{A}_f)/K_f) = \{f : \mathrm{Gl}_2(\mathbb{R})\times\mathrm{Gl}_2(\mathbb{A}_f)/K_f \to \mathbb{C} | f(\underline{u}\underline{g}) = \tau(\underline{u})f(\underline{g})\}$$

which is a module for $\mathrm{Gl}_2(\mathbb{R}) \times \bigotimes' \mathcal{H}_p$.

Let us start from the Harish-Chandra module $\pi_\infty = \mathcal{D}^+_\lambda$ and a homomorphism $\pi_f = \otimes'\pi_p : \otimes'\mathcal{H}_p \to F$ from the unramified Hecke algebra to F. Here $F/\mathbb{Q}$ is a finite extension of $\mathbb{Q}$. We assume it comes with an embedding $\iota : F \hookrightarrow \mathbb{C}$, i.e. we also may consider it as a subfield of $\mathbb{C}$.

We still assume for simplicity that $K_f = \mathrm{Gl}_2(\hat{\mathbb{Z}})$. The results on Whittaker-models imply that we have a unique Whittaker-model

$$\mathcal{W}(\pi) = \mathcal{W}(\pi_\infty) \otimes \mathbb{C}h^{(0)}_{\pi_f} \subset \mathcal{C}_\tau(\mathrm{Gl}_2(\mathbb{R}) \times \mathrm{Gl}_2(\mathbb{A}_f)/K_f)$$

for our isomorphism class $\pi = \pi_\infty \times \pi_f$. Here of course $h^{(0)}_{\pi_f} = \otimes h^{(0)}_{\pi_p}$.

We return to Theorem 4.2.1. On the space $\mathcal{C}^{(2)}_\infty(\Gamma\backslash G(\mathbb{R}), \omega^{(0)})$ we have the action of the unramified Hecke algebra. To see this action we start from the observation that the map $\mathrm{Gl}_2(\mathbb{Q}) \to \mathrm{Gl}_2(\mathbb{A}_f)/K_f$ (Chap. III , 1.5) is surjective, and hence

$$\mathrm{Gl}_2(\mathbb{Z})\backslash\mathrm{Gl}_2(\mathbb{R}) \xrightarrow{\sim} \mathrm{Gl}_2(\mathbb{Q})\backslash\mathrm{Gl}_2(\mathbb{R}) \times \mathrm{Gl}_2(\mathbb{A}_f)/K_f, \tag{4.97}$$

which implies

$$\mathcal{C}^{(2)}_\infty(\mathrm{Gl}_2(\mathbb{Z})\backslash\mathrm{Gl}_2(\mathbb{R})) = \mathcal{C}^{(2)}_\infty(\mathrm{Gl}_2(\mathbb{Q})\backslash\mathrm{Gl}_2(\mathbb{R}) \times \mathrm{Gl}_2(\mathbb{A}_f)/K_f), \tag{4.98}$$

and the space on the right is a $\mathrm{Gl}_2(\mathbb{R}) \times \bigotimes' \mathcal{H}_p$-module. Now we consider the $\pi = \pi_\infty \times \pi_f$ isotypical submodule $\mathcal{C}^{(2)}_\infty(\mathrm{Gl}_2(\mathbb{Q})\backslash\mathrm{Gl}_2(\mathbb{R}) \times \mathrm{Gl}_2(\mathbb{A}_f)/K_f)(\pi) \subset \mathcal{C}^{(2)}_\infty(\mathrm{Gl}_2(\mathbb{Q})\backslash\mathrm{Gl}_2(\mathbb{R}) \times \mathrm{Gl}_2(\mathbb{A}_f)/K_f)$.

We have the following famous theorem which in the case $\Gamma = \mathrm{Sl}_2(\mathbb{Z})$ is due to Hecke:

Theorem 4.2.4. *If $\mathcal{C}^{(2)}_\infty(\mathrm{Gl}_2(\mathbb{Q})\backslash\mathrm{Gl}_2(\mathbb{R}) \times \mathrm{Gl}_2(\mathbb{A}_f)/K_f)(\pi) \neq 0$ then have a canonical isomorphism*

$$\mathcal{F} : \mathcal{W}(\pi) \xrightarrow{\sim} \mathcal{C}^{(2)}_\infty(\mathrm{Gl}_2(\mathbb{Q})\backslash\mathrm{Gl}_2(\mathbb{R}) \times \mathrm{Gl}_2(\mathbb{A}_f)/K_f)(\pi). \tag{4.99}$$

Furthermore π occurs with multiplicity one.

Proof. We give the inverse of $\mathcal{F}$. Given a function

$$h \in \mathcal{C}_\infty^{(2)}(\mathrm{Gl}_2(\mathbb{Q})\backslash \mathrm{Gl}_2(\mathbb{R}) \times \mathrm{Gl}_2(\mathbb{A}_f)/K_f)(\pi), y$$

we define

$$h^\dagger((g_\infty, \underline{g}_f)) = \int_{U(\mathbb{Q})\backslash U(\mathbb{A})} h(\underline{u}\underline{g})\overline{\tau(\underline{u})} d\underline{u}. \tag{4.100}$$

It is clear that $h^\dagger(g_\infty, \underline{g}_f) \in \mathcal{W}(\pi)$. It follows from the theory of automorphic forms that h is actually in the space of cusp forms, which means that the constant Fourier coefficient $\int_{U(\mathbb{Q})\backslash U(\mathbb{A})} h(\underline{u}\underline{g}) d\underline{u} = 0$, hence our Fourier expansion yields ((4.96), evaluated at $u = 0$)

$$h(\underline{g}) = \sum_{a \in \mathbb{Q}^\times} \int_{U(\mathbb{A})/U(\mathbb{Q})} h(\underline{u}\underline{g}) \tau^{[-a]}(\underline{u}) d\underline{u}. \tag{4.101}$$

The measure $d\underline{u}$ is invariant under multiplication by $a \in \mathbb{Q}^\times$, and hence an individual term in the summation is

$$\int_{U(\mathbb{A})/U(\mathbb{Q})} h\left(\begin{pmatrix} 1 & \underline{u} \\ 0 & 1 \end{pmatrix} \underline{g}\right) \tau\left(\begin{pmatrix} 1 & -a\underline{u} \\ 0 & 1 \end{pmatrix}\right) d\underline{u}$$
$$= \int_{U(\mathbb{A})/U(\mathbb{Q})} h\left(\begin{pmatrix} 1 & a^{-1}\underline{u} \\ 0 & 1 \end{pmatrix} \underline{g}\right) \overline{\tau\left(\begin{pmatrix} 1 & \underline{u} \\ 0 & 1 \end{pmatrix}\right)} d\underline{u}$$

Now

$$\begin{pmatrix} 1 & a^{-1}\underline{u} \\ 0 & 1 \end{pmatrix} = \begin{pmatrix} a^{-1} & 0 \\ 0 & 1 \end{pmatrix} \begin{pmatrix} 1 & \underline{u} \\ 0 & 1 \end{pmatrix} \begin{pmatrix} a & 0 \\ 0 & 1 \end{pmatrix}.$$

Since h is invariant under the action of $G(\mathbb{Q})$ from the left we find

$$\int_{U(\mathbb{A})/U(\mathbb{Q})} h\,(\underline{u}\underline{g})\, \tau^{[a]}(\underline{u}) d\underline{u} = h^\dagger\left(\begin{pmatrix} a & 0 \\ 0 & 1 \end{pmatrix} (g_\infty, \underline{g}_f)\right).$$

We evaluate at $\underline{g} = (g_\infty, e)$, then

$$h^\dagger\left(\begin{pmatrix} a & 0 \\ 0 & 1 \end{pmatrix} (g_\infty, e)\right) = h^\dagger\left(\begin{pmatrix} a_\infty & 0 \\ 0 & 1 \end{pmatrix} g_\infty, \begin{pmatrix} \underline{a}_f & 0 \\ 0 & 1 \end{pmatrix}\right).$$

For a fixed g_∞, the function $\underline{g}_f \mapsto h^\dagger(g_\infty, \underline{g}_f)$ is up to a factor equal to $h_{\pi_f}^{(0)} = \bigotimes_p' h_{\pi_p}^{(0)}$, and hence we find

$$h^\dagger\left(\begin{pmatrix} a & 0 \\ 0 & 1 \end{pmatrix} (g_\infty, e)\right) = h^\dagger\left(\begin{pmatrix} a_\infty & 0 \\ 0 & 1 \end{pmatrix} g_\infty, e\right) h_{\pi_f}^{(0)}\left(\begin{pmatrix} \underline{a}_f & 0 \\ 0 & 1 \end{pmatrix}\right)$$

The recursion formulae (4.86), (4.88) imply that $h_{\pi_f}^{(0)}\left(\begin{pmatrix} \underline{a}_f & 0 \\ 0 & 1 \end{pmatrix}\right) = 0$ unless $a \in \mathbb{Z}$.

We restrict our functions to $\mathrm{Gl}_2^+(\mathbb{R})$, i.e. we take $g_\infty \in \mathrm{Gl}_2(\mathbb{R})^+$. Then we know that for $h_\infty \in \mathcal{D}_{\lambda^\vee}^+$ the value $h^\dagger\left(\begin{pmatrix} a_\infty & 0 \\ 0 & 1 \end{pmatrix} g_\infty, e\right) = 0$ if $a_\infty < 0$, and hence

$$h^\dagger\left(\begin{pmatrix} a & 0 \\ 0 & 1 \end{pmatrix} (g_\infty, e)\right) = h^\dagger\left(\begin{pmatrix} a_\infty & 0 \\ 0 & 1 \end{pmatrix} g_\infty, \begin{pmatrix} \underline{a}_f & 0 \\ 0 & 1 \end{pmatrix}\right) = 0 \text{ unless } a > 0, a \in \mathbb{Z},$$

and our Fourier expansion (4.96) becomes

$$h(\underline{g}) = \sum_{a=1}^{\infty} h^{\dagger}\left(\begin{pmatrix} a_{\infty} & 0 \\ 0 & 1 \end{pmatrix} g_{\infty}, e\right) h_{\pi_f}^{(0)}\left(\begin{pmatrix} \underline{a}_f & 0 \\ 0 & 1 \end{pmatrix}\right). \tag{4.102}$$

□

We notice that there is never any problem with convergence. The Whittaker functions $h_{\infty}^{\dagger}$ always decay very rapidly at infinity. We write $g_{\infty} = \begin{pmatrix} 1 & u \\ 0 & 1 \end{pmatrix}\begin{pmatrix} t & 0 \\ 0 & 1 \end{pmatrix} k$ with $k \in K_{\infty}$, then it is easy to see

$$\left| h_{\infty}^{\dagger}\left(\begin{pmatrix} t & 0 \\ 0 & 1 \end{pmatrix} g_{\infty}\right) \right| < P(t)e^{-2\pi t}$$

where $P(t)$ is a Laurent polynomial in t. This implies that the series is really very rapidly converging (see remark below).

Now we choose for the component at infinity the function $h_{\infty}^{\dagger} = \psi_{n+2}^{\dagger}$ and we compute the corresponding holomorphic cusp form h^{Φ} under the Eichler-Shimura isomorphism. We have the formula (4.80)

$$\begin{aligned} h^{\Phi}(z) = h^{\Phi}(x+iy) &= h\left(\begin{pmatrix} y^{\frac{1}{2}} & \frac{x}{y^{\frac{1}{2}}} \\ 0 & y^{-\frac{1}{2}} \end{pmatrix}\right) j\left(\begin{pmatrix} y^{\frac{1}{2}} & \frac{x}{y^{\frac{1}{2}}} \\ 0 & y^{-\frac{1}{2}} \end{pmatrix}, i\right)^{n+2} \\ &= h\left(\begin{pmatrix} y^{\frac{1}{2}} & \frac{x}{y^{\frac{1}{2}}} \\ 0 & y^{-\frac{1}{2}} \end{pmatrix}\right) y^{-\frac{n}{2}-1}, \end{aligned}$$

and hence our Fourier expansion (4.102) becomes

$$h^{\Phi}(z) = y^{-\frac{n}{2}-1} \sum_{a=1}^{\infty} \psi_{n+2}^{\dagger}\left(\begin{pmatrix} ay & ax \\ 0 & 1 \end{pmatrix}\right) h_{\pi_f}^{(0)}\left(\begin{pmatrix} \underline{a} & 0 \\ 0 & 1 \end{pmatrix}\right). \tag{4.103}$$

We have the formula (4.92) for $\phi_{n+2}^{\dagger}$, and then this becomes

$$h^{\Phi}(z) = \sum_{a=1}^{\infty} a^{\frac{n}{2}+1} h_{\pi_f}^{(0)}\left(\begin{pmatrix} a & 0 \\ 0 & 1 \end{pmatrix}\right) e^{2\pi i z a}.$$

This is now the classical Fourier expansion of a holomorphic cusp eigenform of weight $k = n+2$, [62]. The numbers $c(\pi_f, a) = a^{\frac{n}{2}+1} h_{\pi_f}^{(0)}\left(\begin{pmatrix} a & 0 \\ 0 & 1 \end{pmatrix}\right)$ are the Fourier coefficients, and they are also the eigenvalues of the operator T_a - defined by Hecke in [62] - on h^{Φ}. If we apply the Eichler-Shimura isomorphism and interpret h^{Φ} as a cohomology class, then it is an eigenclass in $H_!^1(\Gamma\backslash\mathbb{H}, \tilde{\mathcal{M}}_n \otimes \mathbb{C})$ and for any prime p the number $c(\pi_f, p)$ is the eigenvalue of the operator T_p defined in (3.5).

We briefly come back to the question of convergence. Hecke proved in [62] that

$$|c(\pi_f, a)| \le C a^{n+1+\epsilon} \tag{4.104}$$

and with this estimate the convergence becomes obvious.

Actually there is a much better estimate, which will be discussed in the "probably removed" section.

4.2.4 The L-functions

We still assume that $K_f = \mathrm{Gl}_2(\hat{\mathbb{Z}})$ or, what amounts to the same, that $\Gamma = \mathrm{Sl}_2(\mathbb{Z})$. We start from an eigenspace $H^1_!(\Gamma\backslash\mathbb{H}, \tilde{\mathcal{M}}_\lambda \otimes F)(\pi_f)$, now π_f is simply a homomorphism $\pi_f : \mathcal{H}_{K_f} \to \mathcal{O}_F$. To this homomorphism we attach the *cohomological L-function*

$$L^{\mathrm{coh}}(\pi_f, s) = \prod_p \frac{1}{1 - \pi_p(T_p)p^{-s} + p^{1+n-2s}}, \tag{4.105}$$

here T_p is the Hecke operator defined in (3.5), and it differs from the Hecke operator defined by convolution by a factor $p^{\frac{n}{2}}$ in front. If we expand this product over all primes we get

$$L^{\mathrm{coh}}(\pi_f, s) = \sum_{a=1}^{\infty} \frac{c(\pi_f, a)}{a^s}, \tag{4.106}$$

and this is exactly the L-function Hecke attaches to the cusp form provided by π_f. But we want to stress that this cohomological L-function is defined in purely combinatorial terms (see Section 3.2.1, and Part 2, Chapter 7).

At this moment this L-function is a formal expression, as it is a formal Dirichlet series with coefficients in our field F, which is simply a finite extension of $\mathbb{Q}$. If we assume that $F \subset \mathbb{C}$ then we may interpret s as a complex variable, and the above estimate of the size of the coefficients implies that this series converges absolutely and locally uniformly for $\Re(s) > n+2$. Hence this gives a holomorphic function in this halfspace. But something much better is true. We define the completed L-function

$$\Lambda^{\mathrm{coh}}(\pi_f, s) = \frac{\Gamma(s)}{(2\pi)^s} L^{\mathrm{coh}}(\pi_f, s).$$

For this completed L-function Hecke proved that

Theorem 4.2.5. *The function $\Lambda^{\mathrm{coh}}(\pi_f, s)$ has holomorphic continuation into the entire complex plane and satisfies the functional equation*

$$\Lambda^{\mathrm{coh}}(\pi_f, s) = (-1)^{\frac{n}{2}+1}\Lambda^{\mathrm{coh}}(\pi_f, n+2-s).$$

Proof. We could refer to Hecke, but for some reason we give an outline of the argument. We have the integral representation (Mellin transform)

$$\Lambda^{\mathrm{coh}}(\pi_f, s) = \int_0^\infty \sum_{a=1}^{\infty} c(\pi_f, a)e^{-2\pi a y} y^s \frac{dy}{y} = \int_0^\infty h^\Phi(iy) y^s \frac{dy}{y}.$$

Of course here we have to be courageous (or stupid) enough to exchange integration and summation. But since $e^{-2\pi a y}$ goes rapidly to zero if $y \to \infty$ there is no problem with the upper integration limit ∞. If $\Re(s) \gg 0$, the y^s also tends to zero fast enough, so that we do not have a problem with the lower integration limit. But now we can split the integration into two parts

$$\int_0^\infty \sum_a c(\pi_f, a)e^{-2\pi a y} y^s \frac{dy}{y}$$
$$= \int_0^1 \sum_a c(\pi_f, a)e^{-2\pi a y} y^s \frac{dy}{y} + \int_1^\infty \sum_a c(\pi_f, a)e^{-2\pi a y} y^s \frac{dy}{y},$$

where the second integration converges for all values of s. To handle the first integral we observe that $h^\Phi(-\frac{1}{z}) = z^{n+2}h^\Phi(z)$. Hence we can substitute $y \to \frac{1}{y}$ in the first integral and get

$$\Lambda^{\mathrm{coh}}(\pi_f, s) =$$

$$\sum_a \Big(\frac{1}{(2\pi)^s}\frac{c(\pi_f,a)}{a^s}\Gamma(s,2\pi a) + \frac{(-1)^{\frac{n}{2}+1}}{(2\pi)^{n+2-s}}\frac{c(\pi_f,a)}{a^{n+2-s}}\Gamma(n+2-s,2\pi a)\Big). \tag{4.107}$$

Here $\Gamma(\cdot,\cdot)$ is the incomplete Γ-function. It is defined by $\Gamma(s,A) = \int_A^\infty e^{-y}y^s\frac{dy}{y}$. It has the virtue that for any given value of s it decays rapidly if A goes to infinity.

Therefore we see that $\Lambda^{\mathrm{coh}}(\pi_f, s)$ can be written as a sum of two infinite series which are very rapidly converging. Hence it follows that $\Lambda^{\mathrm{coh}}(\pi_f, s)$ is holomorphic in the entire s plane, and the functional equation also becomes obvious. □

We included the proof of the above theorem, because the above formula also gives us a very effective procedure to compute the numerical value of $\Lambda^{\mathrm{coh}}(\pi_f, s_0)$ with high accuracy. We will come back to this issue in Section 5.3.

4.2.5 The periods

Together with the map $\mathcal{F}$ comes the map $\tilde{\mathcal{F}} = \mathrm{Id}\otimes\mathcal{F}\otimes\mathrm{Id}$

$$\mathrm{Hom}_{\tilde K_\infty}(\Lambda^\bullet(\mathfrak{g}/\tilde{\mathfrak{k}}), \mathcal{W}(\pi)\otimes\mathcal{M}_\lambda)$$
$$\longrightarrow$$
$$\mathrm{Hom}_{\tilde K_\infty}(\Lambda^\bullet(\mathfrak{g}/\tilde{\mathfrak{k}}), \mathcal{C}_\infty(\mathrm{Gl}_2(\mathbb{Q})\backslash(\mathrm{Gl}_2(\mathbb{R})\times\mathrm{Gl}_2(\mathbb{A}_f)/K_f)))\otimes\mathcal{M}_\lambda).$$

The purpose of the following computations is to fix a specific choice of basis elements $\omega_\pm^\dagger \in \mathrm{Hom}_{\tilde K_\infty}(\Lambda^1(\mathfrak{g}/\mathfrak{k}), \mathcal{D}_{\lambda^\vee}{}^\dagger\otimes\mathcal{M}_\lambda)$ (in case A), and basis elements $\omega_{1,2}^\dagger \in \mathrm{Hom}_{\tilde K_\infty}(\Lambda^{1,2}(\mathfrak{g}/\mathfrak{k}), \mathcal{D}_{\lambda^\vee}^\dagger\otimes\mathcal{M}_\lambda)$ (in case B). These "canonical" generators will serve us to define the periods. In case A) we have

$$\mathfrak{g}/\tilde{\mathfrak{k}} \xrightarrow{\sim} \mathbb{Q}\begin{pmatrix}1&0\\0&-1\end{pmatrix}\oplus\mathbb{Q}\begin{pmatrix}0&1\\1&0\end{pmatrix} = \mathbb{Q}H\oplus\mathbb{Q}V = \mathfrak{p}.$$

If we put $P = H + V\otimes i, \bar P = H - V\otimes i \in \mathfrak{g}/\tilde{\mathfrak{k}}\otimes\mathbb{Q}(i)$, then

$$\mathfrak{g}/\tilde{\mathfrak{k}}\otimes\mathbb{Q}(i) = \mathbb{Q}(i)P\oplus\mathbb{Q}(i)\bar P,$$

and

$$e(\phi)Pe(-\phi) = e^{2i\phi}P;\ e(\phi)\bar Pe(-\phi) = e^{-2i\phi}\bar P.$$

On $\mathfrak{p}$ we define an euclidian scalar product $<;>_\mathfrak{p}$ which is proportional to the restriction of the Killing form and is normalised by $<H,H>_\mathfrak{p} = 1$. We use this form to identify $\Lambda^1(\mathfrak{p}) \xrightarrow{\sim} \Lambda^1(\mathfrak{p})^\vee$. Hence we always get $\mathrm{Hom}_{K_\infty}(\Lambda^1(\mathfrak{g}/\mathfrak{k}), ?) = (\Lambda^1(\mathfrak{g}/\mathfrak{k})\otimes ?)^{K_\infty}$.

The module $\mathcal{M}_\lambda \otimes \mathbb{Q}(i)$ decomposes under the action of $\tilde{K}_\infty$ into eigenspaces

$$\mathcal{M}_{\lambda^\vee} \otimes \mathbb{Q}(i) = \bigoplus_{\nu=0}^{n} \mathbb{Q}(i)(X + Y \otimes i)^{n-\nu}(X - Y \otimes i)^\nu, \tag{4.108}$$

where

$$e(\phi)((X + Y \otimes i)^{n-\nu}(X - Y \otimes i)^\nu) = e^{i(n-2\nu)\phi} \cdot (X + Y \otimes i)^{n-\nu}(X - Y \otimes i)^\nu.$$

Then we define the basis elements

$$\omega^\dagger = \frac{1}{2}\bar{P} \otimes {\psi^\dagger}_{n+2} \otimes (X - Y \otimes i)^n \ ; \ \bar{\omega}^\dagger = \frac{1}{2}P \otimes {\psi^\dagger}_{-n-2} \otimes (X + Y \otimes i)^n.$$

We still have our involution $\mathbf{c} \in \tilde{K}^*_\infty$ (see (4.23)) and clearly we have $\mathbf{c}\omega^\dagger = i^n\bar{\omega}^\dagger$ (remember $n \equiv 0 \mod 2$).

Now we put

$$\omega^\dagger_+ = \frac{1}{2}(\omega^\dagger + i^n\bar{\omega}^\dagger) \ ; \ \omega^\dagger_- = \frac{1}{2}(\omega^\dagger - i^n\bar{\omega}^\dagger), \tag{4.109}$$

then these elements

$$\omega^\dagger_\pm = \frac{1}{2}(\omega^\dagger \pm i^n\bar{\omega}^\dagger) \in \mathrm{Hom}_{\tilde{K}_\infty}(\Lambda^1(\mathfrak{g}/\mathfrak{k}), \mathcal{D}^\dagger_{\lambda^\vee} \otimes \mathcal{M}_\lambda)_\pm$$

and they are generators of these one-dimensional spaces. The choice of these generators seems to be somewhat arbitrary, in [55] and [60] we give some motivation for this choice.

There is an alternative way to select these generators $\omega^\dagger_\pm$. We assumed that the centre acts trivially on $\mathcal{M}_\lambda$. Hence $\mathcal{M}_\lambda$ is self dual, let $< \ , \ >_\mathcal{M}$ be the pairing we defined in (4.6).

If we evaluate $\omega^\dagger_\pm$ on the element $H \in \mathfrak{g}/\mathfrak{k} = \mathfrak{p}$ then

$$\omega^\dagger_\pm(H) = \frac{1}{4}(\psi^\dagger_{n+2} \otimes (X - Y \otimes i)^n \pm i^n\psi^\dagger_{-n-2} \otimes (X + Y \otimes i)^n) \in \mathcal{D}^\dagger_\lambda \otimes \mathcal{M}_\lambda.$$

These are functions on $\mathrm{Gl}_2(\mathbb{R})$ with values in $\mathcal{M}_\lambda$. We consider the modified function $g \mapsto g\omega^\dagger_\pm(H)(g)$. We consider the functions $g \mapsto < g\omega^\dagger_\pm(H)(g), X^{n-\nu}Y^\nu >_\mathcal{M}$. If we restrict these scalar-valued functions to the real points of the split torus we get

$$\left\langle \begin{pmatrix} t & 0 \\ 0 & 1 \end{pmatrix} \omega^\dagger_\pm(H) \left(\begin{pmatrix} t & 0 \\ 0 & 1 \end{pmatrix} \right), X^{n-\nu}Y^\nu \right\rangle_\mathcal{M} =$$

$$\left\langle \tfrac{1}{4} \left(\psi^\dagger_{n+2}(\begin{pmatrix} t & 0 \\ 0 & 1 \end{pmatrix}) \otimes (t^{1/2}X - t^{-1/2}Y \otimes i)^n X^{n-\nu}, Y^\nu \right) \right\rangle_\mathcal{M} \pm \tag{4.110}$$

$$\left\langle \tfrac{1}{4} \left(i^n\psi^\dagger_{-n-2}(\begin{pmatrix} t & 0 \\ 0 & 1 \end{pmatrix}) \otimes (t^{1/2}X + t^{-1/2}Y \otimes i)^n \right), X^{n-\nu}Y^\nu \right\rangle_\mathcal{M}.$$

Now let ϵ be a variable which can take the values ± 1. Our formula (4.6) gives us $\langle (X - \epsilon Y \otimes i)^n, X^{n-\nu}Y^\nu \rangle_{\mathcal{M}} = (-\epsilon i)^{n-\nu}$, and combining this with the explicit formula (4.92) for the values of $\psi^\dagger_{\epsilon(n+2)}\left(\begin{pmatrix} t & 0 \\ 0 & 1 \end{pmatrix}\right)$ we get

$$\left\langle \begin{pmatrix} t & 0 \\ 0 & 1 \end{pmatrix} \omega^\dagger_\epsilon(H) \left(\begin{pmatrix} t & 0 \\ 0 & 1 \end{pmatrix}\right), X^{n-\nu}Y^\nu \right\rangle_{\mathcal{M}} = \frac{1}{4}\begin{cases} (-i)^{n-\nu} t^{\nu+1} e^{-2\pi t} & \text{for } t > 0 \\ \epsilon i^{n-\nu}(-t)^{\nu+1} e^{2\pi t} & \text{for } t < 0 \end{cases} \tag{4.111}$$

(Here we use that n is even, but with suitable minor modifications we can also treat the case when n is odd.) Then a straightforward computation yields

$$\begin{aligned} &\int_{T^{\mathrm{ad}}(\mathbb{R})} \left\langle \begin{pmatrix} t & 0 \\ 0 & 1 \end{pmatrix} \omega^\dagger_\epsilon(H) \left(\begin{pmatrix} t & 0 \\ 0 & 1 \end{pmatrix}\right), X^{n-\nu}Y^\nu \right\rangle \frac{dt}{t} \\ &= \frac{(-i)^{n-\nu}}{2} \begin{cases} \frac{\Gamma(\nu+1)}{(2\pi)^{\nu+1}} & \text{if } (-1)^{\frac{n}{2}-\nu} = \mathrm{sg}(\epsilon) \\ 0 & \text{else} \end{cases}. \end{aligned} \tag{4.112}$$

For each choice of the sign $\epsilon = \pm 1$, one of these equations determines the generator $\omega^\dagger_\pm$. This formula will be of importance when we discuss the special values of L-functions.

In case B) we do basically the same, in some sense it is even simpler because K_∞ is maximal compact in this case, i.e. $K_\infty = {K_\infty}^*$. But on the other hand we need some very explicit information about the theory of irreducible representations of K_∞ and also about the decomposition of tensor products of these representations. We will also use some explicit formulas for Bessel functions.

The quotient $\mathfrak{g}/\tilde{\mathfrak{k}}$ is a three-dimensional vector space over $\mathbb{Q}$, the group K_∞ acts by the adjoint representation, and this gives us the standard three-dimensional representation of $K_\infty = U(2)$, which, in addition, is trivial on the center (see 4.1.2). This module is given by the highest weight $2\gamma_c$. We must have $\lambda = n(\gamma + \bar{\gamma}) + \cdots$, if we want $\mathcal{E}^{(2)}(\lambda, w, \Gamma) \neq 0$. Then the formulae 4.50 and (4.51) imply that for $\bullet = 1, 2$

$$\dim_{\mathbb{C}} \mathrm{Hom}_{K_\infty}(\Lambda^\bullet(\mathfrak{g}/\tilde{\mathfrak{k}}), \mathcal{D}^\dagger_{\lambda^\vee} \otimes \mathcal{M}_{\lambda^\vee}) = 1. \tag{4.113}$$

Now we recall that we have defined a structure of an $R = \mathbb{Z}[\frac{1}{2}]$-module on the modules $\mathfrak{g}/\mathfrak{k}, \mathcal{M}_\lambda$ and adapting the concepts in [60],Chap. 8 we can define Harish-Chandra module $\mathcal{D}^\dagger_{\lambda^\vee} = \mathcal{D}^\dagger_{\lambda^\vee R} \otimes \mathbb{C}$ over R. Hence we see that

$$\mathrm{Hom}_{K_\infty}(\Lambda^\bullet(\mathfrak{g}/\tilde{\mathfrak{k}}), \mathcal{D}_{\lambda^\vee} \otimes \mathcal{M}_{\lambda^\vee}) = \mathrm{Hom}_{K_\infty}(\Lambda^\bullet(\mathfrak{g}/\tilde{\mathfrak{k}})_R, \mathcal{D}^\dagger_{\lambda^\vee R} \otimes \mathcal{M}_{\lambda^\vee R}) \otimes \mathbb{C},$$

where here we are a little bit sloppy: the first subscript K_∞ is the compact group, and the second subscript is a smooth group scheme over R. For both choices of $\bullet$ the second term in the above equation is a free R-module of rank 1. We choose generators

$$\omega^{\dagger,\bullet}_\lambda \in \mathrm{Hom}_{K_\infty}(\Lambda^\bullet(\mathfrak{g}/\tilde{\mathfrak{k}})_R, \mathcal{D}_{\lambda^\vee R} \otimes \mathcal{M}_{\lambda^\vee R}).$$

These generators $\omega^{\dagger,1}, \omega^{\dagger,2}$ are well-defined up to an element in $R^\times$.

We may also fix these generators by prescribing values of certain Mellin transforms. To do this we need a little bit of representation theory. Of course we may replace K_∞ by SU(2) because the action of the center on the different modules cancels out. The modules $\mathfrak{g}/\mathfrak{k}\otimes\mathbb{C}$, $\mathcal{D}^\dagger_{\lambda^\vee}$ and $\mathcal{M}_{\lambda^\vee}\otimes\mathbb{C}$ extend naturally to $\mathrm{Sl}_2(\mathbb{C})$-modules, and hence we have to find an explicit generator in

$$\mathrm{Hom}_{\mathrm{Sl}_2(\mathbb{C})}(\mathfrak{g}/\mathfrak{k}\otimes\mathbb{C}, \mathcal{D}^\dagger_{\lambda^\vee}\otimes\mathcal{M}_{n\gamma}\otimes\mathcal{M}_{n\bar{\gamma}}).$$

We have an explicit basis for $\mathfrak{g}/\mathfrak{k}\otimes\mathbb{C}$ (see (4.18)), and our module $\mathcal{M}_{\lambda^\vee} = \mathcal{M}^\flat_{n\gamma}\otimes\mathcal{M}^\flat_{n\gamma}\otimes_{\mathcal{O}}\mathbb{C}$ is given explicitly to us.

Our module $\mathcal{D}^\dagger_{\lambda^\vee}\subset \mathfrak{I}_B^G\lambda_\mathbb{R}\rho^2_\mathbb{R}$, and this last module decomposes into SU(2)-types (see (4.30)). These SU(2)-modules canonically extend to $\mathrm{Sl}_2(\mathbb{C})$-modules, we have

$$\mathfrak{I}_B^G\lambda_\mathbb{R}\rho^2_\mathbb{R} = \bigoplus_{\nu=0}^{\infty}\mathcal{M}_{2\nu\gamma} = \bigoplus_{\nu=0}^{\infty}\mathfrak{I}_B^G\lambda_\mathbb{R}\rho^2_\mathbb{R}(2\nu) \tag{4.114}$$

and

$$(\mathfrak{I}_B^G\lambda_\mathbb{R}\rho^2_\mathbb{R}(2(n+1)))^\dagger = \mathcal{D}^\dagger_{\lambda^\vee}(2(n+1)).$$

Now it is clear that we have the problem to select a specific generator in

$$\mathrm{Hom}_{\mathrm{Sl}_2(\mathbb{C})}(\mathfrak{g}/\mathfrak{k}\otimes\mathbb{C}, \mathcal{D}^\dagger_{\lambda^\vee}(2(n+1))\otimes\mathcal{M}^\flat_{n\gamma}\otimes\mathcal{M}^\flat_{n\bar{\gamma}}\otimes\mathbb{C}).$$

The modules $\mathfrak{g}/\mathfrak{k}\otimes\mathbb{C}, \mathcal{M}^\flat_{n\gamma}, \mathcal{M}^\flat_{n\bar{\gamma}}$ come with an explicit basis (see (4.18)). If we want to write down a specific generator $\omega^{\dagger,\bullet}$, we have to write down a basis of $\mathcal{D}^\dagger_{\lambda^\vee}(2(n+1))$.

Again we start from our exact sequence

$$0\to\mathcal{D}_{\lambda^\vee}\to\mathfrak{I}_B^G\lambda_\mathbb{R}\rho^2_\mathbb{R}\to\mathcal{M}_\lambda\to 0,$$

and we apply the map $\mathcal{F}$ to it, and get an exact sequence of Whittaker modules

$$0\to\mathcal{D}^\dagger_{\lambda^\vee}\to(\mathfrak{I}_B^G\lambda_\mathbb{R}\rho^2_\mathbb{R})^\dagger\to\mathcal{M}_\lambda\to 0.$$

We recall the definition of $\mathfrak{I}_B^G\lambda_\mathbb{R}\rho^2_\mathbb{R}$ as an induced representation, the space of K_∞-invariant vectors is spanned by the spherical function

$$\psi_{\lambda,0}(bk) = \psi_{\lambda,0}\left(\begin{pmatrix} t_1 & u\\ 0 & t_2\end{pmatrix}\begin{pmatrix}\alpha & \beta\\ -\bar{\beta} & \bar{\alpha}\end{pmatrix}\right) = \lambda_\mathbb{R}\rho^2_\mathbb{R}(b).$$

We map the induced representation to its Whittaker-model by

$$\mathcal{F}:\psi\mapsto\left(g\mapsto\int\psi\left(w\begin{pmatrix}1 & x+iy\\ 0 & 1\end{pmatrix}g\right)e^{4\pi ix}dxdy\right), \tag{4.115}$$

our basis element will be $\phi^\dagger_{\lambda,0} = \mathcal{F}(\psi_{\lambda,0})$. A straightforward computation yields

$$\phi^\dagger_{\lambda,0}\left(\begin{pmatrix}t & 0\\ 0 & 1\end{pmatrix}\right) = \mathcal{F}(\psi_{\lambda,0})\left(\begin{pmatrix}t & 0\\ 0 & 1\end{pmatrix}\right) = \int_{-\infty}^{\infty}\frac{t^{n+2}}{(t^2+x^2+y^2)^{n+2}}e^{4\pi ix}dxdy.$$

This function in the variable t is well-known, we have

$$\phi^{\dagger}_{\lambda,0}\left(\begin{pmatrix} t & 0 \\ 0 & 1 \end{pmatrix}\right) = \frac{2\pi^{n+2}}{\Gamma(n+2)} t K_{n+1}(2\pi t)$$

where $K_n(2\pi t)$ is the modified Bessel function ([118], Chap 17.9. Of course $\phi^{\dagger}_{\lambda,0}$ is a function on $G(\mathbb{R}) = \mathrm{Gl}_2(\mathbb{C})$. It is right invariant under K_∞, and

$$\phi^{\dagger}_{\lambda,0}\left(\begin{pmatrix} 1 & x+iy \\ 0 & 1 \end{pmatrix}\begin{pmatrix} t & 0 \\ 0 & 1 \end{pmatrix}\right) = e^{4\pi i x}\phi^{\dagger}_{\lambda,0}\left(\begin{pmatrix} t & 0 \\ 0 & 1 \end{pmatrix}\right)$$

hence it is defined by its restriction to $T^{\mathrm{ad}}(\mathbb{R})_{>0}$.

Starting from this function we construct the desired basis of $\mathcal{D}^{\dagger}_{\lambda^\vee}(2(n+1))$. The Lie-algebra $\mathfrak{g}$ acts on $\mathfrak{I}_B^G \lambda_{\mathbb{R}} \rho_{\mathbb{R}}^2$, and we restrict this action to $\mathfrak{p}$, and it is clear that under this action

$$\mathfrak{p} \otimes \mathfrak{I}_B^G \lambda_{\mathbb{R}} \rho_{\mathbb{R}}^2(2\nu) \to \mathfrak{I}_B^G \lambda_{\mathbb{R}} \rho_{\mathbb{R}}^2(2\nu+2) \oplus \mathfrak{I}_B^G \lambda_{\mathbb{R}} \rho_{\mathbb{R}}^2(2\nu) \oplus \mathfrak{I}_B^G \lambda_{\mathbb{R}} \rho_{\mathbb{R}}^2(2\nu-2),$$

and if we extend this action to the tensor algebra we get a map

$$\mathfrak{U}_{n+1} : \mathfrak{p}^{\otimes(n+1)} \otimes \mathfrak{I}_B^G \lambda_{\mathbb{R}} \rho_{\mathbb{R}}^2(0) \to \bigoplus_{\nu=0}^{n+1} \mathfrak{I}_B^G \lambda_{\mathbb{R}} \rho_{\mathbb{R}}^2(2\nu), \tag{4.116}$$

here we may replace $n+1$ by any positive integer k.

The group K_∞ acts on $\mathfrak{p}^{\otimes(n+1)}$ by the adjoint action, and the above map is of course a K_∞-homomorphism. On the right hand side we can project to the highest K_∞-type $\mathfrak{I}_B^G \lambda_{\mathbb{R}} \rho_{\mathbb{R}}^2(2n+2) = \mathcal{D}^{\dagger}_{\lambda^\vee}(2(n+1))$, i.e. we get a surjective homomorphism

$$\Pi_{n+1} : \mathfrak{p}^{\otimes(n+1)} \otimes \mathfrak{I}_B^G \lambda_{\mathbb{R}} \rho_{\mathbb{R}}^2(0) \to \mathcal{D}^{\dagger}_{\lambda^\vee}(2(n+1)),$$

again we may replace $n+1$ by any positive integer k.

We have the standard surjective homomorphism $\mathfrak{p}^{\otimes(n+1)} \to \mathrm{Sym}^{n+1}(\mathfrak{p})$, let us denote its kernel by I_{n+1}. For any $f \in \mathfrak{I}_B^G \lambda_{\mathbb{R}} \rho_{\mathbb{R}}^2$ and $X', X'' \in \mathfrak{p}$ we have

$$(X'X'' - X''X')f = [X', X'']f.$$

Since the Lie bracket $[X_1', X_2''] \in \mathfrak{k}$ it follows easily that Π_{n+1} vanishes on the kernel I_{n+1}. Hence our homomorphism Π_{n+1} factors over the quotient, i.e.

$$\Pi_{n+1} : \mathrm{Sym}^{n+1}(\mathfrak{p}) \to \mathcal{D}^{\dagger}_{\lambda^\vee}(2(n+1)).$$

We change our notation for the basis of $\mathfrak{p} \otimes \mathbb{C}$ (see (4.18)) and put

$$X_1 = \tfrac{1}{2}\left(\begin{pmatrix} 0 & 1 \\ 1 & 0 \end{pmatrix} - i \otimes \begin{pmatrix} 0 & i \\ -i & 0 \end{pmatrix}\right); X_0 = \tfrac{1}{\sqrt{2}}\left(\begin{pmatrix} 1 & 0 \\ 0 & -1 \end{pmatrix}\right)$$
$$X_{-1} = \tfrac{1}{2}\left(\begin{pmatrix} 0 & 1 \\ 1 & 0 \end{pmatrix} + i \otimes \begin{pmatrix} 0 & i \\ -i & 0 \end{pmatrix}\right).$$

We have the following proposition

Proposition 4.2.1. *The $2n+3$ elements*

$$\{X_1^{n+1}, X_0X_1^n, \ldots, X_0^{n+1}, X_0^nX_{-1}, \ldots, X_{-1}^{n+1}\}$$

form a basis of a K_∞-invariant subspace of $Sym^{n+1}(\mathfrak{p}) \otimes \mathbb{C}$. This subspace is irreducible, it is isomorphic to $\mathcal{M}_{2n+2}$. These basis elements are weight eigenvectors for the action of T_c.

Proof. The representation of the algebraic group K_∞ on $\mathfrak{p}$ extends to a representation of the algebraic group $\mathrm{Sl}_2/\mathbb{C}$ on $\mathfrak{p}\otimes\mathbb{C}$. As such it is isomorphic to the symmetric square $\mathrm{Sym}^2(\mathbb{C}^2)$ of the tautological representation, i.e. to the module $\mathcal{M}_2$ of polynomials $aU^2 + bUV + cV^2$. We get an isomorphism $\mathcal{M}_2 \xrightarrow{\sim} \mathfrak{p}\otimes\mathbb{C}$ by sending $U^2 \mapsto X_1, UV \mapsto X_0, V^2 \mapsto X_{-1}$. Now we see that $\mathrm{Sym}^{2n+2}(\mathbb{C}^+2) \subset \mathrm{Sym}^{n+1}(\mathrm{Sym}^2(\mathbb{C}^2)) = \mathrm{Sym}^{n+1}(\mathfrak{p}\otimes\mathbb{C})$ is an invariant submodule. It has the basis $U^{2n+2-\nu}V^\nu$, and clearly

$$U^{2n+2-\nu}V^\nu = X_1^{n+1-\nu}X_0^\nu \text{ if } \nu \le n+1 \text{ and } X_0^{2n+2-\nu}X_{-1}^{\nu-n-1}\text{else,}$$

and this implies the assertion. □

This implies that the elements

$$\begin{aligned}\{\Pi_{n+1}(X_1^{n+1}\phi^\dagger_{\lambda,0}), \Pi_{n+1}(X_0X_1^n\phi^\dagger_{\lambda,0}), \ldots, \Pi_{n+1}(X_0^nX_1\phi^\dagger_{\lambda,0}),\\ \Pi_{n+1}(X_0^{n+1}\phi^\dagger_{\lambda,0}), \Pi_{n+1}(X_0^nX_{-1}\phi^\dagger_{\lambda,0}), \ldots, \Pi_{n+1}(X_{-1}^{n+1}\phi^\dagger_{\lambda,0})\}\end{aligned} \tag{4.117}$$

form a basis of $\mathcal{D}^\dagger_{\lambda^\vee}(2(n+1))$.

We change our notation slightly. For $m<0$ we put $X_1^m := X_{-1}^{-m}$ and for $0 \le \nu \le 2n+2$ we put $[\nu] = \nu$ if $\nu \le n+1$ and $[\nu] = 2n+2-\nu$ if $\nu \ge n+1$. Then our above basis can be written as

$$\{\ldots, \Pi_{n+1}(X_0^{[\nu]}X_1^{n+1-\nu}\phi^\dagger_{\lambda,0}), \ldots\}_{\nu=0,\ldots,\nu=2n+2}.$$

These are the weight vectors of weight $2(n+1-\nu)\gamma$. We introduce the notation

$$\phi^\dagger_{\lambda,n+1-\nu} := \Pi_{n+1}(X_0^{[\nu]}X_1^{n+1-\nu}\phi^\dagger_{\lambda,0}).$$

These functions $\phi^\dagger_{\lambda,\mu}$ are Whittaker functions, and they satisfy

$$\phi^\dagger_{\lambda,\mu}\left(\begin{pmatrix}1 & x+iy\\ 0 & 1\end{pmatrix} g\right) = e^{4\pi i x}\phi^\dagger_{\lambda,\mu}(g).$$

They are not K_∞-invariant, but they are weight vectors for the torus, and we have

$$\phi^\dagger_{\lambda,\mu}\left(g\begin{pmatrix}e^{2i\varphi} & 0\\ 0 & 1\end{pmatrix}\right) = e^{2\mu i\varphi}\phi^\dagger_{\lambda,\mu}(g),$$

and more generally $\phi^\dagger_{\lambda,\nu}(gk) = \sum_\mu a_{\nu,\mu}(k)\phi^\dagger_{\lambda,\mu}(g)$ where the $a_{\nu,\mu}(k)$ are the matrix coefficients of $\mathcal{M}_{2n+2}$ (the above proposition).

We consider the restriction of the functions $\phi^\dagger_{\lambda,\nu}$ to the maximal torus $T(\mathbb{R})$. Since $\mathfrak{I}_B^G\lambda_\mathbb{R}\rho^2_\mathbb{R}(2\nu)$ has a central character, it suffices to consider the restriction

$$\phi^\dagger_{\lambda,\nu} \to \{z \mapsto \phi^\dagger_{\lambda,\nu}\left(\begin{pmatrix}z & 0\\ 0 & 1\end{pmatrix}\right)\},$$

and we write $z = te^{2i\varphi}$. This means that we map the module $\mathfrak{I}_B^G\lambda_{\mathbb{R}}\rho_{\mathbb{R}}^2$ to its Kirillow realisation $\mathfrak{I}_B^{G,\kappa}\lambda_{\mathbb{R}}\rho_{\mathbb{R}}^2 \subset \mathcal{C}_\infty(\mathbb{C}^\times)$ (see [40] , §2.5.), especially this map is injective.

We express the restriction of these functions $\phi^\dagger_{\lambda,\nu}$ to the torus $T^{\mathrm{ad}}(\mathbb{R})_{>0}$ in terms of Bessel functions. We introduce the notation

$$\mathfrak{I}_B^G[2k] := \bigoplus_{\nu=0}^{k} \mathfrak{I}_B^G\lambda_{\mathbb{R}}\rho_{\mathbb{R}}^2(2\nu).$$

For any Whittaker function $\phi^\dagger \in \mathfrak{I}_B^G[2k]^\dagger$ we have

$$\Pi_{k+1}(X_1\phi^\dagger)(\begin{pmatrix} t & 0\\ 0 & 1\end{pmatrix}) = \frac{\phi^\dagger\left(\begin{pmatrix} t & 0\\ 0 & 1\end{pmatrix}\exp(\epsilon X_1)\right) - \phi^\dagger\left(\begin{pmatrix} t & 0\\ 0 & 1\end{pmatrix}\right)}{\epsilon}.$$

We write $X_1 = \frac{1}{2}\left(\begin{pmatrix} 0 & 2\\ 0 & 0\end{pmatrix} + i\otimes\begin{pmatrix} 0 & -2i\\ 0 & 0\end{pmatrix} + \begin{pmatrix} 0 & -1\\ 1 & 0\end{pmatrix} + i\otimes\begin{pmatrix} 0 & i\\ i & 0\end{pmatrix}\right)$, where the last two matrices are in $\mathfrak{k}$ so they preserve the K_∞-type and

$$\phi^\dagger\left(\begin{pmatrix} t & 0\\ 0 & 1\end{pmatrix}\exp\left(\epsilon(\begin{pmatrix} 0 & 1\\ 0 & 0\end{pmatrix} + \begin{pmatrix} 0 & i\\ 0 & 0\end{pmatrix})\right)\right) =$$

$$\phi^\dagger\left(\exp\left(\epsilon(\begin{pmatrix} 0 & t\\ 0 & 0\end{pmatrix} + \begin{pmatrix} 0 & it\\ 0 & 0\end{pmatrix})\right)\begin{pmatrix} t & 0\\ 0 & 1\end{pmatrix}\right) = \phi^\dagger\left(\begin{pmatrix} 1 & \epsilon(t+it)\\ 0 & 1\end{pmatrix}\begin{pmatrix} t & 0\\ 0 & 1\end{pmatrix}\right)$$
$$= e^{2\pi i\epsilon t}\phi^\dagger\left(\begin{pmatrix} t & 0\\ 0 & 1\end{pmatrix}\right) = \phi^\dagger\left(\begin{pmatrix} t & 0\\ 0 & 1\end{pmatrix}(1+it\epsilon)\right),$$

and hence

$$\Pi_{k+1}(X_1\phi^\dagger)\left(\begin{pmatrix} t & 0\\ 0 & 1\end{pmatrix}\right) = 2it\phi^\dagger\left(\begin{pmatrix} t & 0\\ 0 & 1\end{pmatrix}\right).$$

If $\phi^\dagger$ is a weight vector, i.e. $\phi^\dagger\left(\begin{pmatrix} te^{2\pi i\varphi} & 0\\ 0 & 1\end{pmatrix}\right) = e^{2\pi i\mu\varphi}\phi^\dagger\left(\begin{pmatrix} t & 0\\ 0 & 1\end{pmatrix}\right)$, then $X_1\phi^\dagger$ is also a weight vector with weight $e^{2\pi i(\mu+2)\varphi}$.

This gives us

$$\phi^\dagger_{\lambda,n+1}\left(\begin{pmatrix} t & 0\\ 0 & 1\end{pmatrix}\right) := (X_1^{n+1}\phi^\dagger_{\lambda,0})\left(\begin{pmatrix} t & 0\\ 0 & 1\end{pmatrix}\right) = i^{n+1}\frac{2^{2n+3}\pi^{2n+3}}{\Gamma(n+2)}t^{n+2}K_{n+1}(2\pi t).$$

Since this function is of weight $2n+2$ we can forget the projection Π_{n+1}. We have recursion formulae for the Bessel functions, namely ([118]

$$\begin{aligned} \tfrac{d}{dt}K_n(t) &= -\tfrac{1}{2}(K_{n-1}(t)+K_{n+1}(t)) \\ K_{n+1}(t) &= K_{n-1}(t)+\tfrac{2n}{t}K_n(t). \end{aligned} \tag{4.118}$$

A straightforward calculation yields

$$t\frac{d}{dt}t^\mu K_\nu(2\pi t) = (\mu-\nu)t^\mu K_\nu(2\pi t) - 2\pi t^{\mu+1}K_{\nu-1}(2\pi t). \tag{4.119}$$

Then $\phi^{\dagger}_{\lambda,n+1-\nu}\left(\begin{pmatrix} t & 0 \\ 0 & 1 \end{pmatrix}\right) = \Pi_{n+1}\left(X_0^{[\nu]}X_1^{n+1-\nu}\phi^{\dagger}_{\lambda,0}\left(\begin{pmatrix} t & 0 \\ 0 & 1 \end{pmatrix}\right)\right)$. Hence

$$X_1^{n+1-\nu}\phi^{\dagger}_{\lambda,0}\left(\begin{pmatrix} t & 0 \\ 0 & 1 \end{pmatrix}\right) = i^{n+1-\nu}\frac{(2\pi)^{n+2+|n+1-\nu|}}{\Gamma(n+2)}t^{1+|n+1-\nu|}K_{n+1}(2\pi t).$$

To this we apply $X_0^{[\nu]}$. The operator X_0 is $\sqrt{2}t\frac{d}{dt}$, thus the above formula yields

$$\Pi_{n+1}\left(X_0^{[\nu]}X_1^{n+1-\nu}\phi^{\dagger}_{\lambda,0}\left(\begin{pmatrix} t & 0 \\ 0 & 1 \end{pmatrix}\right)\right) =$$
$$\Pi_{n+1}\left(\cdots + \frac{2^{2n+3}\pi^{2n+3}}{\Gamma(n+2)}t^{n+2}K_{n+1-\nu}(2\pi t)\right),$$

where the dots $\cdots$ are a sum of those terms which are in the image of $\mathfrak{I}_B^{G,\kappa}\lambda_{\mathbb{R}}\rho_{\mathbb{R}}^2[2n]$, hence they vanish under Π_{n+1}. Consequently

$$\phi^{\dagger}_{\lambda,\mu}\left(\begin{pmatrix} t & 0 \\ 0 & 1 \end{pmatrix}\right) = \frac{2^{2n+3}\pi^{2n+3}}{\Gamma(n+2)}t^{n+2}K_{\mu}(2\pi t),$$

where μ runs from $n+1$ to $-n-1$ and, of course, $K_\mu = K_{-\mu}$.

Decompositions of tensor products

If $\lambda_1 = n_1\gamma, \lambda_2 = n_2\gamma$ are two highest weights and if we consider the highest weight modules $\mathcal{M}_{\lambda_1,\mathbb{Q}}, \mathcal{M}_{\lambda_2,\mathbb{Q}}$ then it is a classical theorem that

$$\mathcal{M}_{\lambda_1,\mathbb{Q}} \otimes \mathcal{M}_{\lambda_2,\mathbb{Q}} = \mathcal{M}_{(n_1+n_2)\gamma,\mathbb{Q}} \oplus \mathcal{M}_{(n_1+n_2-2)\gamma,\mathbb{Q}} \oplus \cdots \oplus \mathcal{M}_{(n_1-n_2)\gamma,\mathbb{Q}}$$

where we assume $n_1 \geq n_2$. We put $n = n_1 + n_2$. Our next aim is to give an explicit homomorphism

$$j_{n_1,n_2} : \mathcal{M}^{\flat}_{(n_1+n_2)\gamma} \hookrightarrow \mathcal{M}^{\flat}_{n_1\gamma} \otimes \mathcal{M}^{\flat}_{n_2\gamma},$$

in other words, we want to write explicit tensors for the images of $e^{\flat}_{\mu}$, where $\mu = 0, \ldots, n_1+n_2$. Of course we send the highest weight vector $e^{\flat}_0 \mapsto {'e}^{\flat}_0 \otimes {''e}^{\flat}_0$, this vector is the highest weight vector in the direct summand $\mathcal{M}^{\flat}_{(n_1+n_2)\gamma,\mathbb{Q}} \subset \mathcal{M}_{(n_1+n_2)\gamma,\mathbb{Q}} \oplus \cdots \oplus \mathcal{M}_{(n_1-n_2)\gamma,\mathbb{Q}}$. In terms of the explicit realisation of these modules we can say

$$X^{n_1+n_2} \mapsto {'X}^{n_1} \otimes {''X}^{n_2}.$$

Now we apply the matrix $\begin{pmatrix} 1 & 0 \\ t & 1 \end{pmatrix}$ to it, where we may think of t as an indeterminate. Then we see

$$(X+tY)^{n_1+n_2} \mapsto ({'X}+t\,{'Y})^{n_1} \otimes ({''X}+t\,{''Y})^{n_2}. \tag{4.120}$$

We expand on both sides and find

$$\sum_{\mu=0}^{n_1+n_2}\binom{n_1+n_2}{\mu}t^{\mu}X^{n_1+n_2-\mu}Y^{\mu} \mapsto$$

$$\sum_{\mu=0}^{n_1+n_2} t^{\mu}\Big(\sum_{\mu_1,\mu_2:\mu_1+\mu_2=\mu}\binom{n_1}{\mu_1}{'X}^{n_1-\mu_1}{'Y}^{\mu_1} \otimes \binom{n_2}{\mu_2}{''X}^{n_2-\mu_2}{''Y}^{\mu_2}\Big).$$

We remember the definition of the basis elements $e^\flat_\mu$, where the formula above gives us

$$j_{n_1,n_2} : e^\flat_\mu \mapsto \sum_{\mu_1+\mu_2=\mu} {}'e^\flat_{\mu_1} \otimes {}''e^\flat_{\mu_2}.$$

We apply this to the SU(2)-module

$$(\mathfrak{g}/\mathfrak{k})^\vee_F \otimes \mathcal{M}_{n\gamma} \otimes_F \mathcal{M}_{n\bar{\gamma}}.$$

This module contains a unique copy of $\mathcal{M}^\flat_{2n+2}$. We write

$$(\mathfrak{g}/\mathfrak{k})^\vee_F = F\,{}^0e^\flat_2 \oplus F\,{}^0e^\flat_1 \oplus F\,{}^0e^\flat_0,\ \mathcal{M}_{n_1\gamma,F} = \bigoplus_{\mu_1} F e^\flat_{\mu_1},\ \mathcal{M}_{n_2\gamma,F} = \bigoplus_{\mu_2} F \bar{e}^\flat_{\mu_2}$$

where of course μ_i runs from 0 to n. Then our copy of $\mathcal{M}^\flat_{2n+2}$ comes with the basis

$$\tilde{e}^\flat_\mu = \sum_{\mu_0+\mu_1+\mu_2=\mu} {}^0e^\flat_{\mu_0} \otimes e^\flat_{\mu_1} \otimes \bar{e}^\flat_{\mu_2}.$$

We have the invariant pairing, and this tells us that we can choose as our generator

Korrektur:

$$\omega^{\dagger,1}_\lambda = \sum_{\mu=-n-1}^{n+1} \phi^\dagger_{\lambda,\mu} \otimes \Big(\sum_{\mu_0+\mu_1+\mu_2=n+1+\mu} {}^0e^\flat_{\mu_0} \otimes e^\flat_{\mu_1} \otimes \bar{e}^\flat_{\mu_2} \Big). \tag{4.121}$$

This generator is only determined up to a scalar.

The "canonical" choice of the generator

Again we can fix the generator by requiring that certain Mellin transforms have a prescribed value at certain prescribed arguments.

We do essentially the same as in case A). We can interpret $\omega^{\dagger,1}_\lambda$ as a differential 1-form on $G(\mathbb{R})$ with values in $\mathcal{M}^\flat_\lambda \otimes \mathbb{C}$. We can restrict this 1-form to the torus $T^{\mathrm{ad}}(\mathbb{R})_{>0} = \left\{ \begin{pmatrix} t & 0 \\ 0 & 1 \end{pmatrix} \mid t > 0 \right\}$. We have the "cycles" $e_{\mu_1} \otimes e_{\mu_2} \in \mathcal{M}^\vee_\lambda$. We evaluate $\omega^{\dagger,1}_\lambda(X_0)$ on these "cycles" and get

$$\begin{aligned} \left\langle \omega^{\dagger,1}_\lambda(X_0), e_{\mu_1} \otimes e_{\mu_2} \right\rangle \left(\begin{pmatrix} t & 0 \\ 0 & 1 \end{pmatrix} \right) &= \phi^\dagger_{\lambda,\mu_1+\mu_2-n} \left(\begin{pmatrix} t & 0 \\ 0 & 1 \end{pmatrix} \right) t^{\mu_1+\mu_2} \\ &= c'_n t^{n+2+\mu_1+\mu_2} K_{n-\mu_1-\mu_2}(2\pi t). \end{aligned} \tag{4.122}$$

Later - when we study special values of L-functions - we need to know the value

$$\int_0^\infty \left\langle \omega^{\dagger,1}_\lambda(X_0), e_{\mu_1} \otimes e_{\mu_2} \right\rangle \left(\begin{pmatrix} t & 0 \\ 0 & 1 \end{pmatrix} \right) \frac{dt}{t} = c'_n \int_0^\infty t^{n+2+\mu_1+\mu_2} K_{n-\mu_1-\mu_2}(2\pi t) \frac{dt}{t}.$$

We will also need formulae for the Mellin transforms of these Bessel functions. Here we quote [3, p. 331, 334], and we recall two of them (the second one for later use):

$$\int_0^\infty K_\nu(2\pi t)t^s\frac{dt}{t} = 2^{s-2}(2\pi)^{-s}\Gamma(\frac{s-\nu}{2})\Gamma(\frac{s+\nu}{2})$$
$$\int_0^\infty K_\mu(2\pi t)K_\nu(2\pi t)t^s\frac{dt}{t} = 2^{s-3}(2\pi)^{-s}\prod_{i,j=0}^{1}\Gamma\left(\frac{s+(-1)^i\mu+(-1)^j\nu}{2}\right).$$

The first one gives us

$$\int_0^\infty t^{n+2+\mu_1+\mu_2}K_{n-\mu_1-\mu_2}(2\pi t)\frac{dt}{t} = \frac{\Gamma(n+1)}{4\pi}\frac{\Gamma(\mu+1)}{\pi^{\mu+1}}.$$

We observe that the first factor in front does not depend on μ_1, μ_2. So we renormalise our generator, and for $\mu = -n-1, -n, \ldots, n+1$ we now put

$$\phi^\dagger_{\lambda,\mu}(t) = \frac{4\pi}{\Gamma(n+1)}t^{n+2}K_{n+1-\mu}(t), \tag{4.123}$$

and with this choice of $\phi^\dagger_{\lambda,\nu}$ the $\omega^{\dagger,1}_\lambda$ in (4.121) is now our canonical generator. Now our formula (4.122) becomes

$$\int_0^\infty \left\langle \omega^{\dagger,1}_\lambda(X_0), e_{\mu_1}\otimes e_{\mu_2}\right\rangle\left(\begin{pmatrix} t & 0\\ 0 & 1\end{pmatrix}\right)\frac{dt}{t} = \frac{\Gamma(\mu+1)}{\pi^{\mu+1}}.$$

Hence we may just choose $\mu_1 = \mu_2 = 0$ to nail down $\omega^{\dagger,1}_\lambda$. It is not clear to me whether or not it is a "miracle" that the above relation holds for all values of μ_1, μ_2.

The definition of the periods

The inner cohomology with rational coefficients is a semi-simple module under the action of the Hecke algebra (see Theorem 3.1.1). We find a finite Galois-extension $F/\mathbb{Q}$ such that we get a decomposition into absolutely irreducible modules

$$H^\bullet_!(\Gamma\backslash\mathbb{H}, \tilde{\mathcal{M}}_\lambda\otimes F) = \bigoplus_{\pi_f} H^\bullet_!(\Gamma\backslash\mathbb{H}, \tilde{\mathcal{M}}_\lambda\otimes F)(\pi_f).$$

Since we assume that $\Gamma = \mathrm{Gl}_2(\mathcal{O})$, the π_f are homomorphisms $\pi_f : \mathcal{H}\to\mathcal{O}_F$ (see 3.17). In case A), such an isotypical piece is a direct sum

$$H^\bullet_!(\Gamma\backslash\mathbb{H}, \tilde{\mathcal{M}}_\lambda\otimes F)(\pi_f) = H^1_!(\Gamma\backslash\mathbb{H}, \tilde{\mathcal{M}}_\lambda\otimes F)(\pi_f)_+ \oplus H^1_!(\Gamma\backslash\mathbb{H}, \tilde{\mathcal{M}}_\lambda\otimes F)(\pi_f)_-$$

where both summands are of dimension one over F.

In case B) we get

$$H^\bullet_!(\Gamma\backslash\mathbb{H}, \tilde{\mathcal{M}}_\lambda\otimes F)(\pi_f) = H^1_!(\Gamma\backslash\mathbb{H}, \tilde{\mathcal{M}}_\lambda\otimes F)(\pi_f) \oplus H^2_!(\Gamma\backslash\mathbb{H}, \tilde{\mathcal{M}}_\lambda\otimes F)(\pi_f),$$

and again the summands are one dimensional.

We have defined the module of integral classes $H^\bullet_{!,\,\mathrm{int}}(\Gamma\backslash\mathbb{H},\tilde{\mathcal{M}}^\flat_\lambda\otimes\mathcal{O}_F)\subset H^\bullet_!(\Gamma\backslash\mathbb{H},\tilde{\mathcal{M}}^\flat_\lambda\otimes F)$ (see 2.47), and we consider the intersection

$$H^\bullet_{!,\,\mathrm{int}}(\Gamma\backslash\mathbb{H},\tilde{\mathcal{M}}^\flat_\lambda\otimes\mathcal{O}_F)(\pi_f)_\epsilon = H^\bullet_!(\Gamma\backslash\mathbb{H},\tilde{\mathcal{M}}_\lambda\otimes F)(\pi_f)_\epsilon\cap H^\bullet_{!,\,\mathrm{int}}(\Gamma\backslash\mathbb{H},\tilde{\mathcal{M}}^\flat_\lambda\otimes\mathcal{O}_F),$$

as a locally free $\mathcal{O}_F$-module of rank 1, where

$$\epsilon=\pm,\bullet=1(\text{resp. }\epsilon=1,\bullet\in\{1,2\}).$$

We assume for simplicity that it is actually free, otherwise the formulation of the following becomes slightly more complicated (see below). On the set of π_f which occur in this decomposition, we have an action of the Galois group (see (3.18)) and the Galois action yields canonical isomorphisms

$$\Phi_{\sigma,\tau}: H^\bullet_{!,\,\mathrm{int}}(\Gamma\backslash\mathbb{H},\tilde{\mathcal{M}}^\flat_\lambda\otimes\mathcal{O}_F)({}^\sigma\pi_f)_\epsilon \xrightarrow{\sim} H^\bullet_{!,\,\mathrm{int}}(\Gamma\backslash\mathbb{H},\tilde{\mathcal{M}}^\flat_\lambda\otimes\mathcal{O}_F)({}^\tau\pi_f)_\epsilon.$$

We choose generators ${}^\sigma e^\bullet_\epsilon(\pi_f)$, and a simple argument using Hilbert's Theorem 90 shows that we can assume the consistency condition

$$\Phi_{\sigma,\tau}(e^\bullet_\epsilon({}^\sigma\pi_f)) = e^\bullet_\epsilon({}^\tau\pi_f). \tag{4.124}$$

We get isomorphisms

$$\mathcal{F}^\bullet(\omega^\dagger_\epsilon): \mathcal{W}({}^\sigma\pi_f)\otimes_F\mathbb{C}\xrightarrow{\sim} H^\bullet_\epsilon(\Gamma\backslash\mathbb{H},\mathcal{M}_{\lambda^\vee})({}^\sigma\pi_f)\otimes_F\mathbb{C}$$

which are defined by

$$\mathcal{F}^\bullet(\omega^\dagger_\epsilon): h_{{}^\sigma\pi_f}\mapsto[\mathcal{F}(\omega^\dagger_\epsilon\times h_{{}^\sigma\pi_f})], \tag{4.125}$$

here $\mathcal{F}(\omega^\dagger_\epsilon\times h_{{}^\sigma\pi_f})$ is viewed as a closed $\mathcal{M}_\lambda\otimes\mathbb{C}$-valued differential via the identification (4.75), and $[\cdots]$ is its cohomology class.

Since we assume that π_f is unramified everywhere, we have the canonical basis element $h_f^{(0)}=\prod_p h^{(0)}_{{}^\sigma\pi_p}$ where $h^{(0)}_{{}^\sigma\pi_p}$ is defined by the equality (4.88). Then we obviously have $\sigma(h^{(0)}_{\pi_p})=h^{(0)}_{{}^\sigma\pi_p}$.

We thus define the *periods* by the relation

$$\mathcal{F}(\omega^\dagger_\epsilon)(h^{\dagger(0)}_{{}^\sigma\pi_f}) = \Omega^\bullet(\epsilon\times\pi_f)e^\bullet_\epsilon(\pi_f). \tag{4.126}$$

These periods depend of course on our choice of the "canonical" generator $\omega^\dagger_\epsilon$. We see that the numbers $\Omega^\bullet({}^\sigma\pi_f,\epsilon)$ are well-defined up to an element in $\mathcal{O}_F^\times$.

If $H^\bullet_{!,\,\mathrm{int}}(\Gamma\backslash\mathbb{H},\tilde{\mathcal{M}}^\flat_\lambda\otimes\mathcal{O}_F)(\pi_f)_\epsilon$ is not a free $\mathcal{O}_F$-module, then we can find a covering by two open subsets of $\mathrm{Spec}(\mathcal{O}_F)$, to be denoted U_1,U_2, such that $H^\bullet_{!,\,\mathrm{int}}(\Gamma\backslash\mathbb{H},\tilde{\mathcal{M}}^\flat_\lambda\otimes\mathcal{O}_F(U_i))(\epsilon\times\pi_f)$ is free. We can apply the above procedure, and we get periods $\Omega_1(\epsilon\times\pi_f),\Omega_2(\epsilon\times\pi_f)$. They are well-defined up to an element in $\mathcal{O}_F(U_1)^\times,\mathcal{O}_F(U_2)^\times$ respectively. The ratio of these periods is an element in $\mathcal{O}_F(U_1\cap U_2)^\times$.

Perhaps at this point we should introduce the sheaf $\mathcal{P}$ of periods over F. For any open subset $U\subset\mathrm{Spec}(\mathcal{O}_F)$ we put $\mathcal{P}^*_F(U):=\mathbb{C}^\times/\mathcal{O}_F(U)^\times$, which is a Zariski presheaf on $\mathrm{Spec}(\mathcal{O}_F)$, where the associated sheaf is our *sheaf of periods* $\mathcal{P}_F$.

Now we can interpret the generators $e^\bullet_\epsilon(\pi_f)$ as (unique) sections in the sheaf of generators modulo $\mathcal{O}_F^\times$, and then the equation (4.126) makes sense without the assumption on the class number. These considerations will play a role in the following chapter.

Some little subtleties

We should notice that these periods are defined with respect to the "small" sheaves $\tilde{\mathcal{M}}^\flat_\lambda$. We have $\tilde{\mathcal{M}}^\flat_\lambda \subset \tilde{\mathcal{M}}_\lambda$, and therefore the map

$$H^\bullet_{!,\,\mathrm{int}}(\Gamma\backslash\mathbb{H}, \tilde{\mathcal{M}}^\flat_\lambda \otimes \mathcal{O}_F)(\pi_f)_\epsilon \to H^\bullet_{!,\,\mathrm{int}}(\Gamma\backslash\mathbb{H}, \tilde{\mathcal{M}}_\lambda \otimes \mathcal{O}_F)(\pi_f)_\epsilon \tag{4.127}$$

may not be surjective (the reader should not be puzzled by the fact that we have $\tilde{\mathcal{M}}^\flat_\lambda \otimes F = \tilde{\mathcal{M}}_\lambda \otimes F$). Therefore, if we would work with $\tilde{\mathcal{M}}_\lambda$ instead and define the periods $\Omega^{\bullet,\#}({}^\sigma\pi_f, \epsilon)$ by the same procedure, then we would get the relation

$$\Omega^{\bullet,\#}({}^\sigma\pi_f, \epsilon) = d(\pi_f, \epsilon)\Omega^\bullet({}^\sigma\pi_f, \epsilon)$$

where $d(\pi_\bullet, \epsilon)$ is a non-zero factor in $\mathcal{O}_F$. The primes in these factors are the divisors of the binomial coefficients.

But we could also use the module $H^\bullet(\Gamma\backslash\mathbb{H}, \tilde{\mathcal{M}}^\flat_\lambda \otimes \mathcal{O}_F)_{\,\mathrm{int},!}(\pi_f)_\epsilon$ to define the periods with respect to this module. Again these periods will be integral multiples of the periods $\Omega^\bullet(\pi_f, \epsilon)$.

In Chapter 5 we will discuss rationality results (of Manin and Shimura) which relate these periods to special values of L-functions (see Section 5.3). But we also want to discuss this method not only for cuspidal cohomology classes but also for the Eisenstein cohomology classes, therefore we close this chapter with a brief account of these Eisenstein classes.

4.2.6 The Eisenstein cohomology class

We assume n even, let $\lambda = n\gamma$, and consider the module $\mathcal{M}_\lambda$, i.e. we assume that the center acts trivially. In Section 3.3.5 we claimed the existence of the specific cohomology class $[\mathrm{Eis}(\omega)] \in H^1(\Gamma\backslash\mathbb{H}, \mathcal{M}_n)$. This class was determined by its eigenvalues under the Hecke operators and by its restriction to the boundary. In this section we give a construction of this class on the transcendental level, i.e. we construct a closed differential form which represents the above class and whose restriction to the boundary $H^1(\partial(\Gamma\backslash\mathbb{H}), \mathcal{M}_n \otimes \mathbb{C})$ is a given class ω_n. For the general theory of Eisenstein cohomology we refer to Chapter 9 in Part 2 of this book.

We start from our highest weight module $\mathcal{M}_\lambda$, and we observe that by definition we have an inclusion

$$i_0 : \mathfrak{I}_B^G \lambda_\mathbb{R} \rho_\mathbb{R}^2 \hookrightarrow \mathcal{C}_\infty(\Gamma_\infty^+\backslash G^+(\mathbb{R}))$$

where

$$\Gamma_\infty^+ = \left\{ \begin{pmatrix} t_1 & m \\ 0 & t_1 \end{pmatrix} \mid m \in \mathbb{Z}\,; t_1 = \pm 1 \right\}.$$

Therefore we get an isomorphism

$$H^1(\mathfrak{g}, K_\infty, \mathfrak{I}_B^G \lambda_\mathbb{R} \rho_\mathbb{R}^2 \otimes \mathcal{M}_\lambda \otimes \mathbb{C}) \xrightarrow{\sim} H^1(\Gamma_\infty^+\backslash\mathbb{H}, \mathcal{M}_\lambda \otimes \mathbb{C}) = H^1(\partial(\Gamma\backslash\mathbb{H}), \mathcal{M}_\lambda \otimes \mathbb{C}).$$

The inclusion i_0 sends the module $\mathfrak{I}_B^G \lambda_\mathbb{R} \rho_\mathbb{R}^2$ into a space of functions which are Γ_∞^+-invariant under left translations. Now we apply an infinite summation for $f \in \mathfrak{I}_B^G \lambda_\mathbb{R} \rho_\mathbb{R}^2$ and we put

$$\mathrm{Eis}(f)(x) = \sum_{\gamma \in \Gamma_\infty^+ \backslash \mathrm{Sl}_2(\mathbb{Z})} f(\gamma x). \tag{4.128}$$

and this provides the Eisenstein intertwining operator

$$\mathrm{Eis} : \mathfrak{I}_B^G \lambda_{\mathbb{R}} \rho_{\mathbb{R}}^2 \to \mathcal{C}_\infty(\Gamma \backslash \mathrm{Sl}_2(\mathbb{R})).$$

Of course we have to discuss the convergence of this infinite series. We could quote H. Jacquet: "Let us speak about convergence later", but here is a short interlude discussing this issue.

Interlude: Here is the point: we twist our module, for any complex number $z \in \mathbb{C}$ we consider the induced module

$$\mathfrak{I}_B^G \lambda_{\mathbb{R}}^{w_0} |\rho|^z \subset \mathcal{C}_\infty(\Gamma_\infty^+ \backslash \mathrm{Sl}_2(\mathbb{R}))$$

and again we write down the Eisenstein series. Now it is an elementary exercise to show that the map

$$\begin{pmatrix} a & b \\ c & d \end{pmatrix} \mapsto (c, d)$$

provides a bijection

$$\Gamma_\infty^+ \backslash \mathrm{Sl}_2(\mathbb{Z}) \xrightarrow{\sim} \{(c,d) \in \mathbb{Z} \times \mathbb{Z} \mid (c,d) \text{ coprime}\}/\{\pm 1\} = \mathbb{P}^1(\mathbb{Q}).$$

An element $x \in \mathrm{Sl}_2(\mathbb{R})$ can be written as $x = \begin{pmatrix} t & u \\ 0 & t^{-1} \end{pmatrix} k$ with $k \in K_\infty$. Then for $f \in \mathfrak{I}_B^G \lambda_{\mathbb{R}}^{w_0} |\rho|^z$ we have $f(\gamma x, z) =$

$$\begin{aligned}
&= f\left(\begin{pmatrix} a & b \\ c & d \end{pmatrix} \begin{pmatrix} t & u \\ 0 & t^{-1} \end{pmatrix} k, z\right) \\
&= f\left(\begin{pmatrix} (c^2t2 + (cu + dt^{-1})^2)^{-1/2} & * \\ 0 & (c^2t^2 + (cu + dt^{-1})^2)^{1/2} \end{pmatrix} k(\gamma x), z\right) \\
&= (c^2t^2 + (cv + dt^{-1})^2)^{\frac{n-z}{2}} f(k(\gamma x)k).
\end{aligned}$$

Since $|f(k(\gamma)x)|$ is bounded, the series

$$\mathrm{Eis}(f, z)(x) = \sum_{\gamma \in \Gamma_\infty^+ \backslash \mathrm{Sl}_2(\mathbb{Z})} f(\gamma x, z)$$

is convergent for $\Re(z) \gg 0$, and then it is also holomorphic in z. Selberg and others showed that it can be extended to a meromorphic function in the entire complex plane. This result is now a special case of a theorem of Langlands [77]. If the function $z \mapsto \mathrm{Eis}(f, z)(x)$ is holomorphic at $z = 0$ for all x, then we do not care about convergence and we simply define

$$\mathrm{Eis}(f)(x) = \sum_{\gamma \in \Gamma_\infty^+ \backslash \mathrm{Sl}_2(\mathbb{Z})} f(\gamma x) := \mathrm{Eis}(f, 0)(x).$$

In our special case it is easy to see that the series is convergent at $z = 0$ provided we have $n > 0$, and this is the only case where we will apply this construction.

End of interlude

This construction provides a homomorphism

$$\mathrm{Eis}^1 : H^1(\mathfrak{g}, K_\infty, \mathfrak{I}_B^G \lambda_{\mathbb{R}} \rho_{\mathbb{R}}^2 \otimes \mathcal{M}_\lambda \otimes \mathbb{C}) \to H^1(\Gamma\backslash\mathbb{H}, \mathcal{M}_\lambda \otimes \mathbb{C}). \tag{4.129}$$

Using the isomorphism of Delorme (4.63) we wrote down in (4.72) a distinguished cocycle

$$\begin{aligned}\omega_n = E_+^\vee \otimes Y^n &\in \mathrm{Hom}_{\tilde{K}_\infty^T}(\Lambda^0(\mathfrak{t}/\mathfrak{c}), \mathbb{C}\lambda_{\mathbb{R}}\rho_{\mathbb{R}}^2 \otimes \mathrm{Hom}(\Lambda^1(\mathfrak{u}), \mathcal{M}_{\lambda,\mathbb{C}}))\\ &\subseteq \mathrm{Hom}_{\tilde{K}_\infty}(\Lambda^1(\mathfrak{g}/\tilde{\mathfrak{k}}), \mathfrak{I}_B^G \lambda_{\mathbb{R}} \rho_{\mathbb{R}}^2 \otimes \mathcal{M}_{\lambda,\mathbb{C}})\end{aligned}$$

in the relative Lie algebra complex, giving a generator

$$[\omega_n] \in H^1(\mathfrak{g}, K_\infty, \mathfrak{I}_B^G \lambda_{\mathbb{R}} \rho_{\mathbb{R}}^2 \otimes \mathcal{M}_\lambda \otimes \mathbb{C}),$$

where now $E_+^\vee : \mathbb{Q}H \oplus \mathbb{Q}E_+ \to \mathbb{Q}$ is defined by $E_+^\vee(E_+) = 1$ and $E_+^\vee(H) = 0$. We define

$$\mathrm{Eis}_n := \mathrm{Eis}^1([\omega_n]) \in H^1(\Gamma\backslash\mathbb{H}, \mathcal{M}_\lambda \otimes \mathbb{C}). \tag{4.130}$$

One checks easily that already the differential form $\mathrm{Eis}(\omega_n)$ is an eigenvector for all Hecke operators

$$T_p(\mathrm{Eis}(\omega_n)) = (1 + p^{n+1})\mathrm{Eis}(\omega_n).$$

Proposition 4.2.2. *The restriction of* Eis_n *to* $H^1(\partial(\Gamma\backslash\mathbb{H}), \mathcal{M}_\lambda \otimes \mathbb{C})$ *is the class* $[Y^n]$.

For later purpose we analyse $\mathrm{Eis}^\bullet$ a little bit more carefully. We consider the constant Fourier coefficient $\mathcal{F}^B : \mathcal{C}_\infty(\Gamma\backslash G(\mathbb{R}) \to \mathcal{C}_\infty(\Gamma_\infty^+\backslash G^+(\mathbb{R})$, which is defined by

$$\mathcal{F}^B : (g \mapsto f(g)) \to (g \mapsto \int_{U(\mathbb{Z})\backslash U(\mathbb{R})} f(ug)du).$$

Now we consider the composition

$$\mathcal{F}^B \circ \mathrm{Eis} : \mathfrak{I}_B^G \lambda_{\mathbb{R}} \rho_{\mathbb{R}}^2 \to \mathcal{C}_\infty(\Gamma_\infty^+\backslash G^+(\mathbb{R})) \tag{4.131}$$

Now a standard computation shows that the image of this composition lies in the direct sum $\mathfrak{I}_B^G \lambda_{\mathbb{R}} \rho_{\mathbb{R}}^2 \oplus \mathfrak{I}_B^G \lambda_{\mathbb{R}}^{-1} \subset \mathcal{C}_\infty(\Gamma_\infty^+\backslash G^+(\mathbb{R}))$. Actually we can describe this image more precisely. We remember the intertwining operators $T_\infty^{\mathrm{loc}} : \mathfrak{I}_B^G \lambda_{\mathbb{R}} \rho_{\mathbb{R}}^2 \to \mathfrak{I}_B^G \lambda_{\mathbb{R}}^{-1}$ in section 4.1.4. This intertwining operator sends the finite dimensional top quotient $\mathcal{M}_\lambda$ of $\mathfrak{I}_B^G \lambda_{\mathbb{R}} \rho_{\mathbb{R}}^2$ isomorphically to the finite dimensional submodule $\mathcal{M}_\lambda \subset \mathfrak{I}_B^G \lambda_{\mathbb{R}}^{-1}$.

Then

$$\mathcal{F}^B \circ \mathrm{Eis}(f) = f + c(\lambda) T_\infty^{\mathrm{loc}}(f). \tag{4.132}$$

Especially we see for $f \in \mathcal{D}_\lambda$ that $\mathcal{F}^B \circ \mathrm{Eis}(f) = f$. This implies that the class $[\mathrm{Eis}(\omega_\lambda^+)] \in H^1(\Gamma\backslash\mathbb{H}, \mathcal{M}_\lambda \otimes \mathbb{C})$ restricts to the generator in the cohomology of the boundary whereas $[\mathrm{Eis}(\omega_\lambda^-)] = 0$. This implies that the class $[\mathrm{Eis}(\omega_n)]$ restricts to the generator $[Y^n]$ in the cohomology of the boundary.

The adelic Eisenstein class

The group of units $\mathbb{G}_m(\mathbb{A}) = I_{\mathbb{Q}}$ is called the idele group. On this group we have a very specific character namely the idele norm

$$\underline{x} \mapsto |\underline{x}| = \prod_v |x_v|_v$$

where $|x_\infty|_\infty$ is the usual absolute value and for a finite place we have $|x_p|_p = \frac{1}{p^{n_p}}$ where $x_p = p^{n_p} \cdot u, u \in \mathbb{Z}_p^\times$. The product formula tells us that for a principal idele $a = (a, a, \dots)$ the idele norm is 1. Hence the idele norm provides a character $\alpha : I_{\mathbb{Q}}/\mathbb{Q}^\times \to \mathbb{R}^\times$, this character called the Tate character.

Now we can introduce the adelic versions of our induced representations above, i.e. we translate the notions in (4.1.3) into the adelic language. To any rational character $\lambda \in X(T)$ we define the space of functions

$$I_B^G|\lambda| = \{f \in \mathcal{C}_\infty(G(\mathbb{A})) \mid f(\underline{b}\underline{g}) = |\lambda|(\underline{b})f(\underline{g})\},$$

where of course λ is considered as a character on the torus $T = B/U$, and $\mathcal{C}_\infty$ means that we require that $\underline{g} = (g_\infty, \underline{g}_f) \to f(g_\infty, \underline{g}_f)$ is invariant under right translations under a suitable open subgroup $K_f \subset \mathcal{G}(\hat{\mathbb{Z}})$ and f is $\mathcal{C}_\infty$ in the variable g_∞. This is now a $G(\mathbb{A}) = G(\mathbb{R}) \times G(\mathbb{A}_f)$ module by right translations.

The group $G(\mathbb{Q})$ acts transitively on the quotient $G(\mathbb{A}_f)/\mathcal{G}(\hat{\mathbb{Z}})$. The stabiliser of the identity $\underline{e}_f$ is $\mathcal{G}(\mathbb{Z}) = \mathrm{Gl}_2(\mathbb{Z})$. Hence we get

$$G(\mathbb{Q})\backslash G(\mathbb{A})/\mathcal{G}(\hat{\mathbb{Z}}) = \Gamma\backslash G(\mathbb{R}).$$

We introduce the smaller $(\mathfrak{g}, K_\infty) \times G(\mathbb{A}_f)$ module

$$\mathfrak{I}_B^G|\lambda| := \{f \in I_B^G|\lambda|) \mid \forall\, \underline{g}_f \text{ the map } g_\infty \mapsto f(g_\infty, \underline{g}_f) \text{ is } K_\infty - \text{finite}\}.$$

It is the tensor product of the Harish-Chandra module at the infinite place and the $G(\mathbb{A}_f)$-module $\mathfrak{I}_B^G|\lambda|_f$, here $|\lambda|_f$ is the restriction of $|\lambda|$ to $B(\mathbb{A}_f)$. The space $(\mathfrak{I}_B^G|\lambda|_f)^{\mathcal{G}(\mathbb{Z})}$ is one dimensional and generated by the functions

$$\psi_{\lambda,f}(\underline{g}_f) = \psi_{\lambda,f}(\underline{b}_f k_f) = |\lambda|_f(\underline{b}_f).$$

Now we consider the space $\mathfrak{I}_B^G|\lambda||\rho|^2$. We have seen that this space contains the subspace $\underline{\mathcal{D}}_\lambda = \mathcal{D}_\lambda \otimes \mathfrak{I}_B^G|\lambda|_f|\rho|_f^2$. The space $\mathrm{Hom}_{K_\infty}(\Lambda^1(\mathfrak{g}/\mathfrak{k}), \underline{\mathcal{D}}_\lambda \otimes \mathcal{M}_\lambda)^{\mathcal{G}(\hat{\mathbb{Z}})}$ has the two differential forms

$$\underline{\omega}_\lambda = \tfrac{1}{2}\bar{P} \otimes \psi_{n+2} \otimes (X - i \otimes Y)^n \otimes \psi_{\lambda,f},$$

$$\underline{\bar{\omega}}_\lambda = \tfrac{1}{2}P \otimes \psi_{-n-2} \otimes (X + i \otimes Y)^n \otimes \psi_{\lambda,f}$$

as a basis of this space.

Now the Eisenstein intertwining operator $\mathrm{Eis} : \mathfrak{I}_B^G|\lambda||\rho|^2 \to \mathcal{C}_\infty(G(\mathbb{Q})\backslash G(\mathbb{A}))$ is defined by the summation

$$(g \mapsto f(g)) \mapsto \left(g \mapsto \sum_{a \in B(\mathbb{Q})\backslash G(\mathbb{Q})} f(ag) \right).$$

By the result above that this infinite series is converging. We can also leave it as an exercise to show that the summation converges to a K_∞-finite function. (This Eisenstein operator plays the same role as the operator $\mathcal{F}$ in the previous case, when we considered cuspidal classes). From here we get a map

$$\mathrm{Hom}_{K_\infty}(\Lambda^1(\mathfrak{p}), \underline{\mathcal{D}}_\lambda \otimes \mathcal{M}_\lambda)^{\mathcal{G}(\hat{\mathbb{Z}})} \xrightarrow{\mathrm{Eis}} \mathrm{Hom}_{K_\infty}(\Lambda^1(\mathfrak{p}), \mathcal{C}_\infty(G(\mathbb{Q})\backslash G(\mathbb{A})) \otimes \mathcal{M}_\lambda)^{\mathcal{G}(\hat{\mathbb{Z}})}$$

and hence we get a linear map

$$\mathrm{Eis}^1 : H^1(\mathfrak{g}, K_\infty, \underline{\mathcal{D}}_\lambda \otimes \mathcal{M}_\lambda)^{\mathcal{G}(\hat{\mathbb{Z}})} \to H^1(G(\mathbb{Q})\backslash G(\mathbb{A})/K_\infty \times \mathcal{G}(\hat{\mathbb{Z}}), \tilde{\mathcal{M}}_\lambda \otimes \mathbb{C})$$

the last space is of course equal to $H^1(\Gamma\backslash\mathbb{H}, \tilde{\mathcal{M}}_\lambda \otimes \mathbb{C})$.

To understand the image of this map, we consider the composition of the map Eis^1 and the restriction map to the boundary cohomology. We know that we have to consider the constant Fourier coefficient

We have a brief look at the Eisenstein cohomology in case B). We refer to the final remark at the end of Chapter 2 and retain the notation used there. For our imaginary field we take again $F = \mathbb{Q}(\mathbf{i}), \Gamma = \mathrm{Sl}_2(\mathbb{Z}[\mathbf{i}])$ and $\mathcal{M}_\lambda = \mathcal{M}_{n_1} \otimes \mathcal{M}_{n_2}$. We assume the parity condition $n_1 \equiv n_2 + 2 \equiv 0 \mod 4$. By a rather elementary computation we get

$$H^1(\partial(\Gamma\backslash\mathbb{H}_3), \tilde{\mathcal{M}}_\lambda)_{\text{ int}} = \mathbb{Z}[\mathbf{i}]e_{0,1} \oplus \mathbb{Z}[\mathbf{i}]e_{1,0}.$$

If we extend the scalars to $\mathbb{C}$ we can represent these classes by differential forms. To do so, we apply the two reflections s_1, s_2 to our highest weight $\lambda = n_1\gamma_1 + n_2\gamma_2$ and get the two characters

$$s_1 \cdot \lambda = (-n_1 - 2)\gamma_1 + n_2\gamma_2;\ s_2 \cdot \lambda = n_1\gamma_1 + (-n_2 - 2)\gamma_2.$$

These two characters yield characters $s_i \cdot \gamma : B(\mathbb{R}) \to \mathbb{C}^\times$ and then we have seen that the two classes $E_+^{1,\vee} \otimes e_0^{(1)} \otimes e_{n_2}^{(2)},\ e_{n_1}^{(1)} \otimes E_+^{2,\vee} \otimes e_0^{(2)}$ provide differential forms $e_{s_i\cdot\lambda}$ and cohomology classes in $\mathbb{H}^1(\mathfrak{g}, K_\infty, \mathfrak{I}_B^G s_i \cdot \lambda \otimes \mathcal{M}_\lambda)$ (see Thm. 4.1.2). Since we have $\mathfrak{I}_B^G s_i \cdot \lambda \subset \mathcal{C}_\infty(B(\mathbb{Z})\backslash G(\mathbb{R}))$, we get the two classes $e_{s_i\cdot\lambda} \in H^1(\partial(\Gamma\backslash\mathbb{H}_3), \mathcal{M}_\lambda \otimes \mathbb{C})$. Hence we know

$$H^1(\partial(\Gamma\backslash\mathbb{H}_3), \tilde{\mathcal{M}}_\lambda \otimes \mathbb{C}) = \mathbb{C}e_{s_1\cdot\lambda} \oplus \mathbb{C}e_{s_2\cdot\lambda},$$

but a close inspection shows that the $e_{s_1\cdot\lambda}$ are equal to the $e_{i,j}$.

Again we invoke the theory of Eisenstein series. We assume $n_1 > n_2$ and define the Eisenstein intertwining operator

$$\mathrm{Eis} : \mathfrak{I}_B^G s_1 \cdot \lambda \to \mathcal{C}_\infty(G(\mathbb{Z})\backslash G(\mathbb{R}))\ ; (g \mapsto f(g)) \mapsto \left(g \mapsto \sum_{\alpha \in B(\mathbb{Z})\backslash G(\mathbb{Z})} f(\alpha g) \right).$$

It is not difficult to check that this infinite series is convergent (locally uniformly on compact sets), and we can also define the Eisenstein operator if $n_2 > n_1$ by analytic continuation.

Hence we get the Eisenstein cohomology class

$$\mathrm{Eis}(e_{s_1\cdot\lambda}) \in H^1(\Gamma\backslash\mathbb{H}_3, \mathcal{M}_\lambda \otimes \mathbb{C}).$$

If we restrict again to the boundary we get by a standard computation

$$r(\mathrm{Eis}(e_{s_1\cdot\lambda})) = e_{s_1\cdot\lambda} + c(\lambda)e_{s_2\cdot\lambda}$$

where a priori $c(\lambda)\in\mathbb{C}$. Hence we conclude that $\dim r(H^1(\Gamma\backslash\mathbb{H}_3,\tilde{\mathcal{M}}_\lambda\otimes\mathbb{C}))\geq 1$, but then it follows from Proposition 2.2.3 that this dimension is equal to one. Therefore

$$r(H^1(\Gamma\backslash\mathbb{H}_3,\mathcal{M}_\lambda\otimes\mathbb{C})) = \mathbb{C}\cdot\left(e_{s_1\cdot\lambda}+c(\lambda)e_{s_2\cdot\lambda}\right) = r(H^1(\Gamma\backslash\mathbb{H}_3,\mathcal{M}_\lambda))\otimes_{\mathbb{Q}(\mathbf{i})}\mathbb{C}$$

and then we can conclude that $c(\lambda)\in\mathbb{Q}[\mathbf{i}]$ - actually it lies in $\mathbb{Q}$.

The computation of the constant term of the Eisenstein series yields an explicit and very simple formula for $c(\lambda)$, which we explain next (see for instance [44]): let $\mathcal{I}_{\mathbb{Q}(\mathbf{i})}$ be the group of fractional ideals of $\mathbb{Q}(\mathbf{i})$. We can view our characters $s_i\cdot\lambda\in X^*(T\times\mathbb{Q}[\mathbf{i}])$ also as Hecke characters $s_i\cdot\lambda:\mathcal{I}_{\mathbb{Q}(\mathbf{i})}\to\mathbb{Q}(\mathbf{i})^\times$, where we exploit the fact that every fractional ideal is principal, and define

$$s_1\cdot\lambda(\mathfrak{a}) = s_1\cdot\lambda((\alpha)) = \alpha^{-n_1-2}\bar{\alpha}^{n_2};\ s_2\cdot\lambda(\mathfrak{a}) = s_2\cdot\lambda((\alpha)) = \alpha^{n_1}\bar{\alpha}^{-n_2-2}.$$

Here we need the above parity condition.

To these Hecke characters we attach (completed) Hecke L-functions

$$\Lambda(s_i\cdot\lambda,z) := \frac{\Gamma(z)}{(2\pi)^z}\prod_{\mathfrak{p}}\frac{1}{1-\frac{s_i\cdot\lambda(\mathfrak{p})}{N\mathfrak{p}^z}} = \frac{\Gamma(z)}{(2\pi)^z}\sum_{\mathfrak{a}}\frac{s_i\cdot\lambda(\mathfrak{a})}{N\mathfrak{a}^z},$$

where the $\mathfrak{p}$ run over all prime ideals and $\mathfrak{a}=(a+b\mathbf{i})$ runs over all integral ideals in $\mathbb{Z}[\mathbf{i}]$ and of course $N\mathfrak{a}=(a^2+b^2)$. It is well-known that these Hecke L-functions are meromorphic in the entire complex z-plane, with possible first order poles at $z=0,-1,\dots$

Then our simple formula for the restriction of the Eisenstein class to the boundary cohomology is

$$r(H^1(\Gamma\backslash\mathbb{H}_3,\mathcal{M}_\lambda\otimes\mathbb{C})) = \mathbb{C}\cdot(e_{s_1\cdot\lambda}+\frac{\Lambda(s_1\cdot\lambda,-1)}{\Lambda(s_1\cdot\lambda,0)}e_{s_2\cdot\lambda}). \tag{4.133}$$

Here we have to observe that the two poles at $z=-1$ and $z=0$ cancel out.

We comment a little bit on this result. We put $4k=n_1+n_2+2$ and consider the Hecke character $\phi_k:\mathfrak{a}=(a+b\mathbf{i})\to(a-b\mathbf{i})^{4k}$ and the completed Hecke L-function

$$\Lambda(\phi_k,z) := \frac{\Gamma(z)}{(2\pi)^z}\prod_{\mathfrak{p}}\frac{1}{1-\frac{\phi_k(\mathfrak{p})}{N\mathfrak{p}^z}} = \frac{\Gamma(z)}{(2\pi)^z}\sum_{\mathfrak{a}}\frac{\phi_k(\mathfrak{a})}{N\mathfrak{a}^z} = \frac{\Gamma(z)}{(2\pi)^z}L(\phi_k,z).$$

It is well-known that the series is converging for $\Re(z)>>0$ and can be analytically continued to the entire z-plane. It satisfies the functional equation

$$\Lambda(\phi_k,z) = 2^{4k+1-z}\Lambda(\phi_k,4k+1-z). \tag{4.134}$$

In his paper [68], Hurwitz considered the period integral

$$\Omega = \int_0^1\frac{1}{\sqrt{1-x^4}}dx = 2.62206\dots.$$

and proved that for all values of $k > 0$ the number

$$E_{4k} = \frac{L(\phi_k, 4k)}{\Omega^{4k}} \in \mathbb{Q}.$$

Hurwitz proved many more things about these numbers. His main concern was an analogue of the von Staudt-Clausen theorem.

We now observe that for $s_1 \cdot \lambda = (-n_1 - 2)\gamma_1 + n_2\gamma_2$ we have the equality

$$L(\phi_k, 4k - n_2) = L(s_1 \cdot \lambda, 0)$$

and hence we get

$$\frac{\Lambda(s_1 \cdot \lambda, -1)}{\Lambda(s_1 \cdot \lambda, 0)} = -2\pi \frac{L(s_1 \cdot \lambda, -1)}{L(s_1 \cdot \lambda, 0)} = -2\pi \frac{L(\phi_k, 4k - 1 - n_2)}{L(\phi_k, 4k - n_2)} \in \mathbb{Q} \tag{4.135}$$

and this implies that

$$\pi^{\nu} \frac{L(\phi_k, 4k - \nu)}{\Omega^{4k}} \in \mathbb{Q} \text{ for } \nu = 0, 1, \ldots 4k - 2. \tag{4.136}$$

This has been proven by Damerell [30]. It is one of the first instances of Deligne's conjecture [32]. We come back to this point at the end of this book.

Chapter 5

Applications to Number Theory

5.1 Modular symbols, L-values and denominators of Eisenstein classes.

In this chapter we want to restrict to the case $\Gamma = \mathrm{Sl}_2(\mathbb{Z})$ or $\Gamma = \mathrm{Sl}_2(\mathcal{O})$, where $\mathcal{O}$ is the ring of integers of an imaginary quadratic field. Our coefficient systems will be obtained from the modules $\mathcal{M}_\lambda$. We assume that we have $d = 0$ and hence $n \equiv 0 \mod 2$ in case A), and $d_1 = d_2 = 0,\ n_1 = n_2$ in case B). This has the effect that $\lambda^\vee = \lambda$.

We want to study the pairing

$$H^1_c(\Gamma\backslash X, \tilde{\mathcal{M}}^\flat_\lambda) \times H_1(\Gamma\backslash X, \partial(\Gamma\backslash X), \underline{\mathcal{M}}_\lambda) \to \mathbb{Z}.$$

5.1.1 Modular symbols attached to a torus in $\mathrm{Gl}_2/\mathbb{Q}$

First we construct (relative) cycles in $C_1(\Gamma\backslash X, \underline{\mathcal{M}}_\lambda), C_1(\Gamma\backslash X, \partial(\Gamma\backslash X), \underline{\mathcal{M}}_\lambda)$. Our starting point is a maximal torus $T/\mathbb{Q} \subset G/\mathbb{Q}$, and we assume that it is split over a real quadratic extension $F/\mathbb{Q}$. Then the group of real points

$$T(\mathbb{R}) = \mathbb{R}^\times \times \mathbb{R}^\times$$

acts on $\mathbb{H}$ and $\bar{\mathbb{H}}$ and it has two fixed points $r, s \in \mathbb{P}^1(F)$. There is a unique geodesic (half) circle $\bar{C}_{r,s} \subset \bar{\mathbb{H}}$ joining these two points. Then $T(\mathbb{R})$ acts transitively on $C_{r,s} = \bar{C}_{r,s} \setminus \{r, s\}$. We have two cases:

a) The torus $T/\mathbb{Q}$ is split. Then the two points $r, s \in \mathbb{P}^1(\mathbb{Q})$. Here for instance we can take $r = 0, s = \infty$, then the geodesic circle is the line $\{iy, y > 0\}$ and the torus is the standard diagonal split torus.

b) Here $\{r, s\} \in \mathbb{P}^1(F) \setminus \mathbb{P}^1(\mathbb{Q})$, then r, s are Galois conjugates of each other. Our torus $T/\mathbb{Q}$ is given by a suitable embedding

$$j : R_{F/\mathbb{Q}}(\mathbb{G}_m/F) = T \hookrightarrow \mathrm{Gl}_2/\mathbb{Q}.$$

G. Harder, *Cohomology of Arithmetic Groups*,
Springer Monographs in Mathematics, https://doi.org/10.1007/978-3-032-10378-9_5

In case a) we can choose any reasonable homeomorphism $[0,1] \xrightarrow{\sim} [0,\infty]$ - for instance $x \mapsto x/(1-x)$ - and then we get a one-chain

$$\sigma : [0,1] \xrightarrow{\sim} \bar{C}_{r,s} = \mathbb{R}_{>0} \cup \{0\} \cup \{\infty\}, \sigma(0) = r, \sigma(1) = s \in \partial(\bar{\mathbb{H}}),$$

and for any $m \in \mathcal{M}$ we can consider the image of $\sigma \otimes m \in C_1(\bar{\mathbb{H}}) \otimes \mathcal{M}$ in $C_1(\Gamma\backslash\bar{\mathbb{H}}, \partial(\Gamma\backslash\bar{\mathbb{H}}), \underline{\mathcal{M}})$. By definition this is a relative cycle, and hence we get a (relative) homology class

$$[\bar{C}_{r,s} \otimes m] \in H_1(\Gamma\backslash\bar{\mathbb{H}}, \partial(\Gamma\backslash\bar{\mathbb{H}}), \underline{\mathcal{M}}_\lambda). \tag{5.1}$$

It is easy to see that it does not depend on the choice of σ.

In case b) we have $T(\mathbb{Q}) \xrightarrow{\sim} F^\times$. Then the group $T(\mathbb{Q}) \cap \Gamma$ is a subgroup of finite index in the group of units $\mathcal{O}_F^\times = \{\epsilon_0\} \times \{\pm 1\}$, where ϵ_0 is a fundamental unit. Hence

$$\Gamma_T = T(\mathbb{Q}) \cap \Gamma = \{\epsilon_T\} \times \{\pm 1\} \tag{5.2}$$

where ϵ_T is an element of infinite order. This element ϵ_T induces a translation on $C_{r,s}$. The quotient $C_{r,s}/\Gamma_T$ is a circle. If we pick any point $x \in C_{r,s}$, then $[x, \epsilon_T x] \subset C_{r,s}$ is an interval. As above we can find a $\sigma : [0,1] \xrightarrow{\sim} [x, \epsilon_T x]$, $\sigma(0) = x, \sigma(1) = \epsilon_T x$. As before we can consider the 1-chain $\sigma \otimes m \in C_1(\mathbb{H}) \otimes \mathcal{M}$. Its boundary is the 0-chain $-\{x\} \otimes m + \{\epsilon_T x\} \otimes m$. If we look at the images in $C_\bullet(\Gamma\backslash\mathbb{H}, \underline{\mathcal{M}}_\lambda)$ then

$$\partial_1(\sigma \otimes m) = \sigma(0) \otimes (\epsilon_T^{-1} m - m) = x \otimes (\epsilon_T^{-1} m - m).$$

Hence we see that $\sigma \otimes m$ is a 1-cycle if and only if $m = \epsilon_T m$ and hence $m \in \mathcal{M}^T$. We have constructed homology classes

$$[C_{r,s} \otimes m] \in H_1(\Gamma\backslash\mathbb{H}, \underline{\mathcal{M}}_\lambda) \text{ for all } m \in \mathcal{M}_\lambda^{\langle \epsilon_T \rangle} = \mathcal{M}_\lambda^T.$$

5.1.2 Evaluation of cuspidal classes on modular symbols

The following issue will also be discussed in greater generality and more systematically in Part 2 of this book.

In the following we consider case A). We start from a highest weight $\lambda = n\gamma$. As usual we assume n to be even and $d = 0$. Then $\lambda = \lambda^\vee$, and we consider the two modules $\mathcal{M}_\lambda$ and $\mathcal{M}_\lambda^\flat$. Then we have the pairings

$$\begin{gathered} H^1(\Gamma\backslash\mathbb{H}, \tilde{\mathcal{M}}_\lambda^\flat) \times H_1(\Gamma\backslash\mathbb{H}, \underline{\mathcal{M}}_\lambda) \to \mathbb{Z} \\ H_c^1(\Gamma\backslash\mathbb{H}, \tilde{\mathcal{M}}_\lambda^\flat) \times H_1(\Gamma\backslash\bar{\mathbb{H}}, \partial(\Gamma\backslash\bar{\mathbb{H}}), \underline{\mathcal{M}}_\lambda) \to \mathbb{Z}. \end{gathered} \tag{5.3}$$

These two pairings are non-degenerate if we invert 6 and divide by the torsion on both sides.

We have the surjective homomorphism $H_c^1(\Gamma\backslash\mathbb{H}, \tilde{\mathcal{M}}_\lambda^\flat) \to H_!^1(\Gamma\backslash\mathbb{H}, \tilde{\mathcal{M}}_\lambda^\flat)$, and over a suitably large finite extension $F/\mathbb{Q}$ we have the isotypical decomposition

$$H_!^1(\Gamma\backslash\mathbb{H}, \tilde{\mathcal{M}}_\lambda^\flat \otimes F) = \bigoplus_{\pi_f} H_!^1(\Gamma\backslash\mathbb{H}, \tilde{\mathcal{M}}_\lambda^\flat \otimes F)(\pi_f) \tag{5.4}$$

where the π_f are absolutely irreducible (see Theorem 5.4). We choose an embedding $\iota : F \hookrightarrow \mathbb{C}$. In Section 4.2.5 we constructed the isomorphism

$$\mathcal{F}^1(\omega_\epsilon^\dagger) : \mathcal{W}(\pi_f) \otimes_{F,\iota} \mathbb{C} \to H^1_{\epsilon,!}(\Gamma\backslash\mathbb{H}, \tilde{\mathcal{M}}_\lambda \otimes F)({}^\iota\pi_f).$$

The space $\mathcal{W}(\pi_f)$ is a very explicit space. Since we want to stick to the case $K_f = K_f^{(0)}$, it is of dimension one and is generated by the element

$$h_{\pi_f}^{\dagger,0} = \prod_p h_p^{\dagger,0} \in \prod_p \mathcal{W}(\pi_p) \text{ where } h_p^{\dagger,0}(e) = 1. \tag{5.5}$$

Now we want to compute the value

$$\left\langle \mathcal{F}^1(\omega_\epsilon^\dagger)(h_{\pi_f}^{\dagger,0})), \bar{C}_{r,s} \otimes m) \right\rangle, \tag{5.6}$$

here we assume that the torus is split, i.e. $r, s \in \mathbb{P}^1(\mathbb{Q})$. Then this expression is problematic. The argument $C_{r,s}$ on the right lives in the relative homology group, hence the argument on the left should be in $H^1_c(\Gamma\backslash\mathbb{H}, \mathcal{M}_n \otimes \mathbb{C})$. Of course we can lift the class $\mathcal{F}^1(\omega_\epsilon^\dagger)(h_{\pi_f}^{\dagger,0})$ to a class

$$\widetilde{\mathcal{F}^1(\omega_\epsilon^\dagger)(h_{\pi_f}^{\dagger,0})} \in H^1_c(\Gamma\backslash\mathbb{H}, \tilde{\mathcal{M}}_n \otimes \mathbb{C}).$$

Then

$$\left\langle \widetilde{\mathcal{F}^1(\omega_\epsilon^\dagger \times h_{\pi_f}^{\dagger,0})}, \bar{C}_{r,s} \otimes m \right\rangle$$

makes sense, but the result may depend on the lift. We have:

Proposition 5.1.1. *If $\partial(\bar{C}_{r,s} \otimes m)$ gives the trivial class in the homology group $H_0(\partial(\Gamma\backslash\bar{\mathbb{H}}), \tilde{\mathcal{M}}_\lambda \otimes \mathbb{C})$, then $\left\langle \widetilde{\mathcal{F}^1(\omega_\epsilon^\dagger)(h_{\pi_f}^{\dagger,0})}, \bar{C}_{r,s} \otimes m \right\rangle$ does not depend on the lift, i.e. the value $\left\langle \mathcal{F}^1(\omega_\epsilon^\dagger)(h_{\pi_f}^{\dagger,0}), \bar{C}_{r,s} \otimes m \right\rangle$ is well-defined.*

Proof. This is rather clear, and we refer to the systematic discussion in 2.2.4. □

Now we compute the value of the pairing. We realised the relative homology class by an $\mathcal{M}_\lambda$-valued 1-chain $\sigma \otimes m$. The cohomology class $\mathcal{F}^1(\omega_\epsilon^\dagger)(h_{\pi_f}^{\dagger,0})$ is represented by $\widetilde{\mathcal{F}^1(\omega_\epsilon^\dagger \times h_{\pi_f}^{\dagger,0})}$ (see (4.75), 6.6). We consider the pullback $\sigma^*(\widetilde{\mathcal{F}^1(\omega_\epsilon^\dagger} \times h_{\pi_f}^{\dagger,0}))$, since $\mathcal{F}^1(\omega_\epsilon^\dagger \times h_{\pi_f}^{\dagger,0})$ is rapidly decaying if $x \to 0$ or $x \to 1$. This gives us a 1-form with values in $\mathcal{M}_\lambda \otimes \mathbb{C}$ on the closed interval $[0,1]$.

We claim - under the assumption $[\partial(\overline{C}_{r,s} \otimes m)] = 0$ - that

$$\left\langle \mathcal{F}^1(\omega_\epsilon^\dagger)(h_{\pi_f}^{\dagger,0}), \overline{C}_{r,s} \otimes m \right\rangle = \int_0^1 \left\langle \sigma^*(\widetilde{\mathcal{F}^1(\omega_\epsilon^\dagger \times h_{\pi_f}^{\dagger,0})}), m \right\rangle.$$

We have to be a little bit careful at this point. Of course, our assumption implies that the integral class $[\partial(\overline{C}_{r,s} \otimes m)] \in H_0(\partial(\Gamma\backslash\bar{\mathbb{H}}), \tilde{\mathcal{M}}_\lambda)$ is a torsion class. Let $\delta_{r,s}(m)$ be the order of this torsion class, hence we can write

$$\delta_{r,s}(m)\partial\overline{C}_{r,s} \otimes m = \partial c_{r,s} \text{ with } c_{r,s} \in C_1(\partial(\Gamma\backslash\bar{\mathbb{H}}), \underline{\mathcal{M}_\lambda}).$$

This 1-chain lies in the boundary of the Borel-Serre compactification (see Section 1.2.6). We consider the special case that T is the standard split diagonal torus, this means that $\{r,s\}=\{0,\infty\}$. We pull the cycle $\delta_{r,s}(m)\overline{C}_{r,s}\otimes m-c_{r,s}$ into the interior $\Gamma\backslash\mathbb{H}$ by a simple homotopy, which means we replace it by $\delta_{r,s}(m)[iy_0^{-1},iy_0]\otimes m-\partial c_{r,s}(m)(y_0)$ where $y_0\gg 1$, and $c_{r,s}(m)(y_0)$ is the bounding 1-chain $c_{r,s}$ on the level y_0. Then

$$\delta_{r,s}(m)\left\langle \mathcal{F}^1(\omega_\epsilon^\dagger)(h_{\pi_f}^{\dagger,0}),\overline{C}_{r,s}\otimes m\right\rangle =$$
$$\left\langle \mathcal{F}^1(\omega_\epsilon^\dagger)(h_{\pi_f}^{\dagger,0}),\delta_{r,s}(m)[iy_0^{-1},iy_0]\otimes m - c_{r,s}(y_0)\right\rangle,$$

where now the value on the right hand side is an integral over the moved cycle. Since the differential form $\mathcal{F}^1(\omega_\epsilon^\dagger)(h_{\pi_f}^{\dagger,0})$ is rapidly decreasing if $y_0\to\infty$ we get

$$\delta_{r,s}(m)\left\langle \mathcal{F}^1(\omega_\epsilon^\dagger)(h_{\pi_f}^{\dagger,0}),\overline{C}_{r,s}\otimes m\right\rangle = \lim_{y_0\to\infty}\left\langle \mathcal{F}^1(\omega_\epsilon^\dagger)(h_{\pi_f}^{\dagger,0}),\delta_{r,s}(m)[iy_0^{-1},iy_0]\otimes m\right\rangle.$$

We use the above identification $[0,1]=[0,\infty]$, and our 1-chain is given by the map

$$\sigma:[0,\infty]\to\bar{\mathbb{H}}:t\mapsto\begin{pmatrix} t & 0\\ 0 & 1\end{pmatrix} i = ti\in\bar{\mathbb{H}}, \tag{5.7}$$

especially $\sigma(0)=0$ and $\sigma(\infty)=i\infty$. The group $T(\mathbb{R})$ acts transitively on the open part $C_{0,i\infty}$. This action can be used to trivialize the tangent bundle. The tangent space at $i\in\mathbb{H}$ is identified with the subspace $\mathfrak{p}\subset\mathfrak{g}$ (see 4.2.5), and $\frac{H}{2}$ is a generator of the tangent space of $C_{0,i\infty}$ at one. Using the translations by $T(\mathbb{R})$, we get an invariant vector field on $C_{0,i\infty}$. If we identify $C_{0,i\infty}=\mathbb{R}_{>0}$, it is clear that this vector field is $t\frac{d}{dt}=D^*$.

Now an easy calculation (see 6.6) shows that (here e_f is the identity element in $G(\mathbb{A}_f)$)

$$\mathcal{F}^1(\omega_\epsilon^\dagger\times h_{\pi_f}^{\dagger,0})(D^*)\left(\begin{pmatrix} t & 0\\ 0 & 1\end{pmatrix} e_f\right) =$$
$$\rho_\lambda\left(\begin{pmatrix} t & 0\\ 0 & 1\end{pmatrix}\right)\mathcal{F}^1(\omega_\epsilon^\dagger(\frac{H}{2})\times h_{\pi_f}^{\dagger,0})\left(\begin{pmatrix} t & 0\\ 0 & 1\end{pmatrix},e_f\right),$$

and our integral in the formula above becomes

$$\int_0^\infty\left\langle \rho_\lambda\left(\begin{pmatrix} t & 0\\ 0 & 1\end{pmatrix}\right)\mathcal{F}^1(\omega_\epsilon^\dagger(\frac{H}{2})\times h_{\pi_f}^{\dagger,0})\left(\begin{pmatrix} t & 0\\ 0 & 1\end{pmatrix},e_f\right),m\right\rangle\frac{dt}{t}.$$

Our formulae in 4.2.5 give

$$\omega_\pm^\dagger(\frac{H}{2})=\frac{1}{8}(\psi_{n+2}^\dagger\otimes(X-Y\otimes i)^n\pm i^n\psi_{-n-2}^\dagger\otimes(X+Y\otimes i)^n). \tag{5.8}$$

This is an element in $\mathcal{D}_\lambda^\dagger\otimes\mathcal{M}_\lambda$. We apply $\mathcal{F}^1$ to $\omega_\pm^\dagger(\frac{H}{2})\times h_{\pi_f}^{\dagger,0}$ and evaluate at $\left(\begin{pmatrix} t & 0\\ 0 & 1\end{pmatrix},e_f\right)$. Applying $\mathcal{F}^1$ means that we have to sum over $a\in\mathbb{Q}^\times$, but since $h_{\pi_f}^{\dagger,0}$ is the Whittaker function attached to the unramified spherical function, only the terms with $a\in\mathbb{Z}$ can be non-zero. Hence we get

$$\mathcal{F}^1(\omega_{\pm}^{\dagger}(\tfrac{H}{2}) \times h_{\pi_f}^{\dagger,0})(\begin{pmatrix} t & 0 \\ 0 & 1 \end{pmatrix}, e_f) =$$

$$\frac{1}{8} \sum_{a\in\mathbb{Z};a\neq 0} (\psi_{n+2}^{\dagger}(\begin{pmatrix} at & 0 \\ 0 & 1 \end{pmatrix}) \otimes (X - Y \otimes i)^n \tag{5.9}$$

$$\pm i^n \psi_{-n-2}^{\dagger}(\begin{pmatrix} at & 0 \\ 0 & 1 \end{pmatrix}) \otimes (X + Y \otimes i)^n h_{\pi_f}^{\dagger,0}(\begin{pmatrix} a & 0 \\ 0 & 1 \end{pmatrix}).$$

We have seen that $\psi_{n+2}^{\dagger}\left(\begin{pmatrix} at & 0 \\ 0 & 1 \end{pmatrix}\right) = 0$ if $at < 0$ and $\psi_{n+2}^{\dagger}\left(\begin{pmatrix} -at & 0 \\ 0 & 1 \end{pmatrix}\right) = \psi_{-n-2}^{\dagger}\left(\begin{pmatrix} at & 0 \\ 0 & 1 \end{pmatrix}\right)$, and therefore our Fourier expansion becomes

$$\frac{1}{8}\sum_{a=1}^{\infty} \psi_{n+2}^{\dagger}(\begin{pmatrix} at & 0 \\ 0 & 1 \end{pmatrix}) \otimes ((X - Y \otimes i)^n \pm i^n (X + Y \otimes i)^n) h_{\pi_f}^{\dagger,0}(\begin{pmatrix} a & 0 \\ 0 & 1 \end{pmatrix}).$$

We have

$$\rho_\lambda\left(\begin{pmatrix} t & 0 \\ 0 & 1 \end{pmatrix}\right)((X - Y \otimes i)^n \pm i^n (X + Y \otimes i)^n)$$
$$= \sum_{\nu=0}^{n} \binom{n}{\nu} t^{-\frac{n}{2}+\nu} X^\nu Y^{n-\nu} (i^{n+\nu} \pm i^{-\nu}), \tag{5.10}$$

and we remember that n is even. The last factor is equal to $i^{-\nu}((-1)^{\frac{n}{2}+\nu} \pm 1)$, and this is $i^{-\nu}$ times 2 or 0 or -2, depending on the choices of signs and the parity of $\frac{n}{2}$ and ν. The elements $e_\nu = X^\nu Y^{n-\nu}$ form the dual basis to the basis $(-1)^\nu \binom{n}{n-\nu} X^{n-\nu} Y^\nu$ of $\mathcal{M}_\lambda^\flat$, which implies: if we choose $m = e_{n-\nu}$ in our expression above, then

$$\left\langle \rho_\lambda\left(\begin{pmatrix} t & 0 \\ 0 & 1 \end{pmatrix}\right)((X - Y \otimes i)^n \pm i^n (X + Y \otimes i)^n), m \right\rangle$$
$$= (-1)^\nu (i^{n+\nu} \pm i^{-\nu}) t^{-\frac{n}{2}+\nu}. \tag{5.11}$$

Hence we have to compute

$$(-1)^\nu \frac{i^{n+\nu} \pm i^{-\nu}}{8} \int_0^\infty \sum_{a=1}^{\infty} \psi_{n+2}^{\dagger}(\begin{pmatrix} at & 0 \\ 0 & 1 \end{pmatrix}) t^{-\frac{n}{2}+\nu} h_{\pi_f}^{\dagger,0}(\begin{pmatrix} a & 0 \\ 0 & 1 \end{pmatrix}) \frac{dt}{t}.$$

We remember $\psi_{n+2}^{\dagger}(\begin{pmatrix} t & 0 \\ 0 & 1 \end{pmatrix}) = t^{\frac{n}{2}+1} e^{-2\pi t}$, exchanging summation and integration and after some "innocent" substitutions we get

$$(-1)^\nu \frac{i^{n+\nu} \pm i^{-\nu}}{8} \int_0^\infty \frac{t^{\nu+1}}{(2\pi)^{\nu+1}} e^{-t} \frac{dt}{t} \sum_{a=1}^{\infty} \frac{h_{\pi_f}^{\dagger}(\begin{pmatrix} a & 0 \\ 0 & 1 \end{pmatrix}) a^{\frac{n}{2}}}{a^\nu}.$$

We refer to the discussion of the L-function attached to π_f and get

$$\int_0^\infty \frac{t^{\nu+1}}{(2\pi)^{\nu+1}} e^{-t}\frac{dt}{t}\sum_{a=1}^{\infty}\frac{h^{\dagger}_{\pi_f}(\begin{pmatrix} a & 0\\ 0 & 1\end{pmatrix})a^{\frac{n}{2}+1}}{a^{\nu+1}} = \Lambda^{\mathrm{coh}}(\pi,\nu+1). \tag{5.12}$$

Of course some questions concerning convergence have to be discussed, for this we refer to the proof of Theorem 4.2.5.

In the case that $\nu \neq 0, n$, we know that $\partial(\overline{C}_{0,\infty}\otimes X^{n-\nu}Y^{\nu})$ is a torsion element in $H_0(\partial(\Gamma\backslash\bar{\mathbb{H}}),\underline{\mathcal{M}_\lambda})$ and therefore the value of the integral is also the evaluation of the cohomology class $\mathcal{F}^1(\omega^{\dagger}_{\epsilon})(h^{\dagger,0}_{\pi_f})$ on an integral homology class. We get

$$\left\langle \mathcal{F}^1(\omega^{\dagger}_{\epsilon})(h^{\dagger,0}_{\pi_f}), \overline{C}_{0,\infty}\otimes X^{\nu}Y^{n-\nu})\right\rangle = (-1)^{\nu}\frac{i^{-\nu}\pm i^{n-\nu}}{8}\Lambda^{\mathrm{coh}}(\pi,n+1-\nu). \tag{5.13}$$

In the factor in front of the right side we have the sign $\epsilon=\pm$. We denote this factor by a_ϵ, where this factor is zero unless we have $\epsilon=(-1)^{\frac{n}{2}-\nu}$ (see (4.112)), and then it is simply $\pm\frac{1}{4}$.

If the class number of $\mathcal{O}_F$ is 1 then we defined the periods $\Omega(\epsilon\times\pi_f)$ (see 4.2.5) and then we know that

$$\frac{1}{\Omega(\epsilon\times\pi_f)}\mathcal{F}^1(\omega^{\dagger}_{\epsilon})(h^{\dagger,0}_{\pi_f}) \in H^1(\Gamma\backslash\mathbb{H},\tilde{\mathcal{M}}\otimes\mathcal{O}_F)$$

and hence we can conclude for $\nu\neq 0,n$

$$(1/4)\frac{\delta_{0,\infty}(e_\nu)}{\Omega(\epsilon\times\pi_f)}\Lambda^{\mathrm{coh}}(\pi,n+1-\nu)\in\mathcal{O}_F. \tag{5.14}$$

If the class number is not one, we have to interpret $\Omega(\epsilon\times\pi_f)$ as a section in the sheaf of periods, and $\mathcal{O}_F$ has to be replaced by the monoid of integral ideals in $\mathcal{O}_F$. Notice that the term $\delta_{0,\infty}(e_\nu)$ has only prime factors $\leq n$. We will improve this term after the following discussion of the cases $\nu=0, \nu=n$.

This argument fails for $\nu=0,n$ because $\partial(\overline{C}_{0,\infty}\otimes X^n)=\infty\otimes(X^n-Y^n)$ is not a torsion class in $H_0(\partial(\Gamma\backslash\bar{\mathbb{H}}),\underline{\mathcal{M}_\lambda})$ (see section 3.2.1). We apply the Manin-Drinfeld principle to show that the rationality statement also holds for $\nu=0,n$ but we will get a denominator. We pick a prime p. Then we know that the class $[\partial(\overline{C}_{0,\infty}\otimes X^n)]$ is an eigenclass for T_p, i.e.

$$T_p([\partial(\overline{C}_{0,\infty}\otimes X^n])=(p^{n+1}+1)[\partial(\overline{C}_{0,\infty}\otimes X^n)]. \tag{5.15}$$

This implies that $\partial(T_p([\overline{C}_{0,\infty}\otimes X^n])-(p^{n+1}+1)[\overline{C}_{0,\infty}\otimes X^n])=0$, hence we can apply Proposition 5.1.1 and get that the value of the pairing is equal to the integral against the modular symbol. If we exploit the adjointness formula for the Hecke operator then we get

$$\left\langle T_p([\overline{C}_{0,\infty}\otimes X^n])-(p^{n+1}+1)[\overline{C}_{0,\infty}\otimes X^n],\mathcal{F}^1(\omega^{\dagger}_{\epsilon}\otimes h^{\dagger,0}_{\pi_f})\right\rangle$$
$$=\int_0^\infty\left(\left\langle X^n,\sigma^*(\mathcal{F}^1(\omega^{\dagger}_{\epsilon}\otimes(T_p-(p^{n+1}+1))(h^{\dagger,0}_{\pi_f})))\right\rangle\frac{dt}{t}.$$

We have $T_p(h^{\dagger,0}_{\pi_f}) = a_p h^{\dagger,0}_{\pi_f}$ where $a_p \in \mathcal{O}_F$ and hence we get

$$\Big\langle T_p([\overline{C}_{0,\infty} \otimes X^n]) - (p^{n+1}+1)[\overline{C}_{0,\infty} \otimes X^n], \mathcal{F}^1(\omega^{\dagger}_{\epsilon} \otimes h^{\dagger,0}_{\pi_f})\Big\rangle$$

$$= a_\epsilon(a_p - (p^{n+1}+1))\Lambda^{\mathrm{coh}}(\pi_f, 1).$$

It is again the Manin-Drinfeld principle that tells us that for almost all primes p the number $a_p - (p^{n+1}+1) \neq 0$. Let $Z(n)$ be the ideal in $\mathcal{O}_F$ generated by these numbers. We will see (Theorem 5.1.2) that

$$\mathrm{numerator}(\zeta(-1-n)) \subset Z(n). \tag{5.16}$$

Ribet gives an argument in [94] that even equality yields.

Now we can conclude: *For $\nu = 0$ or n, we have*

$$\frac{1}{4}\frac{Z(n)}{\Omega(\epsilon \times \pi_f)}\Lambda^{\mathrm{coh}}(\pi, n+1-\nu) \subseteq \mathcal{O}_F. \tag{5.17}$$

We want to have an estimate of the denominator ideal of

$$\frac{\Lambda^{\mathrm{coh}}(\pi, n+1-\nu)}{\Omega(\epsilon \times \pi_f)}$$

for all values of ν. For $\nu = 0$, $\nu = n$ we have the estimate $Z(n)$. For the other values of ν we have the $\delta_{0,\infty}(e_\nu)$, but we can do much better. Notice that this denominator ideal is an ideal in $\mathcal{O}_F$. We pick a prime $p \leq n$ which then may divide $\delta_{0,\infty}(e_\nu)$. We work locally at p and replace $\mathbb{Z}$ by $\mathbb{Z}_{(p)}$, the local ring at p. It follows from Proposition 3.3.1 that for $0 < \nu < n$ the torsion element $[\partial(\overline{C}_{0,\infty} \otimes e_\nu)]$ is annihilated by a sufficiently high power of the Hecke operator T_p^m. Hence we can lift $T_p^m(\overline{C}_{0,\infty} \otimes e_\nu)$ to an element $T_p^m(\widetilde{C_{0,\infty} \otimes e_\nu}) \in H_1(\Gamma\backslash\mathbb{H}, \underline{\mathcal{M}}_\lambda \otimes \mathbb{Z}_{(p)})$. We know that

$$\frac{1}{\Omega(\epsilon \times \pi_f)}\Big\langle \mathcal{F}^1(\omega^{\dagger}_{\epsilon})(h^{\dagger,0}_{\pi_f}), T_p^m(\widetilde{C_{0,\infty} \otimes e_\nu})\Big\rangle \in \mathcal{O}_F \otimes \mathbb{Z}_{(p)}.$$

Again we can use the adjointness property of T_p and we get

$$\frac{1}{\Omega(\epsilon \times \pi_f)}\Big\langle \mathcal{F}^1(\omega^{\dagger}_{\epsilon})(h^{\dagger,0}_{\pi_f}), \overline{C}_{0,\infty} \otimes e_\nu\Big\rangle$$

$$= a_\epsilon \frac{\pi_f(T_p)^m}{\Omega(\epsilon \times \pi_f)}\Lambda^{\mathrm{coh}}(\pi, n+1-\nu) \in \mathcal{O}_F \otimes \mathbb{Z}_{(p)}.$$

We consider the ideal $\mathfrak{n}(p,\nu,\pi_f) = (\delta_{0,\infty}(e_\nu), \pi_f(T_p)^m) \subset \mathcal{O}_F \otimes \mathbb{Z}_{(p)}$ with m choosen minimal. This ideal may be much larger than $\delta_{0,\infty}(e_\nu)$. We put $\mathfrak{n}(\nu,\pi_f) = \prod_p \mathfrak{n}(p,\nu,\pi_f)$ for $\nu \neq 0, n$ and for convenience $\mathfrak{n}(n,\pi_f) = \mathfrak{n}(0,\pi_f) = Z(n)$.

Then we get the final result:

Theorem 5.1.1. *For any π_f which occurs in (5.4) and any $\nu = 0, \ldots, n$ the ideal*

$$\frac{\mathfrak{n}(\nu,\pi_f)}{\Omega(\epsilon \times \pi_f)}\Lambda^{\mathrm{coh}}(\pi, n+1-\nu)$$

is an integral ideal in $\mathcal{O}_F$. *The primes* $\mathfrak{p}$ *dividing* $\mathfrak{n}(\nu,\pi_f)$ *lie over primes* $p \leq n$. *Furthermore for* $0<\nu<n$ *these primes are not ordinary for* π_f, *i.e if* $\mathfrak{p}$ *divides* $\mathfrak{n}(\nu,\pi_f)$ *then* $\pi_f(T_p) \equiv 0 \mod \mathfrak{p}$.

These rationality results go back to Manin and Shimura. In principle we may say that the integrality assertion also goes back to these authors, but here we have to take into account the fine tuning of the periods.

It is clear that this is compatible with the action of the Galois group $\mathrm{Gal}(F/\mathbb{Q})$, as for $\sigma \in \mathrm{Gal}(F/\mathbb{Q})$ we have

$$\sigma\left(\frac{1}{\Omega(\epsilon\times\pi_f)}\Lambda^{\mathrm{coh}}(\pi,n+1-\nu)\right)=\frac{1}{\Omega(\epsilon\times^{\sigma}\pi_f)}\Lambda^{\mathrm{coh}}({}^{\sigma}\pi,n+1-\nu). \quad (5.18)$$

There is still a slightly different way to look at the theorem above. For each choice of $\epsilon=\pm$ we can look at the array of numbers

$$(\Lambda^{\mathrm{coh}}(\pi,n+1-\nu),\dots)_{\nu=0,\dots n;(-1)^{\frac{n}{2}-\nu}=\epsilon}.$$

It is easy to see that not all of the entries can be zero, hence we can project the arrays to a point $\mathbf{\Lambda}(\epsilon,\pi_f)$ in the projective space $\mathbb{P}^{d(\epsilon,n)}(\mathbb{C})$. Then a slightly weakened form of our results asserts

$$\mathbf{\Lambda}(\epsilon,\pi_f)\in\mathbb{P}^{d(\epsilon,n)}(F)=\mathbb{P}^{d(\epsilon,n)}(\mathcal{O}_F) \text{ and } \sigma(\mathbf{\Lambda}(\epsilon,\pi_f))=\mathbf{\Lambda}(\sigma(\epsilon,\pi_f)).$$

In this formulation we do not see the period. But now we can fix the period as a section in the period sheaf: we require that the arrays of ideals

$$(\dots,\frac{\mathfrak{n}(\nu,\pi_f)}{\Omega(\epsilon\times\pi_f)}\Lambda^{\mathrm{coh}}(\pi,n+1-\nu),\dots)_{\nu=0,\dots n;(-1)^{\frac{n}{2}-\nu}=\epsilon}$$

is an array of integral and coprime ideals. This period is not necessarily equal to the period we defined earlier, but they may only differ at primes $\mathfrak{p}$ dividing $\mathfrak{n}(\nu,\pi_f)$.

We pay a lot of attention to the careful choice of the periods, because we conjecture that the factorisation of the numbers $\frac{\mathfrak{n}(\nu,\pi_f)}{\Omega(\epsilon\times\pi_f)}\Lambda^{\mathrm{coh}}(\pi,n+1-\nu)$ has influence on the structure of the integral cohomology of some other groups. For example we expect that "large" prime ideals $\mathfrak{p}\subset\mathcal{O}_F$ which divide an ideal $\frac{\mathfrak{n}(\nu,\pi_f)}{\Omega(\epsilon\times\pi_f)}\Lambda^{\mathrm{coh}}(\pi,n+1-\nu)$ will also divide the denominator of an Eisenstein class on the symplectic group. Here it is an issue what "large" means. For instance if we want to check certain conjectures experimentally we need to work with "small large" primes.

5.1.3 Evaluation of Eisenstein classes on capped modular symbols

In the following we consider cohomology with coefficients in $\tilde{\mathcal{M}}_\lambda$, we assume that $\lambda=n\gamma$, i.e. we assume that λ is trivial on the centre. We have seen that

$$H^1(\Gamma\backslash\mathbb{H},\tilde{\mathcal{M}}_\lambda\otimes\mathbb{Q}) = H^1_!(\Gamma\backslash\mathbb{H},\tilde{\mathcal{M}}_\lambda\otimes\mathbb{Q})\oplus\mathbb{Q}\mathrm{Eis_n} \tag{5.19}$$

where $\mathrm{Eis_n}$ is defined in 4.130 and characterized by the two conditions

$$r(\mathrm{Eis_n}) = [Y^n] \text{ and } T_p(\mathrm{Eis_n}) = (p^{n+1}+1)\mathrm{Eis_n} \tag{5.20}$$

for all Hecke operators T_p. In our special situation it suffices to check the second condition for $p = 2$. Earlier, we raised the question of determining the denominator of the class $\mathrm{Eis_n}$, i.e. we want to determine the smallest integer $\Delta(n) > 0$ such that $\Delta(n)\mathrm{Eis_n}$ becomes an integral class. In Chapter 3 we discussed an algorithm that computes this denominator in a given case, but we want to see a general expression for this denominator.

To achieve this goal we compute the evaluation of $\mathrm{Eis_n}$ on the first homology group, i.e we compute the value $\langle c,\mathrm{Eis_n}\rangle$ for $c \in H_1(\Gamma\backslash\mathbb{H},\underline{\mathcal{M}}_\lambda)$. We have the exact sequence

$$\begin{aligned} H_1(\partial(\Gamma\backslash\bar{\mathbb{H}}),\underline{\mathcal{M}}_\lambda) &\xrightarrow{j} H_1(\Gamma\backslash\bar{\mathbb{H}},\underline{\mathcal{M}}_\lambda) \longrightarrow \\ \to H_1(\Gamma\backslash\bar{\mathbb{H}},\partial(\Gamma\backslash\bar{\mathbb{H}}),\underline{\mathcal{M}}_\lambda) &\xrightarrow{\partial} H_0(\partial(\Gamma\backslash\bar{\mathbb{H}}),\underline{\mathcal{M}}_\lambda). \end{aligned}$$

It follows from the construction of $\mathrm{Eis_n}$ that $\langle c,\mathrm{Eis_n}\rangle \in \mathbb{Z}$ for all the elements in the image of j. Therefore we only have to compute the values $\langle\tilde{c}_\nu,\mathrm{Eis_n}\rangle$, where $\tilde{c}_\mu$ are lifts of a system of generators $\{c_\mu\}$ of $\ker(\partial)$.

In our special case it is not difficult to see that the elements $\overline{C}_{0,\infty}\otimes e_\nu$, where $\nu = 0, 1\dots, n$, form a set of generators of $H_1(\Gamma\backslash\bar{\mathbb{H}},\partial(\Gamma\backslash\bar{\mathbb{H}}),\underline{\mathcal{M}}_\lambda)$. We observe:

The boundary of the element $\overline{C}_{0,\infty}\otimes e_\nu$ for $\nu = 0$ and n is an element of infinite order in $H_0(\partial(\Gamma\backslash\bar{\mathbb{H}}),\underline{\mathcal{M}}_\lambda)$.

The boundary of an element $\overline{C}_{0,\infty}\otimes e_\nu$ with $0<\nu<n$ are torsion elements in $H_0(\partial(\Gamma\backslash\bar{\mathbb{H}}),\underline{\mathcal{M}}_\lambda)$. This implies

Proposition 5.1.2. *The elements $\overline{C}_{0,\infty}\otimes m \in H_1(\Gamma\backslash\bar{\mathbb{H}},\partial(\Gamma\backslash\bar{\mathbb{H}}),\underline{\mathcal{M}}_\lambda)$ with $\partial(C_{0,\infty}\otimes m) = 0$ are of the form*

$$c = \overline{C}_{0,\infty}\otimes\left(\sum_{\nu=1}^{\nu=n-1} a_\nu e_\nu\right) \quad \textit{with } a_\nu\in\mathbb{Z}.$$

Now it seems to be tempting to choose for our generators above the $\overline{C}_{0,\infty}\otimes e_\nu$, but this is not possible because $\partial(\overline{C}_{0,\infty}\otimes e_\nu)$ is not necessarily zero, it is only a torsion element. So we see that it is not clear how to find a suitable system of generators.

To overcome this difficulty we use the Hecke operators. If we want to determine the denominator $\Delta(n)$ we can localize, i.e. for each prime p we have to determine the highest power $p^{d(n,p)}$ which divides $\Delta(n)$. As usual we write $d(n,p) = \mathrm{ord}_p(\Delta(n))$. We replace the ring $\mathbb{Z}$ by its localization $\mathbb{Z}_{(p)}$ and replace all our cohomology and homology groups by the localized groups. Thus, we have to find a set of generators $\{\dots,c_\nu\dots\}_\nu \subset H_1(\Gamma\backslash\bar{\mathbb{H}},\partial(\Gamma\backslash\bar{\mathbb{H}}),\underline{\mathcal{M}}_\lambda\otimes\mathbb{Z}_{(p)})$ and compute the denominator of $\langle\tilde{c}_\nu,\mathrm{Eis}_n\rangle$.

It follows from Proposition 3.3.1 that for $c = \overline{C}_{0,\infty} \otimes \left(\sum_{\nu=1}^{\nu=n-1} a_\nu e_\nu\right)$ with $a_i \in \mathbb{Z}_{(p)}$, the torsion element $\partial(c)$ is annihilated by a sufficiently high power of the Hecke operator T_p^m, and hence we see that $T_p^m(c)$ can be lifted to an element $\widetilde{T_p^m(c)} \in H_1(\Gamma\backslash\mathbb{H}, \underline{\mathcal{M}}_\lambda \otimes \mathbb{Z}_{(p)})$. Now if c itself can already be lifted to a $\tilde{c} \in H_1(\Gamma\backslash\mathbb{H}, \underline{\mathcal{M}}_\lambda \otimes \mathbb{Z}_{(p)})$, we have modulo $\mathbb{Z}_{(p)}$

$$\left\langle \widetilde{T_p^m(c)}, \mathrm{Eis}_n \right\rangle \equiv \langle T_p^m(\tilde{c}), \mathrm{Eis}_n \rangle = \langle \tilde{c}, T_p^m(\mathrm{Eis}_n) \rangle = (p^{n+1}+1)^m \langle \tilde{c}, \mathrm{Eis}_n \rangle$$

and hence $\min\left(\mathrm{ord}_p \left\langle \widetilde{T_p^m(c)}, \mathrm{Eis}_n \right\rangle, 0\right) = \min\left(\mathrm{ord}_p \langle \tilde{c}, \mathrm{Eis}_n \rangle, 0\right)$. Hence we get

Proposition 5.1.3. *If ν runs from 1 to $n-1$ and if $T_p^m(\widetilde{C_{0,\infty} \otimes e_\nu})$ is any p-integral lift of $T_p^m(\overline{C}_{0,\infty} \otimes e_\nu)$ then*

$$d(n,p) = -\min_\nu\left(0, \mathrm{ord}_p \left\langle T_p^m(\widetilde{C_{0,\infty} \otimes e_\nu}), \mathrm{Eis}_n \right\rangle\right).$$

Proof. This is now obvious. □

5.1.4 The capped modular symbol

The case A)

To calculate the denominator of Eis_n we have to compute $\left\langle T_p^m(\widetilde{C_{0,\infty} \otimes e_\nu}), \mathrm{Eis}_n \right\rangle$ modulo $\mathbb{Z}_{(p)}$ for various p. At this point some meditation is in order. Our cohomology class Eis_n is represented by a closed differential form $\mathrm{Eis}(\omega_n)$ (see Section 4.2.6), and this differential form lives on $\Gamma\backslash\mathbb{H}$, hence it provides a cohomology class for $\Gamma\backslash\mathbb{H}$. But we know that the inclusion provides an isomorphism

$$H^1(\Gamma\backslash\mathbb{H}, \tilde{\mathcal{M}}_\lambda^\flat) \xrightarrow{\sim} H^1(\Gamma\backslash\bar{\mathbb{H}}, \tilde{\mathcal{M}}_\lambda^\flat),$$

and since $T_p^m(\widetilde{C_{0,\infty} \otimes e_\nu}) \in H_1(\Gamma\backslash\bar{\mathbb{H}}, \underline{\mathcal{M}}_\lambda)$ we can evaluate the cohomology class $\mathrm{Eis}(\omega_n)$ on the cycle. But we want to get the value $\left\langle T_p^m(\widetilde{C_{0,\infty} \otimes e_\nu}), \mathrm{Eis}_n \right\rangle$ by integration of the differential form against the cycle. This is problematic, because the cycle has non-trivial support in $\partial(\Gamma\backslash\bar{\mathbb{H}})$, and on this circle at infinity the differential form is not really defined.

There are certainly several ways out of this dilemma. The Borel-Serre boundary is a circle $\mathbb{H}_\infty = \Gamma_\infty\backslash\mathbb{R}$ where $\Gamma_\infty = \{\pm\mathrm{Id}\} \times \langle T_\infty \rangle$ and $T_\infty = \begin{pmatrix} 1 & 1 \\ 0 & 1 \end{pmatrix}$. The cycle is the sum of two 1-chains:

$$T_p^m(\widetilde{C_{0,\infty} \otimes e_\nu}) = T_p^m\left(\overline{C}_{0,\infty} \otimes e_\nu\right) + \mathcal{Z}_\partial(\nu)$$

where $\mathcal{Z}_\partial(\nu)$ is a p-integral 1-chain on the Borel-Serre boundary, well defined only up to a p-integral cycle, with

$$\partial \mathcal{Z}_\partial(\nu) = -T_p^m\left(\{0\}_\infty \otimes (e_\nu - w_0 e_\nu)\right).$$

One possibility is to deform the cycle $T_p^m(\widetilde{C_{0,\infty} \otimes e_\nu})$ and "pull" it into the interior $\Gamma\backslash\mathbb{H}$. For this we note that the 1-chain $T_p^m\left(\overline{C}_{0,\infty} \otimes e_\nu\right)$ consists of a union of compactified geodesics in $\Gamma\backslash\bar{\mathbb{H}}$, each geodesic starting and ending at some point in $\Gamma_\infty\backslash\mathbb{H}_\infty$. By truncating the geodesics when crossing the Farey circle $S\left(t_0^{-1},\infty\right)$ for some large chosen t_0, and moving $\mathcal{Z}_\partial(\nu)$ from $\mathbb{H}_\infty$ onto the Farey circle by the geodesic action, we get a homologous cycle

$$T_p^m(\widetilde{C_{0,\infty}\otimes e_\nu})(t_0) = \left(T_p^m(C_{0,\infty}\otimes e_\nu)\right)(t_0) + \mathcal{Z}_\partial(\nu)(t_0)$$

in $\Gamma\backslash\mathbb{H}$. So we get

$$\left\langle T_p^m(\widetilde{C_{0,\infty}\otimes e_\nu}), \mathrm{Eis}_n\right\rangle = \int_{\left(T_p^m(C_{0,\infty}\otimes e_\nu)\right)(t_0)+\mathcal{Z}_\partial(\nu)(t_0)} \mathrm{Eis}(\omega_n). \tag{5.21}$$

The value of this integral does not depend on t_0, and we easily check that for both summands the limit for $t_0\to\infty$ exists. We get a sum of integrals of a closed 1-form against two 1-chains. We consider the first summand and use the selfadjointness of the Hecke operator. Then this first summand is equal to

$$\begin{aligned}
&(1+p^{n+1})^m \int_{C_{0,\infty}} \langle \mathrm{Eis}(\omega_n), e_\nu\rangle_{\mathcal{M}} \\
=&(1+p^{n+1})^m \int_0^\infty \langle \mathrm{Eis}(\omega_n), e_\nu\rangle_{\mathcal{M}} \left(t\frac{\partial}{\partial t}\right)\left(\begin{pmatrix} t_\infty & 0\\ 0 & 1\end{pmatrix}\right)\frac{dt_\infty}{t_\infty} \\
=&(1+p^{n+1})^m \int_0^\infty \left\langle \begin{pmatrix} t_\infty & 0\\ 0 & 1\end{pmatrix}\mathrm{Eis}(\omega_n(\frac{H}{2}))\left(\begin{pmatrix} t_\infty & 0\\ 0 & 1\end{pmatrix}\right), e_\nu\right\rangle_{\mathcal{M}} \frac{dt_\infty}{t_\infty}.
\end{aligned}$$

The factor in front is irrelevant (at least for $\mathrm{ord}_p(m)$ large), because we are only interested in the value mod $\mathbb{Z}_{(p)}$.

We know that $B(\mathbb{Z})\backslash\mathcal{G}(\mathbb{Z}) = B(\mathbb{Q})\backslash G(\mathbb{Q})$. The adelic definition of the Eisenstein intertwining operator becomes (we choose the standard system of representatives for $B(\mathbb{Q})\backslash G(\mathbb{Q})$)

$$\mathrm{Eis}(\underline{\omega}_n(\frac{H}{2}))(\underline{x}) = \sum_{a\in\mathbb{Q}^\times} \underline{\omega}_n(\frac{H}{2})(\begin{pmatrix} 1 & 0\\ a & 1\end{pmatrix}\underline{x}) + \underline{\omega}_n(\frac{H}{2})(\underline{x}) + \underline{\omega}_n(\frac{H}{2})(w_0\underline{x}).$$

Now we use the identification $T(\mathbb{A})/T(\mathbb{Q}) = \mathbb{R}^\times_{>0}\times T(\hat{\mathbb{Z}})$ and evaluate

$$\begin{aligned}
&\int_{T(\mathbb{A})/T(\mathbb{Q})} \left\langle \begin{pmatrix} t_\infty & 0\\ 0 & 1\end{pmatrix}\mathrm{Eis}(\underline{\omega}_n(\tfrac{H}{2}))\left(\begin{pmatrix} \underline{t} & 0\\ 0 & 1\end{pmatrix}\right), e_\nu\right\rangle_{\mathcal{M}} \frac{d\underline{t}}{\underline{t}} = \\
&\int_{T(\mathbb{A})/T(\mathbb{Q})} \left\langle \begin{pmatrix} t_\infty & 0\\ 0 & 1\end{pmatrix}\textstyle\sum_{a\in\mathbb{Q}^\times}\underline{\omega}_n(\tfrac{H}{2})\left(\begin{pmatrix} 1 & 0\\ a & 1\end{pmatrix}\begin{pmatrix} \underline{t} & 0\\ 0 & 1\end{pmatrix}\right), e_\nu\right\rangle_{\mathcal{M}} \frac{d\underline{t}}{\underline{t}},
\end{aligned} \tag{5.22}$$

the second and the third term vanish.

We look at an individual summand, and for $a>0$ we write $a=\prod p^{\alpha_p}$

$$\left\langle \begin{pmatrix} t_\infty & 0\\ 0 & 1\end{pmatrix}\omega_n(\frac{H}{2})\left(\begin{pmatrix} 1 & 0\\ a & 1\end{pmatrix}\begin{pmatrix} t_\infty & 0\\ 0 & 1\end{pmatrix}\right), e_\nu\right\rangle_{\mathcal{M}} \prod_p \omega_{f,p}\left(\begin{pmatrix} 1 & 0\\ p^{\alpha_p} & 0\end{pmatrix}\right).$$

Clearly

$$\omega_{f,p}(\begin{pmatrix} 1 & 0 \\ p^{\alpha_p} & 1 \end{pmatrix}) = \begin{cases} 1 & \text{if } \alpha_p \geq 0 \\ p^{\alpha_p(n+2)} & \text{if } \alpha_p < 0 \end{cases}. \tag{5.23}$$

We want to compute the factor at infinity

$$\left\langle \begin{pmatrix} t_\infty & 0 \\ 0 & 1 \end{pmatrix} \omega_n(\frac{H}{2}) \left(\begin{pmatrix} 1 & 0 \\ a & 1 \end{pmatrix} \begin{pmatrix} t_\infty & 0 \\ 0 & 1 \end{pmatrix} \right), e_\nu \right\rangle_{\mathcal{M}}. \tag{5.24}$$

We invoke the Iwasawa decomposition

$$\begin{pmatrix} t_\infty & 0 \\ at_\infty & 1 \end{pmatrix} = \begin{pmatrix} a^{-1} & 0 \\ 0 & 1 \end{pmatrix} \begin{pmatrix} \frac{at_\infty}{\sqrt{1+a^2t_\infty^2}} & y \\ 0 & \sqrt{1+a^2t_\infty^2} \end{pmatrix} \begin{pmatrix} \frac{1}{\sqrt{1+a^2t_\infty^2}} & \frac{-at_\infty}{\sqrt{1+a^2t_\infty^2}} \\ \frac{at_\infty}{\sqrt{1+a^2t_\infty^2}} & \frac{1}{\sqrt{1+a^2t_\infty^2}} \end{pmatrix}$$

and make the substitution $at_\infty = \tau_\infty$ and call the second and third factor $b(\tau_\infty), k(\tau_\infty)$. Then

$$\omega_n(\frac{H}{2}) \left(\begin{pmatrix} t_\infty & 0 \\ at_\infty & 1 \end{pmatrix} \right) = a^{-n/2-1} (\frac{\tau_\infty}{1+\tau_\infty^2})^{n/2+1} \omega_n(\frac{H}{2})(k(\tau_\infty)).$$

Since ω_n is K_∞-invariant we can move $k(\tau_\infty)$ by right translation and we find

$$\begin{aligned} \omega_n(\frac{H}{2})(k(\tau_\infty)) &= k(\tau_\infty)^{-1} \omega_n \left(\mathrm{Ad}(k(\tau_\infty))(\frac{H}{2}) \right) (1) \\ &= E_+^\vee \left(\mathrm{Ad}(k(\tau_\infty))(\frac{H}{2}) \right) \; k(\tau_\infty)^{-1}(Y^n) \\ &= \frac{2\tau_\infty}{1+\tau_\infty^2} \left(\frac{\tau_\infty}{\sqrt{1+\tau_\infty^2}} X + \frac{1}{\sqrt{1+\tau_\infty^2}} Y \right)^n \end{aligned}$$

and hence

$$\omega_n(\frac{H}{2}) \left(\begin{pmatrix} t_\infty & 0 \\ at_\infty & 1 \end{pmatrix} \right) = 2a^{-n/2-1} (\frac{\tau_\infty}{1+\tau_\infty^2})^{n/2+2} \left(\sum_{\mu=0}^{n} \binom{n}{\mu} \frac{\tau_\infty^\mu}{(1+\tau_\infty^2)^{n/2}} X^\mu Y^{n-\mu} \right).$$

We still have to apply $\begin{pmatrix} t_\infty & 0 \\ 0 & 1 \end{pmatrix}$ to this expression and we get

$$\begin{pmatrix} t_\infty & 0 \\ 0 & 1 \end{pmatrix} \omega_n(\tfrac{H}{2}) \left(\begin{pmatrix} t_\infty & 0 \\ at_\infty & 1 \end{pmatrix} \right) =$$

$$2 \sum_{\mu=0}^{n} \binom{n}{\mu} a^{-1-\mu} \frac{\tau_\infty^{2\mu+2}}{(1+\tau_\infty^2)^{n+2}} X^\mu Y^{n-\mu}.$$

We have to pair this expression with e_ν, then it is clear that this pairing picks only one summand - namely the summand $\mu = n - \nu$ - and we get

$$\left\langle \begin{pmatrix} t_\infty & 0 \\ 0 & 1 \end{pmatrix} \omega_n(\frac{H}{2}) \left(\begin{pmatrix} t_\infty & 0 \\ at_\infty & 1 \end{pmatrix} \right), e_\nu \right\rangle_{\mathcal{M}} = (-1)^\nu 2a^{\nu-n-1} \frac{\tau_\infty^{2(n+1-\nu)}}{(1+\tau_\infty^2)^{n+2}}.$$

Here we have to remember $< \binom{n}{n-\nu} X^{n-\nu}Y^\nu, X^\nu Y^{n-\nu} >_{\mathcal{M}} = (-1)^\nu$. Also we note that the expression changes its sign by $(-1)^{\nu+1}$ if we change the sign of a, so that the sum over all $a \in \mathbb{Q}^\times$ is vanishing for even ν. For positive $a \in \mathbb{Q}^\times$ the individual term in the summation is given by (again we write $a = \prod_p p^{\alpha_p}$)

$$(-1)^\nu 2 \frac{\tau_\infty^{2(n+1-\nu)}}{(1+\tau_\infty^2)^{n+2}} \prod_p p^{(\nu-n-1)\alpha_p} \omega_{f,p}\begin{pmatrix} 1 & 0 \\ p^{\alpha_p} & 1 \end{pmatrix}.$$

We interchange the integration and summation in (5.22). Then we get for a fixed positive a that the factor at the infinite place is

$$\int_0^\infty \frac{(at_\infty)^{2(n+1-\nu)}}{(1+(at_\infty)^2)^{n+2}} \frac{dt_\infty}{t_\infty} = \int_0^\infty \frac{\tau_\infty^{2(n+1-\nu)}}{(1+\tau_\infty^2)^{n+2}} \frac{d\tau_\infty}{\tau_\infty} = \frac{1}{2} \frac{\Gamma(\nu+1)\Gamma(n+1-\nu)}{\Gamma(n+2)}.$$

Now we sum over a or better over the arrays of exponents $\underline{\alpha} = \{\alpha_2, \dots, \alpha_p, \dots\}$. This is the product over all primes of the expressions

$$\sum_{\alpha_p=-\infty}^{\infty} p^{(-n-1+\nu)\alpha_p} \omega_{f,p}\begin{pmatrix} 1 & 0 \\ p^{\alpha_p} & 0 \end{pmatrix} = \sum_{\alpha_p=0}^{\infty} p^{(-n-1+\nu)\alpha_p} + \sum_{\beta_p=1}^{\infty} p^{-\beta_p(\nu+1)} =$$

$$\frac{1-p^{-n-2}}{(1-p^{-\nu-1})(1-p^{\nu-n-1})}.$$

Hence taking the product over all places yields

$$\int_{T(\mathbb{A})/T(\mathbb{Q})} \left\langle \begin{pmatrix} t_\infty & 0 \\ 0 & 1 \end{pmatrix} \mathrm{Eis}(\underline{\omega}_n(\tfrac{H}{2})) \left(\begin{pmatrix} \underline{t} & 0 \\ 0 & 1 \end{pmatrix}\right), e_\nu \right\rangle_{\mathcal{M}} =$$

$$\begin{cases} -2 \dfrac{\Gamma(\nu+1)\Gamma(n+1-\nu)}{\Gamma(n+2)} \dfrac{\zeta(\nu+1)\zeta(n+1-\nu)}{\zeta(n+2)} & \text{for } \nu \text{ odd} \\ 0 & \text{for } \nu \text{ even.} \end{cases}$$

Applying the functional equation and some identity for special values of the Gamma-function finally we get

$$\int_{T(\mathbb{A})/T(\mathbb{Q})} \left\langle \begin{pmatrix} t_\infty & 0 \\ 0 & 1 \end{pmatrix} \mathrm{Eis}(\underline{\omega}_n(\frac{H}{2})) \left(\begin{pmatrix} \underline{t} & 0 \\ 0 & 1 \end{pmatrix}\right), e_\nu \right\rangle_{\mathcal{M}} \frac{d\underline{t}}{\underline{t}} = -\frac{\zeta(-\nu)\zeta(\nu-n)}{\zeta(-1-n)}.$$

We consider the second term $\int_{\mathcal{Z}_\partial(\nu)} \mathrm{Eis}(\omega_n)$. As a first step we calculate $T_p^m(\{0\}_\infty \otimes e_\nu)$ and write it explicitly as a boundary of some 1-chain in $\mathbb{H}_\infty$ for $0 < \nu < n$ and m large. For selfadjointness to hold we have to use a normalization of the Hecke operators for homology which corresponds to our normalization of u_α in Section 3.1.2 for cohomology. On the level of singular chains this means that the coset representative of T_p acts on the coefficients as it does on $\mathcal{M}_{n\gamma+\frac{n}{2}\det} = \mathcal{M}_n$.

We write the double coset T_p as a union of left cosets

$$\Gamma\left(\begin{pmatrix} p & 0 \\ 0 & 1 \end{pmatrix}\right)\Gamma = \Gamma\left(\begin{pmatrix} p & 0 \\ 0 & 1 \end{pmatrix}\right) \cup \bigcup_{i=1}^{p-1} \Gamma\left(\begin{pmatrix} 1 & i \\ 0 & p \end{pmatrix}\right),$$

the representatives chosen in such a way that they preserve the boundary component $\mathbb{H}_\infty$. Then the general formula for the Hecke operators is

$$T_p\left(\{r\}_\infty \otimes m\right) = \sum_{i=0}^{p-1} \left\{ \begin{pmatrix} 1 & i \\ 0 & p \end{pmatrix} r \right\}_\infty \otimes \begin{pmatrix} 1 & i \\ 0 & p \end{pmatrix} m + \{pr\}_\infty \otimes \begin{pmatrix} p & 0 \\ 0 & 1 \end{pmatrix} m.$$

Setting

$$\begin{aligned} U_p\left(\{r\}_\infty \otimes m\right) &= \sum_{i=0}^{p-1} \left\{ \begin{pmatrix} 1 & i \\ 0 & p \end{pmatrix} r \right\}_\infty \otimes \begin{pmatrix} 1 & i \\ 0 & p \end{pmatrix} m \\ V_p\left(\{r\}_\infty \otimes m\right) &= \{pr\}_\infty \otimes \begin{pmatrix} p & 0 \\ 0 & 1 \end{pmatrix} m, \end{aligned}$$

we have

$$\begin{aligned} V_p U_p\left(\{r\}_\infty \otimes m\right) &= p^{1+n}\left(\{r\}_\infty \otimes m\right) \\ V_p\left(\{0\}_\infty \otimes e_\nu\right) &= p^\nu\left(\{0\}_\infty \otimes e_\nu\right). \end{aligned}$$

Here we are calculating with chains in $\mathbb{H}_\infty$ modulo the action of Γ_∞. Then

$$T_p^m\left(\{0\}_\infty \otimes e_\nu\right) = U_p^m\left(\{0\}_\infty \otimes e_\nu\right) + c_1 U_p^{m-1}\left(\{0\}_\infty \otimes e_\nu\right) + c_2 U_p^{m-2}\left(\{0\}_\infty \otimes e_\nu\right) + \ldots$$

where $c_i \equiv p^{i\nu}$ modulo p^{n+1}.

Now we have for all m

$$U_p^m\left(\{0\}_\infty \otimes e_\nu\right) = \sum_{j_1,\ldots,j_m=0}^{p-1} \left\{ \frac{j_1 + j_2 p + \ldots + j_m p^{m-1}}{p^m} \right\}_\infty \otimes \begin{pmatrix} 1 & j_m \\ 0 & p \end{pmatrix} \cdots \begin{pmatrix} 1 & j_1 \\ 0 & p \end{pmatrix} e_\nu.$$

Putting $\underline{j} = j_1 + j_2 p + \ldots + j_m p^{m-1}$ the sum becomes

$$U_p^m\left(\{0\}_\infty \otimes e_\nu\right) = \sum_{j_1,\ldots,j_m=0}^{p-1} \left\{ \frac{\underline{j}}{p^m} \right\}_\infty \otimes X^\nu \left(\underline{j}X + p^m Y\right)^{n-\nu}.$$

For large m we can find $P \in \mathcal{M}_n \otimes \mathbb{Z}_{(p)}$ such that

$$X^\nu \left(\underline{j}X + p^m Y\right)^{n-\nu} = \partial P := P(X,Y) - P(X, -X+Y).$$

This is so because the term Y^n does not appear, X^n can be written in that form and the coefficients of the other terms are highly divisble by p. Then we can write the zero-cycle

$$\left\{ \frac{\underline{j}}{p^m} \right\}_\infty \otimes X^\nu \left(\underline{j}X + p^m Y\right)^{n-\nu}$$

as the boundary of the 1-chain

$$-\left[\frac{\underline{j}}{p^m}, \frac{\underline{j}}{p^m} + 1 \right]_\infty \otimes P.$$

We will now use the following formula which we will prove later

$$\int_{[a,b]_\infty \otimes P} \mathrm{Eis}(\omega_n) = \lim_{t_0 \to \infty} \int_{[it_0+a, it_0+b]\otimes P} \mathrm{Eis}(\omega_n) = \int_{-b}^{-a} P(1,u)du.$$

So our 1-chain gives the contribution

$$-\int_{-\frac{j}{p^m}-1}^{-\frac{j}{p^m}} P(1,u)du = \tilde{P}(-\frac{j}{p^m}-1) - \tilde{P}(-\frac{j}{p^m})$$

where $\tilde{P}$ is a primitive function for the polynomial $P(1,u)$. We normalize it to $\tilde{P}(0) = 0$. Since $\tilde{Q}(u) = \tilde{P}(u) - \tilde{P}(u-1)$ is a primitive polynomial for $P(1,u) - P(1,u-1) = \left(\underline{j} + p^m u\right)^{n-\nu}$ we get

$$\tilde{Q}(u) = \frac{1}{(n-\nu+1)p^m}\left(\underline{j}+p^m u\right)^{n-\nu+1} + \tilde{Q}(0) - \frac{\underline{j}^{n-\nu+1}}{(n-\nu+1)p^m}$$

and hence

$$-\int_{-\frac{j}{p^m}-1}^{-\frac{j}{p^m}} P(1,u)du = -\tilde{Q}(0) + \frac{\underline{j}^{n-\nu+1}}{(n-\nu+1)p^m} = \tilde{P}(-1) + \frac{\underline{j}^{n-\nu+1}}{(n-\nu+1)p^m}.$$

If $P(1,u) = a_0 + a_1 u + \ldots$ we have

$$\tilde{P}(-1) = -a_0 + \frac{a_1}{2} - \frac{a_2}{3} + \ldots .$$

Since $P(1,u) - P(1,u-1) = \left(\underline{j} + p^m u\right)^{n-\nu}$, we have $a_1 = \underline{j}^{n-\nu}$, and the coefficients a_i for $i > 1$ are highly divisible by p for m large, and hence $\tilde{P}(-1) \in \mathbb{Z}_{(p)}$ for $p > 2$. Since $\underline{j}$ runs over the integers modulo p^m, we have to consider sums

$$\mathcal{S}_m(\mu) = \frac{1}{p^m \mu} \sum_{j=0}^{p^m-1} j^\mu$$

and to look at their values modulo $\mathbb{Z}_{(p)}$. It's well known that $\mathcal{S}_m(\mu)$ converges for $m \to \infty$ p-adically to $-\zeta(1-\mu)$. Since $p^{n+1}\zeta(1-\mu)$ is p-adic integral for $0 < \mu < n$, we get for a 1-chain $\mathcal{Z}$ in $\mathbb{H}_\infty$ with $\partial(\mathcal{Z}) = T_p^m(\{0\}_\infty \otimes e_\nu)$ and large m

$$\int_{\mathcal{Z}} \mathrm{Eis}(\omega_n) \equiv \sum_{i=0}^{m} p^{i\nu} \mathcal{S}_{m-i}(n-\nu+1) \equiv \frac{1}{p^\nu - 1}\zeta(\nu - n) \text{ modulo } \mathbb{Z}_{(p)}.$$

Since $\partial \mathcal{Z}_\partial(\nu) = -T_p^m(\{0\}_\infty \otimes (e_\nu - w_0 e_\nu))$ we have

$$\int_{\mathcal{Z}_\partial(\nu)} \mathrm{Eis}(\omega_n) \equiv \frac{1}{1-p^\nu}\zeta(\nu-n) + \frac{1}{1-p^{n-\nu}}\zeta(-\nu) \text{ modulo } \mathbb{Z}_{(p)}.$$

Here we are using that $\zeta(-\nu)$ is vanishing for ν even.

Before going on we have to pay off a debt and indicate the proof of the formula for

$$I := \int_{[a,b]_\infty \otimes P} \mathrm{Eis}(\omega_n) = \lim_{y \to \infty} \int_{[iy+a, iy+b]\otimes P} \mathrm{Eis}(\omega_n).$$

For that we lift the 1-chain $\sigma_y : t \mapsto iy + a + t(b-a) \in \mathbb{H}$ to the group

$$\widetilde{\sigma}_y := \begin{cases} [0,1] \to \mathrm{Gl}_2(\mathbb{R}) \\ t \mapsto \begin{pmatrix} y & a+t(b-a) \\ 0 & 1 \end{pmatrix} \end{cases}$$

and have $\widetilde{\sigma}_{y*}(\frac{\partial}{\partial t})_{|t=t_0} = \widetilde{\sigma}_y(t_0)_*(\frac{b-a}{y}E_+)$. We get

$$\begin{aligned} I &= \lim_{y\to\infty} \int_0^1 \left\langle \mathrm{Eis}(\omega_n) \left(\begin{pmatrix} y & a+t(b-a) \\ 0 & 1 \end{pmatrix}_* \left(\frac{b-a}{y} E_+ \right) \right), P \right\rangle dt \\ &= \lim_{y\to\infty} \frac{b-a}{y} \int_0^1 \langle \mathrm{Eis}(\omega_n)\,(E_+)\,(\widetilde{\sigma}_y(t))\,, \widetilde{\sigma}_y(t)^{-1} P \rangle\, dt \\ &= \lim_{y\to\infty} \frac{b-a}{y} \int_0^1 \langle (\mathcal{F} \circ \mathrm{Eis})\,(\omega_n)\,(E_+)\,(\widetilde{\sigma}_y(t))\,, \widetilde{\sigma}_y(t)^{-1} P \rangle\, dt \\ &= \lim_{y\to\infty} \frac{b-a}{y} \int_0^1 \langle (\omega_n + T_w\omega_n)\,(E_+)\,(\widetilde{\sigma}_y(t))\,, \widetilde{\sigma}_y(t)^{-1} P \rangle\, dt. \end{aligned}$$

Here $\mathcal{F}$ is the constant term, and we have $\mathcal{F} \circ \mathrm{Eis} = T_1 + T_w$ with intertwining operators $T_1 = \mathrm{id}$ and T_w. Plugging in

$$\begin{aligned} \omega_n\,(E_+)\,(\widetilde{\sigma}_y(t)) &= y^{\frac{n}{2}+1} Y^n \\ (T_w\omega_n)\,(E_+)\,(\widetilde{\sigma}_y(t)) &= c y^{-\frac{n}{2}} Y^n \end{aligned}$$

for some constant c and

$$\begin{aligned} \widetilde{\sigma}_y(t)^{-1} P = \widetilde{\sigma}_y(t)^{-1} &\left(\sum_{\mu=0}^{n} a_\mu X^\mu Y^{n-\mu} \right) \\ &= \left(y^{-\frac{n}{2}} \sum_{\mu=0}^{n} a_\mu\,(-a - t(b-a))^{n-\mu} \right) X^n + \dots \end{aligned}$$

we get

$$\begin{aligned} I &= \lim_{y\to\infty} \frac{b-a}{y} \int_0^1 (y + c y^{-n}) \left(\sum_{\mu=0}^{n} a_\mu\,(-a - t(b-a))^{n-\mu} \right) dt \\ &= \int_{-b}^{-a} P(1,t) dt. \end{aligned}$$

Putting together the contributions from the integral over the modular symbol and the integral over the boundary chain we get for large m the following congruence modulo $\mathbb{Z}_{(p)}$

$$\begin{aligned} \left\langle T_p^m(\widetilde{C_{0,\infty} \otimes e_\nu}), \mathrm{Eis}_n \right\rangle \equiv &- (1+p^{n+1})^m \frac{\zeta(-\nu)\zeta(\nu-n)}{\zeta(-1-n)} \\ &+ \frac{1}{1-p^\nu}\zeta(\nu-n) + \frac{1}{1-p^{n-\nu}}\zeta(-\nu). \end{aligned} \tag{5.25}$$

For studying possible cancellations between the denominators of the various summands we have need for the following

Proposition 5.1.4. *For $p > 2$ a prime, $a, b > 1$ integers with $p-1 \mid a$ and $\zeta_{(p)}(s) := (1-p^{-s})\zeta(s)$ we have the following congruences*

$$\zeta_{(p)}(1-b) \in \begin{cases} \mathbb{Z}_{(p)} & \textit{for } p-1 \not\mid b \\ -\frac{p-1}{p}\frac{1}{b} + \mathbb{Z}_{(p)} & \textit{for } p-1 \mid b \end{cases}$$

and

$$\zeta_{(p)}(1-b)-\zeta_{(p)}(1-(a+b)) \in \begin{cases} pa\mathbb{Z}_{(p)} & p-1 \not\mid b \\ -\frac{p-1}{p}\left(\frac{1}{b}-\frac{1}{a+b}\right)+\frac{a}{b}\mathbb{Z}_{(p)} & p-1 \mid b, \frac{a}{b} \in \mathbb{Z}_{(p)} \end{cases}.$$

Proof. While the first three congruences are well known as the von Staudt-Clausen theorem and the Kummer congruences, we will prove the last one in some detail. For $\mathrm{ord}_p(a) = \mathrm{ord}_p(b)$ the last congruence follows immediately from the second one (von Staudt-Clausen theorem), and so we assume $\mathrm{ord}_p(a) > \mathrm{ord}_p(b)$. For positive integers m divisible by $p-1$, the zeta value $\zeta_{(p)}(1-m)$ is interpolated by the p-adic zeta function $\zeta_p(s)$, which is a fraction

$$\zeta_p(s) = \frac{g(\kappa_0^s - 1)}{h(\kappa_0^s - 1)}$$

with $\kappa_0 := 1+p$, $h(T) = 1 - \frac{\kappa_0}{1+T}$, and $g(T) \in \mathbb{Z}_p[[T]]$ an element of the Iwasawa algebra. For elements of the Iwasawa algebra Kummer-like congruences are obvious, and we get

$$(\kappa_0^b - 1)\zeta_{(p)}(1-b) \equiv (\kappa_0^{a+b}-1)\zeta_{(p)}(1-(a+b)) \text{ modulo } pa\mathbb{Z}_{(p)}.$$

So we get congruences modulo $pa\mathbb{Z}_{(p)}$

$$\begin{aligned}(\kappa_0^b - 1)\left(\zeta_{(p)}(1-b) - \zeta_{(p)}(1-(a+b))\right) &\equiv \kappa_0^b(\kappa_0^a - 1)\zeta_{(p)}(1-(a+b)) \\ &\equiv \kappa_0^b(\kappa_0^a-1)\left(-\tfrac{p-1}{p}\tfrac{1}{a+b}\right).\end{aligned}$$

Therefore we consider

$$R := \kappa_0^b(\kappa_0^a - 1)\left(-\frac{p-1}{p}\frac{1}{a+b}\right) - (\kappa_0^b - 1)\left(-\frac{p-1}{p}\frac{a}{(a+b)b}\right),$$

and since $\kappa_0^b - 1 \in pb\mathbb{Z}_{(p)}^\times$, we have to show that $R \in pa\mathbb{Z}_{(p)}$. Now we are calculating in $\mathbb{Q}_p$, setting $a = mb$ with $m \in \mathbb{Z}_p$, $q_0 := \kappa_0^b$ and $\alpha := q_0 - 1$:

$$\begin{aligned}-\frac{p(a+b)}{p-1}R = q_0(q_0^m - 1) - (q_0-1)\frac{a}{b} &= (1-q_0)\left(m - q_0\frac{q_0^m-1}{q_0-1}\right) \\ &= -\alpha\left(m - (1+\alpha)\sum_{i=1}^{\infty}\binom{m}{i}\alpha^{i-1}\right) \\ &= \alpha\left(\sum_{i=2}^{\infty}\binom{m}{i}\alpha^{i-1} + \sum_{i=1}^{\infty}\binom{m}{i}\alpha^i\right) \\ &= \sum_{i=2}^{\infty}\binom{m+1}{i}\alpha^i.\end{aligned}$$

Since $i-2 \geq \mathrm{ord}_p(i!)$ for $i \geq 2$ and $\alpha \in pb\mathbb{Z}_p$, each summand is contained in $p^2ab\mathbb{Z}_p$, and hence R is contained in $\frac{pab}{a+b}\mathbb{Z}_p = pa\mathbb{Z}_p$. □

Now we are ready to discuss the proof of

Theorem 5.1.2. *Assume* $\Gamma = \mathrm{Sl}_2(\mathbb{Z}), n > 0$ *even. Then up to power of* 2 *the nominator* $\delta(n)$ *of* $\zeta(-1-n)$ *and the denominator* $\Delta(n)$ *of the Eisenstein class in* $H^1(\Gamma\backslash\mathbb{H}, \tilde{\mathcal{M}}_n)$ *coincide.*

Proof. The proof goes prime by prime. First we prove that $\Delta(n)|\ \delta(n)$ in $\mathbb{Z}_{(p)}$ for $p > 2$. So we have to prove that $\delta(n)\left\langle T_p^m(\widetilde{C_{0,\infty} \otimes e_\nu}), \mathrm{Eis}_n\right\rangle \in \mathbb{Z}_{(p)}$ for $\nu = 2, \ldots, n-1$. There are different cases:

Case I. $p-1 \nmid \nu+1$, $p-1 \nmid n-\nu+1$. Then the zeta values $\zeta(\nu-n)$ and $\zeta(-\nu)$ are p-adic integral and the claim follows from formula (5.25).

For the remaining cases we set $a = \nu+1$ and $b = n+1-\nu$ and we are assuming $p-1|a$. Then up to a p-adic unit the right side of (5.25) is

$$\begin{aligned} &-\frac{\zeta_{(p)}(1-a)\zeta_{(p)}(1-b)}{\zeta_{(p)}(1-(a+b))} \\ &+\frac{1}{(1+p^{n+1})^m(1-p^{n+1})}\left(\zeta_{(p)}(1-b)+\zeta_{(p)}(1-a)\right). \end{aligned} \tag{5.26}$$

Since $p^{n+1}\zeta_{(p)}(1-b)$ and $p^{n+1}\zeta_{(p)}(1-a)$ are both p-adic integral, this is modulo $\mathbb{Z}_{(p)}$ equal to

$$-\frac{\zeta_{(p)}(1-a)\zeta_{(p)}(1-b)}{\zeta_{(p)}(1-(a+b))} + \zeta_{(p)}(1-b) + \zeta_{(p)}(1-a). \tag{5.27}$$

By combining the first and the third term we get

$$-\frac{\zeta_{(p)}(1-a)\left(\zeta_{(p)}(1-b) - \zeta_{(p)}(1-(a+b))\right)}{\zeta_{(p)}(1-(a+b))} + \zeta_{(p)}(1-b). \tag{5.28}$$

We apply Proposition 5.1.4 to this expression and we get:

Case II. $p-1|\ \nu+1$ and $p-1 \nmid n+1-\nu$. Then the remaining contribution from the boundary $\zeta_{(p)}(1-b)$ is p-integral, as well as the nominator of the fraction in (5.28). So the claim follows. The same works when $p-1 \nmid \nu+1$ and $p-1|\ n+1-\nu$ by exchanging a and b.

Case III. $p-1|\ \nu+1$ and $p-1|n+1-\nu$. By symmetry we may assume that $\mathrm{ord}_p(a) \geq \mathrm{ord}_p(b)$. By Proposition 5.1.4 the first summand in (5.28) is contained in

$$\begin{aligned} -\left(-\tfrac{p-1}{p}\tfrac{1}{a} + \mathbb{Z}_{(p)}\right)\left(-\tfrac{p-1}{p}\left(\tfrac{1}{b} - \tfrac{1}{a+b}\right) + \tfrac{a}{b}\mathbb{Z}_{(p)}\right)\left(-\tfrac{p-1}{p}\tfrac{1}{a+b}\right)^{-1}\left(1+p(a+b)\mathbb{Z}_{(p)}\right) \\ \subseteq \tfrac{p-1}{p}\tfrac{1}{b} + \mathbb{Z}_{(p)}. \end{aligned}$$

Thus it cancels with the remaining contribution from the boundary $\zeta_{(p)}(1-b)$ modulo $\mathbb{Z}_{(p)}$. So the term (5.28) is p-adic integral and this finish the proof of $\Delta(n)|\ \delta(n)$.

The divisibility in the other direction would follow if we can show that there is some ν such that $\zeta(\nu-n)$ and $\zeta(-\nu)$ are both p-adic units. For large n this strategy actually works due to a theorem of Carlitz. We look at the numbers $\zeta(-\nu)\zeta(\nu-n) \mod p$, here we leave out the ν for which $(p-1)|\ \nu+1$ or $(p-1)|\ n+1-\nu$. We claim that there exist indices ν_1 such that $\zeta(-\nu_1)\zeta(\nu_1-n) \not\equiv 0 \mod p$. We call such a pair (ν_1, n) a regular pair, actually regularity only depends on the residue class $\mod (p-1)$ of (ν_1, n). We look at the index of irregularity $i(p)$ of p, this is the number of even numbers k with $2 \leq k \leq p-3$ such that $p|\zeta(1-k) = -\frac{B_k}{k}$. It is expected that this index of irregularity is small if we compare it to p and then the above claim is obvious. But it seems to be difficult -maybe even impossible- to prove an estimate for the index which comes close to the probabilistic expectations.

In his paper [27], L. Carlitz proved a very crude estimate for the index of irregularity. His estimate says that

$$i(p) < \frac{p+3}{4} - \frac{\log(2)}{\log(p)}\frac{p-1}{4},$$

and this implies that $i(p) < \frac{p-3}{4} - 2$ provided $p > 100$.

If we now assume $n > p$ then the estimate above shows that not all the $[\frac{p-3}{4}]$ numbers $\zeta(-\nu)\zeta(\nu-n)$ can be divisible by p, and hence there is a ν_0 such that the p-part of the denominator of $\frac{\zeta(-\nu_0)\zeta(\nu_0-n)}{\zeta(-1-n)}$ is precisely the p-part of the nominator of $\zeta(-1-n)$.

We remove the assumption $p < n$. We need a little maneuver. In Section 3.3.11 we explain how the fact $p^\alpha||\Delta(n)$ is reflected in the structure of the Hecke-module $H^1_{\text{ord}}(\Gamma\backslash\mathbb{H}, \tilde{\mathcal{M}}^\flat_n \otimes \mathbb{Z}_p)$, the number p^α is the order of an extension class. But this order is already visible in the structure of the Hecke-module $H^1_{\text{ord}}(\Gamma\backslash\mathbb{H}, \tilde{\mathcal{M}}^\flat_n \otimes \mathbb{Z}_p/p^k)$, provided k is large enough. We can clearly find an integer $n' > p$ with $n' \equiv n \mod (p-1)p^k$. Then for n' the denominator of the Eisenstein class is also visible in the Hecke module $H^1_{\text{ord}}(\Gamma\backslash\mathbb{H}, \tilde{\mathcal{M}}^\flat_{n'} \otimes \mathbb{Z}/p^k\mathbb{Z})$. As claimed in Section 3.3.6 we prove in [50] that we have an isomorphism of Hecke-modules

$$H^1_{\text{ord}}(\Gamma\backslash\mathbb{H}, \tilde{\mathcal{M}}^\flat_n \otimes \mathbb{Z}/p^k\mathbb{Z}) \xrightarrow{\sim} H^1_{\text{ord}}(\Gamma\backslash\mathbb{H}, \tilde{\mathcal{M}}^\flat_{n'} \otimes \mathbb{Z}/p^k\mathbb{Z}).$$

Hence we see that the Eisenstein classes for n and n' have the same denominators, but by the Kummer congruence the nominators of $\zeta(-1-n)$ and $\zeta(-1-n')$ also coincide for large k. So if the theorem is true for n', it is also true for n. This concludes the proof of Theorem 5.1.2. □

A weaker version of this theorem has been proved by Haberland [42]. I have announced this result in my paper with R. Pink [59]. In his diploma thesis C. Kaiser proved a more general result for some congruence subgroups ([71]). $\Gamma_1(p)$. Very recently H. Bekki and S.Sakamoto [8] gave a very detailed proof.

H. Maennel considered in his doctoral thesis the case of Hilbert modular varieties and applied it to give a new proof of a theorem of Colmez on the residuum of the p-adic zeta-function for totally real number fields, assuming the Leopold conjecture for p [82].

We may also ask for a generalisation of this theorem to groups of higher rank.

Roughly speaking it is our general expectation that there is a connection between the prime factorisation of certain special values of L-functions and denominators of Eisenstein classes.

On the one hand, a first example is discussed my article in [24], where I consider the cohomology of the group $\mathrm{Sp}_2(\mathbb{Z})$ with coefficients in a very specific coefficient system. I make a conjecture about the denominator of an Eisenstein class. But it seems that the method to determine the denominator - integrating the Eisenstein class against a cycle - fails.

On the other hand, we know that the denominators create congruences between eigenvalues of Hecke operators acting on the cohomology of different groups. In the special case above, these congruences have been verified in v.d. Geer's article in [24] for a finite number of Hecke operators. In many other cases the congruences have been checked experimentally - always for some finite number of Hecke operators - [10], [11], but these authors never check the denominators. Of course, it is very likely that we have a denominator if there are congruences. There is abundant experimental evidence for the congruences.

Later, G. Chenevier and J. Lannes proved the congruences for all Hecke operators T_p in some cases [28]. They use algebraic modular forms, i.e. they consider cases where the symmetric space is a point, and if I understand they use some kind of denominator argument.

More recently the congruences have proved in many cases in [1],[2], the authors even propose a strategy to prove the conjecture in greater generality. But as far as I see they do not prove the denominator conjecture,

If we want to check experimentally the conjectures about the denominators, we have to upgrade our program with Gangl, We have to write algorithms for $\mathrm{Sp}_2(\mathbb{Z})$ or some other groups, see also [54]. Several attempts have been made, but this seems to be a very difficult task. The complexes to be dealt with become very big, and also the entries in the matrices will become huge. On the other hand we know some algorithms which compute the L values very effectively (see [35])and hence we can provide some interesting values of ℓ which are not so big and we hope that some mod ℓ version of the algorithm might work. (see Chap. 3).

We address these questions again at the end of the book.

The case B)

We may discuss the denominator issue also in case B). Here the situation is slightly different (see also [7]). We go back to the end of Chapter 4. Even if we start from the integral class $e_{s_1\cdot\lambda}$, we cannot expect that the restriction of the Eisenstein class is integral, simply because the number $c(\lambda)$ may not be integral (see (4.133)). We have to multiply the Eisenstein class by the denominator $\delta(\lambda)$ of $c(\lambda)$, and then the restriction $r(\delta(\lambda)\mathrm{Eis}(e_{s_1\cdot\lambda})) \in H^1(\partial(\Gamma\backslash\mathbb{H}_3), \tilde{\mathcal{M}}_\lambda)$ is integral. This class generates a direct summand in the boundary cohomology and it now plays the role of the class ω_n in case A). I think that now it is reasonable to ask

What is the denominator $\Delta(\lambda)$ of the class $\delta(\lambda)\mathrm{Eis}(e_{s_1\cdot\lambda})$? Does this denominator tell us something about the structure of the cohomology as a Hecke-module?

In our discussion we retain the notations and conventions of 4.2.6., in particular, $\lambda = n_1\gamma_1 + n_2\gamma_2$ and for $4k = n_1 + n_2 + 2$. We modify the period Ω^{4k} by a rational factor N_k such that the array

$$\left\{\dots, \pi^\nu \frac{L(\phi_k, 4k-\nu)}{\Omega^{4k} N_k}, \dots\right\}_{\nu=1,\dots,4k-3}.$$

is an array of coprime integers. To simplify the notation we put

$$L^{\mathrm{ar}}(\phi_k, \nu) := \frac{\pi^{4k-\nu} L(\phi_k, \nu)}{\Omega^{4k} N_k},$$

and call this the *arithmetic part* of the L-value. If now

$$\Delta(\lambda) = \gcd(L^{\mathrm{ar}}(\phi_k, 4k - n_2 - 1), L^{\mathrm{ar}}(\phi_k, 4k - n_2)),$$

then

$$\delta(\lambda) = \frac{1}{\Delta(\lambda)} L^{\mathrm{ar}}(\phi_k, 4k - n_2).$$

This says that the above restriction is equal to

$$r(\delta(\lambda)\mathrm{Eis}(e_{s_1\cdot\lambda})) = \frac{1}{\Delta(\lambda)} L^{\mathrm{ar}}(\phi_k, 4k - n_2) e_{s_1\cdot\lambda} + \frac{1}{\Delta(\lambda)} L^{\mathrm{ar}}(\phi_k, 4k - n_2 - 1) e_{s_2\cdot\lambda}.$$

We want to compute (or better to estimate) the denominator. We apply the techniques from case A), namely testing $\delta(\lambda)\mathrm{Eis}(e_{s_1\cdot\lambda})$ against certain modular symbols (see Section 5.1.4). We have a certain supply of (capped) modular symbols $\mathcal{MS}$, these $\tilde{\mathfrak{z}} \in \mathcal{MS}$ yield homology classes $\tilde{\mathfrak{z}} \in H_1(\Gamma\backslash\mathbb{H}_3, \tilde{\mathcal{M}}_\lambda \otimes \mathbb{Q})$ which have a "small" denominator $\mathfrak{n}(\tilde{\mathfrak{z}})$. Here small means that $\mathfrak{n}(\tilde{\mathfrak{z}})$ is an integer which in its prime factorisation has only primes $p < 4k$ (here it is very desirable to have more precise information). Ignoring contributions from the boundary we can evaluate for $\tilde{\mathfrak{z}} \in \mathcal{MS}$ the integral

$$\int_{\tilde{\mathfrak{z}}} \mathrm{Eis}(e_{s_1\cdot\lambda}) = \frac{\mathcal{L}(\tilde{\mathfrak{z}})}{L^{\mathrm{ar}}(\phi_k, 4k - n_2)},$$

here $\mathcal{L}(\tilde{\mathfrak{z}})$ is a number which is given as an expression in terms of special L-values. We formulate an assumption, which can be verified with very high probability by a quick evaluation of L-values:

For any (not too small) prime ℓ, we can find a $\tilde{\mathfrak{z}} \in \mathcal{MS}$ such that $\ell \not| \mathcal{L}(\tilde{\mathfrak{z}})$.

For the class $\mathfrak{n}(\tilde{\mathfrak{z}})\tilde{\mathfrak{z}}$ we get, by ignoring the boundary terms,

$$\langle \delta(\lambda)\mathrm{Eis}(e_{s_1\cdot\lambda}), \mathfrak{n}(\tilde{\mathfrak{z}})\tilde{\mathfrak{z}} \rangle = \frac{\mathfrak{n}(\tilde{\mathfrak{z}})}{\Delta(\lambda)} \times \text{ unit in } \mathbb{Z}_{(\ell)},$$

and we conclude

$$\ell | \frac{\Delta(\lambda)}{\mathfrak{n}(\tilde{\mathfrak{z}})} \implies \ell | D(\lambda).$$

Hence we see again that primes which divide L-values also may divide denominators of Eisenstein classes.

We call a prime ℓ large (with respect to $4k$) if $\ell > 4k$, otherwise it is small. For large primes we know that they do not divide $\mathfrak{n}(\tilde{\mathfrak{z}})$. A highest weight λ is called unitary if $n_1 = n_2$ $(= 2k-1)$. If this is the case the functional equation (see 4.134) tells us that $L^{\mathrm{ar}}(\phi_k, 4k - n_2 - 1) = 2 \times L^{\mathrm{ar}}(\phi_k, 4k - n_2)$, and hence we get

$$\textit{for a unitary } \lambda \textit{ we have } \Delta(\lambda) = L^{\mathrm{ar}}(\phi_k, 4k - n_2),$$

and since the value of the L^{ar} function will be very large, we see a very large denominator if k becomes large.

But if λ is not unitary we should not expect "large" denominators. The two integers $L^{\mathrm{ar}}(\phi_k, 4k-n_2-1), L^{\mathrm{ar}}(\phi_k, 4k-n_2)$ will become very large if k grows, but they are not entangled by the functional equation. Hence it should be a rare event if a large prime ℓ divides both of them. I produced a table of values $\Delta(\lambda)$ for $k = 1, \dots, 20$, and I did not find any large prime ℓ dividing a $\Delta(\lambda)$ for a non-unitary λ. I found some small primes which divide $\Delta(\lambda)$ but then I did not investigate the influence of $\mathfrak{n}(\tilde{\mathfrak{z}})$, hence I do not know whether they divide the denominator.

This leads us to the second half of the question above.

If λ is not unitary we know that the inner cohomology with rational coefficients is trivial, i.e. we have $H_!^\bullet(\Gamma\backslash\mathbb{H}_3, \tilde{\mathcal{M}}_\lambda \otimes \mathbb{Q}) = 0$. We will see in the next chapter that this phenomenon happens quite frequently (see 6.2.1). More generally we can say that $H_!^\bullet(\Gamma\backslash X, \tilde{\mathcal{M}}_\lambda \otimes \mathbb{Q}) = 0$ if λ is not conjugate self-dual. N. Bergeron and A. Venkatesh [9] proposed to look at the torsion of the cohomology, and they formulated some conjectures which in a certain sense say that the torsion of the cohomology becomes very large. In our special case we have a theorem by W. Müller and F. Rochon which says:

Let $\Gamma \subset \mathrm{Sl}_2(\mathbb{Z}[\mathbf{i}])$ be a torsion-free congruence subgroup, then for any $\epsilon > 0$ we find an $n(\epsilon)$ such that for all $n_1 > n(\epsilon)$ and $\lambda = n_1\gamma_1$ we have

$$(1-\epsilon)e^{\frac{\mathrm{vol}(\Gamma\backslash\mathbb{H}_3)}{2\pi}n_1^2} < \# H^2_{\mathrm{tor}}(\Gamma\backslash\mathbb{H}_3), \tilde{\mathcal{M}}_\lambda) < (1+\epsilon)e^{\frac{\mathrm{vol}(\Gamma\backslash\mathbb{H}_3)}{\pi}n_1^2}$$

This is Thm. 1.11 in [87]. The authors claim that they can also prove a similar result for weights λ where also $n_2 \neq 0$. I believe that the assumption of torsion freeness of Γ is not necessary and I expect that $\# H^2(\Gamma\backslash\mathbb{H}_3), \tilde{\mathcal{M}}_\lambda)_{\mathrm{tors}}$ has some kind of exponential growth in the variable λ even for $\Gamma = \mathrm{Sl}_2(\mathbb{Z}[\mathbf{i}])$. But these methods only give some information about the archimedean size of the torsion, whereas we do not get information about the primes dividing $\# H^2(\Gamma\backslash\mathbb{H}_3, \tilde{\mathcal{M}}_\lambda)_{\mathrm{tors}}$.

Small primes will occur in $\# H^2(\Gamma\backslash\mathbb{H}_3), \tilde{\mathcal{M}}_\lambda)_{\mathrm{tors}}$, even with high multiplicity. This is very plausible because if we apply the same arguments as in Section 3.3.1, we get an analogous statement to Proposition 3.3.1. This implies that

$H^1(\partial(\Gamma\backslash\mathbb{H}_3),\tilde{\mathcal{M}}_\lambda)_{\text{tors}}$ is a finite abelian group, and its order is only divisible by small primes. Then the image of $H^1(\partial(\Gamma\backslash\mathbb{H}_3),\tilde{\mathcal{M}}_\lambda)_{\text{tors}}$ under the boundary operator in the fundamental exact sequence provides a notable contribution to $H^2_c(\Gamma\backslash\mathbb{H}_3,\tilde{\mathcal{M}}_\lambda)_{\text{tors}}$. But if the ℓ-torsion in $H^2_c(\Gamma\backslash\mathbb{H}_3,\tilde{\mathcal{M}}_\lambda)_{\text{tors}}$ is non-zero then it is also non-zero in $H^2(\Gamma\backslash\mathbb{H}_3,\tilde{\mathcal{M}}_\lambda)_{\text{tors}}$.

To get further information we have to analyse the following diagram

$$\begin{array}{ccccc} & & H^1(\partial(\Gamma\backslash\mathbb{H}_3),\tilde{\mathcal{M}}_\lambda)_{\text{ tors}} & & \\ & & \downarrow & & \\ H^1(\Gamma\backslash\mathbb{H}_3,\tilde{\mathcal{M}}_\lambda) & \xrightarrow{r} & H^1(\partial(\Gamma\backslash\mathbb{H}_3),\tilde{\mathcal{M}}_\lambda) & \xrightarrow{\delta_1} & H^2_c(\Gamma\backslash\mathbb{H}_3) \\ \downarrow & & \downarrow\pi_1 & & \\ H^1(\Gamma\backslash\mathbb{H}_3,\tilde{\mathcal{M}}_\lambda)_{\text{ int}} & \xrightarrow{r\text{ int}} & H^1(\partial(\Gamma\backslash\mathbb{H}_3),\tilde{\mathcal{M}}_\lambda)_{\text{ int}} & & \\ & & \cup & & \\ & & \mathbb{Z}[\mathbf{i}](r(\delta(\lambda)\text{Eis}(e_{s_1\cdot\lambda}))) & & \\ & & \cup & & \\ & & \mathbb{Z}[\mathbf{i}](L^{\text{ar}}(\phi_k,4k-n_2)r(\text{Eis}(e_{s_1\cdot\lambda}))) & & \end{array} . \tag{5.29}$$

But let us assume we found a large prime ℓ which divides $\Delta(\lambda)$ for a non-unitary λ. We tensorize the above diagram by the semi-local ring $\mathbb{Z}[\mathbf{i}]\otimes\mathbb{Z}_{(\ell)}$. Then π_1 becomes an isomorphism, hence the class $[\delta(\lambda)\text{Eis}(e_{s_1\cdot\lambda})]$ is an element in $H^1(\partial(\Gamma\backslash\mathbb{H}_3),\tilde{\mathcal{M}}_\lambda)\otimes\mathbb{Z}_{(\ell)}$. Then our assumption above implies that

$$\mathbb{Z}_{(\ell)}[\mathbf{i}](r(\delta(\lambda)\text{Eis}(e_{s_1\cdot\lambda})))/\mathbb{Z}_{(\ell)}[\mathbf{i}](L^{\text{ar}}(\phi_k,4k-n_2)\text{Eis}(e_{s_1\cdot\lambda}))=\mathbb{Z}_{(\ell)}[\mathbf{i}]/\Delta(\lambda)\mathbb{Z}_{(\ell)}[\mathbf{i}].$$

Therefore the boundary operator δ_1 yields an injection

$$\mathbb{Z}_{(\ell)}[\mathbf{i}]/\Delta(\lambda)\mathbb{Z}_{(\ell)}[\mathbf{i}]\hookrightarrow H^2_c(\Gamma\backslash\mathbb{H}_3,\tilde{\mathcal{M}}_\lambda).$$

We have constructed an ℓ-torsion class, which owes its existence to some divisibility properties of special L-values.

I remind the reader that as long as λ is not unitary, for $k\le 20$ I did not find an $\ell|\Delta(\lambda)$ with $\ell>4k$.

The situation changes dramatically if λ is unitary. In this case $\Delta(\lambda)$ will become very large. For example if we take $k=20$ and λ unitary then the two primes

$$\ell=27006373,\ell=12621663529147$$

are divisors of $\Delta(\lambda)$. The map $r_{\text{ int}}$ is not necessarily injective anymore. It can happen that the image of $r_{\text{ int}}$ is larger than $\mathbb{Z}[\mathbf{i}](L^{\text{ar}}(\phi_k,4k-n_2)\text{Eis}(e_{s_1\cdot\lambda}))$. To be more precise we take such a prime ℓ and localise our diagram (5.29) at ℓ, then the element $[r(\text{Eis}(\delta(\lambda)e_{s_1\cdot\lambda}))]\in H^1(\partial(\Gamma\backslash\mathbb{H}_3),\tilde{\mathcal{M}}_\lambda)\otimes\mathbb{Z}_{(\ell)}$ has two options

i) *The boundary operator δ_1 maps it to a non-zero ℓ-torsion class.*

ii) *The boundary operator sends it to zero and hence this class is in the image of r.*

In case ii) we get $H^1_!(\Gamma\backslash\mathbb{H}_3),\tilde{\mathcal{M}}_\lambda)\neq 0$. Our previous discussion of case A) also yields that we can find an eigenclass which is congruent to the Eisenstein class mod ℓ.

Now I believe that the case i) almost never occurs. At this point we should return remorsefully to our algorithm (2.22) and carry out some experimental computations. Actually H. Sengun provides some data in his paper [97] in Section 5.1, where he gives a complete list of large primes which divide $\#H^2(\Gamma\backslash\mathbb{H}_3, \tilde{\mathcal{M}}_\lambda)_{\text{tors}}$ for $k = 1, \dots, 13$ (his large primes may be smaller than ours). A simple computation using Poincaré duality and some simple exact sequences shows that for a large prime

$$\ell | \#H^2(\Gamma\backslash\mathbb{H}_3, \tilde{\mathcal{M}}_\lambda)_{\text{tors}} \iff \ell | \#H^2_c(\Gamma\backslash\mathbb{H}_3, \tilde{\mathcal{M}}_\lambda)_{\text{tors}}.$$

Now we can compare this list with the list of large primes dividing $\Delta(\lambda)$, and we see that these two lists do not have any members in common. This shows that for these few values of k we always have option ii). It would be nice if Sengun's list could be extended up to $k = 20$, his parameter n and our parameter k are related by $2k = n + 1$.

5.2 The Deligne-Eichler-Shimura theorem

In this section I try to explain very briefly some results which are specific for Gl_2 and a few other low dimensional algebraic groups. These results concern representations of the Galois group $\mathrm{Gal}(\bar{\mathbb{Q}}/\mathbb{Q})$ which can be attached to irreducible constituents Π_f in the cohomology. These results are very deep and reaching a better understanding and more general versions of these results is a fundamental task of the subject treated in these notes. The first cases have been tackled by Eichler and Shimura, then Ihara made some contributions and finally Deligne proved a general result for $\mathrm{Gl}_2/\mathbb{Q}$.

We start from the group $G = \mathrm{Gl}_2/\mathbb{Q}$, this is now a reductive group and its centre is isomorphic to $\mathbb{G}_m/\mathbb{Q}$. Its group of real points is $\mathrm{Gl}_2(\mathbb{R})$ and the centre $\mathbb{G}_m(\mathbb{R})$ considered as a topological group has two components, the connected component of the identity is $\mathbb{G}_m(\mathbb{R})^{(0)} = \mathbb{R}^\times_{>0}$. Now we enlarge the maximal compact connected subgroup $\mathrm{SO}(2) \subset \mathrm{Gl}_2(\mathbb{R})$ to the group $K_\infty = SO(2) \cdot \mathbb{G}_m(\mathbb{R})^{(0)}$. The resulting symmetric space $X = \mathrm{Gl}_2(\mathbb{R})/K_\infty$ is now a union of a upper and a lower half plane: We write $X = \mathbb{H}_+ \cup \mathbb{H}_-$.

We choose a positive integer $N > 2$ and consider the congruence subgroup $\Gamma(N) \subset \mathrm{Gl}_2(\mathbb{Q})$. We modify our symmetric space: This modification may look a little bit artificial at this point, it will be justified in the next chapter and is in fact very natural. At this point I want to avoid to use the language of adeles.

We replace the symmetric space by

$$X = (\mathbb{H}_+ \cup \mathbb{H}_-) \times \mathrm{Gl}_2(\mathbb{Z}/N\mathbb{Z}).$$

On this space we have an action of $\Gamma = \mathrm{Gl}_2(\mathbb{Z})$, on the second factor it acts via the homomorphism $\mathrm{Gl}_2(\mathbb{Z}) \to \mathrm{Gl}_2(\mathbb{Z}/N\mathbb{Z})$ by translations from the left. Again we look at the quotient of this space by the action of $\mathrm{Gl}_2(\mathbb{Z})$. This quotient space will have several connected components. The group $\mathrm{Gl}_2(\mathbb{Z})$ contains the group $\mathrm{Sl}_2(\mathbb{Z})$ as a subgroup of index two, because the determinant of an element is ± 1. The element $\begin{pmatrix} 1 & 0 \\ 0 & -1 \end{pmatrix}$ interchanges the upper and the lower half plane and hence we see

$$\mathrm{Gl}_2(\mathbb{Z})\backslash X = \mathrm{Gl}_2(\mathbb{Z})\backslash((\mathbb{H}_+ \cup \mathbb{H}_-) \times \mathrm{Gl}_2(\mathbb{Z}/N\mathbb{Z})) = \mathrm{Sl}_2(\mathbb{Z})\backslash(\mathbb{H}_+ \times \mathrm{Gl}_2(\mathbb{Z}/N\mathbb{Z})),$$

the connected components of $\mathbb{H}_+ \times \mathrm{Gl}_2(\mathbb{Z}/N\mathbb{Z})$ are indexed by the elements of $\mathrm{Gl}_2(\mathbb{Z}/N\mathbb{Z})$. The stabiliser of such a component is the full congruence subgroup

$$\Gamma(N) = \{\gamma = \begin{pmatrix} a & b \\ c & d \end{pmatrix} | a, d \equiv 1 \mod N, b, c \equiv 0 \mod N\},$$

this group is torsion free because we assumed $N > 2$.

The image of the natural homomorphism $\mathrm{Sl}_2(\mathbb{Z}) \to \mathrm{Gl}_2(\mathbb{Z}/N\mathbb{Z})$ is the subgroup $\mathrm{Sl}_2(\mathbb{Z}/N\mathbb{Z})$ (strong approximation), therefore the quotient by this subgroup is $(\mathbb{Z}/N\mathbb{Z})^\times$.

We choose as system of representatives for the determinant the matrices $t_a = \begin{pmatrix} a & 0 \\ 0 & 1 \end{pmatrix}, a \in (\mathbb{Z}/N\mathbb{Z})^\times$. Then we get an isomorphism

$$\mathcal{S}_N = \mathrm{Gl}_2(\mathbb{Z})\backslash(\mathbb{H} \times \mathrm{Gl}_2(\mathbb{Z}/N\mathbb{Z})) \xrightarrow{\sim} (\Gamma(N)\backslash\mathbb{H}) \times (\mathbb{Z}/N\mathbb{Z})^\times.$$

We consider the corresponding cohomology groups with compact support $H^\bullet_c(\mathcal{S}_N, \tilde{\mathcal{M}}_n)$, and the groups $H^\bullet(\mathcal{S}_N, \tilde{\mathcal{M}}_n)$ and $H^\bullet(\partial\mathcal{S}_N, \tilde{\mathcal{M}}_n)$. Again we have the fundamental long exact sequence and we define $H^\bullet_!(\mathcal{S}_N, \tilde{\mathcal{M}}_n)$ as before.

To any prime p which does not divide N we can again attach Hecke operators

$$T_{p^r} = T\left(\begin{pmatrix} p^r & 0 \\ 0 & 1 \end{pmatrix}, u_{\begin{pmatrix} p^r & 0 \\ 0 & 1 \end{pmatrix}}\right).$$

These operators are associated to double cosets, they satisfy a recursion formula, and we have $\mathcal{H}_p = \mathbb{Z}[T_p]$. Thus we have a Hecke algebra $\mathcal{H}_p$ for the primes $p|N$, but this will not be commutative anymore. We get an action of a larger Hecke algebra

$$\mathcal{H}_N^{\text{large}} = \bigotimes_p{}' \mathcal{H}_p.$$

We apply 3.1.1 and find a finite normal extension $F/\mathbb{Q}$ such that we get an isotypical decomposition

$$H^\bullet_!(\mathcal{S}_N, \tilde{\mathcal{M}}_n \otimes F) = \bigoplus_{\pi_f} H^\bullet_!(\mathcal{S}_N, \tilde{\mathcal{M}}_n \otimes F)(\pi_f) \tag{5.30}$$

where $\pi_f = \otimes' \pi_p$, and the π_p are isomorphism types of absolutely irreducible $\mathcal{H}_p$ modules. For $p \not| N$ this $\mathcal{H}_p$-module is a one-dimensional F-vector space $H_{\pi_p} = F$ and π_p is simply a homomorphism $\pi_p : \mathcal{H}_p \to \mathcal{O}_F$. If $p|N$ then the $\mathcal{H}_p$-module is $F^{d(\pi_p)}$ with $d(\pi_p) \geq 1$, and the theory of semi-simple algebras tells us that the map $\mathcal{H}_p \otimes F \to \mathrm{End}_F(H_{\pi_p})$ is surjective. Hence we know the isomorphism type π_p once we know the kernel $I(\pi_p)$ of this map.

Now we have some input from the theory of automorphic forms

Theorem 5.2.1. *The isomorphism type π_f is determined by its restriction to the central subalgebra $\otimes_{p \not| N} \mathcal{H}_p$. Under the action of the group $\pi_0(\mathrm{Gl}_2(\mathbb{R})) = \{\pm 1\}$ it decomposes into two eigenspaces*

$$H^\bullet_!(\mathcal{S}_N, \tilde{\mathcal{M}}_n \otimes F)(\pi_f) = H^\bullet_!(\mathcal{S}_N, \tilde{\mathcal{M}}_n \otimes F)_+(\pi_f) \oplus H^\bullet_!(\mathcal{S}_N, \tilde{\mathcal{M}}_n \otimes F)_-(\pi_f),$$

and these two eigenspaces are absolutely irreducible of type π_f. (These assertions are summarised under "strong multiplicity one")

Of course we have the action of the Galoisgroup $\mathrm{Gal}(F/\mathbb{Q})$ on the cohomology groups $H^\bullet_!(\mathcal{S}_N, \tilde{\mathcal{M}}_n \otimes F)$ and it is clear that this induces an action on the isomorphism types π_f. For $\sigma \in \mathrm{Gal}(F/\mathbb{Q})$ we have

$$\sigma(H^\bullet_!(\mathcal{S}_N, \tilde{\mathcal{M}}_n \otimes F)(\pi_f)) = H^\bullet_!(\mathcal{S}_N, \tilde{\mathcal{M}}_n \otimes F)(\sigma(\pi_f)).$$

I want to discuss some applications.

A) To any isotypical component π_f we can attach an automorphic L-function

$$L(\pi_f, s) = \prod_p L(\pi_p, s)$$

where for $p \not| N$ we define

$$L(\pi_p, s) = \frac{1}{1 - \pi_p(T_p)p^{-s} + p^{n+1}\omega(\pi_p)(p)p^{-2s}}$$

and for $p|N$ we have

$$L(\pi_p, s) = \begin{cases} \frac{1}{1-p^{n+1}\omega(\pi_p)(p)p^{-s}} & \text{if } \pi_p \text{ is a Steinberg module} \\ 1 & \text{else} \end{cases}$$

This L-function, which is defined as an infinite product, is holomorphic for $\Re(s) >> 0$, it can be written as the Mellin transform of a holomorphic cusp form F of weight $n + 2$, and this implies that

$$\Lambda(\pi, s) = \frac{\Gamma(s)}{(2\pi)^s} L(\pi_f, s)$$

has a holomorphic continuation into the entire complex plane and satisfies a functional equation

$$\Lambda(\pi_f, s) = W(\pi_f)N(\pi_f)^{s-1-n/2}\Lambda(\pi_f, n + 2 - s)$$

Here $W(\pi_f)$ is the root number, it can be computed from the π_p where $p|N$, its value is ± 1, the number $N(\pi_f)$ is the conductor of π_f, it is a positive integer, whose prime factors are contained in the set of prime divisors of N.

Now we exploit the fact that the disjoint union of Riemann surfaces $\Gamma(N)\backslash X$ is in fact the space of complex points of the moduli scheme $M_N \to \mathrm{Spec}(\mathbb{Z}[1/N])$. This has been explained at several places in the literature. I refer to the second edition of my book [48] Section 5.2.5 where I try to explain that the functor associating to schemes $S \to \mathrm{Spec}(\mathbb{Z}[1/N])$ isomorphism classes of elliptic curves over S with full level-N structure is representable, provided $N \geq 3$. More precisely we have a smooth quasiprojective scheme $M_N \to \mathrm{Spec}(\mathbb{Z}[1/N])$ with one dimensional fibers and we have the universal elliptic curve

$$\pi : \mathcal{E} \to M_N$$

with full level N-structure (e_1, e_2) where $e_i : M_N \to \mathcal{E}$ are sections which yield a pair of generators of the group of N-division points. The group $\mathrm{Gl}_2(\mathbb{Z}/NZ)$ acts on the group of N-division points, this gives an action of $\mathrm{Gl}_2(\mathbb{Z}/NZ)$ on M_N. We can define the moduli stack $M_1 \to \mathrm{Spec}(\mathbb{Z})$ of elliptic curves without level structure. For any $N \geq 3$ we have $M_1 \times \mathrm{Spec}(Z[\frac{1}{N}]) = M_N/\mathrm{Gl}_2(\mathbb{Z}/N\mathbb{Z})$.

On $\mathcal{E}$ we have the constant ℓ-adic sheaf $\mathbb{Z}_\ell$. For $i = 0, 1, 2$ we can consider the ℓ- adic sheaves $R^i\pi_*(\mathbb{Z}_\ell)$ on M_N. We have the spectral sequence

$$E_2^{p,q} = H^p(M_N \times \bar{\mathbb{Q}}, R^q\pi_*(\mathbb{Z}_\ell)) \Rightarrow H^n(\mathcal{E} \times \bar{\mathbb{Q}}, \mathbb{Z}_\ell).$$

We can take the fibered product of the universal elliptic curve

$$\mathcal{E}^{(n)} = \mathcal{E} \times_{M_N} \mathcal{E} \times \cdots \times_{M_N} \mathcal{E} \xrightarrow{\pi_n} M_N$$

where n is the number of factors. This gives us a more general spectral sequence

$$E_2^{p,q} = H^p(M_N \times \bar{\mathbb{Q}}, R^q\pi_{N,*}(\mathbb{Z}_\ell)) \Rightarrow H^n(\mathcal{E}^{(n)} \times \bar{\mathbb{Q}}, \mathbb{Z}_\ell).$$

The stalk $R^q\pi_{n,*}(\mathbb{Z}_\ell)_y$) of the sheaf $R^q\pi_{n,*}(\mathbb{Z}_\ell)$ in a geometric point y of M_N is the q-th cohomology $H^q(\mathcal{E}_y^{(n)}, \mathbb{Z}_\ell)$, and this can be computed using the Kuenneth formula

$$H^q(\mathcal{E}_y^{(n)}, \mathbb{Z}_\ell) \xrightarrow{\sim} \bigoplus_{a_1+a_2+\cdots+a_n=q} H^{a_1}(\mathcal{E}_y, \mathbb{Z}_\ell) \otimes H^{a_2}(\mathcal{E}_y, \mathbb{Z}_\ell) \cdots \otimes H^{a_n}(\mathcal{E}_y, \mathbb{Z}_\ell),$$

where the $a_i \in \{0, 1, 2\}$. We have $H^0(\mathcal{E}_y, \mathbb{Z}_\ell) = \mathbb{Z}_\ell(0)$, and $H^2(\mathcal{E}_y, \mathbb{Z}_\ell) = \mathbb{Z}_\ell(-1)$, and the most interesting factor is $H^1(\mathcal{E}_y, \mathbb{Z}_\ell)$ which is a free $\mathbb{Z}_\ell$-module af rank 2.

This tells us that the sheaf decomposes into a direct sum according to the type of Kuenneth summands. We also have an action of the symmetric group S_n which is obtained from the permutations of the factors in $\mathcal{E}^{(n)}$ which also permutes the Kuenneth summands. We are mainly interested in the case $q = n$ and then we have the special summand where $a_1 = \cdots = a_n = 1$. This summand is invariant under S_n and contains a summand on which S_n acts by the signature character $\sigma : S_n \to \{\pm 1\}$. This defines a unique subsheaf $R^n\pi_{n,*}(\mathbb{Z}_\ell)(\sigma) \subset R^n\pi_{n,*}(\mathbb{Z}_\ell)$ and hence we get an inclusion

$$H^1(M_N \times \bar{\mathbb{Q}}, R^n\pi_{n,*}(\mathbb{Z}_\ell)(\sigma)) \hookrightarrow H^{n+1}(\mathcal{E}^{(n)} \times \bar{\mathbb{Q}}, \mathbb{Z}_\ell)$$

and we can do the same thing for the cohomology with compact supports.

Now a meticulous analysis of the definitions yields

A) The restriction of the étale sheaf $R^n\pi_{n,*}(\mathbb{Z}_\ell)(\sigma)$ on $M_N \times \mathbb{C}$ to the topological space $\mathcal{S}_N = M_N(\mathbb{C})$ is isomorphic to $\mathcal{M}_n[-n] \otimes \mathbb{Z}_\ell$. Then the comparison theorem gives us

$$H^1(M_N(\mathbb{C}), \mathcal{M}_n[-n] \otimes \mathbb{Z}_\ell) \xrightarrow{\sim} H^1(M_N \times \bar{\mathbb{Q}}, R^n\pi_{n,*}(\mathbb{Z}_\ell)(\sigma))$$

B) The Hecke operators T_p for $p \not| N$ are coming from algebraic correspondences $T_p \subset M_N \times M_N$ and induce endomorphisms

$$T_p : H^1(M_N \otimes \bar{\mathbb{Q}}, R^n\pi_{n,*}(\mathbb{Z}_\ell)(\sigma)) \to H^1(M_N \otimes \bar{\mathbb{Q}}, R^n\pi_{n,*}(\mathbb{Z}_\ell)(\sigma))$$

which commute with the action of $\mathrm{Gal}(\bar{\mathbb{Q}}/\mathbb{Q})$ *on the cohomology.*

This gives us the structure of a $\mathrm{Gal}(\bar{\mathbb{Q}}/\mathbb{Q}) \times \mathcal{H}_\Gamma$*-module on the cohomology group* $H^1(M_N(\mathbb{C}), \mathcal{M}_n[-n] \otimes \mathbb{Z}_\ell)$.

C) The operation of the Galois group on $H^1(M_N(\mathbb{C}), \mathcal{M}_n \otimes \mathbb{Z}_\ell)$ is unramified outside N and ℓ therefore we have the conjugacy class Φ_p^{-1} of the inverse of the Frobenius for all $p \not| N$ as endomorphism of $H^1(M_N(\mathbb{C}), \mathcal{M}_n \otimes \mathbb{Q}_\ell)$.

We choose our normal extension $F/\mathbb{Q}$ and a prime $\mathfrak{l}$ above ℓ. Then an isotypical component $H^1(M_N(\mathbb{C}), \mathcal{M}_n \otimes F_\mathfrak{l})(\pi_f)$ is a Galois module. Let H_{π_f} be a vector space over F which is an irreducible $\mathcal{H}_\Gamma$-module of isomorphism type π_f. Then $W(\pi_f) = \mathrm{Hom}_{\mathcal{H}_\Gamma}(H_{\pi_f} \otimes F_\mathfrak{l}, H^1_!(M_N(\mathbb{C}), \mathcal{M}_n \otimes F_\mathfrak{l}))$ is a Galois module which is unramified outside N and ℓ

We now apply our decomposition (5.30) to the cohomology $H^1_!(M_N(\mathbb{C}), \mathcal{M}_n \otimes Z_\ell)$, as a module under this large Hecke algebra. Then the isotypical summands will be invariant under the Galois group.

Theorem 5.2.2. *(Deligne) For all primes $p \not| N\ell$*

$$tr(\Phi_p^{-1}|W(\pi_f)) = \pi_p(T_p), \det(\Phi_p^{-1}|W(\pi_f)) = p^{n+1}\omega(\pi_p)(p)$$

This theorem is much deeper than the previous ones. This requires some tools from algebraic geometry. We have to consider the reduction $M_N \times \mathrm{Spec}(\mathbb{F}_p)$ and to look at the reduction of the Hecke operator T_p modulo p.

We conclude by giving a few applications.

A) To our modular cusp form $\Delta(z)$ we attach the Hecke L-function

$$L(\Delta, s) = \int_0^\infty \Delta(iy) y^s \frac{dy}{y} = \frac{\Gamma(s)}{(2\pi)^s} \sum_{n=1}^\infty \frac{\tau(n)}{n^s} = \frac{\Gamma(s)}{(2\pi)^s} \prod_p \frac{1}{1 - \tau(p)p^{-s} + p^{11-2s}},$$

the product expansion has been discovered by Ramanujan and has been proved by Mordell and Hecke.

Now it is in any textbook on modular forms that the transformation rule

$$\Delta(-\frac{1}{z}) = z^{12}\Delta(z)$$

implies that $L(\Delta, s)$ defines a holomorphic function in the entire s plane and satisfies the functional equation

$$L(\Delta, s) = L(\Delta, 12 - s).$$

This function $L(\Delta, s)$ is the prototype of an automorphic L-function. The above theorem shows that it is equal to a "motivic" L-function. We gave some vague explanations of what this possibly means: We can interpret the projective system $(\mathcal{M}/\ell^n\tilde{\mathcal{M}})_{et}$ as the ℓ–adic realization of a motive:

$$\mathcal{M} = \mathrm{Sym}^{10} R^1\pi_*\mathbb{Z}.$$

(All this is a translation of Deligne's reasoning into a more sophisticated language.)

It is a general hope that "motivic" L-functions $L(M, s)$ have nice properties as functions in the variable s (meromorphicity, control of the poles, functional equation). So far the only cases, in which one could prove such nice properties are cases where one could identify the "motivic" L-function to an automorphic L function. The greatest success of this strategy is Wiles' proof of the Shimura-Taniyama-Weil conjecture, but also the Riemann ζ-function is a motivic $L-$ function and Riemann's proof of the functional equation follows exactly this strategy.

B) But we also have a flow of information in the opposite direction. In 1973 Deligne proved the Weil conjectures, which in this case say that the two roots of the quadratic equation

$$x^2 - \tau(p)x + p^{11} = 0$$

have absolute value $p^{11/2}$, i.e. they have the same absolute value. This implies the famous Ramanujan conjecture

$$\tau(p) \leq 2p^{11/2}$$

and for more than 50 years this has been a brain-teaser for mathematicians working in the field of modular forms.

The ℓ-adic Galois representation in the first non-trivial case

Again we consider the module $\mathcal{M} = \mathcal{M}_{10}$. We choose a prime ℓ and for some reason let us assume $\ell > 7$. Then we can consider the cohomology groups

$$H^1(\Gamma\backslash\mathbb{H}, \tilde{\mathcal{M}}/\ell^n\tilde{\mathcal{M}})$$

and the projective limit

$$H^1(\Gamma\backslash\mathbb{H}, \tilde{\mathcal{M}}\otimes\mathbb{Z}_\ell) = \lim_{\leftarrow} H^1(\Gamma\backslash\mathbb{H}, \tilde{\mathcal{M}}/\ell^n\tilde{\mathcal{M}}).$$

We know that

$$H^1_{et}(M_1 \times_{\mathrm{Spec}(\mathbb{Z})} \bar{\mathbb{Q}}, (\tilde{\mathcal{M}}/\ell^n\tilde{\mathcal{M}})_{et}) \xrightarrow{\sim} H^1(\Gamma\backslash\mathbb{H}, \tilde{\mathcal{M}}/\ell^n\tilde{\mathcal{M}}).$$

On the étale cohomology groups we have an action of the Galois group, hence we get an action

$$\rho_\ell^{(n)} : \mathrm{Gal}(\bar{\mathbb{Q}}/\mathbb{Q}) \to \mathrm{Gl}(H^1(\Gamma\backslash\mathbb{H}, (\tilde{\mathcal{M}}/\ell^n\tilde{\mathcal{M}})_{et})).$$

From Galois theory we get a finite normal extension $K_\ell^{(n)}/\mathbb{Q}$ which is defined by $\mathrm{Gal}(\bar{\mathbb{Q}}/K_\ell^{(n)}) = \ker(\rho^{(n)})$. The representation $\rho_\ell^{(n)}$ is unramified outside ℓ, and this means that the finite extension $K_\ell^{(n)}/\mathbb{Q}$ is unramified outside ℓ.

From the fundamental exact sequence we get a diagram (where we use the abbreviation $M_\ell := \tilde{\mathcal{M}}_\lambda \otimes \mathbb{Z}_\ell$)

$$\begin{array}{ccccccccc}
 & & 0 & & & & & & \\
 & & \downarrow & & & & & & \\
 & & H^0(\partial(\Gamma\backslash\mathbb{H}), M_\ell) & & & & & & \\
 & & \downarrow & & & & & & \\
0 & \rightarrow & H_c^1(\Gamma\backslash\mathbb{H}, M_\ell) & & & & & & \\
 & & \downarrow & & & & & & \\
0 & \rightarrow & H_!^1(\Gamma\backslash\mathbb{H}, M_\ell) & \rightarrow H^1(\Gamma\backslash\mathbb{H}, M_\ell) & \rightarrow & H^1(\partial(\Gamma\backslash\mathbb{H}), M_\ell) & \rightarrow & 0. \\
 & & \downarrow & & & & & & \\
 & & 0 & & & & & &
\end{array}$$

The vertical and the horizontal sequence are exact sequences of Hecke×Galois-modules. Here we may replace $\mathbb{Z}_\ell$ by $\mathbb{Z}/\ell^n\mathbb{Z}$. We computed these Hecke modules, the cohomology $H^1(\Gamma\backslash\mathbb{H}, \tilde{\mathcal{M}} \otimes \mathbb{Z}_\ell)$ is free of rank 3 and $H^1(\partial(\Gamma\backslash\mathbb{H}), \tilde{\mathcal{M}} \otimes \mathbb{Z}_\ell)$ is free of rank one. We get the two Galois representations

$$\rho_! : \mathrm{Gal}(\bar{\mathbb{Q}}/\mathbb{Q}) \rightarrow \mathrm{Gl}(H_!^1(\Gamma\backslash\mathbb{H}, \tilde{\mathcal{M}} \otimes \mathbb{Z}_\ell)), \text{ and } \rho_\partial : \mathrm{Gal}(\bar{\mathbb{Q}}/\mathbb{Q}) \rightarrow \mathbb{Z}_\ell^\times.$$

The ℓ-adic Tate character $\alpha_\ell : \mathrm{Gal}(\bar{\mathbb{Q}}/\mathbb{Q}) \rightarrow \mathbb{Z}_\ell^\times$ is defined by the following rule: for all $\sigma \in \mathrm{Gal}(\bar{\mathbb{Q}}/\mathbb{Q})$ and all ℓ^n-th roots of unity $\zeta \in \bar{\mathbb{Q}}$ we have $\sigma(\zeta) = \zeta^{\alpha_\ell(\sigma)}$. Then it is not difficult to see (or well-known) that $\rho_\partial = \alpha_\ell^{-11}$. The representation ρ_∂ is the ℓ-adic realisation of the Tate motive $\mathbb{Z}(-11)$. (for a slightly more precise explanation I refer to [56] on my home-page). On $\mathbb{Z}_\ell(-1) = H^2(\mathbb{P}^1 \times \bar{\mathbb{Q}}, \mathbb{Z}_\ell)$ the Galois group acts by the Tate-character α_ℓ^{-1}.

For the representation $\rho_!$ the above theorem of Deligne gives

$$\det(\mathrm{Id} - \rho(\Phi_p^{-1})t | H_!^1(\Gamma\backslash\mathbb{H}, \tilde{\mathcal{M}} \otimes \mathbb{Z}_\ell)) = 1 - \tau(p)t + p^{11}t^2.$$

We also have $\det(\rho(\sigma)) = \alpha_\ell^{-11}(\sigma)$ and we can ask what the image of $\mathrm{Gal}(\bar{\mathbb{Q}}/\mathbb{Q})$ in $\mathrm{Gl}(H_!^1(\Gamma\backslash\mathbb{H}, \tilde{\mathcal{M}} \otimes \mathbb{Z}_\ell)) = \mathrm{Gl}_2(\mathbb{Z}_\ell)$ is. This question is discussed in [109]. If $\ell \neq 691$ then the Hecke algebra induces a splitting (Manin-Drinfeld principle)

$$H^1(\Gamma\backslash\mathbb{H}, \tilde{\mathcal{M}} \otimes \mathbb{Z}_\ell) = H_!^1(\Gamma\backslash\mathbb{H}, \tilde{\mathcal{M}} \otimes \mathbb{Z}_\ell)) \oplus \mathbb{Z}_\ell$$

where T_p acts by multiplication by $p^{11} + 1$ on the second summand.

Now Swinnerton-Dyer showed in [109] that for $\ell \neq 23, 691$ the image of the Galois group under $\rho_!$ is as large possible - it is the inverse image of $(\mathbb{F}_\ell^\times)^{11}$.

From now on we choose $\ell = 691$. Then we get a diagram of Hecke modules (we use the abbreviation $M_{10} := \tilde{\mathcal{M}}_{10} \otimes \mathbb{Z}/\ell\mathbb{Z})$).

$$\begin{array}{ccccccccc}
 & & 0 & & & & & & \\
 & & \downarrow & & & & & & \\
 & & H^0(\partial(\Gamma\backslash\mathbb{H}), M_{10}) & & & & & & \\
 & & \downarrow & & & & & & \\
0 & \rightarrow & H_c^1(\Gamma\backslash\mathbb{H}, M_{10}) & & & & & & \\
 & & \downarrow & & & & & & \\
0 & \rightarrow & H_!^1(\Gamma\backslash\mathbb{H}, M_{10}) & \rightarrow H^1(\Gamma\backslash\mathbb{H}, M_{10}) & \rightarrow & H^1(\partial(\Gamma\backslash\mathbb{H}), M_{10}) & \rightarrow & 0 \\
 & & \downarrow & & & & & & \\
 & & 0 & & & & & &
\end{array}$$

We learned in the previous section that we have an action of $\mathrm{Gal}(\bar{\mathbb{Q}}/\mathbb{Q})$ on this diagram, and this action of the Galois group commutes with the action of the Hecke algebra. The two modules $H^0(\partial(\Gamma\backslash\mathbb{H}), \tilde{\mathcal{M}} \otimes \mathbb{Z}/\ell\mathbb{Z}), H^1(\partial(\Gamma\backslash\mathbb{H}), \tilde{\mathcal{M}} \otimes \mathbb{Z}/\ell\mathbb{Z})$ are isomorphic to $\mathbb{Z}/\ell\mathbb{Z}$ and a Hecke operator T_p acts by the eigenvalue $p^{11}+1 \mod \ell$. The module $H^1_!(\Gamma\backslash\mathbb{H}, \tilde{\mathcal{M}}\otimes\mathbb{Z}/\ell\mathbb{Z}) = \mathbb{Z}/\ell\mathbb{Z}\oplus\mathbb{Z}/\ell\mathbb{Z}$ and the Hecke operator acts by the eigenvalue $\tau(p)$.

The Galois group acts on $H^0(\partial(\Gamma\backslash\mathbb{H}), \tilde{\mathcal{M}} \otimes \mathbb{Z}/\ell\mathbb{Z})$ resp. $H^1(\partial(\Gamma\backslash\mathbb{H}), \tilde{\mathcal{M}} \otimes \mathbb{Z}/\ell\mathbb{Z})$ by α_ℓ^0 resp. α_ℓ^{-11}, where here α_ℓ denotes the reduction of the Tate character mod ℓ. We also know that we have the inclusion of Galois modules

$$j : \mathbb{Z}/\ell\mathbb{Z}(-11) \hookrightarrow H^1_!(\Gamma\backslash\mathbb{H}, \tilde{\mathcal{M}} \otimes \mathbb{Z}/\ell\mathbb{Z}).$$

We are going to determine the two Galois modules $H^1_c(\Gamma\backslash\mathbb{H}, \tilde{\mathcal{M}} \otimes \mathbb{Z}/\ell\mathbb{Z})$ and $H^1(\Gamma\backslash\mathbb{H}, \tilde{\mathcal{M}} \otimes \mathbb{Z}/\ell\mathbb{Z})$. There is a perfect pairing with values in $\mathbb{Z}/\ell\mathbb{Z}(-11)$ between them, hence we have to study only one of them, say the Galois-module $H^1(\Gamma\backslash\mathbb{H}, \tilde{\mathcal{M}} \otimes \mathbb{Z}/\ell\mathbb{Z})$.

From the above considerations it follows that we have a basis e_1, e_0, e_{-1} of this module such that $\sigma \in \mathrm{Gal}(\bar{\mathbb{Q}}/\mathbb{Q})$ acts by the matrix

$$\rho(\sigma) = \begin{pmatrix} \alpha_\ell(\sigma)^{-11} & u_{12}(\sigma) & u_{13}(\sigma) \\ 0 & 1 & u_{23}(\sigma) \\ 0 & 0 & \alpha_\ell(\sigma)^{-11} \end{pmatrix} \in B(\mathbb{Z}/\ell\mathbb{Z}). \tag{5.31}$$

We want to describe the image of the Galois group in $B(\mathbb{Z}/\ell\mathbb{Z})$. Let $T^{(1)}(\mathbb{Z}/\ell\mathbb{Z})$ be the torus

$$\begin{pmatrix} t & 0 & 0 \\ 0 & 1 & 0 \\ 0 & 0 & t \end{pmatrix}; t \in (\mathbb{Z}/\ell\mathbb{Z})^\times$$

and let $U(\mathbb{Z}/\ell\mathbb{Z})$ be the unipotent radical in $B(\mathbb{Z}/\ell\mathbb{Z})$. Then I claim

Theorem 5.2.3. *The image of the Galois group under the map ρ is $B^1(\mathbb{Z}/\ell\mathbb{Z}) := T^{(1)}(\mathbb{Z}/\ell\mathbb{Z}) \ltimes U(\mathbb{Z}/\ell\mathbb{Z})$.*

Proof. For the quotient $B^1(\mathbb{Z}/\ell\mathbb{Z})/U(\mathbb{Z}/\ell\mathbb{Z}) = T^{(1)}(\mathbb{Z}/\ell\mathbb{Z})$, and the resulting map $\mathrm{Gal}(\bar{\mathbb{Q}}/\mathbb{Q}) \to T^{(1)}(\mathbb{Z}/\ell\mathbb{Z})$ is surjective. We have to show that the restriction map $\mathrm{Gal}(\bar{\mathbb{Q}}/\mathbb{Q}(\zeta_\ell)) \to U(\mathbb{Z}/\ell\mathbb{Z})$ is surjective. The center of $U(\mathbb{Z}/\ell\mathbb{Z})$ is the group $U_{1,3}(\mathbb{Z}/\ell\mathbb{Z})$ where $u_{1,2} = u_{2,3} = 0$. Let $V(\mathbb{Z}/\ell\mathbb{Z})$ be the quotient $U(\mathbb{Z}/\ell\mathbb{Z})/U_{1,3}(\mathbb{Z}/\ell\mathbb{Z})$. It suffices to show that $\mathrm{Gal}(\bar{\mathbb{Q}}/\mathbb{Q}(\zeta_\ell)) \to V(\mathbb{Z}/\ell\mathbb{Z})$ is surjective because the commutators of the elements in $V(\mathbb{Z}/\ell\mathbb{Z})$ fill up the elements in $U_{1,3}(\mathbb{Z}/\ell\mathbb{Z})$. Then it becomes clear that it suffices to find a $\sigma \in \mathrm{Gal}(\bar{\mathbb{Q}}/\mathbb{Q}(\zeta_\ell))$ with $u_{1,2}(\sigma), u_{2,3}(\sigma) \not\equiv 0 \mod \ell$, because we still have an action of $T^{(1)}(\mathbb{Z}/\ell\mathbb{Z})$ by conjugation on the image of the Galois group.

Now we apply the Eichler-Shimura congruence relation which says that for any p we have

$$\rho(\Phi_p^{-1})^2 - T_p\rho(\Phi_p^{-1}) + p^{11} Id = 0.$$

We have seen that the Hecke operator with respect to our basis is given by

$$T_p \equiv \begin{pmatrix} p^{n+1}+1 & 0 & t^{(p)} \\ 0 & p^{n+1}+1 & 0 \\ 0 & 0 & p^{n+1}+1 \end{pmatrix} \mod \ell \tag{5.32}$$

(See Theorem 3.3.7 , the roles of ℓ and p are exchanged) and if we are courageous enough to do computations with 3×3 matrices we find for $\sigma = \Phi_p$

$$\begin{pmatrix} 0 & 0 & u_{12}(\Phi_p)u_{23}(\Phi_p) + (-1+p^{11})u_{13}(\Phi_p) - p^{11}t^{(p)} \\ 0 & 0 & 0 \\ 0 & 0 & 0 \end{pmatrix} = 0.$$

Hence the right upper entry must be zero. If $p \equiv 1 \mod 691$ this means that $u_{12}\Phi_p)u_{23}\Phi_p) - p^{11}t^{(p)} = 0$ and this implies: if $t^{(p)} \neq 0$ then both entries $u_{12}(\Phi_p)$ and $u_{23}(\Phi_p) \neq 0$. But we have seen that $t^{(p)} = 0$ implies the stronger congruence $\tau(p) \equiv p^{11}+1 \mod 691^2$ (see Proposition 3.3.2). But now the prime $p_1 = 6911$ is congruent to $1 \mod 691$ but $\tau(p_1)$ is not congruent to $p_1^{11}+1$ modulo 691^2. Hence $u_{12}(\Phi_{p_1}) \neq 0, u_{23}(\Phi_{p_1}) \neq 0$. The claim follows. □

By definition $K_\ell^{(1)}/\mathbb{Q}$ is the normal extension of $\mathbb{Q}$ such that

$$\mathrm{Gal}(K_\ell^{(1)}/\mathbb{Q}) = T^{(1)}(\mathbb{Z}/\ell\mathbb{Z}) \ltimes U(\mathbb{Z}/\ell\mathbb{Z}) = B^{(1)}(\mathbb{Z}/\ell\mathbb{Z}),$$

this extension is unramified outside ℓ. It contains the field of ℓ-th roots of unity, i.e. $\mathbb{Q}(\zeta_\ell) \subset K_\ell^{(1)}$. The Galois group $\mathrm{Gal}(K_\ell^{(1)}/\mathbb{Q}(\zeta_\ell)) = U(\mathbb{Z}/\ell\mathbb{Z})$. This group has a center $U_{13}(\mathbb{Z}/\ell\mathbb{Z}) = \mathbb{Z}/\ell\mathbb{Z}$, this is also the center of the larger group $\mathrm{Gal}(K_\ell^{(1)})/\mathbb{Q}$. We define the subfield $K_\ell^{(1,0)}$ by requiring that $\mathrm{Gal}(K_\ell^{(1,0)}/\mathbb{Q}) = \mathrm{Gal}(K_\ell^{(1)}/\mathbb{Q}(\zeta_\ell))/U_{13}(\mathbb{Z}/\ell\mathbb{Z})$. Then $K_\ell^{(1,0)}/\mathbb{Q}$ is the composite of two cyclic extensions $K_\ell^{(1,!)}/\mathbb{Q}(\zeta_\ell)$ and $K_\ell^{(1,\partial)}/\mathbb{Q}(\zeta_\ell)$. These two extensions have the faithful two-dimensional representations

$$\begin{aligned} \rho_! : \mathrm{Gal}(K_\ell^{(1,!)}/\mathbb{Q}) &\to \mathrm{Gl}(H_!^1(\Gamma\backslash\mathbb{H}, \tilde{\mathcal{M}} \otimes \mathbb{Z}/\ell\mathbb{Z})) \\ \sigma \mapsto \rho_!(\sigma) &= \begin{pmatrix} 1 & u_{23}(\sigma) \\ 0 & \alpha_\ell(\sigma)^{-11} \end{pmatrix} \\ \rho_\partial : \mathrm{Gal}(K_\ell^{(1,\partial)}/\mathbb{Q}) &\to \mathrm{Gl}(H^1(\Gamma\backslash\mathbb{H}, \tilde{\mathcal{M}} \otimes \mathbb{Z}/\ell\mathbb{Z}), \mathbb{Z}/\ell\mathbb{Z}e_1) \\ \sigma \mapsto \rho_\partial(\sigma) &= \begin{pmatrix} \alpha_\ell(\sigma)^{-11} & u_{12}(\sigma) \\ 0 & 1 \end{pmatrix}. \end{aligned} \tag{5.33}$$

We consider the restriction of the action of $\mathrm{Gal}(\bar{\mathbb{Q}}/\mathbb{Q})$ on $H_!^1(\Gamma\backslash\mathbb{H}, \tilde{\mathcal{M}}\otimes\mathbb{Z}_\ell)$ to the local Galois group $\mathrm{Gal}(\bar{\mathbb{Q}}_\ell/\mathbb{Q}_\ell)$. It is a general theorem, that $H_{!,\mathrm{ord}}^1(\Gamma\backslash\mathbb{H}, \tilde{\mathcal{M}}\otimes\mathbb{Z}_\ell)$ is an ordinary Galois module for $\mathrm{Gal}(\bar{\mathbb{Q}}_\ell/\mathbb{Q}_\ell)$, which means in our case that it has a filtration

$$\{(0)\} \subset \mathbb{Z}_\ell(0) \subset H_!^1(\Gamma\backslash\mathbb{H}, \tilde{\mathcal{M}} \otimes \mathbb{Z}_\ell) \text{ with } H_!^1(\Gamma\backslash\mathbb{H}, \tilde{\mathcal{M}} \otimes \mathbb{Z}_\ell)/\mathbb{Z}_\ell(0) = \mathbb{Z}_\ell(-11).$$

Here $\mathbb{Z}_\ell(0) = H_!^1(\Gamma\backslash\mathbb{H}, \tilde{\mathcal{M}} \otimes \mathbb{Z}_\ell)^{I_\ell}$, where I_ℓ is the inertia group. Therefore we get a direct sum decomposition for the $\mathrm{Gal}(\bar{\mathbb{Q}}_\ell/\mathbb{Q}_\ell)$-module

$$H_!^1(\Gamma\backslash\mathbb{H}, \tilde{\mathcal{M}} \otimes \mathbb{Z}/\ell\mathbb{Z}) = \mathbb{Z}/\ell\mathbb{Z}(0) \oplus \mathbb{Z}/\ell\mathbb{Z}(-11),$$

and this implies that $K_\ell^{(1,!)}/\mathbb{Q}(\zeta_\ell)$ is also unramified at ℓ, hence it is an unramified extension.

This unramified extension has been constructed by Ribet in [94]. It has also been constructed in [59]. At the end of [59] we raise the question for a decomposition law. This means that for any prime p we want to find a rule to determine the conjugacy class of $\rho(\Phi_p), \rho_!(\Phi_p), \dots$. If $p \not\equiv 1 \mod \ell$ then the two conjugacy classes $\rho_!(\Phi_p), \rho_\partial(\Phi_p)$ are semi-simple and determined by their eigenvalues. But if $p \equiv 1 \mod \ell$ and $\tau(p) \equiv 2 \mod \ell$ then $\rho_!(\Phi_p)$ is unipotent, and we do not know whether $\rho_!(\Phi_p)$ is the identity or not. If we look at the entire cohomology and not only at the cuspidal part we get something slightly better:

Theorem 5.2.4. *If $p \equiv 1 \mod \ell$ and if the horizontal long exact sequence of $\mathbb{Z}[T_p]$-modules splits (see 3.76), then p splits completely either in the field $K_\ell^{(1,!)}$ or in the field $K_\ell^{(1,\partial)}$.*

If $p \equiv 1 \mod \ell$ and if the horizontal long exact sequence of $\mathbb{Z}[T_p]$-modules does not split then in both field extensions $K_\ell^{(1,!)}/\mathbb{Q}(\zeta_\ell)$ and $K_\ell^{(1,\partial)}/\mathbb{Q}(\zeta_\ell)$ the primes above p are inert at the primes above p.

The density of primes which satisfy the congruences $p \equiv 1 \mod 691$ and $\tau(p) \equiv p^{11} + 1 \mod 691^2$ is bounded below by $\frac{1}{238568}$.

Proof. For the last assertion we invoke Čebotarev's celebrated density theorem ([88], Thm. 13.4). This theorem says that the density of primes for which $p \equiv 1 \mod \ell$, and $u_{12}(\Phi_p) \cdot u_{23}(\Phi_p) = 0$ is equal to the cardinality of the subset of elements of $U(\mathbb{Z}/\ell\mathbb{Z})$ with $u_{12} \cdot u_{23} = 0$ divided by the order of $B^1(\mathbb{Z}/\ell\mathbb{Z})$. By an immediate calculation the density is equal to $\frac{2\ell-1}{(\ell-1)\ell^2}$. □

For the curious reader: the first such prime is $p = 3178601$. We leave it as an exercise for the reader to find out whether it splits completely in $K_\ell^{(1,!)}$ or in $K_\ell^{(1,\partial)}$. It is the 228759 -th prime and this is well matched with the above value for the density of these primes.

Finally we have a brief look at the action of the Galois group on $H^1_!(\Gamma\backslash\mathbb{H}, \tilde{\mathcal{M}} \otimes \mathbb{Z}_\ell)$. Again we choose a basis e_1, e_0, e_{-1}, where the element e_1 maps to a generator in the boundary cohomology, and e_0, e_{-1} form a basis of $H^1_!(\Gamma\backslash\mathbb{H}, \tilde{\mathcal{M}} \otimes \mathbb{Z}_\ell)$. We assume that this basis reduces mod ℓ to the basis which we denoted by the same letters. Then

$$\rho(\sigma) = \begin{pmatrix} \alpha_\ell(\sigma)^{-11} & u_{12}(\sigma) & u_{13}(\sigma) \\ 0 & a(\sigma) & b(\sigma) \\ 0 & \ell c(\sigma) & d(\sigma) \end{pmatrix} \in \mathrm{Gl}_3(\mathbb{Z}_\ell), \tag{5.34}$$

where $a(\sigma) \equiv 1 \mod \ell, d(\sigma) \equiv \alpha^{-11}(\sigma) \mod \ell$.

We claim that there is a σ with $c(\sigma) \not\equiv 0 \mod \ell$.

For a prime p and the Frobenius Φ_p we get $a(\Phi_p^{-1})d(\Phi_p^{-1}) - \ell b(\Phi_p^{-1})c(\Phi_p^{-1}) = p^{11}$ and $\tau(p) = a(\Phi_p^{-1}) + d(\Phi_p^{-1})$. Now a straightforward calculation shows that for a prime $p \equiv 1 \mod \ell$ which in addition satisfies $c(\Phi_p) \equiv 0 \mod \ell$ we must have $\tau(p) \equiv p^{11} + 1 \mod \ell^2$. But $p = 1 + 10 * 691 = 6911$ does not satisfy this congruence, hence $c(\Phi_{6911}) \not\equiv 0 \mod \ell$.

The cohomology $H^1(\Gamma\backslash\mathbb{H}, \tilde{\mathcal{M}} \otimes \mathbb{Z}_\ell)$ has the submodule $H^1_!(\Gamma\backslash\mathbb{H}, \tilde{\mathcal{M}} \otimes \mathbb{Z}_\ell) \oplus \mathbb{Z}_\ell e_1^\dagger$, where $e_1^\dagger = \ell f_{10}^\dagger$ (see (3.58)). This submodule is determined by T_2. Therefore it is invariant under the action of the Galois group with respect to the basis

$e_1^\dagger, e_0, e_{-1}$, the Galois action is given by

$$\rho^\dagger(\sigma) = \begin{pmatrix} \alpha_\ell(\sigma)^{-11} & 0 & 0 \\ 0 & a(\sigma) & b(\sigma) \\ 0 & \ell c(\sigma) & d(\sigma) \end{pmatrix} \in \mathrm{Gl}_3(\mathbb{Z}_\ell), \tag{5.35}$$

where we still have $\alpha_\ell(\sigma)^{-11} = \det \begin{pmatrix} a(\sigma) & b(\sigma) \\ \ell c(\sigma) & d(\sigma) \end{pmatrix}$. It is clear from the above considerations that the image of the Galois group under $\rho^\dagger$ is given by those matrices in $\mathrm{Gl}_3(\mathbb{Z}_\ell)$ which satisfy the conditions above.

But we want to know the image of the Galois group with respect to our basis e_1, e_0, e_{-1}. For this we write $e_1 = \frac{x_0 e_{-1} + e_1^\dagger}{\ell}$ and then clearly

$$\rho(\sigma) = \begin{pmatrix} \alpha_\ell(\sigma)^{-11} & c(\sigma)x_0 & \frac{d(\sigma)-\alpha_\ell(\sigma)^{-11}}{\ell} x_0 \\ 0 & a(\sigma) & b(\sigma) \\ 0 & \ell c(\sigma) & d(\sigma) \end{pmatrix} \in \mathrm{Gl}_3(\mathbb{Z}_\ell).$$

We put

$$a(x_0, \sigma) = \alpha_\ell^{11}(\sigma)(u_{12}(x_0,\sigma)e_0 + u_{13}(x_0,\sigma)e_{-}1),$$

then $\sigma \mapsto a(x_0, \sigma)$ is a 1-cocycle with values in $H^1_!(11) := H^1_!(\Gamma\backslash\mathbb{H}, \tilde{\mathcal{M}} \otimes \mathbb{Z}_\ell) \otimes \mathbb{Z}_\ell(11)$. We can compute its cohomology class $[\underline{v}] \in H^1(\mathrm{Gal}(\bar{\mathbb{Q}}/\mathbb{Q}, H^1_!(11))$. We start from the exact sequence of Galois modules

$$0 \to H^1_!(11) \to H^1(11) \to \mathbb{Z}_\ell(0) \to 0,$$

This yields a long exact sequence in Galois cohomology. The element e_1 yields a generator in $H^0(\mathrm{Gal}(\bar{\mathbb{Q}}/\mathbb{Q}), \mathbb{Z}_\ell(0))$ and clearly $\delta(\tilde{e_1}) = \underline{v}$.

Now we can say that the image of the Galois group under ρ consists of the matrices

$$\{\begin{pmatrix} x & u_{12} & u_{13} \\ 0 & a & b \\ 0 & c & d \end{pmatrix} \mid ad - bc = x; \ell u_{12} = c; \ell u_{13} = d - x\} \subset \mathrm{Gl}_3(\mathbb{Z}_\ell).$$

It is the cocycle condition which makes this into a subgroup. Now it is not difficult to see that this is the group of $\mathbb{Z}_\ell$-valued points of a smooth group scheme $\mathcal{I}^{(1)}/\mathbb{Z}_\ell$ with a map to $\mathrm{Gl}_2/\mathbb{Z}_\ell$, which is an isomorphism in the generic fibre.

We can say that we have constructed a Galois-extension $K_\ell^{(\infty)}/\mathbb{Q}$ which is unramified outside ℓ, and we have an isomorphism

$$\rho_\ell : \mathrm{Gal}(K_\ell^{(\infty)}/\mathbb{Q}) \xrightarrow{\sim} \mathcal{I}^{(1)}(\mathbb{Z}_\ell).$$

We also consider the finite extensions with $\mathrm{Gal}(K_\ell^{(r)}/\mathbb{Q}) \xrightarrow{\sim} \mathcal{I}^{(1)}(\mathbb{Z}/\ell^r\mathbb{Z})$ and for $r = 1$ we have $\mathcal{I}^{(1)}(\mathbb{Z}/\ell\mathbb{Z}) = B^{(1)}(\mathbb{Z}/\ell\mathbb{Z})$.

Here we see the prototype of a very general strategy to get insight into the structure of the Galois group $\mathrm{Gal}(\bar{\mathbb{Q}}/\mathbb{Q})$. We see that values of certain L-functions at certain special arguments (in this case the Riemann $\zeta(s)$-function at $s = -11$) has influence on the structure of certain cohomology groups of arithmetic groups, and these Hecke modules contain some information on the structure of the Galois group. We can construct certain interesting extensions of $\mathbb{Q}$ with controlled ramification. On the other hand we see that there is also a flow of information from the Galois side back to the structure of the Hecke modules.

This connection between the congruence for the values of $\tau(p)$ and the structure of the Galois group was observed by Serre in his paper [101] and proved later by Deligne [31]. Later this relationship was exploited by Ribet in [94] and by many other authors.

Here I want to point out that there is a change of paradigm between our approach and the approach by the authors mentioned above. These authors mostly look at the space of holomorphic cusp forms, what essentially amounts to the inner cohomology $H^1_!(\Gamma\backslash\mathbb{H}, \tilde{\mathcal{M}}_n \otimes \mathbb{C})$ (by Eichler-Shimura isomorphism). Then these authors get the consequences for the structure of the Galois group from the congruences.

In our approach we consider the module $H^1(\Gamma\backslash\mathbb{H}, \tilde{\mathcal{M}}_n)$ and exploit our knowledge of the denominator of the Eisenstein class.

We explain this approach for a general choice of ℓ, we also keep an eye on the computational aspects. We start from any irregular prime ℓ and an even integer $n > 0$ such that $\ell;||\,\zeta(-1-n)$. In Section 3.3.8 we introduced the Hecke modules $H^1(\Gamma\backslash\mathbb{H}, \tilde{\mathcal{M}}_n \otimes \mathbb{Z}/\ell)\{\bar{\pi}_f^{\mathrm{Eis}}\}$ or $H^1(\Gamma\backslash\mathbb{H}, \tilde{\mathcal{M}}_n \otimes \mathbb{Z}_\ell)\{\bar{\pi}_f^{\mathrm{Eis}}\}$. Deligne's theorem tells us that we have an action of the Galois group on these modules. This action commutes with the action of the Hecke algebra $\mathcal{H}$. More precisely, there is a finite normal extension $\mathbb{Q}_{\mathrm{Eis}}(\mathcal{M}_n \otimes \mathbb{F}_\ell)/\mathbb{Q}$ and an injective homomorphism

$$\rho_{\ell,1} : (\mathrm{Gal}(\mathbb{Q}_{\mathrm{Eis}}(\mathcal{M}_n \otimes \mathbb{F}_\ell)/\mathbb{Q})) \to \mathrm{Gl}_{\mathcal{H}}(H^1(\Gamma\backslash\mathbb{H}, \tilde{\mathcal{M}}_n \otimes \mathbb{Z}/\ell)\{\bar{\pi}_f^{\mathrm{Eis}}\}).$$

We can replace ℓ by higher powers ℓ^r and get representations $\rho_{\ell,r}$, and in the limit we get an injection

$$\rho_\ell : \mathrm{Gal}(\mathbb{Q}_{\mathrm{Eis}}(\mathcal{M}_n \otimes \mathbb{Z}_\ell)/\mathbb{Q}) \to \mathrm{Gl}_{\mathcal{H}}(H^1(\Gamma\backslash\mathbb{H}, \tilde{\mathcal{M}}_n \otimes \mathbb{Z}_\ell)\{\bar{\pi}_f^{\mathrm{Eis}}\}).$$

This is the representation in the above theorem of Deligne. This theorem of Deligne also asserts that the extensions $\mathbb{Q}_{\mathrm{Eis}}(\mathcal{M}_n \otimes \mathbb{Z}_\ell)/\mathbb{Q}$ is unramified outside ℓ.

We want to understand the image We make some further assumptions, for instance we assume that $m(\{\bar{\pi}_f^{\mathrm{Eis}}\}) = 1$, We expect it to be true most of the times, but we know that this is not always true (see Section 3.3.8). Assuming that $m(\{\bar{\pi}_f^{\mathrm{Eis}}\}) = 1$, then we can proceed as in our toy example.

We know that $H^1(\Gamma\backslash\mathbb{H}, \tilde{\mathcal{M}}_n \otimes \mathbb{Z}_\ell)\{\bar{\pi}_f^{\mathrm{Eis}}\}$ is a free $\mathbb{Z}_\ell$-module of rank 3. We can find generators e_1, e_0, e_{-1} of this module such that

a) $(\mathbb{Z}/\ell^\delta)e_{-1}$ is the submodule in Theorem 3.3.1.

b) the two elements e_{-1}, e_0 generate $H^1_!(\Gamma\backslash\mathbb{H}, \tilde{\mathcal{M}}_n \otimes \mathbb{Z}_\ell)\{\bar{\pi}_f^{\mathrm{Eis}}\}$.

We recall that with respect to this basis, the Hecke operator $T_p \mod \ell$ is of the form

$$T_p = \begin{pmatrix} p^{n+1}+1 & 0 & t^{(p)} \\ 0 & p^{n+1}+1 & 0 \\ 0 & 0 & p^{n+1}+1 \end{pmatrix} \mod \ell^\delta.$$

In our example above we needed the assumption

$$\textit{We can find a prime } p_1 \equiv 1 \mod \ell \textit{ such that } t^{(p_1)} \not\equiv 0 \mod \ell. \tag{5.36}$$

This assumption is almost certainly always true, but I do not have a proof. If this condition is fulfilled then we can conclude that $u_{12}(\Phi_{p_1}^{-1})$ and $u_{23}(\Phi_{p_1}^{-1})$ are both non zero. This means: If we restrict our representation to $\mathrm{Gal}(\bar{\mathbb{Q}}/\mathbb{Q}(\zeta_\ell))$ we get to nonzero homomorphism

$$u'_{12}, u'_{23} : \mathrm{Gal}(\bar{\mathbb{Q}}/\mathbb{Q}(\zeta_\ell)) \to \mathbb{Z}/\ell\mathbb{Z}.$$

These two homomorphisms give rise to cyclic field extensions $K^{1,!}/\mathbb{Q}(\zeta_\ell)$ and $K^{1,\partial}/\mathbb{Q}(\zeta_\ell)$ of degree ℓ. If now $\alpha_\ell^{-1-n} \neq \alpha_\ell^{1+n}$, i.e. $\ell - 1 \nmid 2n+2$, then these two homomorphisms are independent and this implies that the homomorphism

$$\mathrm{Gal}(\bar{\mathbb{Q}}/\mathbb{Q}(\zeta_\ell)) \to U(\mathbb{Z}/\ell\mathbb{Z})/U_{13}(\mathbb{Z})$$

is surjective. But we can look at commutators and produce non zero elements in $U_{13}(\mathbb{Z}/\ell\mathbb{Z})$ and see that $\mathrm{Gal}(\bar{\mathbb{Q}}/\mathbb{Q}(\zeta_\ell)) \to U(\mathbb{Z}/\ell\mathbb{Z})$ is surjective. If now $d = gcd(n+1, \ell-1)$ and

$$\mathcal{I}^{(d)}(\mathbb{Z}/\ell) = \left\{ \begin{pmatrix} x^d & u_{1,2} & u_{1,3} \\ 0 & 1 & u_{2,3} \\ 0 & 0 & x^d \end{pmatrix} \mid x \in (\mathbb{Z}/\ell)^\times, u_{ij} \in \mathbb{Z}/\ell \right\}.$$

Using the same arguments as in the proof of Thm. 5.2.4 we get an isomorphism

$$\mathrm{Gal}(\mathbb{Q}_{\mathrm{Eis}}(\mathcal{M}_n \otimes \mathbb{Z}/\ell)/\mathbb{Q}) \xrightarrow{\sim} \mathcal{I}^{(d)}(\mathbb{Z}/\ell)$$

provided we have (5.36) and $(\ell - 1) \nmid 2(n+1)$.

For given values of n and ℓ we can verify (5.36), but this verification might be very expensive bcause we have the consider primes $p_1 \equiv 1 \mod \ell$. Already in the toymodel we needed the tables for $\tau(p)$ provided by Mathematica.

Now we will show that this result provides an indication on how to verify (5.36). A direct computation shows that the function

$$\begin{aligned} t : \mathcal{I}^{(d)}(\mathbb{Z}/\ell) &\to \mathbb{F}_\ell \\ \gamma &\mapsto u_{1,2}u_{2,3} + (-1 + x^d)u_{1,3} \end{aligned}$$

is constant on the conjugacy classes in $\mathcal{I}^{(d)}$. Again we invoke the Chebotareff density theorem. For $\gamma \in \mathcal{I}^{(d)}(\mathbb{Z}/\ell)$ we denote its conjugacy class by C_γ and its centraliser by Z_γ. Then the theorem says that the density of primes p for which $\Phi_p \in C_\gamma$ is $1/\#Z_\gamma$. I leave to the reader to analyse the conjugacy classes and to check that

$$d(\{p \mid \Phi_p \in C_\gamma \, ; \, t(\gamma) \neq 0\}) = \ell d(\{p \mid \Phi_p \in C_\gamma \, ; \, t(\gamma) = 0\})$$

(this can easily be made more precise). But this suggests that the probability to find $t(\Phi_p) = t^{(p)} = 0$ is roughly $\frac{1}{\ell}$, hence very small. So we must be really very unlucky if we do not find a $t^{(p)} \neq 0$ for some very small prime $p = 2, 3$. Since we can write algorithms which compute T_p for small values of p, we should be able to verify this in practice (in a given case). In the six cases $n = 10, 14, 16, 18, 20, 24$

the Hecke module $H^1_!(\Gamma\backslash\mathbb{H},\tilde{\mathcal{M}}_n)$ is of rank 2 and for any $\ell|\zeta(-1-n)$ our program with Gangl yields $t^{(2)} \not\equiv 0 \mod \ell$, hence T_2 suffices to verify (5.36) in a few seconds in these cases.

But we have to find primes $p \equiv 1 \mod \ell$ with $t^{(p)} \not\equiv 1 \mod \ell$. In [75] the authors prove effective versions of the Chebotareff density theorem, and using their results we can show that there is a computable constant $c_0(n,\ell)$ such that we find a $p < c_0(n,\ell)$ which satisfies (5.36) if there is any. But the constant is huge and not of practical use. Actually already the condition $p \equiv 1 \mod \ell$ forces us to use large Hecke operators.

I want to show that there is an elegant way to verify (5.36) in a given case, which with very high probability also works in practice.

We look at the (two) homomorphisms

$$u'_{1,2} \times u'_{2,3} : \mathrm{Gal}(\bar{\mathbb{Q}}/\mathbb{Q}(\zeta_\ell)) \to \mathbb{F}_\ell \times \mathbb{F}_\ell.$$

In our considerations above the decisive step was to verify that the image is 2-dimensional, i.e. this homomorphism is surjective. Assume it is not surjective. This is only possible if there exists a $\lambda \in \mathbb{F}_\ell$ such that $u'_{1,2} = \lambda u'_{2,3}$ or $\lambda u'_{1,2} = u'_{2,3}$. In any case it is clear that the image of $\mathrm{Gal}(\bar{\mathbb{Q}}/\mathbb{Q}(\zeta_\ell))$ lies in an abelian group $U_\lambda(\mathbb{F}_\ell) \times U_{13}(\mathbb{F}_\ell)$. Then we get for the larger Galois group

$$\rho_{\ell,1} : \mathrm{Gal}(\bar{\mathbb{Q}}/\mathbb{Q}) \to \left(T^{(1)}(\mathbb{F}_\ell)^d)\right) \ltimes (U_\lambda(\mathbb{F}_\ell)) \times U_{13}(\mathbb{F}_\ell)).$$

Our goal is to rule out that this happens simply by checking a few small Hecke operators $T_2, T_3, \dots T_{p_0}$.

The nasty case is $\lambda \neq 0$. We have seen already that $\lambda \neq 0$ can only happen if $\ell - 1 \mid 2(n+1)$. Let assume that even $n+1 = \frac{\ell-1}{2}$ and $\ell \mid \zeta(\frac{1-\ell}{2})$. This is an extremely unlikely event, the probability for this to happen is $1/\ell$. It never happens for $\ell < 10^5$.

Again we invoke the congruence relation

$$\lambda u_{12}(\Phi_p)^2 + (-1 + p^{n+1})u_{13}(\Phi_p) - p^{n+1}t^{(p)} = 0.$$

Now we may try to compute some $t^{(p_i)}$ for some small primes $p_1, \dots, p_r$ for which in addition p_i is a square $\mod \ell$. Then $p_i^{n+1} \equiv 1 \mod \ell$, and we get $\lambda u_{12}(\Phi_{p_i})^2 = t^{(p_i)}$. Hence $t^{(p_i)}/\lambda$ must be a square $\mod \ell$, and this implies that the ratios $t^{(p_1)}/t^{(p_i)}$ must be squares. With a little bit of luck we will quickly find an index i for which this is not the case, provided we still can compute the $t^{(p)}$ for such a large value of n.

We consider the case $\lambda = 0$ or $\lambda = \infty$. In this case we have a very elegant procedure which in a given case rules this possibility out. We see that for one of the two entries $u_{12} = 0$ or $u_{23} = 0$ and hence the Eichler-Shimura congruence relation gives us (in both cases)

$$(-1 + p^{n+1})u_{13}(\Phi_p) = p^{n+1}t^{(p)}. \tag{5.37}$$

On the other hand it is clear that under our assumptions, the map

$$\begin{aligned} &u_{1,3} : \mathrm{Gal}(\mathbb{Q}_{\mathrm{Eis}}(\mathcal{M}_n \otimes \mathbb{F}_\ell)/\mathbb{Q}) \to \mathbb{F}_\ell \\ &\sigma \mapsto u_{1,3}(\sigma) \end{aligned}$$

is a homomorphism. We accordingly define the subfield $L \subset \mathbb{Q}_{\mathrm{Eis}}(\mathcal{M}_n \otimes \mathbb{F}_\ell)$ by $\mathrm{Gal}(\mathbb{Q}_{\mathrm{Eis}}(\mathcal{M}_n \otimes \mathbb{F}_\ell)/L) = \ker(u_{1,3})$. Then we get a homomorphism

$$\bar{u}_{1,3} : \mathrm{Gal}(L/\mathbb{Q}) \hookrightarrow \mathbb{F}_\ell.$$

Here either $L = \mathbb{Q}$ or $L/\mathbb{Q}$ is a cyclic extension of order ℓ which is unramified outside ℓ. Then class field theory tells us that $L/\mathbb{Q}$ is actually the unique cyclic subfield of degree ℓ in $\mathbb{Q}(\zeta_{\ell^2})$. The Galois group $\mathrm{Gal}(L/\mathbb{Q}) = \mathbb{F}_\ell$, and for any prime $p \neq \ell$ the Frobenius is given by a number $v_\ell(p)$ in $\mathbb{F}_\ell$, this number is defined by

$$p^{\ell-1} = 1 + \ell v_\ell(p) \mod \ell^2.$$

Hence there is a number $\lambda_{n,\ell} \in \mathbb{F}_\ell$ such that for all $p \neq \ell$

$$u_{1,3}(\Phi_p) = \lambda_{n,\ell} v_\ell(p)$$

Now we assume that we found a prime $p_0 (\neq \ell)$ such that $t^{(p_0)} \neq 0 \mod \ell$, then we get for any other prime p the relation

$$\frac{t^{(p)}}{t^{(p_0)}} = \frac{(1-p^{-n-1})v_\ell(p)}{(1-p_0^{-n-1})v_\ell(p_0)} \mod \ell \tag{5.38}$$

We look at some other small primes and check the relation (5.38). Almost certainly we will find a small prime such that (5.38) fails and we have verified that our Hecke module is not degenerated. For small values of n, ℓ this should be rather effective. We must be extremely unlucky if can not rule out the regular behavior of $p \mapsto t^{(p)} \mod \ell$. For the first six cases $n = 10, 14, 16, 18, 20, 24$, we checked that (5.38) fails to be true.

Here we have just discussed the unramified cohomology, it is interesting enough. We did not allow ramification because we wanted to avoid the discussion of the arithmetic properties of local intertwining operators at the ramified primes. I explain briefly what should happen if we have ramification.

We resume the consideration from the end of Chapter 3. Our arithmetic group $\mathrm{Sl}_2(\mathbb{Z})$, is denoted by Γ_0, and we pick an integer $N > 1$ and a character $\chi : (\mathbb{Z}/N\mathbb{Z})^\times \to \mathbb{C}^\times$. Let $\Gamma(N)$ be the full congruence subgroup and $\Gamma_0(N)$ the subgroup with $c \equiv 0 \mod N$. Then χ can be viewed as a character $\chi : \Gamma_0(N) \to \mathbb{C}^\times$. Let $F = \mathbb{Q}[\chi((\mathbb{Z}/N\mathbb{Z})^\times)]$ be the field which is generated by the values of the character χ. Now we pick a prime ℓ, and a prime $\mathfrak{l} \subset \mathcal{O}_F$ which divides ℓ. Let $F_\mathfrak{p}$ be the completion of F at $\mathfrak{l}$, let $\mathcal{O}_\mathfrak{l}$ be the ring of integers in $F_\mathfrak{p}$. We put $\mathbb{F}_\mathfrak{l} = \mathcal{O}_\mathfrak{l}/\mathfrak{l}$, We consider the induced representation

$$\hat{\mathcal{V}}_\chi := \{f : \mathrm{Sl}_2(\mathbb{Z}) \to \mathcal{O}_\ell \mid f(bg) = \chi(b)f(g)\ \forall\, b \in \Gamma_0(N)\}$$

and we get the exact sequence in cohomology

$$H^1_!(\Gamma_0\backslash\mathbb{H}, \widetilde{\mathcal{M}_n \otimes \hat{\mathcal{V}}_\chi}) \to H^1(\Gamma_0\backslash\mathbb{H}, \widetilde{\mathcal{M}_n \otimes \hat{\mathcal{V}}_\chi}) \xrightarrow{r} H^1(\partial(\Gamma_0\backslash\mathbb{H}), \widetilde{\mathcal{M}_n \otimes \hat{\mathcal{V}}_\chi}).$$

This is now an exact sequence of Hecke $\times$ Galois-modules. The Hecke algebra will not be commutative anymore, the local factors $\mathcal{H}_p$ for $p|N$ are "mildly" non-commutative.

Now in a first step we investigate the structure $H^1(\partial(\Gamma_0\backslash\mathbb{H}),\widetilde{\mathcal{M}_n\otimes\hat{\mathcal{V}}_\chi})$ as Hecke $\times$ Galois-module.

If $\pi_{f,\chi}$ is an irreducible direct summand of $H^1(\partial(\Gamma_0\backslash\mathbb{H}),\widetilde{\mathcal{M}_n\otimes\hat{\mathcal{V}}_\chi})$, and $\bar{\pi}_{f,\chi}$ its reduction $\mod \mathfrak{l}$, then we are in the position to define the generalised eigenspace $H^1(\Gamma_0\backslash\mathbb{H},\widetilde{\mathcal{M}_n\otimes\hat{\mathcal{V}}_\chi}\otimes\mathbb{F}_\mathfrak{l})\{\bar{\pi}_{f,\chi}\}$. and the module (see 3.92) $H^1(\Gamma_0\backslash\mathbb{H},\widetilde{\mathcal{M}_n\otimes\hat{\mathcal{V}}_\chi})\{R_\mathfrak{l}^{-1}(\bar{\pi}_{f,\chi})\})$. We get the exact sequence

$$H^1_!(\Gamma_0\backslash\mathbb{H},\widetilde{\mathcal{M}_n\otimes\hat{\mathcal{V}}_\chi})\{R_\mathfrak{l}^{-1}(\bar{\pi}_{f,\chi})\}\to H^1(\Gamma_0\backslash\mathbb{H},\widetilde{\mathcal{M}_n\otimes\hat{\mathcal{V}}_\chi}))\{R_\mathfrak{l}^{-1}(\bar{\pi}_{f,\chi})\}\to$$
$$\to H^1(\partial(\Gamma_0\backslash\mathbb{H}),\widetilde{\mathcal{M}_n\otimes\hat{\mathcal{V}}_\chi}))\{\bar{\pi}_{f,\chi}\}\to 0 \tag{5.39}$$

This sequence may not split, it splits if we tensorise by $F_\mathfrak{l}$. Hence we see a denominator $\Delta(\bar{\pi}_{f,\chi})$ of this splitting.

Now we have to prove a accurate extension of Theorem 5.1.2. to this case. Here I mean by accurate the following: We want to know that $\mathfrak{l}\mid L(\chi,-1-n)$ implies $\mathfrak{l}\mid\Delta(\bar{\pi}_{f,\chi})$, but there may be some exceptions for small primes ℓ. This set of exceptional primes should be kept as small as possible.

We know that $\mathfrak{l}\mid\Delta(\bar{\pi}_{f,\chi})$ implies that the multiplicity $m_!(\bar{\pi}_{f,\chi})>0$ and we can look for examples where this multiplicity is >1. If ℓ is not too large we have some chances to find such examples. Again we can test the weak multiplicity one conjecture, a counterexample would be very interesting.

Let $K/F_\mathfrak{l}$ be a normal finite extension such that we have a decomposition into absolutely irreducible Hecke-modules

$$H^1_!(\Gamma_0\backslash\mathbb{H},\widetilde{\mathcal{M}_n\otimes\hat{\mathcal{V}}_\chi}\otimes K)\{R_\mathfrak{l}^{-1}(\bar{\pi}_{f,\chi})\}=\bigoplus_{\pi_f\in R_\mathfrak{l}^{-1}(\bar{\pi}_{f,\chi})}H(\pi_f). \tag{5.40}$$

The Galois group $\mathrm{Gal}(K/F_\mathfrak{l})$ is acting on $R_\mathfrak{l}^{-1}(\bar{\pi}_{f,\chi})$ and according to the Maeda conjecture we can ask whether or not this action is transitive. We discussed this issue in the unramified case where we had this one example with $\ell=547$. In this case we have multiplicity 2 and in Section 3.3.10 we performed some - maybe very speculative - computations and found that we always have weak multiplicity one. (Remember that K depends on the number $n=484+\alpha\cdot 546 \mod 547^2$)

Finally we have to study the interaction between the Galois module and the Hecke module structure. We can ask for an extension of the Theorem 5.2.3 to this more general situation. Here we should also consider the case of higher multiplicity, this is not carried out in the unramified example.

Of course there is a much greater challenge waiting for us, we have to generalise our considerations to higher dimensional groups. I mentioned already that for groups of higher rank our method to determine the denominator does not work anymore, we do not have Theorem 5.1.2 in higher rank. But - at least in principle - we can compute the denominator if we write an algorithm which computes the cohomology and some Hecke operators $T_2^\chi,T_3^\chi\dots$. Here it would be already nice if we could transfer the approach we made for $\ell = 691, \mathcal{M}_{10}$ in Chapter 3 to the case $\ell = 41, \mathcal{M}_{4\gamma_1+7\gamma_2}$ in [51]. I also refer to my paper [54].

Once we have algorithms which work in the special case in [51] we can try to say something about the size of the Galois group action on $H^3(\Gamma_0\backslash\mathbb{H}_2, \mathcal{M}_{4\gamma_1+7\gamma_2} \otimes \mathbb{F}_{41})$ We have to transfer the considerations following Theorem 5.2.3 to this case.

Chapter 6

Analytic Methods

6.1 Representation theory and the de-Rham complex

6.1.1 Rational representations

We start from a reductive group $G/\mathbb{Q}$. We assume for simplicity that the semi-simple component $G^{(1)}/\mathbb{Q}$ is quasisplit. There is a unique minimal finite normal extension $F/\mathbb{Q}, F \subset \mathbb{C}$, such that $G^{(1)} \times_{\mathbb{Q}} F$ becomes split. If $T^{(1)}/\mathbb{Q} \subseteq G^{(1)}/\mathbb{Q}$ is a maximal torus which is contained in a Borel subgroup $B/\mathbb{Q}$ we let $U/\mathbb{Q}$ be its unipotent radical. Then the Galois group $\mathrm{Gal}(\bar{\mathbb{Q}}/\mathbb{Q})$ acts on $X^*(T^{(1)} \times_{\mathbb{Q}} F)$. It acts by permutations on the set of positive roots $\pi_G \subset X^*(T^{(1)} \times_{\mathbb{Q}} F)$ corresponding to $B/\mathbb{Q}$. This action factors over the quotient $\mathrm{Gal}(F/\mathbb{Q})$. Then it also acts on the set of dominant weights. Since our group is quasisplit we find for any dominant weight λ an absolutely irreducible $G \times_{\mathbb{Q}} F$-module $\mathcal{M}_\lambda$:

$$r : G \times_{\mathbb{Q}} F \to \mathrm{Gl}(\mathcal{M}_\lambda).$$

This representation is characterised by the following two properties - the space of invariants under the action of U is one-dimensional, i.e. ${\mathcal{M}_\lambda}^U = Fe_\lambda$; the torus acts on this one-dimensional space by the character λ, i.e. $te_\lambda = \lambda(t)e_\lambda$. We say that e_λ is a *highest weight vector*. Since we assumed that $\mathbb{Q} \subset F \subset \bar{\mathbb{Q}} \subset \mathbb{C}$ we get the extension

$$r_{\mathbb{C}} : (G \times_{\mathbb{Q}} F) \times_F \mathbb{C} \to \mathrm{Gl}(\mathcal{M}_\lambda \otimes_F \mathbb{C}).$$

Given such an absolutely irreducible rational representation, we can construct two new representations. We can form the dual $\mathcal{M}^\vee_{\lambda,\mathbb{C}} = \mathrm{Hom}_{\mathbb{C}}(\mathcal{M}_\lambda, \mathbb{C})$ and the complex conjugate $\overline{\mathcal{M}}_{\mathbb{C}}$ of our module $\mathcal{M}_\lambda$. On the dual module we have the contragredient representation $r^\vee$, which is defined by $\phi(r_{\mathbb{C}}(g)(v)) = r^\vee_{\mathbb{C}}(g^{-1})(\phi)(v)$.

To get the rational representation on the conjugate module $\bar{\mathcal{M}} \otimes_F \mathbb{C}$, we recall its definition: as abelian groups we have $\mathcal{M} \otimes_F \mathbb{C} = \bar{\mathcal{M}} \otimes_F \mathbb{C}$, but the action of the scalars is conjugated - we write this as $z \cdot_c m = \bar{z}m$. Then the identity gives us an identification

$$\mathrm{End}_{\mathbb{C}}(\mathcal{M} \otimes_F \mathbb{C}) = \mathrm{End}_{\mathbb{C}}(\bar{\mathcal{M}}_\lambda \otimes_F \mathbb{C}).$$

G. Harder, *Cohomology of Arithmetic Groups*,
Springer Monographs in Mathematics, https://doi.org/10.1007/978-3-032-10378-9_6

Now we define an action $\bar{r}_{\mathbb{C}}$ on $\bar{\mathcal{M}}_\lambda \otimes_F \mathbb{C}$: for $g \in G(\mathbb{C})$ we put

$$\bar{r}_{\mathbb{C}}(g)m = r_{\mathbb{C}}(\bar{g})m,$$

where $g \mapsto \bar{g}$ is the complex conjugation on $G(\mathbb{C})$. This defines an action of the abstract group $G(\mathbb{C})$, but this is in fact obtained from a rational representation. Therefore $\mathcal{M}_\lambda^\vee \otimes_F \mathbb{C}$ and $\overline{\mathcal{M}}_\lambda \otimes_F \mathbb{C}$ both are given by a highest weight. The highest weight of $\mathcal{M}_\lambda^\vee$ is $-w_0(\lambda)$. Here w_0 is the unique element $w_0 \in W$, which sends the system of positive roots Δ^+ into the system $\Delta^- = -\Delta^+$. Since G is quasisplit the highest weight of $\bar{\mathcal{M}}_\lambda \otimes_F \mathbb{C}$ is $c(\lambda)$ where c, viewed as an element in $\mathrm{Gal}(\mathbb{C}/\mathbb{R}) \to \mathrm{Gal}(F/\mathbb{Q})$, is the complex conjugation acting on $X^*(T \times_{\mathbb{Q}} F)$. So we may say that $\bar{\mathcal{M}}_\lambda = \mathcal{M}_{c(\lambda)}$.

We will call the module $\mathcal{M}_\lambda$ - *conjugate-autodual* or simply *c-autodual* if

$$c(\lambda) = -w_0(\lambda). \tag{6.1}$$

If our group $G/\mathbb{Q}$ is split, then c acts trivially on the character module, and the condition becomes $\lambda = -w_0(\lambda)$. If now, in addition, the element w_0 acts by -1 on the character module, then every λ is conjugate-autodual.

In the following few sections (until 6.2.2) we will always assume that our local system (resp. the corresponding representation) is a local system in $\mathbb{C}$-vector spaces (resp. a $\mathbb{C}$-vector space $\mathcal{M}_\lambda$). Therefore we will suppress the factor $\otimes\mathbb{C}$.

Now we choose an arithmetic subgroup $\Gamma \subset G(\mathbb{R})$ and we will use transcendental methods to investigate the cohomology $H^\bullet(\Gamma\backslash X, \tilde{\mathcal{M}}_\lambda)$.

6.1.2 Harish-Chandra modules and $(\mathfrak{g}, K_\infty)$-cohomology

In the following we assume that $G/\mathbb{Q}$ is semi simple, it is always easy to extend a concept or an assertion to the case of reductive groups. We consider the group of real points $G(\mathbb{R})$. It has the Lie algebra $\mathfrak{g}$, for $X_1, X_2 \in \mathfrak{g}$ let $B(X_1, X_2)$ be the Killing form. Since B is non degenerate we may also consider the Killing form as an element in the symmetric tensors $\mathrm{Sym}^2(\mathfrak{g}) \subset \mathfrak{g} \otimes \mathfrak{g}$. Inside the Lie algebra $\mathfrak{g}$ we have the Lie algebra $\mathfrak{k}$ of the group K_∞. We define the notion of a $(\mathfrak{g}, K_\infty)$-module: this is a $\mathbb{C}$-vector space V together with an action of $\mathfrak{g}$ and an action of the group K_∞. We have certain assumptions of consistency:

i) The action of K_∞ is differentiable, which means it induces an action of $\mathfrak{k}$, the derivative of the group action.

ii) The action of $\mathfrak{g}$ restricted to $\mathfrak{k}$ is the derivative of the action of K_∞.

iii) For $k \in K_\infty, X \in \mathfrak{g}$ and $v \in V$ we have

$$(\mathrm{Ad}(k)X)v = k(X(k^{-1}v)).$$

Inside V we have have the subspace of K_∞-finite vectors. A vector v is called K_∞*-finite* if the $\mathbb{C}$-subspace generated by all translates kv is finite dimensional, i.e. v lies in a finite dimensional K_∞-invariant subspace. The K_∞-finite vectors form a subspace $V^{(K_\infty)}$, and it is obvious that $V^{(K_\infty)}$ is invariant under the action of $\mathfrak{g}$, hence it is a $(\mathfrak{g}, K_\infty)$-submodule of V. We call a $(\mathfrak{g}, K_\infty)$-module a Harish-Chandra module if $V = V^{(K_\infty)}$.

For such a $(\mathfrak{g}, K_\infty)$-module we can write down a complex

$$\begin{aligned} &\mathrm{Hom}_{K_\infty}(\Lambda^\bullet(\mathfrak{g}/\mathfrak{k}), V) := \\ 0 \quad &\to V^{K_\infty} \to \mathrm{Hom}_{K_\infty}(\Lambda^1(\mathfrak{g}/\mathfrak{k}), V) \quad \to \mathrm{Hom}_{K_\infty}(\Lambda^2(\mathfrak{g}/\mathfrak{k}), V) \to \cdots, \end{aligned} \tag{6.2}$$

where the differential is given by

$$\begin{aligned} d\omega(X_0, X_1, \ldots, X_p) &= \sum_{i=0}^{p} (-1)^i X_i \omega(X_0, \ldots, \hat{X}_i, \ldots, X_p) \\ &+ \sum_{0 \le i < j \le p} (-1)^{i+j} \omega([X_i, X_j], X_0, \ldots, \hat{X}_i, \ldots, \hat{X}_j, \ldots). \end{aligned} \tag{6.3}$$

A few comments are in order. We have the inclusions

$$\mathrm{Hom}_{K_\infty}(\Lambda^\bullet(\mathfrak{g}/\mathfrak{k}), V) \subset \mathrm{Hom}(\Lambda^\bullet(\mathfrak{g}/\mathfrak{k}), V) \subset \mathrm{Hom}(\Lambda^\bullet(\mathfrak{g}), V).$$

The above differential defines the structure of a complex for the rightmost term, and we have to verify that the leftmost term is a subcomplex, which is not so difficult.

We define the $(\mathfrak{g}, K_\infty)$-cohomology as the cohomology of this complex, i.e.

$$H^\bullet(\mathfrak{g}, K_\infty, V) = H^\bullet(\mathrm{Hom}_{K_\infty}(\Lambda^\bullet(\mathfrak{g}/\mathfrak{k}), V)) \tag{6.4}$$

It is clear that the map

$$H^\bullet(\mathfrak{g}, K_\infty, V^{(K_\infty)}) \to H^\bullet(\mathfrak{g}, K_\infty, V)$$

is an isomorphism.

If we have two $(\mathfrak{g}, K_\infty)$-modules V_1, V_2 and form the algebraic tensor product $W = V_1 \otimes V_2$, then we have a natural structure of a $(\mathfrak{g}, K_\infty)$-module on W : the group K_∞ acts via the diagonal and $U \in \mathfrak{g}$ acts by the Leibniz rule $U(v_1 \otimes v_2) = Uv_1 \otimes v_2 + v_1 \otimes Uv_2$. If both modules are Harish-Chandra modules then the tensor product is also a Harish-Chandra module. Of course any finite dimensional rational representation of the algebraic group also yields a Harish-Chandra module.

6.1.3 The representation theoretic de-Rham isomorphism

For us, the $(\mathfrak{g}, K_\infty)$-module $\mathcal{C}_\infty(\Gamma\backslash G(\mathbb{R}))$ - this being the space of functions which are $\mathcal{C}_\infty$ in the variable g_∞ - is one of the most important $(\mathfrak{g}, K_\infty)$-modules. On these functions the group $G(\mathbb{R})$ acts by translations from the right. Since our functions are $\mathcal{C}_\infty$ we also get an action of the Lie algebra $\mathfrak{g}$. Hence this is also a $(\mathfrak{g}, K_\infty)$-module.

We define the action of the Hecke algebra $\mathcal{H}_\Gamma$ on $\mathcal{C}_\infty(\Gamma\backslash G(\mathbb{R}))$. This Hecke algebra is generated by operators T_α where α runs through the elements of $G(\mathbb{Q})$. For $\alpha \in G(\mathbb{Q}), f \in \mathcal{C}_\infty(\Gamma\backslash G(\mathbb{R}))$ we put

$$T_\alpha(f)(g) = \sum_{\eta \in \alpha\Gamma\alpha^{-1} \cap \Gamma\backslash\Gamma} f(\alpha^{-1}\eta g).$$

We define the Hecke operators

$$T_\alpha \otimes \alpha : \mathcal{C}_\infty(\Gamma\backslash G(\mathbb{R})) \otimes \mathcal{M}_\lambda \to \mathcal{C}_\infty(\Gamma\backslash G(\mathbb{R})) \otimes \mathcal{M}_\lambda,$$

hence $\mathcal{C}_\infty(\Gamma\backslash G(\mathbb{R}) \otimes \mathcal{M}_\lambda)$ becomes a Hecke $\times$ $(\mathfrak{g}, K_\infty)$-module.

Let us assume for the moment that Γ is torsion-free. Then we can define the *sheaf* $\Omega^\bullet(\tilde{\mathcal{M}}_\lambda)$ *of de-Rham complexes with values in* $\tilde{\mathcal{M}}_\lambda$: For any given small open subset $V \subset \Gamma\backslash X$, the complex of sections is the de-Rham complex $\Omega^\bullet(V) \otimes \tilde{\mathcal{M}}_\lambda(V)$ (see for instance [48], 4.10). If V is small enough, a section in $\Omega^\bullet(V) \otimes \mathcal{M}_\lambda$ can be written as $\omega = \sum_\nu \omega_\nu \otimes m_\nu$ where $m_1, m_2, \ldots, m_k$ is a basis of $\mathcal{M}_\lambda$, and then the differential is given by $\sum_\nu d\omega_\nu \otimes m_\nu$, because the m_ν are locally constant, in other words: we have a flat connection.

We have the following fundamental fact:

Proposition 6.1.1. *For semisimple G we have canonical isomorphisms of complexes*

$$\mathrm{Hom}_{K_\infty}(\Lambda^\bullet(\mathfrak{g}/\mathfrak{k}), \mathcal{C}_\infty(\Gamma\backslash G(\mathbb{R})) \otimes \mathcal{M}_\lambda) \xrightarrow{\sim} (\Omega^\bullet(X) \otimes \mathcal{M}_\lambda)^\Gamma \xrightarrow{\sim} \Omega^\bullet(\Gamma\backslash X, \tilde{\mathcal{M}}_\lambda),$$

being compatible with the action of the Hecke algebra.

This is clear, we have to look carefully at the definitions. We have the projection map

$$q_\infty : G(\mathbb{R}) \to G(\mathbb{R})/K_\infty = X,$$

let $x_0 \in X$ be the image of the identity $e \in G(\mathbb{R})$. The differential $D_{q_\infty}(e)$ maps the Lie algebra $\mathfrak{g}$, which is the tangent space of $G(\mathbb{R})$ at e, to the tangent space T_{X,x_0}. This provides the identification $T_{X,x_0} \xrightarrow{\sim} \mathfrak{g}/\mathfrak{k}$.

To a given ω we want to produce an element $\tilde{\omega}$ in the de-Rham complex $\Omega^\bullet(\Gamma\backslash X, \tilde{\mathcal{M}}_\lambda)$. We pick a point $x \in \Gamma\backslash X$ and a point $\tilde{x} \in X$ lying over this point, we find an element $g_\infty \in G(\mathbb{R})$ such that $g_\infty x_0 = \tilde{x}$. Our still to be defined form $\tilde{\omega}$ can be evaluated at a p-tuple $(\mathfrak{t}_0, \ldots, \mathfrak{t}_{p-1})$ of tangent vectors at $\tilde{x}$, and the result has to be an element in the stalk $\tilde{\mathcal{M}}_{\mathbb{C},x}$. The derivative $D_{g_\infty^{-1}}$ of the left translation by g_∞^{-1} provides an isomorphism from the tangent space at $\tilde{x}$ and the tangent space at x_0 which is identified with $\mathfrak{g}/\mathfrak{k}$. Hence we can evaluate the given ω at our p-tuple of tangent vectors, we get as value

$$\omega\big(D_{g_\infty^{-1}}(\mathfrak{t}_0), \ldots, D_{g_\infty^{-1}}(\mathfrak{t}_{p-1})\big)(g_\infty) \in \mathcal{M}_\lambda. \tag{6.5}$$

This does not yet give the evaluation of $\tilde{\omega}$. The value has to be an element in $\tilde{\mathcal{M}}_{\lambda_x}$. Hence we define

$$\tilde{\omega}(\mathfrak{t}_0, \ldots, \mathfrak{t}_{p-1})(\tilde{x}) = g_\infty \omega(D_{g_\infty^{-1}}(\mathfrak{t}_0), \ldots, D_{g_\infty^{-1}}(\mathfrak{t}_{p-1}))(g_\infty). \tag{6.6}$$

Now we leave it as an exercise to the reader to verify that for $\tilde{x}$ this value does not depend on the choice of g_∞ because ω is K_∞-invariant. Hence we can consider the family of values $\{\tilde{\omega}(\mathfrak{t}_0, \ldots, \mathfrak{t}_{p-1})(\gamma\tilde{x})\}_{\gamma\in\Gamma}$, and this family is indeed an element in $\tilde{\mathcal{M}}_{\lambda x}$.

Now it is clear that $\omega \mapsto \tilde{\omega}$ gives the isomorphism we want ([20]).

Actually in this context we do not really need that Γ is torsion-free. If this is not the case then $\tilde{\mathcal{M}}_\lambda$ is only an orbilocal system. We have to define the sheaf

$\Omega^\bullet(\tilde{\mathcal{M}}_\lambda)$ accordingly. In any case the left hand side of the complex is always defined, we simply take the left hand side as definition of the de-Rham complex.

We recall that the de-Rham complex ([48], Sect. 4.8.) computes the cohomology, and therefore we can rewrite the de-Rham isomorphism

$$H^\bullet(\Gamma\backslash X, \tilde{\mathcal{M}}_\lambda) \xrightarrow{\sim} H^\bullet\big(\mathrm{Hom}_{K_\infty}(\Lambda^\bullet(\mathfrak{g}/\mathfrak{k}), \mathcal{C}_\infty(\Gamma\backslash G(\mathbb{R})) \otimes \mathcal{M}_\lambda\big). \tag{6.7}$$

From now on the complex $\mathrm{Hom}_{K_\infty}(\Lambda^\bullet(\mathfrak{g}/\mathfrak{k}), \mathcal{C}_\infty(\Gamma\backslash G(\mathbb{R})) \otimes \mathcal{M}_\lambda)$ will also be called the de-Rham complex.

By the same token we can compute the cohomology with compact supports

$$H^\bullet_c(\Gamma\backslash X, \tilde{\mathcal{M}}_\lambda) \xrightarrow{\sim} H^\bullet\big(\mathrm{Hom}_{K_\infty}(\Lambda^\bullet(\mathfrak{g}/\mathfrak{k}), \mathcal{C}_{c,\infty}(\Gamma\backslash G(\mathbb{R})) \otimes \mathcal{M}_\lambda\big) \tag{6.8}$$

where $\mathcal{C}_{c,\infty}(\Gamma\backslash G(\mathbb{R}))$ are the $\mathcal{C}_\infty$-functions with compact support. These isomorphisms are also valid if we drop the assumption that Γ is torsion-free.

The Poincaré duality on the cohomology is induced by the pairing on the de-Rham complexes:

Proposition 6.1.2. *If* $\omega_1 \in \mathrm{Hom}_{K_\infty}(\Lambda^\bullet(\mathfrak{g}/\mathfrak{k}), \mathcal{C}_\infty(\Gamma\backslash G(\mathbb{R})) \otimes \mathcal{M})$ *is a closed form and* $\omega_2 \in \mathrm{Hom}_{K_\infty}(\Lambda^\bullet(\mathfrak{g}/\mathfrak{k}), \mathcal{C}_{\infty,c}(\Gamma\backslash G(\mathbb{R})) \otimes \mathcal{M}^\vee)$ *is a closed form with compact support in complementary degree, then the value of the cup product pairing of the classes* $[\omega_1] \in H^p(\Gamma\backslash X, \tilde{\mathcal{M}}_\lambda), [\omega_2] \in H^{d-p}_c(\Gamma\backslash X, \tilde{\mathcal{M}}^\vee_\lambda)$ *is given by*

$$\langle [\omega_1] \cup [\omega_2] \rangle = \int_{\Gamma\backslash X} \langle \omega_1 \wedge \omega_2 \rangle \,.$$

(Reference Book Vol. !)

6.1.4 Input from representation theory of real reductive groups

Let us consider an arbitrary irreducible $(\mathfrak{g}, K_\infty)$-module V. Let $\hat{K}_\infty$ be the set of irreducible continuous representations that are finite dimensional. We also assume that for any $\vartheta \in \hat{K}_\infty$ the multiplicity of ϑ in V is finite (we say that V is *admissible*). Then we can extend the action of the Lie-algebra $\mathfrak{g}$ to an action of the universal enveloping algebra $\mathfrak{U}(\mathfrak{g})$ on V and we can restrict this action to an action of the centre $\mathfrak{Z}(\mathfrak{g})$. The structure of this centre is well-known by a theorem of Harish-Chandra - it is a polynomial algebra in $r = \mathrm{rank}(G)$ variables. Here the rank is the absolute rank, i.e. the dimension of a maximal torus in $G/\mathbb{Q}$ (see Chap. 4 Sect. 4).

Clearly this centre respects the decomposition into K_∞-types. Since these K_∞-types come with finite multiplicity, we can apply the standard argument, which proves the Lemma of Schur. Hence $\mathfrak{Z}(\mathfrak{g})$ has to act on V by scalars, thus we get a homomorphism $\chi_V : \mathfrak{Z}(\mathfrak{g}) \to \mathbb{C}$ which is defined by

$$zv = \chi_V(z)v.$$

This homomorphism is called the *infinitesimal character* of V.

A fundamental theorem of Harish-Chandra asserts that for a given infinitesimal character there exist only finitely many isomorphism classes of irreducible, admissible $(\mathfrak{g}, K_\infty)$-modules with this infinitesimal character.

For any rational finite-dimensional representation $r : G/\mathbb{Q} \to \mathrm{Gl}(\mathcal{M}_\lambda)$ we can consider $\mathcal{M}_\lambda \otimes \mathbb{C}$ as $(\mathfrak{g}, K_\infty)$-module. If $\mathcal{M}_\lambda$ is absolutely irreducible with highest weight λ then it also has an infinitesimal character $\chi_{\mathcal{M}_\lambda} = \chi_\lambda$.

Wigner's lemma: *Let V be an irreducible, admissible $(\mathfrak{g}, K_\infty)$-module, let $\mathcal{M} = \mathcal{M}_\lambda$, a finite dimensional, absolutely irreducible rational representation. Then $H^\bullet(\mathfrak{g}, K_\infty, V \otimes \mathcal{M}_\mathbb{C}) = 0$ unless we have*

$$\chi_V(z) = \chi_{\mathcal{M}^\vee}(z) = \chi_{\lambda^\vee}(z) \text{ for all } z \in \mathfrak{Z}(\mathfrak{g}).$$

Since we also know that the number of isomorphism classes of irreducible, admissible $(\mathfrak{g}, K_\infty)$-modules with a given infinitesimal character is finite, we can conclude that for a given absolutely irreducible rational module $\mathcal{M}_\lambda$, the number of isomorphism classes of irreducible, admissible $(\mathfrak{g}, K_\infty)$-modules V with $H^\bullet(\mathfrak{g}, K_\infty, V \otimes \mathcal{M}_\mathbb{C}) \neq 0$ is finite.

The proof of Wigner's lemma is very elegant. We have $\mathcal{M}\otimes V = \mathrm{Hom}(\mathcal{M}^\vee, V)$ and hence we have

$$H^0(\mathfrak{g}, K_\infty, \mathcal{M} \otimes V) = \mathrm{Hom}(\mathcal{M}^\vee, V)^{(\mathfrak{g},K_\infty)} = \mathrm{Hom}_{(\mathfrak{g},K_\infty)}(\mathcal{M}^\vee, V).$$

In [20] , Chap. I, 2.4, it is shown, that the category of $(\mathfrak{g}, K_\infty)$-modules has enough injective and projective elements (see [20], I. 2.5). If I is an injective $(\mathfrak{g}, K_\infty)$-module then $\mathcal{M} \otimes I$ is also injective because for any $(\mathfrak{g}, K_\infty)$-module A we have $\mathrm{Hom}(A, \mathcal{M} \otimes I) = \mathrm{Hom}(A \otimes \mathcal{M}^\vee, I)$. Hence an injective resolution $0 \to V \to I^0 \to I^1 \to \cdots$ yields an injective resolution $0 \to \mathcal{M}\otimes V \to \mathcal{M}\otimes I^0 \to \mathcal{M}\otimes I^1 \to \cdots$, and from this we get, using that $(\mathfrak{g}, K_\infty)$-cohomology is a derived functor,

$$H^q(\mathfrak{g}, K_\infty, \mathcal{M} \otimes V) = \mathrm{Ext}^q_{(\mathfrak{g},K_\infty)}(\mathcal{M}^\vee, V).$$

Any $z \in \mathfrak{Z}(\mathfrak{g})$ induces an endomorphism of $\mathcal{M}_\lambda$ and V. Since $\mathrm{Ext}^\bullet$ is functorial in both variables, we see that z induces endomorphisms z_1 (via the action on $\mathcal{M}_\lambda$) and z_2 (via the action on V) on $\mathrm{Ext}^q_{(\mathfrak{g},K_\infty)}(\mathcal{M}^\vee, V)$. We show that $z_1 = z_2$. This is clear by definition for $\mathrm{Ext}^0_{(\mathfrak{g},K_\infty)}(\mathcal{M}^\vee, V) = \mathrm{Hom}_{(\mathfrak{g},K_\infty)}(\mathcal{M}^\vee, V)$: for $z \in \mathfrak{Z}(\mathfrak{g})$ and $\phi \in \mathrm{Hom}_{(\mathfrak{g},K_\infty)}(\mathcal{M}^\vee, V), m \in \mathcal{M}_\lambda$ we have $z_1\phi(m) = \phi(zm) = z_2(\phi(m))$. To prove it for an arbitrary q we use dévissage and induction. We embed V into an injective $(\mathfrak{g}, K_\infty)$-module I and get an exact sequence

$$0 \to V \to I \to I/V \to 0.$$

From this we get

$$\mathrm{Ext}^{q-1}_{(\mathfrak{g},K_\infty)}(\mathcal{M}_\lambda, I/V) = \mathrm{Ext}^q_{(\mathfrak{g},K_\infty)}(\mathcal{M}_\lambda, V) \text{ for } q > 0.$$

Now by induction we know $z_1 = z_2$ on the left hand side, so it also holds on the right hand side.

If now $\chi_V \neq \chi_{\mathcal{M}^\vee}$ then we can find a $z \in \mathfrak{Z}(\mathfrak{g})$ such that $\chi_{\mathcal{M}^\vee}(z) = 0$ and $\chi_V(z) = 1$. Therefore $z_1 = 0$ and $z_2 = 1$ on all $\mathrm{Ext}^q_{(\mathfrak{g},K_\infty)}((\mathfrak{g}, K_\infty)(\mathcal{M}_\lambda, V))$.

Since we know that $z_1 = z_2$ we see that the identity on $\mathrm{Ext}^q_{(\mathfrak{g},K_\infty)}(\mathcal{M}_\lambda, V)$ is equal to zero, and this implies the assertion.

On the universal enveloping algebra $\mathfrak{U}(\mathfrak{g})$ we have an antiautomorphism given by $u \mapsto {}^t u$ which is induced by the antiautomorphism $X \mapsto -X$ on the Lie algebra $\mathfrak{g}$. If V is an admissible $(\mathfrak{g}, K_\infty)$-module, then we can form the dual module $V^\vee$, and if we denote the pairing between $V, V^\vee$ by $\langle\ ,\ \rangle_V$ then

$$\langle Uv, \phi\rangle_V = \langle v, {}^t U\phi\rangle_V \text{ for all } U \in \mathfrak{U}(\mathfrak{g}), v \in V, \phi \in V^\vee.$$

If V is irreducible, then it has an infinitesimal character and we get

$$\chi_{V^\vee}(z) = \chi_V({}^t z).$$

This applies to finite-dimensional and to infinite-dimensional $(\mathfrak{g}, K_\infty)$-modules.

6.1.5 Representation theoretic Hodge-theory

Let V be a Hilbert space over $\mathbb{C}$, denote the hermitian scalar product by $\langle\ ,\ \rangle_V$. We consider irreducible unitary representations $G(\mathbb{R}) \to U(V)$. We know from the work of Harish-Chandra:

1) If we fix an isomorphism class ϑ of irreducible representations of K_∞, then the isotypical subspace $\dim_\mathbb{C} H(\vartheta) \le \dim(\vartheta)^2$, i.e. ϑ occurs at most with multiplicity $\dim(\vartheta)$.

2) The direct sum $\sum_{\vartheta\in\hat{K}_\infty} H(\vartheta) = H^{(K_\infty)} \subset H$ is dense in H and it is an admissible irreducible Harish-Chandra module.

We call an irreducible $(\mathfrak{g}, K_\infty)$-module unitary, if it is isomorphic to such an $H^{(K_\infty)}$.

For any rational irreducible $G/\mathbb{C}$-module $\mathcal{M}_\lambda$, Vogan and Zuckerman [113] give a finite list of certain irreducible, admissible $(\mathfrak{g}, K_\infty)$-modules $A_\mathfrak{q}(\lambda)$, for which $H^\bullet(\mathfrak{g}, K_\infty, A_\mathfrak{q}(\lambda)\otimes\mathcal{M}_\lambda) \neq 0$, and they compute these cohomology groups. This list contains all unitary, irreducible $(\mathfrak{g}, K_\infty)$-modules, which have non-trivial cohomology with coefficients in $\mathcal{M}_\lambda$.

For the following we refer to [20] Chap. II , §1-2. We apply the methods of Hodge theory to compute the cohomology groups $H^\bullet(\mathfrak{g}, K_\infty, V \otimes \mathcal{M}_\lambda)$ for a unitary $(\mathfrak{g}, K_\infty)$-module V. This means we have a positive-definite scalar product $\langle\ ,\ \rangle_V$ on V, for which the action of K_∞ is unitary, and for $U \in \mathfrak{g}$ and $v_1, v_2 \in V$ we have $\langle Uv_1, v_2\rangle_V + \langle v_1, Uv_2\rangle_V = 0$.

We assume that $\mathcal{M}_\lambda$ is conjugate-autodual. In the next step we introduce for all p a hermitian form on $\mathrm{Hom}_{K_\infty}(\Lambda^p(\mathfrak{g}/\mathfrak{k}), V\otimes\mathcal{M}_\lambda)$. To do this we construct a hermitian form on $\mathcal{M}_\lambda$.

The following considerations are only true modulo the centre. We consider the Lie algebra and its complexification $\mathfrak{g}_\mathbb{C} = \mathfrak{g} \otimes \mathbb{C}$. On this complex vector space we have the complex conjugation $^- : U \mapsto \bar{U}$. We rediscover $\mathfrak{g}$ as the set of fixed points under $^-$. We also have the Cartan involution Θ which is the involution which has $\mathfrak{k}$ as its fixed point set. Then we get the Cartan decomposition

$$\mathfrak{g} = \mathfrak{k} \oplus \mathfrak{p} \text{ where } \mathfrak{p} \text{ is the } -1 \text{ eigenspace of } \Theta.$$

The Killing form is negative definite on $\mathfrak{k}$ and positive-definite on $\mathfrak{p}$. We have for the Lie bracket $[\mathfrak{p},\mathfrak{p}] \subset \mathfrak{k}$. We consider the invariants under $^{-} \circ \Theta$, which is the Lie algebra $\mathfrak{g}_c = \mathfrak{k} \oplus (\sqrt{-1} \otimes \mathfrak{p})$. On this real Lie algebra, the Killing form is negative definite and $\mathfrak{g}_c$ is the Lie algebra of an algebraic group $G_c/\mathbb{R}$. This group is an $\mathbb{R}$-form of $G/\mathbb{R}$, i.e. $G_c \otimes_{\mathbb{R}} \mathbb{C} \xrightarrow{\sim} G \otimes_{\mathbb{R}} \mathbb{C}$. The group $G_c(\mathbb{R})$ of real points is compact (this is the *compact form* of G). We still have the representation $G_c/\mathbb{R} \to \mathrm{Gl}(\mathcal{M}_\lambda)$ which is irreducible, and hence we find a hermitian form $\langle\, ,\, \rangle_\lambda$ on $\mathcal{M}_\lambda$ which is invariant under $G_c(\mathbb{R})$ and which is unique up to a scalar.

This form satisfies the equations

$$\begin{gathered}\langle Ym_1, m_2\rangle_\lambda + \langle m_1, Ym_2\rangle_\lambda = 0 \text{ for all } m_1, m_2 \in \mathcal{M}_\lambda, Y \in \mathfrak{k},\\ \text{which is the invariance under } K_\infty \text{ and}\\ \langle Ym_1, m_2\rangle_\lambda = \langle m_1, Ym_2\rangle_\lambda \text{ for all } m_1, m_2 \in \mathcal{M}_\lambda, Y \in \mathfrak{p},\\ \text{which is the invariance under } \sqrt{-1} \otimes \mathfrak{p}.\end{gathered}$$

Now we define a hermitian metric on $V \otimes \mathcal{M}_\lambda$, we simply take the tensor product $\langle\, ,\, \rangle_V \otimes \langle\, ,\, \rangle_\lambda = \langle\, ,\, \rangle_{V\otimes\lambda}$. Finally we define the (hermitian) scalar product on $\mathrm{Hom}_{K_\infty}(\Lambda^\bullet(\mathfrak{g}/\mathfrak{k}), V \otimes \mathcal{M}_\lambda)$. We choose an orthonormal (with respect to the Killing form) basis $E_1, E_2, \ldots, E_d$ on $\mathfrak{p}$, where we identify $\mathfrak{g}/\mathfrak{k} \xrightarrow{\sim} \mathfrak{p}$. Then a form $\omega \in \mathrm{Hom}_{K_\infty}(\Lambda^p(\mathfrak{g}/\mathfrak{k}), V \otimes \mathcal{M}_\lambda)$ is given by its values $\omega(E_I) \in V \otimes \mathcal{M}_\lambda$, where $I = \{i_1, i_2, \ldots, i_p\}$ runs through the ordered subsets of $\{1, 2, \ldots, d\}$ with p elements. For $\omega_1, \omega_2 \in \mathrm{Hom}_{K_\infty}(\Lambda^p(\mathfrak{g}/\mathfrak{k}), V \otimes \mathcal{M}_\lambda)$ we put

$$\langle \omega_1, \omega_2\rangle = \sum_{I,|I|=p} \langle \omega_1(E_I), \omega_2(E_I)\rangle_{V\otimes\lambda}\,. \tag{6.9}$$

Now we can define an adjoint operator

$$\delta : \mathrm{Hom}_{K_\infty}(\Lambda^p(\mathfrak{g}/\mathfrak{k}), V \otimes \mathcal{M}_\lambda) \to \mathrm{Hom}_{K_\infty}(\Lambda^{p-1}(\mathfrak{g}/\mathfrak{k}), V \otimes \mathcal{M}_\lambda), \tag{6.10}$$

which can be defined by a straightforward calculation. We simply write a formula for δ: for an element E_i we define $E_i^*(v \otimes m) = -E_i v \otimes m + v \otimes E_i m$. Then we can define δ by the following formula:

Evaluating $\delta(\omega)$ on $E_J = (E_{i_1}, \ldots, E_{i_{p-1}})$ where $J = \{i_1, \ldots, i_{p-1}\}$. We put

$$\delta(\omega)(E_J) = \sum_{i \notin J} (-1)^{p(i, J\cup\{i\})} E_i^* \omega(E_{J\cup\{i\}}),$$

where $p(i, J \cup \{i\})$ denotes the position of i in the ordered set $J \cup \{i\}$. With this definition we get for a pair of forms $\omega_1 \in \mathrm{Hom}_{K_\infty}(\Lambda^{p-1}(\mathfrak{g}/\mathfrak{k}), V \otimes \mathcal{M}_\lambda)$ and $\omega_2 \in \mathrm{Hom}_{K_\infty}(\Lambda^p(\mathfrak{g}/\mathfrak{k}), V \otimes \mathcal{M}_\lambda)$ (see [20]

$$\langle d\omega_1, \omega_2\rangle = \langle \omega_1, \delta\omega_2\rangle\,. \tag{6.11}$$

We define the Laplacian $\Delta = \delta d + d\delta$. Then ([20] , II , Thm. 2.5)

$$\langle \Delta\omega, \omega\rangle \geq 0 \text{ and we have equality if and only if } d\omega = 0, \delta\omega = 0. \tag{6.12}$$

Inside $\mathfrak{Z}(\mathfrak{g})$ we have the Casimir operator C which is induced by the Killing form (see 1.1.4): Since $G/\mathbb{Q}$ is semi simple the Killing form provides an identification $\mathfrak{g} \xrightarrow{\sim} \mathfrak{g}^\vee$ and therefore we can view the Killing form as an element

$C \in \mathfrak{g} \otimes \mathfrak{g}$ which is invariant under the adjoint action and hence central in the universal envelloping algebra.

Kuga's lemma: *The action of the Casimir operator and the Laplace operator on* $\mathrm{Hom}_{K_\infty}(\Lambda^p(\mathfrak{g}/\mathfrak{k}), V \otimes \mathcal{M}_\lambda)$ *are related by the identity*

$$\Delta = -C \otimes \mathrm{Id} + \chi_\lambda(C)\mathrm{Id}.$$

If the $(\mathfrak{g}, K_\infty)$*-module* V *is irreducible, then* Δ *acts by multiplication by the scalar* $-\chi_V(C) + \chi_\lambda(C)$.

For a proof we refer to [20], Chap. II, Thm. 2.5.

This has the following consequence (See also [20]):

If V *is an irreducible unitary* $(\mathfrak{g}, K_\infty)$*-module and if* $\mathcal{M}_\lambda$ *is an irreducible representation with highest weight* λ *then*

$$H^\bullet(\mathfrak{g}, K_\infty, V \otimes \mathcal{M}_\mathbb{C}) = \begin{cases} 0 & \text{if } \chi_V(C) - \chi_\lambda(C) \neq 0 \\ \mathrm{Hom}_{K_\infty}(\Lambda^\bullet(\mathfrak{g}/\mathfrak{k}), V \otimes \mathcal{M}_\lambda) & \text{if } \chi_V(C) - \chi_\lambda(C) = 0 \end{cases}.$$

This only applies to unitary $(\mathfrak{g}, K_\infty)$-modules, but for these it is much stronger: it says that under the assumption $\chi_V(C) = \chi_\lambda(C)$ we have $\chi_V = \chi_{\lambda^\vee}$ (we only have to test the Casimir operator), and it says that all the differentials in the complex are zero.

6.2 Automorphic forms and cohomology

6.2.1 Input from the theory of automorphic forms

We apply this to the square integrable functions on $\Gamma\backslash G(\mathbb{R})$. Because of the presence of a non-trivial center C, we have to consider functions which transform in a certain way under the action of the center. We may assume that the coefficient system $\mathcal{M}_\lambda$ has a central character, and this central character defines a character ζ_λ on the maximal $\mathbb{Q}$-split torus $S \subset C$. This character can be evaluated on $S^0(\mathbb{R})$, this being the connected component of the identity of the real valued points of S. It follows from [18] that the quotient $S^0(\mathbb{R})\Gamma\backslash G(\mathbb{R})$ has finite volume. We define the space of functions $\mathcal{C}_\infty(\Gamma\backslash G(\mathbb{R}), \zeta_\infty^{-1})$ to be the subspace of those $\mathcal{C}_\infty$-functions which satisfy $f(z_\infty \underline{g}) = \zeta_\infty^{-1}(z_\infty) f(\underline{g})$ for all $z_\infty \in S^0(\mathbb{R})$. Let $G^{(1)}$ be the semi-simple component of G, then $C' := G/G^{(1)}$ is also a torus with maximal $\mathbb{Q}$-split subtorus $S' \subset C'$. We get an isogeny $d_C : C \to C'$ which induces an isomorphism $S^0(\mathbb{R}) \xrightarrow{\sim} S'^{,0}(\mathbb{R})$. Therefore we get a character $\zeta'_\infty : S'^{,0}(\mathbb{R}) \to \mathbb{R}^\times_{>0}$, and by extending to $C'(\mathbb{R})$ we get also a character $\zeta'_\infty : G(\mathbb{R}) \to \mathbb{R}^\times_{>0}$. Its restriction to $S^0(\mathbb{R})$ is ζ_∞. If now $f \in \mathcal{C}_\infty(\Gamma\backslash G(\mathbb{R}), \zeta_\infty^{-1})$ then

$$f(\underline{g})\zeta'_\infty(\underline{g}) \in \mathcal{C}_\infty(\Gamma S^0(\mathbb{R})\backslash G(\mathbb{R})).$$

We say that $f \in \mathcal{C}_\infty(\Gamma\backslash G(\mathbb{R}), \zeta_\infty^{-1})$ is square integrable if

$$\int_{\Gamma S^0(\mathbb{R})\backslash G(\mathbb{R})} |f(\underline{g})\zeta'_\infty(\underline{g})|^2 d\underline{g} < \infty, \tag{6.13}$$

and this allows us to define the Hilbert space $L^2(\Gamma\backslash G(\mathbb{R}),\zeta_\infty^{-1})$. We clearly have $\zeta_\infty'^{-1} \in L^2(\Gamma\backslash G(\mathbb{R}),\zeta_\infty^{-1})$.

The group $G(\mathbb{R})$ acts on $\mathcal{C}_\infty(\Gamma\backslash G(\mathbb{R}),\zeta_\infty^{-1})$ by right translations, and hence we get, by differentiating, an action of the universal enveloping algebra $\mathfrak{U}(\mathfrak{g})$ on it. We define by $\mathcal{C}_\infty^{(2)}(\Gamma\backslash G(\mathbb{R}),\zeta_\infty^{-1})$ the subspace of functions f for which Uf is square integrable for all $U \in \mathfrak{U}(\mathfrak{g})$.

This allows us to define a subcomplex of the de-Rham complex

$$\mathrm{Hom}_{K_\infty}(\Lambda^\bullet(\mathfrak{g}/\mathfrak{k}),\mathcal{C}_\infty^{(2)}(\Gamma\backslash G(\mathbb{R}),\zeta_\infty^{-1})\otimes\mathcal{M}_\lambda). \tag{6.14}$$

We will not work with this complex because its cohomology may show some bad behavior (see remark below).

We do something less sophisticated, we simply define $H^\bullet_{(2)}(\Gamma\backslash X,\tilde{\mathcal{M}}_\lambda) \subset H^\bullet(\Gamma\backslash X,\tilde{\mathcal{M}}_\lambda)$ to be the image of the cohomology of the complex (6.14) in the cohomology. Hence $H^\bullet_{(2)}(\Gamma\backslash X,\tilde{\mathcal{M}}_\lambda)$ is the space of cohomology classes which can be represented by square integrable forms.

Remark: Some authors also define L^2-de-Rham complexes, using the above complex (6.14), and then they take suitable completions to get complexes of Hilbert spaces. These complexes also give cohomology groups which run under the name of L^2-cohomology. These L^2-cohomology groups are related but not necessarily equal to our $H^\bullet_{(2)}(\Gamma\backslash X,\tilde{\mathcal{M}}_\lambda)$. The L^2-cohomology groups can be infinite dimensional.

The Hilbert space $L^2(\Gamma\backslash G(\mathbb{R}),\zeta_\infty^{-1})$ is a module for $G(\mathbb{R})\times\mathcal{H}_\Gamma$. The group $G(\mathbb{R})$ acts by unitary transformations, and the algebra $\mathcal{H}_{K_f}$ is self-adjoint.

Let us assume that $H = H_{\pi_\infty\times\pi_f}$ is an irreducible unitary $G(\mathbb{R})\times\mathcal{H}$-module, and assume that we have an inclusion of this $G(\mathbb{R})\times\mathcal{H}$-module

$$j : H \hookrightarrow L^2(\Gamma\backslash G(\mathbb{R}),\zeta_\infty^{-1}).$$

It follows from the finiteness results in 6.1.5 that this induces an inclusion into the space of square-integrable $\mathcal{C}_\infty$ functions

$$H^{(K_\infty)} \hookrightarrow \mathcal{C}_\infty^{(2)}(\Gamma\backslash G(\mathbb{R}),\zeta_\infty^{-1})^{(K_\infty)}.$$

We consider the $(\mathfrak{g},K_\infty)$-cohomology of this module with coefficients in our irreducible module $\mathcal{M}_\lambda$, we furthermore assume $\chi_V(C) = \chi_\lambda(C)$. We have $H^\bullet(\mathfrak{g},K_\infty,H\otimes\mathcal{M}_\lambda) = \mathrm{Hom}_{K_\infty}(\mathfrak{g},K_\infty,H^{(K_\infty)}\otimes\mathcal{M}_\lambda)$ and get

$$H^\bullet(\mathfrak{g},K_\infty,H^{(K_\infty)}\otimes\mathcal{M}_\mathbb{C}) \xrightarrow{j^\bullet} H^\bullet(\mathfrak{g},K_\infty,\mathcal{C}_\infty(\Gamma\backslash G(\mathbb{R}),\zeta_\infty^{-1})^{(K_\infty)}\otimes\mathcal{M}_\lambda).$$

This suggests that we try to "decompose" $\mathcal{C}_\infty(\Gamma\backslash G(\mathbb{R}),\zeta_\infty^{-1})^{(K_\infty)}$ into irreducibles and then investigate the contributions of the irreducible summands to the cohomology. Essentially we follow the strategy of [17] and [16], but instead of working with complexes of Hilbert spaces we work with complexes of $\mathcal{C}_\infty$ forms and modify the arguments accordingly.

It has been shown by Langlands, that we have a decomposition into a discrete and a continous spectrum

$$L^2(\Gamma\backslash G(\mathbb{R})) = L^2_{\text{disc}}(\Gamma\backslash G(\mathbb{R})) \oplus L^2_{\text{cont}}(\Gamma\backslash G(\mathbb{R})),$$

where $L^2_{\text{disc}}(\Gamma\backslash G(\mathbb{R}))$ is the closure of the sum of all irreducible closed subspaces occuring in $L^2(\Gamma\backslash G(\mathbb{R}))$ and where $L^2_{\text{cont}}(\Gamma\backslash G(\mathbb{R}))$ is the complement.

The discrete spectrum $L^2_{\text{disc}}(\Gamma\backslash G(\mathbb{R}))$ contains as a subspace the *cuspidal spectrum* $L^2_{\text{cusp}}(\Gamma\backslash G(\mathbb{R}))$:

A function $f \in L^2(\Gamma\backslash G(\mathbb{R}))$ is called a *cusp form* if for all proper parabolic subgroups $P/\mathbb{Q} \subset G/\mathbb{Q}$, with unipotent radical $U/\mathbb{Q}$ and $\Gamma_P = \Gamma \cap P(\mathbb{R}), \Gamma_U = \Gamma \cap U(\mathbb{R})$ the integral

$$\mathcal{F}^P(f)(g) = \int_{\Gamma_U\backslash U_P(\mathbb{R})} f(ug)du = 0.$$

This means that the integral is defined for almost all g and zero for almost all g. The function $\mathcal{F}^P(f)(g)$, which is an almost everywhere defined function on $\Gamma\backslash G(\mathbb{R})$, is called the constant Fourier coefficient of f along $P/\mathbb{Q}$. The cuspidal spectrum is the intersection of all the kernels of the $\mathcal{F}^P$.

If our group is anisotropic, then it does not have any proper parabolic subgroup, and in this case we have $L^2_{\text{cusp}}(\Gamma\backslash G(\mathbb{R})) = L^2_{\text{disc}}(\Gamma\backslash G(\mathbb{R})) = L^2(\Gamma\backslash G(\mathbb{R}))$.

For any unitary $G(\mathbb{R}) \times \mathcal{H}$-module $H_\pi = H_{\pi_\infty} \otimes H_{\pi_f}$ we put

$$W_{\text{cusp}}(\pi) := \text{Hom}_{G(\mathbb{R})\times\mathcal{H}}(H_\pi, L^2_{\text{cusp}}(\Gamma\backslash G(\mathbb{R}))).$$

We can ignore the $\mathcal{H}$-module structure and define

$$W_{\text{cusp}}(\pi_\infty) = \text{Hom}_{G(\mathbb{R})}(H_{\pi_\infty}, L^2_{\text{cusp}}(\Gamma\backslash G(\mathbb{R}))).$$

Then the dimension of $W_{\text{cusp}}(\pi_\infty)$ is the multiplicity $m_{\text{cusp}}(\pi_\infty)$. It has been shown by Gelfand-Graev and Langlands that

$$m_{\text{cusp}}(\pi_\infty) = \sum_{\pi_f} \dim(W_{\text{cusp}}(\pi_\infty \times \pi_f)).$$

We get a decomposition into isotypical subspaces

$$L^2_{\text{cusp}}(\Gamma\backslash G(\mathbb{R})) = \overline{\bigoplus_{\pi_\infty\times\pi_f} (L^2_{\text{cusp}}(\Gamma\backslash G(\mathbb{R})(\pi_\infty \times \pi_f))},$$

where $L^2_{\text{cusp}}(\Gamma\backslash G(\mathbb{R}))(\pi_\infty\times\pi_f)$ is defined to be the image of $W_{\text{cusp}}(\pi_\infty\times\pi_f)\otimes H_\pi$ in $L^2_{\text{cusp}}(\Gamma\backslash G(\mathbb{R}))$.

The cuspidal spectrum has a complement in the discrete spectrum, this is the *residual spectrum* $L^2_{\text{res}}(\Gamma\backslash G(\mathbb{R}))$. It is called residual spectrum, because the irreducible subspaces contained in it are obtained by residues of Eisenstein series.

Again we define $W_{\text{res}}(\pi) = \text{Hom}_{G(\mathbb{R})\times\mathcal{H}}(H_\pi, L^2_{\text{res}}(\Gamma\backslash G(\mathbb{R})))$ (resp. $W_{\text{res}}(\pi_\infty) = \text{Hom}_{G(\mathbb{R})}(H_{\pi_\infty} L^2_{\text{cusp}}(\Gamma\backslash G(\mathbb{R}))))$, and it is a deep theorem of Langlands that $m_{\text{res}}(\pi_\infty) = \dim(W_{\text{res}}(\pi_\infty)) < \infty$. Hence we get a decomposition

$$L^2_{\mathrm{res}}(\Gamma\backslash G(\mathbb{R})) = \overline{\bigoplus_{\pi_\infty\times\pi_f} (L^2_{\mathrm{res}}(\Gamma\backslash G(\mathbb{R}))(\pi_\infty \times \pi_f)}.$$

If our group $G/\mathbb{Q}$ is isotropic, then the one dimensional space of constants is in the residual (discrete) spectrum but not in the cuspidal spectrum.

Langlands has given a description of the continuos spectrum using the theory of Eisenstein series. We have a decomposition

$$L^2_{\mathrm{cont}}(\Gamma\backslash G(\mathbb{R})) = \overline{\bigoplus_{\Sigma} \tilde{H}^+_P(\pi_\Sigma)}. \tag{6.15}$$

We briefly explain this decomposition following [17]. The Σ are so called cuspidal data, these are pairs (P, π_Σ) where P is a proper parabolic subgroup and π_Σ is a representation of $M(\mathbb{R}) = P(\mathbb{R})/U(\mathbb{R})$ occurring in the discrete spectrum $L^2_{\mathrm{disc}}(\Gamma_M\backslash M(\mathbb{R}))$.

Let $M^{(1)}/\mathbb{Q}$ be the semi-simple part of M and recall that $C/\mathbb{Q}$ was the center of $G/\mathbb{Q}$. We consider the character module $X^*(P) = \mathrm{Hom}(M/C{\cdot}M^{(1)}, \mathbb{G}_m)$. The elements $X^*(P)\otimes\mathbb{C}$ provide homomorphisms $\gamma\otimes z : M(\mathbb{R})/C(\mathbb{R})M^{(1)}(\mathbb{R}) \to \mathbb{C}^\times$ (see (4.19)). The module $X^*(P)\otimes\mathbb{Q}$ comes with a canonical basis which is given by the dominant fundamental weights γ_μ which are trivial on $M^{(1)}$. We define

$$\Lambda_\Sigma = X^*(P) \otimes i\mathbb{R} = \{\sum_\mu \gamma_\mu \otimes it_\mu | t_\mu \in \mathbb{R}\}$$

which is a group of unitary characters. For $\sigma \in \Lambda_\Sigma$ we define the unitarily induced representation

$$I^G_P\pi_\Sigma \otimes \sigma := \mathrm{Ind}^{G(\mathbb{R})}_{P(\mathbb{R})}\pi_\Sigma \otimes (\sigma + \rho_P),$$

where of course $p \in P(\mathbb{R}), g \in G(\mathbb{R})$ and $\rho_P \in X^*(P) \otimes \mathbb{Q}$ is the half sum of the roots in the unipotent radical of P. This gives us a unitary representation of $G(\mathbb{R})$. Let d_Σ be the Lebesgue measure on Λ_Σ, then we can form the direct integral of unitary representations

$$H_P(\pi_\Sigma) = \int_{\Lambda_\Sigma} I^G_P\pi_\Sigma \otimes \sigma \, d_\Sigma\sigma.$$

The theory of Eisenstein series [77], [61] gives us a homomorphism of $G(\mathbb{R})\times\mathcal{H}_\Gamma$-modules

$$\mathrm{Eis}_P(\pi_\Sigma) : H_P(\pi_\Sigma) \to L^2_{\mathrm{cont}}(\Gamma\backslash G(\mathbb{R})).$$

Let us put

$$\Lambda^+_\Sigma = \{\sum_\mu \gamma_\mu \otimes it_\mu | t_\mu \geq 0\},$$

then the restriction

$$\mathrm{Eis}_P(\pi_\Sigma) : H^+_P(\pi_\Sigma) = \int_{\Lambda^+_\Sigma} I^G_P\pi_\Sigma \otimes \sigma \, d_\Sigma\sigma \to L^2_{\mathrm{cont}}(\Gamma\backslash G(\mathbb{R}))$$

is an isometric embedding. The image will be denoted by $\tilde{H}_P^+(\pi_\Sigma)$, these spaces are the elementary subspaces in [17]. Two such elementary subspaces $\tilde{H}_P^+(\pi_\Sigma), \tilde{H}_{P_1}^+(\pi_{\Sigma_1})$ are either orthogonal to each other or they are equal. We get the above decomposition if we sum over a suitable set of representatives of cuspidal data.

Now we are ready to discuss the contribution of the continuous spectrum to the cohomology. If we have a closed square integrable form

$$\omega \in \mathrm{Hom}_{K_\infty}(\Lambda^p(\mathfrak{g}/\mathfrak{k}), \mathcal{C}_\infty^{(2)}(\Gamma\backslash G(\mathbb{R})) \otimes \mathcal{M}_\lambda),$$

then we can decompose it

$$\omega = \omega_{\mathrm{cusp}} + \omega_{\mathrm{res}} + \omega_{\mathrm{cont}},$$

all summands are $\mathcal{C}_\infty^{(2)}$ and closed.

Proposition 6.2.1. *The cohomology class* $[\omega_{\mathrm{cont}}]$ *is trivial.*

Proof. We use the (now standard) argument in Hodge theory, but this time we apply it to a continuous spectrum instead of a discrete one. We follow Borel-Casselman and prove their Lemma 5.5 [16] in our context. We may assume that ω_{cont} lies in one of the summands, i.e. $\omega_{\mathrm{cont}} = \mathrm{Eis}(\int_{\Lambda_\Sigma} \omega^\vee(\sigma) d_\Sigma\sigma)$ where $\omega^\vee(\sigma) \in \mathrm{Hom}_{K_\infty}(\Lambda^p(\mathfrak{g}/\mathfrak{k}), I_P^G\pi_\Sigma \otimes \sigma \otimes \mathcal{M}_\lambda)$ is the Fourier transform of ω_{cont} in L^2 (theorem of Plancherel). As it stands, the expression $\int_{\Lambda_\Sigma} \omega^\vee(\sigma) d_\Sigma\sigma$ does not make sense because the integrand is in L^2 and not necessarily in L^1. If we choose a symmetric positive definite quadratic form $h(\sigma) = \sum_{\nu,\mu} b_{\nu,\mu} t_\nu t_\mu$ and a positive real number τ then the function

$$h_\tau(\sigma) = (1 + \tau h(\sigma)^m)^{-1} \in L^2(\Lambda_\Sigma).$$

Then $\omega^\vee(\sigma) h_\tau(\sigma)$ is in L^1 and by definition

$$\lim_{\tau\to 0} \int_{\Lambda_\Sigma} \omega^\vee(\sigma) h_\tau(\sigma) d_\Sigma\sigma = \int_{\Lambda_\Sigma} \omega^\vee(\sigma) d_\Sigma\sigma.$$

where the convergence is in the L^2 sense. Since

$$\omega_{\mathrm{cont}} \in \mathrm{Hom}_{K_\infty}(\Lambda^p(\mathfrak{g}/\mathfrak{k}), I_P^G\pi_\Sigma \otimes \sigma \otimes \mathcal{M}_\lambda)$$

we get that $\omega^\vee(\sigma)$ has the following property:

For any polynomial $P(\sigma) = \sum a_{\underline{\mu}} t^{\underline{\mu}}$ in the variables t_μ and with real coefficients, the section

$$\omega^\vee(\sigma) P(\sigma) \text{ is square integrable.} \tag{6.16}$$

This follows from the well-known rules that differentiating a function provides multiplication by the variables for the Fourier transform.

The Lemma of Kuga implies

$$\Delta(\omega^\vee(\sigma)) = (\chi_\sigma(C) - \chi_\lambda(C))\omega^\vee(\sigma)$$

and if $\sigma = \sum \gamma_\mu \otimes it_\mu$ the eigenvalue is

$$\chi_\sigma(C) - \chi_\lambda(C) = \sum a_{\nu,\mu} t_\nu t_\mu + \sum b_\mu t_\mu + c_{\pi_\Sigma} - c_\lambda, \tag{6.17}$$

where c_{π_Σ} is the eigenvalue of the Casimir operator of $M^{(1)}$ on π_Σ. If the $t_\mu \in \mathbb{R}$ then this expression is always ≤ 0. Especially we see that the quadratic form on the right hand side is negative definite. This implies that for $\sigma \in \Lambda_F$ the expression $\chi_\sigma(C) - \chi_\lambda(C)$ assumes a finite number of maximal values, all of them ≤ 0, and hence

$$V_\Sigma = \{\sigma | \chi_\sigma(C) - \chi_\lambda(C) = 0\}$$

is a finite set of points. This set has measure zero, since we assumed that P was a proper parabolic subgroup. The σ for which $H^\bullet(\mathfrak{g}, K_\infty, H_{\Lambda_\Sigma}(\sigma) \otimes \mathcal{M}_\mathbb{C}) \neq 0$ is finite. We choose a $\mathcal{C}_\infty$ function $h_\Sigma(\sigma)$ which is positive, which takes value 1 in a small neighbourhood of V_Σ, which takes values ≤ 1 in a slightly larger neighbourhood and which is zero outside this second neighbourhood. Then we write

$$\omega_{\text{cont}} = \text{Eis}\left(\int_{\Lambda_\Sigma^+} h_\Sigma(\sigma)\omega^\vee(\sigma) d_\Sigma\sigma\right) + \text{Eis}\left(\int_{\Lambda_\Sigma^+} (1 - h_\Sigma(\sigma))\omega^\vee(\sigma) d_\Sigma\sigma\right).$$

We have $d\omega^\vee(\sigma) = 0$, and hence we get

$$\Delta(1 - h_\Sigma(\sigma))\omega^\vee(\sigma) = d\big((\chi_\sigma(C) - \chi_\lambda(C))(1 - h_\Sigma(\sigma))\delta\omega^\vee(\sigma)\big).$$

This implies that

$$\begin{aligned} &\text{Eis}\left(\int_{\Lambda_\Sigma^+} (1 - h_\Sigma(\sigma))\omega^\vee(\sigma) d_\Sigma\sigma\right) \\ =&d\text{Eis}\left(\int_{\Lambda_\Sigma^+} (1 - h_\Sigma(\sigma))(\chi_\sigma(C) - \chi_\lambda(C))^{-1}\delta\omega^\vee(\sigma) d_\Sigma\sigma\right). \end{aligned}$$

It is clear that the integrand in the second term $\int_{\Lambda_\Sigma^+}(1 - h_\Sigma(\sigma))(\chi_\sigma(C) - \chi_\lambda(C))^{-1}\delta\omega^\vee(\sigma)$ still satisfies (6.16), and then our well-known rules above imply that $\psi = \text{Eis}(\int_{\Lambda_\Sigma^+}(1 - h_\Sigma(\sigma))(\chi_\sigma(C) - \chi_\lambda(C))^{-1}\delta\omega^\vee(\sigma) d_\Sigma\sigma)$ is $\mathcal{C}_\infty^2$. Therefore the second term in our above formula is a boundary.

$$\omega_{\text{cont}} = \int_{\Lambda_\Sigma} h_\Sigma(\sigma)\omega(\sigma) d_\Sigma\sigma + d\psi.$$

This is true for any choice of h_Σ. Hence the scalar product $\langle \omega - d\psi, \omega - d\psi\rangle$ can be made arbitrarily small. Then we claim that the cohomology class $[\omega] \in H^\bullet(\text{Hom}_{K_\infty}(\Lambda^p(\mathfrak{g}/\mathfrak{k}), \mathcal{C}_\infty(\Gamma\backslash G(\mathbb{R}) \otimes \mathcal{M}_\lambda)))$ must be zero. This needs a tiny final step.

We invoke Poincaré duality: a cohomology class in $[\omega] \in H^p(\Gamma\backslash X, \tilde{\mathcal{M}}_\lambda)$ is zero if and only if the value of the pairing with any class $[\omega_2] \in H_c^{d-p}(\Gamma\backslash X, \tilde{\mathcal{M}}_\lambda^\vee)$ is zero. But the (absolute) value $[\omega] \cup [\omega_2]$ of the cup product can be given by an integral (see Prop. 6.1.2). Therefore it can be estimated by the norm $\langle \omega - d\psi, \omega - d\psi\rangle$ (Cauchy-Schwarz inequality) and hence must be zero. □

As usual we denote by $\widehat{G(\mathbb{R})}$ the unitary spectrum. For us it is simply the set of isomorphism classes of unitary irreducible representations of $G(\mathbb{R})$. Given $\tilde{\mathcal{M}}_\lambda$, we define

$$\mathrm{Coh}_2(\lambda) = \{\pi_\infty \in \widehat{G(\mathbb{R})} \mid H^\bullet(\mathfrak{g}, K_\infty, H_{\pi_\infty} \otimes \mathcal{M}_\lambda) \neq 0\}. \tag{6.18}$$

The theorem of Harish-Chandra says that this set is finite. Let

$$\begin{aligned} H_{\mathrm{Coh}_2(\lambda)} &= \bigoplus_{\pi:\pi_\infty \in \mathrm{Coh}_2(\lambda)} L^2_{\mathrm{disc}}(\Gamma\backslash G(\mathbb{R}))(\pi_\infty \times \pi_f) \\ &= \bigoplus_{\pi:\pi_\infty \in \mathrm{Coh}_2(\lambda)} W_{\mathrm{disc}}(\pi) \otimes H_\pi, \end{aligned} \tag{6.19}$$

the theorem of Gelfand-Graev and Langlands asserts that this is a finite sum of irreducible modules. This space decomposes again into $H^{\mathrm{cusp}}_{\mathrm{Coh}_2(\lambda)} \oplus H^{\mathrm{res}}_{\mathrm{Coh}_2(\lambda)}$.

Then we get the following theorem which is due to Borel, Garland, Matsushima and Murakami

Theorem 6.2.1. *a) The map*

$$H^\bullet(\mathfrak{g}, K_\infty, H^{(K_\infty)}_{\mathrm{Coh}_2(\lambda)} \otimes \mathcal{M}_\lambda) = \mathrm{Hom}_{K_\infty}(\Lambda^\bullet(\mathfrak{g}/\mathfrak{k}), H^{(K_\infty)}_{\mathrm{Coh}_2(\lambda)} \otimes \mathcal{M}_\lambda) \to H^\bullet_{(2)}(\Gamma\backslash X, \tilde{\mathcal{M}}_\lambda)$$

is surjective. Especially, we have that the image contains $H^\bullet_!(\Gamma\backslash X, \tilde{\mathcal{M}}_\lambda)$.

b) (Borel) The homomorphism

$$H^\bullet(\mathfrak{g}, K_\infty, H^{\mathrm{cusp},(\mathrm{K}_\infty)}_{\mathrm{Coh}_2(\lambda)} \otimes \mathcal{M}_\lambda) \to H^\bullet(\Gamma\backslash X, \tilde{\mathcal{M}}_\lambda)$$

is injective.

In [17] Prop. 5.6., they do not consider the above space $H^\bullet_{(2)}(\Gamma\backslash X, \tilde{\mathcal{M}}_\lambda)$, hence we added a new small insight to this proposition by claiming that this space is the image.

In general, the homomorphism

$$H^\bullet(\mathfrak{g}, K_\infty, H^{\mathrm{res}}_{\mathrm{Coh}_2(\lambda)} \otimes \mathcal{M}_\lambda) \to H^\bullet(\Gamma\backslash X, \tilde{\mathcal{M}}_\lambda)$$

is not injective. We come back to this issue in the next section.

If we denote by $H^\bullet_{\mathrm{cusp}}(\Gamma\backslash X, \tilde{\mathcal{M}}_\lambda)$ the image of the homomorphism in b), then we get a filtration of the cohomology by four subspaces

$$\begin{aligned} H^\bullet_{\mathrm{cusp}}(\Gamma\backslash X, \tilde{\mathcal{M}}_\lambda) &\subset H^\bullet_!(\Gamma\backslash X, \tilde{\mathcal{M}}_\lambda) \\ &\subset H^\bullet_{(2)}(\Gamma\backslash X, \tilde{\mathcal{M}}_\lambda) \subset H^\bullet(\Gamma\backslash X, \tilde{\mathcal{M}}_\lambda). \end{aligned} \tag{6.20}$$

We get the representation theoretic Hodge decomposition

$$\bigoplus_{\pi_\infty} W_{\mathrm{cusp}}(\pi_\infty) \otimes H^\bullet(\mathfrak{g}, K_\infty, H_{\pi_\infty} \otimes \mathcal{M}_\lambda) \xrightarrow{\sim} H^\bullet_{\mathrm{cusp}}(\Gamma\backslash X, \tilde{\mathcal{M}}_\lambda).$$

If we replace the subscript $_{\text{cusp}}$ by the subscript $_!$, then the corresponding map is still surjective but may be not injective.

We want to point out that our space $H^\bullet_{(2)}(\Gamma\backslash X, \tilde{\mathcal{M}}_\lambda)$ is not the space denoted by the same symbol in the paper [16]. They define L^2 cohomology as the complex of square integrable forms, i.e. ω and $d\omega$ have to be square integrable. But then a closed form ω which is in L^2 gives the trivial class in their cohomology if we can write $\omega = d\psi$ for ψ also square integrable. In our definition we do not have that restriction on ψ.

6.2.2 The semi-simplicity of the inner cohomology

Now we assume again that our representation $\tilde{\mathcal{M}}_\lambda$ is defined over some number field F, and we consider it as a subfield of $\mathbb{C}$. In other words, we have a representation $r: G \times F \to \mathrm{Gl}(\mathcal{M}_\lambda)$. We have defined $H^\bullet_!(\Gamma\backslash X, \tilde{\mathcal{M}})$ as a finite-dimensional F-vector space, and Theorem 3.1.1 in Chapter 3 asserts that this is a semi-simple module under the Hecke algebra. The following argument shows that this is an easy consequence of our results above.

Using the decomposition in (6.19) of $H_{\mathrm{Coh}_2(\lambda)}$ into a finite direct sum of irreducible $G(\mathbb{R}) \times \mathcal{H}_{K_f}$-modules, we get

$$H^\bullet(\mathfrak{g}, K_\infty, H_{\mathrm{Coh}_2(\lambda)} \otimes \mathcal{M}_\lambda) = \bigoplus_{\pi_\infty \times \pi_f \in \hat{H}_{\mathrm{Coh}_2(\lambda)}} (\mathrm{Hom}_{K_\infty}(\Lambda^\bullet(\mathfrak{g}/\mathfrak{k}), H_{\pi_\infty} \otimes \mathcal{M}_\lambda) \otimes H_{\pi_f})^{m(\pi_\infty \times \pi_f)}.$$

This is a decomposition of the left-hand side into irreducible $\mathcal{H}_{K_f}$-modules. Now we have the surjective map

$$H^\bullet(\mathfrak{g}, K_\infty, H_{\mathrm{Coh}_2(\lambda)} \otimes \mathcal{M}_\lambda) \to H^\bullet_{(2)}(\Gamma\backslash X, \tilde{\mathcal{M}}_\lambda \otimes \mathbb{C}),$$

hence it follows that $H^\bullet_{(2)}(\Gamma\backslash X, \tilde{\mathcal{M}}_\lambda \otimes \mathbb{C})$ is a semi-simple $\mathcal{H}_{K_f}$-module and hence also $H^\bullet_!(\Gamma\backslash X, \tilde{\mathcal{M}}_\lambda)$ is a semi-simple $\mathcal{H}_{K_f}$-module.

Friendship

We touch upon a question which comes up naturally in this context. Assume we have a non-zero isotypical submodule $H^\bullet_!(\Gamma\backslash X, \mathcal{M}_\lambda)(\pi_f)$. Then we know that there is a unitary $(\mathfrak{g}, K_\infty)$-module H_{π_∞} with $\pi_\infty \in \mathrm{Coh}(\lambda)$ such that we can embed $H_{\pi_\infty} \times H_{\pi_f}$ into $L^2_{\mathrm{disc}}(\Gamma\backslash G(\mathbb{R}))$. The interesting question is:

Given π_f, what are the possible choices for π_∞?

We can formulate this differently. We recall that

$$W_?(\pi_\infty \otimes \pi_f) = \mathrm{Hom}_{G(\mathbb{R}) \times \mathcal{H}_{K_f}}(H_{\pi_\infty} \otimes H_{\pi_f}, L^2_?(\Gamma\backslash G(\mathbb{R})),$$

where ? can be cusp or (2) resp. disc, then we get the surjective map

$$\bigoplus_{\pi_\infty} W_?(\pi_\infty \times \pi_f) \otimes H_{\pi_\infty} \otimes H_{\pi_f} \to H^\bullet_?(\Gamma\backslash X, \tilde{\mathcal{M}}_\lambda)(\pi_f),$$

which is an isomorphism if ? is cusp. The friends of π_f are those π_∞ for which $W_?(\pi_\infty \times \pi_f) \neq 0$.

This question may become very delicate, and we will not discuss it profoundly. (As J. Arthur put it: π_f looks around and asks "Who is my friend?".) In principle we give a complete answer to these questions in the low-dimensional cases discussed in Section 4.1.5, i.e $G/\mathbb{R} = \mathrm{Gl}_2/\mathbb{R}$ and $G/\mathbb{R} = R_{\mathbb{C}/\mathbb{R}}(\mathrm{Gl}_2/\mathbb{C})$.

In Section 6.1.5 we mentioned the Vogan-Zuckerman classification of unitary representations with non-trivial cohomology. We mentioned already that Vogan and Zuckerman construct a family of irreducible $(\mathfrak{g}, K_\infty)$-modules $A_\mathfrak{q}(\lambda)$ for which they show $H^\bullet(\mathfrak{g}, K_\infty, A_\mathfrak{q}(\lambda) \otimes \mathcal{M}_{\lambda^\vee}) \neq 0$, and they explicitly compute $H^\bullet(\mathfrak{g}, K_\infty, A_\mathfrak{q}(\lambda) \otimes \mathcal{M}_{\lambda^\vee})$. Moreover, they show that any irreducible unitary module V with $H^\bullet(\mathfrak{g}, K_\infty, V \otimes \mathcal{M}_{\lambda^\vee}) \neq 0$ is isomorphic to an $A_\mathfrak{q}(\lambda)$. Later on it was proved hat all $A_\mathfrak{q}(\lambda)$ are actually unitary representations.

We give a very cursory description of their construction. To simplify notations, we assume G to be semi simple in the following. Let $T^c/\mathbb{R}$ be a maximal torus in $K_\infty/\mathbb{R}$. Then it is clear that its centraliser $T/\mathbb{R}$ is a maximal torus in $G/\mathbb{R}$. In Section 1.1.2 we introduced the one-dimensional torus $\mathbb{S}^1/\mathbb{R}$ and we chose an isomorphism $i_0 : \mathbb{S}^1 \times_\mathbb{R} \mathbb{C} \xrightarrow{\sim} \mathbb{G}_m/\mathbb{C}$. We consider cocharacters $\chi : \mathbb{S}^1/\mathbb{R} \to T^c/\mathbb{R}$. Using the identification i_0 we can also view this as a cocharacter $\chi : \mathbb{G}_m \to T^{(c)} \times_\mathbb{R} \mathbb{C}$. Let $\Delta \subset X^*(T \times \mathbb{C})$ be the system of roots, let U_α be the corresponding one parameter subgroups, then $U_\alpha \xrightarrow{\sim} \mathbb{G}_a/\mathbb{C}$. We define the subsets $\Delta_0(\chi) := \{\alpha| < \chi, \alpha >= 0\}, \Delta_+(\chi) := \{\alpha| < \chi, \alpha >\geq 0\}$.

The centraliser $Z_\chi \subset G/\mathbb{R}$ is a reductive sub group, the root system of its semi simple component is $\Delta_0(\chi)$. This centraliser is a Levi sub group of the parabolic $P_\chi/\mathbb{C} \subset G \times_\mathbb{R} \mathbb{C}$, this parabolic has as its unipotent radical the group

$$U_\chi = \prod_{\alpha \in \Delta_+(\chi) \setminus \Delta_0(\chi)} U_\alpha,$$

this group is not defined over $\mathbb{R}$, its complex conjugate $\bar{P}_\chi$ is a opposite parabolic subgroup, we get it if we replace i_0 by i_0^{-1} or if we replace χ by its complex conjugate $\bar{\chi} = \chi^{-1}$.

The Lie-algebra $\mathfrak{q} = \mathrm{Lie}(P_\chi/\mathbb{C})$ will be the first parameter is the $\mathfrak{q}$ in $A_\mathfrak{q}(\lambda)$. We will also denote these modules by $A_\chi(\lambda)$, i.e. $A_\chi(\lambda) := A_\mathfrak{q}(\lambda)$ if χ and $\mathfrak{q}$ are related as above.

It is clear that the above sets of roots are stable under the Cartan involution Θ (attached to K_∞) and hence $\Theta(\mathfrak{q}) = \mathfrak{q}$.

We look at the action of Θ on $\Delta(\mathfrak{u}_\chi) = \Delta_+(\chi) \setminus \Delta_0(\chi)$. For any α in this set we have either $\alpha = \Theta\alpha$ or $\alpha \neq \Theta\alpha$. In the first case it clear that $\Theta : U_\alpha \to U_\alpha$ is the identity if $U_\alpha \subset K_\infty \times \mathbb{C}$ and multiplication by -1 if $U_\alpha \not\subset K_\infty \times \mathbb{C}$. Let us put $\Delta(\mathfrak{u}_\chi)_{nc} := \{\alpha \in \Delta(\mathfrak{u}_\chi) | \Theta\alpha = \alpha;\ \Theta\alpha \text{ acts by } -1\}$ In the second case it is clear that for the Lie algebras $\mathfrak{u}_\alpha \oplus \mathfrak{u}_{\Theta\alpha}$ decomposes into a one dimensional $+$ and a one dimensional $-$ eigenspace. This tells us that

$$R_\chi := \dim((\mathfrak{p} \otimes \mathbb{C} \cap \mathfrak{u}_\chi) = \#\Delta(\mathfrak{u}_\chi)_{nc} + \frac{1}{2}\#\{\alpha \in \Delta(\mathfrak{u}_\chi) | \alpha \neq \Theta\alpha\}$$

The second datum is a highest weight $\lambda \in X^*(T \times_\mathbb{R} \mathbb{C})$, and it has to satisfy two conditions

a) The weight λ is c-autodual (see (6.1)) i.e. $c(\lambda) = -w_0\lambda$.

b) The highest weight λ is trivial on the semi-simple part $Z_\chi^{(1)}$, or, what amounts to the same, λ extends to a character $\lambda : P_\chi \to \mathbb{G}_m/\mathbb{C}$.

Under these conditions Vogan and Zuckerman show that there is a (unique up to isomorphism) Harish-Chandra module $A_{\mathfrak{q}}(\lambda)$ which has non trivial cohomology:

$$H^\bullet(\mathfrak{g}, K_\infty, A_{\mathfrak{q}}(\lambda)\otimes \mathcal{M}_\lambda) = \mathrm{Hom}_{Z_\chi(\mathbb{R})\cap K_\infty}(\Lambda^{\bullet - R_\chi}(\mathfrak{z}_\chi \cap \mathfrak{k}), \mathbb{C}\lambda)$$

where of course $\mathfrak{z}_\chi$ is the Lie algebra of Z_χ, this is the $\mathfrak{l}$ in [113].

We have two extreme cases. In the first case the cocharacter χ is trivial, then the centraliser is the entire group $G\times_{\mathbb{R}}\mathbb{C}/\mathbb{C}$ and then condition b) implies that $\lambda = 0$. This implies that $\mathcal{M}_{\lambda^\vee}$ is the trivial representation, $\mathfrak{q} = \mathfrak{g}$ and $A_{\mathfrak{q}}(0)$ is the trivial one-dimensional $(\mathfrak{g}, K_\infty)$-module. The number $R_\chi = 0$ and the cohomology is

$$H^\bullet(\mathfrak{g}, K_\infty, A_{\mathfrak{q}}(0), \mathbb{C}) = \mathrm{Hom}_{K_\infty}(\Lambda^\bullet(\mathfrak{g}/\mathfrak{k})\otimes\mathbb{C})$$

Here we mention that that $R_\chi = 0$ only holds if χ is trivial. For instance for the group $G = \mathrm{Sl}_n$ we have $R_\chi \geq n-1$ if $\chi \neq 1$ hence we get in this case a vanishing result

If $G/\mathbb{R} = \mathrm{Sl}_n/\mathbb{R}$ and $\chi\neq 1$ then

$$H^q(\mathfrak{g}, K_\infty, A_{\mathfrak{q}}(\lambda)) = 0 \textit{ for } q < n-1$$

In the second case χ is regular, this means that $Z_\chi = T$ and P_χ is a Borel sub group. Now we have the only constraint on λ is a). In this case the $A_{\mathfrak{q}}(\lambda) = A_{\mathfrak{b}}(\lambda)$ are the so called *tempered representations* (see [20], IV, 3.6.). In this case the number R_χ is close to $\frac{1}{2}\dim X$ and hence the non zero cohomology is concentrated near the middle degrees.

The regular cocharacters $\chi \in X_*(T_1^c)\otimes\mathbb{R}$ lie in the complement of finitely many hyperplanes, hence the set $(X_*(T_1^c)\otimes\mathbb{R})^{(0)}$ of regular characters is a finite union of connected components. It is clear from the description that the module $A_{\mathfrak{q}}(\lambda)$ does not change if χ varies inside a connected component. Finally we have the action of the real Weyl group $W(\mathbb{R}) = N(T)(\mathbb{R})/T(\mathbb{R})$ on $X_*(T_1^c)\otimes\mathbb{R}$, and again it is clear that the isomorphism type does not change if we conjugate χ by an element in $W(\mathbb{R})$. Hence we can say that the tempered $A_{\mathfrak{p}}(\lambda)$ are parametrised by $\pi_0((X_*(T_1^c)\otimes\mathbb{R})^{(0)})/W(\mathbb{R})$.

We have a brief look at the case that $G^{(1)}/\mathbb{R}$ has a compact maximal torus T_1^c, i.e. $T = T_1^c$. This case played an important role in the section on the Gauss-Bonnet formula. Then

$$T_1^c\times_{\mathbb{R}}\mathbb{C}\subset {K_\infty}^{(1)}\times_{\mathbb{R}}\mathbb{C}\subset G^{(1)}\times_{\mathbb{R}}\mathbb{C},$$

hence T_1^c is a maximal torus in both reductive groups. In addition, we have the two (absolute) Weyl groups $W_{K_\infty} = W(\mathbb{R}) = N_{K_\infty}(T)(\mathbb{C})/T(\mathbb{C})$ and $W_G = N_G(T)(\mathbb{C})/T(\mathbb{C})$. The big Weyl group W_G acts simply transitively on the set of connected components of $(X_*(T_1^c)\otimes\mathbb{R})^{(0)}$, thus, $W_G = \pi_0((X_*(T_1^c)\otimes\mathbb{R})^{(0)})$ once we choose a base point $[\chi_0]\in\pi_0((X_*(T_1^c)\otimes\mathbb{R})^{(0)})$. Therefore we get a family

$$\{A_{w\chi_0}(\lambda)\}_{w\in W_{K_\infty}\backslash W_G},$$

and the results of Vogan and Zuckerman assert:

These repesentations are unitary, they are pairwise non-isomorphic, and they are the Harish-Chandra modules attached to the discrete series representations of $G(\mathbb{R})$. The cohomology groups are given by

$$H^q(\mathfrak{g}, K_\infty, A_{w\chi_0}(\lambda) \otimes \mathcal{M}_{\lambda^\vee}) = \begin{cases} \mathbb{C} & \textit{if } q = \frac{d}{2} \\ 0 & \textit{else} \end{cases},$$

where d denotes the (real) dimension of the symmetric space attached to $G(\mathbb{R})$.

Now it is clear that for a regular highest weight λ the condition b) forces the cocharacter χ to be regular.

We come back to the question raised above. Assume λ is regular and that we have an isotypical component $H^\bullet(\Gamma\backslash X, \tilde{\mathcal{M}}_{\lambda^\vee})(\pi_f)$. Then the possible "friends" are the $A_\chi(\lambda)$ with χ regular. Hence we get

$$\begin{aligned} H^\bullet_!(\Gamma\backslash X, \tilde{\mathcal{M}}_{\lambda^\vee})(\pi_f) &= \bigoplus_{w \in W_K\backslash W_G} H^{\frac{d}{2}}(\mathfrak{g}, K_\infty, A_{w\chi_0}(\lambda) \otimes \mathcal{M}_{\lambda^\vee})^{m(w\chi_0 \times \pi_f)} \otimes H_{\pi_f} \\ &= \bigoplus_{w \in W_K\backslash W_G} \mathbb{C}^{m(w\chi_0 \times \pi_f)} \otimes H_{\pi_f}, \end{aligned}$$

where $m(w\chi_0 \times \pi_f)$ is the multiplicity of $A_{w\chi_0}(\lambda) \times \pi_f$ in $L^2_{\mathrm{disc}}(\Gamma\backslash G(\mathbb{R})$. (In Arthurs' words : if λ is regular then the only friends of a $\pi_f \in \mathrm{Coh}(H^\bullet_!(\Gamma\backslash X, \tilde{\mathcal{M}}_\lambda))$ are the $w\chi_0$.)

If we refrain from decomposing into isotypical subspaces then we get a simpler formula

$$H^\bullet_!(\Gamma\backslash X, \tilde{\mathcal{M}}_\lambda) = \bigoplus_{w \in W_K\backslash W_G} \mathbb{C}^{m(w\chi_0)}$$

where of course $m(w\chi_0)$ is the multiplicity of $A_{w\chi_0}(\lambda)$ in $L^2_{\mathrm{disc}}(\Gamma\backslash G(\mathbb{R}))$. Actually we know that $A_{w\chi_0}(\lambda)$ must even lie in the cuspidal spectrum (see [115]). In principle we already used this fact because we tacitly used the theorem of Borel (see Thm. 6.2.1.b).

6.2.3 Cuspidal cohomology vs. inner cohomology

Now we remember that in the previous sections we made the convention (see end of 6.1.1) that our coefficient systems $\mathcal{M}_\lambda$ are $\mathbb{C}$-vector spaces. We now revoke this convention and recall that the coefficient systems $\mathcal{M}_\lambda$ should be replaced by $\mathcal{M}_\lambda \otimes_F \mathbb{C}$, where F is some number field over which $\mathcal{M}_\lambda$ is defined. Then in the above list (6.20) of four subspaces in the cohomology, the second and the fourth subspaces have a natural structure of F-vector spaces and they have a combinatorial definition, whereas the first and third subspaces need some input from analysis in their definition. In other words, if we replace $\mathcal{M}_\lambda$ in (6.20) by $\mathcal{M}_\lambda \otimes_F \mathbb{C}$ then (6.20) can be written as

$$\begin{aligned} H^\bullet_{\mathrm{cusp}}(\Gamma\backslash X, \tilde{\mathcal{M}}_\lambda \otimes_F \mathbb{C}) &\subset H^\bullet_!(\Gamma\backslash X, \tilde{\mathcal{M}}_\lambda) \otimes_F \mathbb{C} \\ &\subset H^\bullet_{(2)}(\Gamma\backslash X, \tilde{\mathcal{M}}_\lambda \otimes_F \mathbb{C}) \subset H^\bullet(\Gamma\backslash X, \tilde{\mathcal{M}}_\lambda) \otimes_F \mathbb{C}. \end{aligned}$$

It is a very important question to understand the discrepancy between the first two steps. If λ is regular then it follows from the results of [81] that in fact

$$H^\bullet_{\mathrm{cusp}}(\Gamma\backslash X, \tilde{\mathcal{M}}_\lambda \otimes_F \mathbb{C}) = H^\bullet_!(\Gamma\backslash X, \tilde{\mathcal{M}}_\lambda) \otimes_F \mathbb{C},$$

but without the assumption of λ being regular this is not true for interesting reasons.

Of course we should also take the action of the Hecke algebra into account. If π_f is the isomorphism type of an absolutely irreducible Hecke module which is defined over F, then we can consider

$$H^\bullet_{\mathrm{cusp}}(\Gamma\backslash X, \tilde{\mathcal{M}}_\lambda \otimes_F \mathbb{C})(\pi_f) \subset H^\bullet_!(\Gamma\backslash X, \tilde{\mathcal{M}}_\lambda)(\pi_f) \otimes_F \mathbb{C}$$

and compare these two modules. We will say that π_f is *strongly inner* if we have equality in the above.

We come back to this issue in part 2 of this book.

At this point we encounter an interesting problem. We recall that we have the three subspaces

$$\begin{aligned} H^\bullet_{\mathrm{cusp}}(\Gamma\backslash X, \tilde{\mathcal{M}}_\lambda \otimes \mathbb{C}) &\subset H^\bullet_!(\Gamma\backslash X, \tilde{\mathcal{M}}_\lambda) \otimes \mathbb{C} \\ &\subset H^\bullet_{(2)}(\Gamma\backslash X, \tilde{\mathcal{M}}_\lambda \otimes \mathbb{C}) \subset H^\bullet(\Gamma\backslash X, \tilde{\mathcal{M}}_\lambda) \otimes \mathbb{C}, \end{aligned}$$

note the positions of the tensor symbol $\otimes$. The first and the third space are only defined after we tensorize the coefficient system by $\mathbb{C}$, whereas the second and the fourth cohomology groups are by definition F vector spaces tensorized by $\mathbb{C}$.

Now the question is whether the first and the third space also have a natural F-vector space structure. Of course we get a positive answer, if the Manin-Drinfeld principle holds. All the vector spaces are of course modules under the Hecke algebra and we can look at their spectra

$$\begin{aligned} &\Sigma(H^\bullet_{\mathrm{cusp}}(\Gamma\backslash X, \tilde{\mathcal{M}}_\lambda \otimes \mathbb{C})) =: \Sigma_{\mathrm{cusp}} &\quad &\Sigma(H^\bullet_!(\Gamma\backslash X, \tilde{\mathcal{M}}_\lambda \otimes \mathbb{C})) =: \Sigma_!, \\ &\Sigma(H^\bullet_{(2)}(\Gamma\backslash X, \tilde{\mathcal{M}}_\lambda \otimes \mathbb{C})) =: \Sigma_{(2)} &\quad &\Sigma(H^\bullet(\Gamma\backslash X, \tilde{\mathcal{M}}_\lambda \otimes \mathbb{C})) =: \Sigma. \end{aligned}$$

If now for instance $\Sigma_{\mathrm{cusp}} \cap (\Sigma_! \setminus \Sigma_{\mathrm{cusp}}) = \emptyset$ then we can define $H^\bullet_{\mathrm{cusp}}(\Gamma\backslash X, \tilde{\mathcal{M}}_\lambda) \subset H^\bullet_!(\Gamma\backslash X, \tilde{\mathcal{M}}_\lambda)$ as the subspace which is the sum of the isotypical components in Σ_{cusp}.

If this is the case, we say that the cuspidal cohomology is *intrinsically definable* and we get a canonical decomposition

$$H^\bullet_!(\Gamma\backslash X, \tilde{\mathcal{M}}_\lambda) = H^\bullet_{\mathrm{cusp}}(\Gamma\backslash X, \tilde{\mathcal{M}}_\lambda) \oplus H^\bullet_{!,\mathrm{noncusp}}(\Gamma\backslash X, \tilde{\mathcal{M}}_\lambda).$$

The classical Manin-Drinfeld principle refers to the two spectra $\Sigma_! \subset \Sigma$. If it is true in this case we get a decomposition

$$H^\bullet(\Gamma\backslash X, \tilde{\mathcal{M}}_\lambda) = H^\bullet_!(\Gamma\backslash X, \tilde{\mathcal{M}}_\lambda) \oplus H^\bullet_{\mathrm{Eis}}(\Gamma\backslash X, \tilde{\mathcal{M}}_\lambda),$$

the canonical complement is called the Eisenstein cohomology (see Chap. 2, 2.2.3 and Chap. 3 Section 5).

6.2.4 Consequences

Vanishing Theorems

If V is an irreducible unitary $(\mathfrak{g}, K_\infty)$-module, then we have that $\bar{V} \xrightarrow{\sim} V^\vee$ and this implies for the infinitesimal character

$$\overline{\chi_V(z)} = \chi_{V^\vee}(z) \text{ for all } z \in \mathfrak{Z}(\mathfrak{g}).$$

Combining this with Wigner's lemma we can conclude

If V is an irreducible unitary $(\mathfrak{g}, K_\infty)$-module, $\mathcal{M}_\lambda$ an irreducible rational representation, and if

$$H^\bullet(\mathfrak{g}, K_\infty, V \otimes \mathcal{M}_\lambda) \neq 0$$

then $\chi_{\mathcal{M}_\lambda^\vee}(z) = \chi_{\mathcal{M}_\lambda}({}^t z) = \chi_{\bar{\mathcal{M}}_\lambda}(z)$.

In other words: for an unitary irreducible $(\mathfrak{g}, K_\infty)$-module V, the cohomology with coefficients in an irreducible rational representation $\mathcal{M}_\lambda$ vanishes, unless we have $\mathcal{M}_\lambda^\vee \xrightarrow{\sim} \bar{\mathcal{M}}_\lambda$, or in terms of highest weights, unless $-w_0(\lambda) = c(\lambda)$ (see 3.1.1).

If we combine this with the considerations following Wigner's lemma we get the result

Corollary 6.2.1. *If $\mathcal{M}$ is an absolutely irreducible rational representation and if $\mathcal{M}_\lambda^\vee$ is not isomorphic to $\bar{\mathcal{M}}_\lambda$ then*

$$H^\bullet_{(2)}(\Gamma\backslash X, \tilde{\mathcal{M}}_\lambda) = 0.$$

Hence we also have $H^\bullet_!(\Gamma\backslash X, \tilde{\mathcal{M}}_\lambda) = 0$.

We will discuss examples for this result below.

The group $G/\mathbb{Q} = \mathrm{Sl}_2/\mathbb{Q}$

Let us consider the group $G/\mathbb{Q} = \mathrm{Sl}_2/\mathbb{Q}$. We have tautological representations $\mathrm{Sl}_2 \hookrightarrow \mathrm{Gl}(\mathbb{Q}^2) = \mathrm{Gl}(V)$, and we get all irreducible representations via the symmetric powers $\mathcal{M}_n = \mathrm{Sym}^n(V)$ of V.

In this case the Vogan-Zuckerman list is very short. It is discussed in Chapter 4 for the groups $\mathrm{Sl}_2(\mathbb{R})$ and $\mathrm{Sl}_2(\mathbb{C})$, where both groups are considered as real Lie-groups.

In the case $\mathrm{Sl}_2(\mathbb{R})$ we have the trivial module $\mathbb{C}$ and for any integer $k \geq 2$ we have two irreducible unitarizable $(\mathfrak{g}, K_\infty)$-modules $\mathcal{D}_k^\pm$ (the discrete series representations). These are the only $(\mathfrak{g}, K_\infty)$-modules which have non-trivial cohomology with coefficients in a rational representation. If we now pick one of our rational representations $\mathcal{M}_n$, then the non-vanishing cohomology groups are

$$H^q(\mathfrak{g}, K_\infty, \mathcal{M}_n \otimes \mathbb{C}) = \mathbb{C} \text{ for } n = 0, q = 0, 2$$
$$H^q(\mathfrak{g}, K_\infty, \mathcal{D}_k^\pm \otimes \mathcal{M}_n \otimes \mathbb{C}) = \mathbb{C} \text{ for } n = k-2, q = 1.$$

The trivial $(\mathfrak{g}, K_\infty)$-module $\mathbb{C}$ occurs with multiplicity one in $L^2(\Gamma\backslash G(\mathbb{R}))$, hence we get for the trivial coefficient system a contribution

$$H^\bullet(\mathfrak{g}, K_\infty, \mathbb{C}) = H^0(\mathfrak{g}, K_\infty, \mathbb{C}) \oplus H^2(\mathfrak{g}, K_\infty, \mathbb{C}) = \mathbb{C} \oplus \mathbb{C} \to H^\bullet_{(2)}(\Gamma\backslash X, \mathbb{C}).$$

This map is injective in degree 0 and zero in degree 2.

For the modules $\mathcal{D}_k^{\pm}$ we have to determine the multiplicities $m^{\pm}(k)$ of these modules in the discrete spectrum of $L^2(\Gamma\backslash G(\mathbb{R}))$. A simple argument using complex conjugation tells us $m^+(k) = m^-(k)$. Now we have the fundamental observation made by Gelfand and Graev, which links representation theory to automorphic forms:

We have an isomorphism

$$\mathrm{Hom}_{(\mathfrak{g},K_\infty)}(\mathcal{D}_k^+, L^2_{\mathrm{disc}}(\Gamma\backslash G(\mathbb{R}))) \xrightarrow{\sim} S_k(\Gamma\backslash\mathbb{H})$$

where $S_k(\Gamma\backslash\mathbb{H})$ is the space of holomorphic cusp forms of weight k and level Γ.

We have explained how we get, starting from a holomorphic cusp form f of weight k, an inclusion

$$\Phi_f : \mathcal{D}_k^+ \hookrightarrow L^2_{\mathrm{disc}}(\Gamma\backslash G(\mathbb{R}))$$

and how this map $f \mapsto \Phi_f$ establishes the above isomorphism. This gives us the famous Eichler-Shimura isomorphism

$$S_k(\Gamma\backslash\mathbb{H}) \oplus \overline{S_k(\Gamma\backslash\mathbb{H})} \xrightarrow{\sim} H^1_!(\Gamma\backslash X, \tilde{\mathcal{M}}_{k-2}).$$

The group $G/\mathbb{Q} = R_{F/\mathbb{Q}}(\mathrm{Sl}_2/F)$.

For any finite extension $F/\mathbb{Q}$ we may consider the base restriction $G/\mathbb{Q} = R_{F/\mathbb{Q}}(\mathrm{Sl}_2/F)$ (see Chap-II, 1.1.1). Here we want to consider the special case of $F/\mathbb{Q}$ being imaginary quadratic. In this case we have $G \otimes \mathbb{C} = \mathrm{Sl}_2 \times \mathrm{Sl}_2/\mathbb{C}$, where the factors correspond to the two embeddings of F into $\mathbb{C}$. The rational irreducible representations are tensor products of irreducible representations of the two factors $\mathcal{M}_\lambda = \mathcal{M}_{k_1} \otimes \mathcal{M}_{k_2}$, where again $\mathcal{M}_k = \mathrm{Sym}^k(\mathbb{C}^2)$. These representations are defined over F.

In this case we discuss the Vogan-Zuckerman list in Chapter 4, but here we want to discuss a particular aspect of this. We observe that

$$\mathcal{M}_\lambda^\vee \xrightarrow{\sim} \mathcal{M}_{k_1} \otimes \mathcal{M}_{k_2}, \bar{\mathcal{M}}_\lambda = \mathcal{M}_{k_2} \otimes \mathcal{M}_{k_1},$$

and hence our corollary above yields for any choice of the congruence group Γ.

$$H_!^\bullet(\Gamma\backslash X, \mathcal{M}) = 0 \text{ if } k_1 \neq k_2.$$

In Chapter 2 we discussed the special examples in low dimensions. We take $F = \mathbb{Q}[i]$ and $\Gamma = \mathrm{Sl}_2(\mathbb{Z}[\mathbf{i}])$ which amounts to taking the standard maximal compact subgroup $K_f = \mathrm{Sl}_2(\hat{\mathcal{O}}_F)$. If now for instance $k_1 > 0$ and $k_2 = 0$, then we get $H_!^\bullet(\Gamma\backslash X, \mathcal{M}_\lambda) = 0$. Hence we have by definition $H^\bullet(\Gamma\backslash X, \tilde{\mathcal{M}}) = H^\bullet_{\mathrm{Eis}}(\Gamma\backslash X, \tilde{\mathcal{M}})$ and we have complete control over the Eisenstein cohomology in this case. Hence we know the cohomology in this case if we apply the analytic methods.

On the other hand, in Chapter 2 we wrote an explicit complex of finite-dimensional vector spaces, which computes the cohomology. It is not clear to me how we can read off from this complex the structure of the cohomology groups.

We get another example where this phenomenon happens, if we consider the group $\mathrm{Sl}_n/\mathbb{Q}$ for $n > 2$. In 2.1.2 we described the simple roots $\alpha_1, \alpha_2, \ldots, \alpha_{n-1}$.

Accordingly we have the fundamental highest weights $\omega_1, \dots, \omega_{n-1}$. The element $-w_0$ (see 6.1.1) has the effect of reversing the order of the weights. Hence we see that for $\lambda = \sum n_i \omega_i$ we have

$$H_!^\bullet(\Gamma\backslash X, \mathcal{M}_\lambda) = 0$$

unless we have $-w_0(\lambda) = \lambda$, and this means $n_i = n_{n-1-i}$.

The algebraic K-theory of number fields

We briefly recall the definition of the K-groups of an algebraic number field $F/\mathbb{Q}$. There are many references, and we refer to the graduate text book [95].

We consider the group $\mathrm{Gl}_n(\mathcal{O}_F)$, which has a classifying space BG_n. We can pass to the limit $\lim\limits_{n\to\infty} \mathrm{Gl}_n(\mathcal{O}_F) = \mathrm{Gl}(\mathcal{O}_F) = \mathrm{G}$ and let BG be its classifying space. Quillen invented a procedure to modify this space to another space BG^+, whose fundamental group is now abelian, but which has the same homology and cohomology as BG. Then Quillen defines the algebraic K-groups as

$$K_i(\mathcal{O}_F) = \pi_i(\mathrm{BG}^+).$$

The space BG^+ is an H-space, which means that we have a multiplication $m : \mathrm{BG}^+ \times \mathrm{BG}^+ \to \mathrm{BG}^+$ which has a two-sided identity element. We also have the diagonal $\Delta : \mathrm{BG}^+ \to \mathrm{BG}^+ \times \mathrm{BG}^+$. Then by the Milnor-Moore theorem we know that the Hurewicz homomorphism $\pi_i(\mathrm{BG}^+) \otimes \mathbb{Q} \to H_i(\mathrm{BG}, \mathbb{Q})$ provides an isomorphism

$$\pi_i(\mathrm{BG}^+) \otimes \mathbb{Q} \xrightarrow{\sim} \{x \in H_i(\mathrm{BG}, \mathbb{Q}) \mid \Delta_i(x) = 1 \otimes x + x \otimes 1\}.$$

The submodule on the righthand side is called the module of primitive elements $P(H_i(\mathrm{BG}, \mathbb{Q}))$.

Obviously $P(H_i(\mathrm{BG}, \mathbb{Q})) \simeq P(H_i(\mathrm{G}, \mathbb{Q})) \simeq \lim\limits_{n\to\infty} P(H_i(\mathrm{Gl}_n(\mathcal{O}_F), \mathbb{Q}))$. We will see later on that for a fixed degree i the homology groups $H_i(\mathrm{Gl}_N(\mathcal{O}_F), \mathbb{Q})$ become stable for large N and the same holds for their submodules of primitive elements. By going to the dual we get the module of undecomposable elements of the cohomology algebra

$$UH^\bullet(\mathrm{Gl}_N(\mathcal{O}_F), \mathbb{Q}) := H^\bullet(\mathrm{Gl}_N(\mathcal{O}_F), \mathbb{Q})/H^{\geq 1}(\mathrm{Gl}_N(\mathcal{O}_F), \mathbb{Q}) \cdot H^{\geq 1}(\mathrm{Gl}_N(\mathcal{O}_F), \mathbb{Q}).$$

Since the map $\mathrm{Sl}_N(\mathcal{O}_F) \times \mathcal{O}_F^\times \to \mathrm{Gl}_N(\mathcal{O}_F)$ has finite kernel and a normal image of finite index, it induces an isomorphism in group cohomology with rational coefficients. By applying a Künneth theorem and Dirichlet's structure theorem about units of rings of integers in number fields, we get for $i \geq 2$:

$$UH^i(\mathrm{Sl}_N(\mathcal{O}_F), \mathbb{Q}) \simeq UH^i(\mathrm{Gl}_N(\mathcal{O}_F), \mathbb{Q}).$$

On the other hand,

$$H^i(\mathrm{Sl}_N(\mathcal{O}_F), \mathbb{Q}) \simeq H^i(\mathrm{Sl}_N(\mathcal{O}_F)\backslash X_F^{(N)}, \mathbb{Q})$$

with the symmetric space

$$X_F^{(N)} = \mathrm{Sl}_N(F \otimes \mathbb{R})/K_\infty = \prod_{i=1}^{r_1} \mathrm{Sl}_N(\mathbb{R})/\mathrm{SO}(N) \times \prod_{i=1}^{r_2} \mathrm{Sl}_N(\mathbb{C})/\mathrm{SU}(N).$$

In the years around 1970, various people proved square integrability criteria for the cohomology classes of arithmetic groups ([39], [17],...). In our situation here, these results imply that in low degrees $i < \frac{N}{4}$

$$H^i(\mathrm{Sl}_N(\mathcal{O}_F)\backslash X_F^{(N)}, \mathbb{C}) = H^i_{(2)}(\mathrm{Sl}_N(\mathcal{O}_F)\backslash X_F^{(N)}, \mathbb{C}).$$

On the other hand it follows from the Vogan-Zuckerman classification ([113]) that the only irreducible unitary $(\mathfrak{g}, K_\infty)$-modules $A_\mathfrak{q}$, for which $H^i(\mathfrak{g}, K_\infty, A_\mathfrak{q}) \neq 0$ for some $i \leq N/4$ are one dimensional, i.e. $\lambda = 0, \mathfrak{q} = \mathfrak{g}$.

Hence we see that in low degrees $i \leq N/4$ the natural map

$$H^i(\mathfrak{g}, K_\infty, \mathbb{C}) \to H^i(\Gamma\backslash X, \mathbb{C})$$

is an isomorphism (the proof of injectivity requires some additional reasoning). Now $H^q(\mathfrak{g}, K_\infty, \mathbb{C}) = \mathrm{Hom}_{K_\infty}(\Lambda^\bullet(\mathfrak{g}/\mathfrak{k}), \mathbb{C})$, and obviously this last complex is isomorphic to the complex $\Omega^\bullet(X)^{G(\mathbb{R})}$ of $G(\mathbb{R})$-invariant forms on the symmetric space $G(\mathbb{R})/K_\infty$. Now the complex $\Omega^\bullet(X)^{G(\mathbb{R})}$ of invariant differential forms (all differentials are zero) does not change if we replace the group $G/\mathbb{R}$ by its compact form $G_c/\mathbb{R}$, and then we get the complex of invariant forms on the compact twin of our symmetric space

$$X_c = \prod_{i=1}^{r_1} \mathrm{SU}_N(\mathbb{R})/\mathrm{SO}(N) \times \prod_{i=1}^{r_2} (\mathrm{SU}(N) \times \mathrm{SU}(N))/\mathrm{SU}(N),$$

and hence

$$H^\bullet(\mathfrak{g}, K_\infty, \mathbb{C}) = \Omega^\bullet(X_c)^{G_c(\mathbb{R})} = H^\bullet(X_c, \mathbb{C}).$$

The cohomology of the topological spaces like the one on the right hand side has been computed by Borel in the early days of his career (c.f. [12]):

$$H^\bullet(\mathrm{SU}(2N+1)/\mathrm{SO}(2N+1), \mathbb{C}) = \Lambda^\bullet(\mathbb{C}e_5 \oplus \cdots \oplus \mathbb{C}e_{4N+1})$$
$$H^\bullet(\mathrm{SU}(N), \mathbb{C}) = \Lambda^\bullet(\mathbb{C}f_3 \oplus \cdots \oplus \mathbb{C}f_{2N-1}).$$

Here the $e_{4\nu+1}$(resp. $f_{2\mu+1}$) are non-zero classes, and we can choose them to be rational classes, compatible with the restriction map associated to the symmetric space for N to that for $N-1$. They also give us a dual basis for the primitive elements. In particular we see the announced stability property of the (co)homology groups. Putting together we get for the ranks of the K-groups $K_i(\mathcal{O}_F) \otimes \mathbb{Q}$ with $i \geq 2$ the following result of Borel:

$$\mathit{rank}\ K_i(\mathcal{O}_F) \otimes \mathbb{Q} = \begin{cases} r_1 + r_2 & \text{if } i = 4k+1 \\ r_2 & \text{if } i = 4k+3 \\ 0 & \text{else.} \end{cases}$$

6.2.5 The Eisenstein cohomology program

We return to our fundamental exact sequence

$$0 \to H^\bullet_!(\Gamma\backslash X, \tilde{\mathcal{M}}) \to H^\bullet(\Gamma\backslash X, \tilde{\mathcal{M}}) \xrightarrow{r} H^\bullet(\overset{\bullet}{\mathcal{N}}(\Gamma\backslash X), \tilde{\mathcal{M}}) \to \cdots. \qquad (6.21)$$

We know that $H^\bullet_!(\Gamma\backslash X, \tilde{\mathcal{M}}) \subset H^\bullet_{(2)}(\Gamma\backslash X, \tilde{\mathcal{M}})$ and we gained some understanding of the latter space using analytic methods. We have seen that cohomology classes $[\omega] \in H^\bullet_{(2)}(\Gamma\backslash X, \tilde{\mathcal{M}} \otimes \mathbb{C})$ can be represented by harmonic forms ω.

The goal of the Eisenstein cohomology is to gain some understanding of the the restriction map r, in particular we want to describe the image of this restriction map r. This leads us to very difficult questions. Again we have to use analytic tools (Langlands theory of Eisenstein series), and we touched this subject only very superficially in this part 1.

If we want to describe the image of r as a subspace of $H^\bullet(\overset{\bullet}{\mathcal{N}}(\Gamma\backslash X), \tilde{\mathcal{M}})$, it might be better to understand the boundary cohomology first.

For the computation of $H^\bullet(\overset{\bullet}{\mathcal{N}}(\Gamma\backslash X), \tilde{\mathcal{M}}) = H^\bullet(\partial(\Gamma\backslash \bar{X}), \tilde{\mathcal{M}})$ we invoke the spectral sequences (2.41), (2.42), their $E_2^{p,q}$-terms are the cohomology of the complex

$$\cdots \to \bigoplus_{[P]:d(P)=p+1} H^q(\Gamma_P\backslash X, \tilde{\mathcal{M}}) \xrightarrow{d_1^{p,q}} \bigoplus_{[Q]:d(Q)=p+2} H^q(\Gamma_Q\backslash X, \tilde{\mathcal{M}}) \to \cdots . \tag{6.22}$$

We can go one step further and employ the spectral sequence (2.31) and invoke Kostant's theorem to decompose the cohomology of the fiber into weight spaces. We explain this briefly in the following subsection.

The cohomology of unipotent groups

We now drop the subscript $_P$, and we know that the group scheme $U/\mathbb{Q}$ is a unipotent group scheme. Let $A = A(U)$ be its affine algebra (see Section 1.1.1). Then $U/\mathbb{Q}$ has a filtration by normal subgroup schemes

$$U_0 = \{e\} \subset U_1 \subset U_2 \subset \cdots \subset U_{m-1} \subset U_m = U$$

such that $U_i/U_{i-1} \xrightarrow{\sim} \mathcal{G}_a$. The subgroup $\Gamma_U \subset U(\mathbb{Q})$ is Zariski dense, more precisely, we know the following: if we define $\Gamma_i = U_i(\mathbb{Q}) \cap \Gamma$ then we see $\Gamma_i/\Gamma_{i-1} \xrightarrow{\sim} \mathbb{Z} \subset U_i/U_{i-1}(\mathbb{R}) \xrightarrow{\sim} \mathbb{R}$.

We consider the category of $U/\mathbb{Q}$-modules $\mathrm{Mod}_{\mathrm{U}}$ (see 1.1.1). We choose the action of U on A by right translations on A. It follows from Frobenius reciprocity that the $U/\mathbb{Q}$-module A is an injective module in Mod_U. This implies that we get an injective resolution of the $U/\mathbb{Q}$-module $\mathbb{Q}$ by

$$0 \to \mathbb{Q} \to A \to (A/\mathbb{Q}) \otimes A \to \cdots = 0 \to \mathbb{Q} \to I^0 \to I^1 \to \cdots$$

and hence

$$\begin{aligned} H^q(U, \mathcal{M}) &= H^q(0 \to (I^1 \otimes \mathcal{M})^U \to (I^2 \otimes \mathcal{M})^U \to \cdots) \\ &= H^q((I^\bullet \otimes \mathcal{M})^U). \end{aligned} \tag{6.23}$$

It is clear that the functor $\mathcal{M} \to \mathcal{M}^U$ is equal to $\mathcal{M} \to \mathcal{M}^{\Gamma_U}$. (Our $\mathbb{Z}$- module $\mathcal{M}$ above is now a $\mathbb{Q}$-vector space, i.e. we consider coefficient systems with rational coefficients.) To conclude that the cohomology groups $H^q(U, \mathcal{M})$ and

$H^q(\Gamma_U, \mathcal{M})$ coincide we have to prove that injective $U/\mathbb{Q}$-modules are acyclic as Γ_U-modules. Actually it is enough to check this for the $U/\mathbb{Q}$-module A, since the resolution $I^\bullet \otimes \mathcal{M}$ shows, that any $U/\mathbb{Q}$-module has an injective resolution consisting of sums of copies of A. For $U = \mathcal{G}_a$ this means we have to prove the vanishing of $H^1(\mathbb{Z}, \mathbb{Q}[X]) = \mathbb{Q}[X]/(1-T)\mathbb{Q}[X]$, where $TP(X) = P(X+1)$. But this is obviously true since $1 - T$ maps polynomials of degree $\leq n+1$ to polynomials of degree $\leq n$, and for dimension reasons the map is surjective. A (nontrivial) general unipotent group U has a central subgroup $\mathcal{G}_a$ with quotient $\overline{U}$, and by the Hochschild-Serre spectral sequence and using induction we have

$$H^q(\Gamma_U, A(U)) \simeq H^q(\Gamma_{\overline{U}}, A(U)^{\mathbb{Z}}) \simeq H^q(\Gamma_{\overline{U}}, A(\overline{U})) = 0 \text{ for } q > 0.$$

Hence we have

$$H^q(U, \mathcal{M}) = H^q(\Gamma_U, \mathcal{M}) = H^q(\Gamma_U \backslash U(\mathbb{R}), \mathcal{M}). \tag{6.24}$$

Since $U/\mathbb{Q}$ is the unipotent radical of the parabolic group $P/\mathbb{Q}$, the parabolic group $P/\mathbb{Q}$ acts via the adjoint action on the modules $I^m \otimes \mathcal{M}$. This action respects the submodules $(I^m \otimes \mathcal{M})^U$, and $U/\mathbb{Q}$ acts trivially on $(I^m \otimes \mathcal{M})^U$. This implies that the modules $(I^m \otimes \mathcal{M})^U$ are $M/\mathbb{Q} = (P/U)/\mathbb{Q}$-modules. The group $M/\mathbb{Q}$ is reductive, and we know that the category of $M/\mathbb{Q}$-modules is semi-simple. This implies that we can decompose

$$(I^\bullet \otimes \mathcal{M})^U = \mathbb{H}^\bullet(U, \mathcal{M}) \oplus ACI(I^\bullet)^U \tag{6.25}$$

where the first summand is a complex of $M/\mathbb{Q}$-modules in which all the differentials are zero, and the second is an acyclic complex of $M/\mathbb{Q}$-modules. Hence

$$H^\bullet(U, \mathcal{M}) = H^\bullet(\Gamma_U, \mathcal{M}) \xrightarrow{\sim} \mathbb{H}^\bullet(U, \mathcal{M}). \tag{6.26}$$

We get a "smaller" resolution from the (algebraic) de-Rham complex of differential forms. On the smooth affine scheme $U/\mathbb{Q}$ we have the sheaves of differential forms $\Omega^p_U = \Lambda^p \Omega^1_U$ (see [49], 7.5) and we have the de-Rham complex

$$\Omega(U)^\bullet : 0 \to \Omega^0(U) \to \Omega^1(U) \to \Omega^2(U) \to \cdots,$$

where $\Omega^p(U) = \Omega^p_U(U)$ is the module of global sections and $A = \Omega^0(U)$. These modules of differentials are as A-modules freely generated by the invariant differential forms. Hence they are injective. Since our unipotent group scheme $U/\mathbb{Q}$ is isomorphic to the affine space $\mathbb{A}^d$ (as an affine scheme) we easily see that this de-Rham complex is acyclic and hence provides a resolution of $\mathbb{Q}$. As before, we get the cohomology by taking the complex $(\Omega^p(U) \otimes \mathcal{M})^U$ of invariants under the action of $U/\mathbb{Q}$. Since a $U/\mathbb{Q}$-invariant differential form with values in $\mathcal{M}$ is determined by its value at the identity e, the complex of invariants under $U/\mathbb{Q}$ becomes

$$\mathrm{Hom}(\Lambda^\bullet \mathfrak{u}, \mathcal{M}) : 0 \to \mathcal{M} \to \mathrm{Hom}(\mathfrak{u}, \mathcal{M}) \to \mathrm{Hom}(\Lambda^2 \mathfrak{u}, \mathcal{M}) \to \ldots, \tag{6.27}$$

and the cohomology of this complex is the Lie-algebra cohomology $H^\bullet(\mathfrak{u}, \mathcal{M})$. We still have the action of $P/\mathbb{Q}$ on $\mathfrak{u}$ by the adjoint action, hence we get an action of P on $\mathrm{Hom}(\Lambda^\bullet \mathfrak{u}, \mathcal{M})$. We conclude

Theorem 6.2.2. *We have isomorphism of $M/\mathbb{Q}$-modules*

$$H^\bullet(U, \mathcal{M}) \xrightarrow{\sim} H^\bullet(\mathfrak{u}, \mathcal{M}) \xrightarrow{\sim} \mathbb{H}^\bullet(\mathfrak{u}, \mathcal{M}).$$

We note that this result, combined with the identification (6.24), gives an algebraic proof of van Est's Lie theoretic isomorphism [111].

A famous theorem due to Kostant [73] yields a description of the $M/\mathbb{Q}$-module $\mathbb{H}^\bullet(\mathfrak{u},\mathcal{M})$. Let $\lambda \in X^*(T)$ be the highest weight of $\mathcal{M}$, i.e. we have $\mathcal{M}=\mathcal{M}_\lambda$. The set

$$W^P = \{w \in W \mid \alpha \in \Delta_M^+,\ w^{-1}(\alpha) \in \Delta^+\}$$

is the set of Kostant representatives for $W^M\backslash W$. For any $w \in W^P$ we define the element

$$\omega_w = (\Lambda_{\alpha\in\Delta_U; w^{-1}\alpha<0}\, u_\alpha^\vee) \otimes e_{w\lambda}.$$

It is clear that this is the highest weight vector of an irreducible M-module of highest weight $w\cdot\lambda = w(\lambda+\rho)-\rho$. One checks that this module appears with multiplicity one in the full Lie algebra cohomology complex, hence it is annulated by the differential and gives an irreducible submodule

$$\mathbb{H}^\bullet(\mathfrak{u},\mathcal{M}_\lambda)_{w\cdot\lambda} \subset \mathbb{H}^\bullet(\mathfrak{u},\mathcal{M}_\lambda).$$

Now the theorem of Kostant [73] says

Theorem 6.2.3.

$$\mathbb{H}^\bullet(\mathfrak{u},\mathcal{M}) = \bigoplus_{w\in W^P} \mathbb{H}^\bullet(\mathfrak{u},\mathcal{M})_{w\cdot\lambda} = \bigoplus_{w\in W^P} \mathbb{H}^{l(w)}(\mathfrak{u},\mathcal{M})_{w\cdot\lambda},$$

where we have to be aware that the summand $\mathbb{H}^\bullet(\mathfrak{u},\mathcal{M})_{w\cdot\lambda}$ *sits in degree* $l(w)$.

Proof. Rather clear after the preparation. □

Since the differentials in the complex $\mathbb{H}^\bullet(\mathfrak{u},\mathcal{M}_\lambda)$ are zero, the spectral sequence attached to the fibration

$$\Gamma_P\backslash X^P(\underline{c}_{\pi'}) \longrightarrow \Gamma_M\backslash X^M(r(\underline{c}_{\pi'}))$$

degenerates, and we get

$$H^n(\Gamma_P\backslash X^P(\underline{c}_{\pi'}),\tilde{\mathcal{M}}) \xrightarrow{\sim} \bigoplus_{w\in W^P} H^{n-l(w)}(\Gamma_M\backslash X^M(r(\underline{c}_{\pi'})),\widetilde{\mathbb{H}^\bullet(\mathfrak{u},\mathcal{M})_{w\cdot\lambda}}).$$

This is the decomposition of the cohomology of the boundary stratum $\partial_P(\Gamma\backslash X)$ into weight spaces. We insert this decomposition into the spectral sequence and get a more precise description of the $E_1^{p,q}$ page (cf. 6.22):

$$\xrightarrow{d_1^{p-1,q}} \bigoplus_{[P]:d(P)=p+1}\bigoplus_{w\in W^P} H^{q-l(w)}(\Gamma_{M_P}\backslash X^{M_P},\widetilde{\mathbb{H}^{l(w)}(\mathfrak{u}_P,\mathcal{M})_{w\cdot\lambda}}) \xrightarrow{d_1^{p,q}}$$

$$\xrightarrow{d_1^{p,q}} \bigoplus_{[Q]:d(Q)=p+2}\bigoplus_{v\in W^Q} H^{q-l(v)}(\Gamma_{M_Q}\backslash X^{M_Q},\widetilde{\mathbb{H}^{l(v)}(\mathfrak{u}_Q,\mathcal{M})_{v\cdot\lambda}}) \xrightarrow{d_1^{p+1,q}}$$

Here the differentials $d_1^{p,q}$ are $\pm$ sums of restriction maps

$$H^{q-l(w)}(\Gamma_{M_P}\backslash X^{M_P},\mathbb{H}^{l(w)}\widetilde{(\mathfrak{u}_P,\mathcal{M})}_{w\cdot\lambda}) \to H^{q-l(v)}(\Gamma_{M_Q}\backslash X^{M_Q},\mathbb{H}^{l(v)}\widetilde{(\mathfrak{u}_Q,\mathcal{M})}_{v\cdot\lambda})$$

where $v = s_i w$ and s_i is a reflection at a simple root.

It becomes clear that the computation of the cohomology of this complex, i.e. the computation of the $E_2^{p,q}$ term, requires that we understand the above problem for all Levi quotients M. The computation of the various differentials $d_2^{p,q} : E_2^{p,q} \to E_2^{p+2,q-1}$ will become a very delicate issue. The computation of the higher pages of the spectral sequence will be even more difficult.

If the highest weight λ is regular then then the spectral sequence should degenerate at $E_2^{p,q}$ level, i.e. $d_2^{p,q} = 0$. This should follow from the results in [81], but if λ is not regular this will not be true.

I am convinced that a thorough study of the deeper pages in the spectral sequence and of the map r will be very rewarding. We get interesting applications to number theory. We can prove rationality (-algebraicity-) results for special values of L-functions (divided by a well chosen period). In certain cases these normalised L-values $L^{\mathrm{ar}}(\pi_f,\nu) = \frac{L(\pi_f,\nu)}{\Omega_{\epsilon(\nu)}}$ *carry some relevant arithmetic information. It may tell us something about the structure of some cohomology groups as Hecke modules (for instance the denominator of Eisenstein classes) and as a consequence something about the structure of the Galois group.*

This program has been executed in special cases in this book. We discussed the rationality of normalised special L-values in a specific situation at the end of Chapter 4 (see (4.136)). In the papers [55] and [60] we address this rationality issue for some higher rank cases. Several authors [55] ,[60],[4],[67] use the method of construction rational cohomology classes from Eisenstein series to prove rationality results for special values of L functions. To be a little bit more precise: To a cohomology eigenclass σ_f in the cohomology of the Levi of a maximal parabolic they attach a cohomological L function $L(\sigma_f, s)$ and prove that for integers $\nu, \nu+1 \in \{n_(\sigma_\infty)\dots n_+(\sigma_\infty))$ the ratio

$$\Omega(\sigma_f)^{\epsilon(\nu)}\frac{L(\sigma_f,\nu)}{L(\sigma_f,\nu+1)} \in \bar{\mathbb{Q}}$$

Here $\Omega(\sigma_f)$ is a well chosen period in $\mathbb{C}^\times$ which is well defined up to a unit in some $\mathcal{O}_F^\times$. Such a result is of course non tautological only if the length of the interval of critical arguments is > 3. We treat one instance of such a result in Chapter 4 (see 4.135)

So far the authors do not go to deeper pages of the spectral sequence. In the note [58] I consider the group $\mathrm{Sl}_4(\mathbb{Z})$ and carry out some speculative computations which indicate that $d_2^{\bullet,\bullet} \neq 0$. Hence we get $E_2^{\bullet,\bullet} \neq E_3^{\bullet,\bullet}$, and this will imply a rationality relation between special values of the Riemann ζ-function which of course will be well-known. We also learn from this example that the Weyl group combinatorics will become very involved and we can hope that some computational experiments may help us to figure out some nice general patterns (if there are any).

Once we have proved such rationality (-better algebraicity-) result, we can ask the next question: Do the ratios of special values carry any interesting information? Do they tell us something about the structure of the cohomology as module under the Hecke-algebra?

In Chapter 5 we show that at least in some special cases the answer is yes: We have Theorem 5.1.2 for the group $\mathrm{Sl}_2(\mathbb{Z})$. For the group $\mathrm{Sl}_2(\mathbb{Z}[\mathbf{i}])$ I discuss a situation where we may look for a possible analogue to Theorem 5.1.2.

But I think that the proof given in this book is a "wrong" proof, I think that the argument checking the denominator by evaluating the Eisenstein class on a suitable modular symbol does not extend to higher rank cases. It is just a kucky accident.

In my papers [51] and [54] I describe a strategy which might explain the implication

$$\ell | L\text{-value} \Longrightarrow \ell | \text{denominator}.$$

I construct some mixed Tate motives in the cohomology of Shimura varieties [84]. If certain - to some extent plausible and also accepted- assumptions concerning these mixed Tate motives are fulfilled, we have some speculative arguments which prove this implication. One of these assumptions is that these mixed Tate motives are not "exotic", this concept is explained in (see [84]).

We do not know whether there are any exotic mixed Tate motives at all. on the other hand it seems to be an insurmountable difficulty to prove that exotic mixed Tate motives do not exist.

Therefore I come back to the option to verify this implication experimentally. To achieve this we have to write algorithms which do the same as what we did in Chapter 3 for $\mathrm{Gl}_2(\mathbb{Z})$ for some cases of higher rank. One should verify this above assertion in some cases with the help of a computer. In principle we have to do the same for some groups of higher rank as what we did for $\ell = 691$ in Chapter 3. A first case might be the case discussed in [51] for the prime $\ell = 41$. Here we must be aware that this may become very expensive, the complexes become very big and the matrix entries will be huge. At this point it may be helpful to work with the cohomology $\mod \ell$ or $\mod \ell^2$.

Of course the outcome of such experiments could be a falsification of the assumption about mixed Tate-Motives.

finis-operis

Bibliography

[1] Atobe, Hiraku; Chida, Masataka; Ibukiyama, Tomoyoshi; Katsurada, Hidenori; Yamauchi, Takuya *Harder's conjecture I* J. Math. Soc. Japan 75 (2023), no. 4, 1339–1408.

[2] H. Atobe, M. Chida, T. Ibukiyama, H. Katsurada, T. Yamauchi *Harder's conjecture II*, J. Math. Soc. Japan 75 (2023), 1309-1378.

[3] BATEMAN MANUSCRIPT PROJECT

[4] C. Bhagwat & A. Raghuram *Eisenstein cohomology for orthogonal groups and the special values L-functions for* $O(2n) \times GL_1$, to appear in Journal of the Institute of Mathematics at Jussieu.

[5] K.A. Behrend, *Semi-stability of reductive group schemes over curves,* Math. Ann. 301 (1995), 281–305.

[6] J. Bellaïche, R. Pollack, *Congruences with Eisenstein series and* μ *-invariants*, Compos. Math. 155 (2019), 863-901.

[7] T. Berger, *Denominators of Eisenstein cohomology classes for* Gl_2 *over imaginary quadratic fields,* Manuscripta Math 125 (2008), 427-470.

[8] H. Bekki, S.Sakamoto, *Harder's denominator problem for* $SL_2(\mathbb{Z})$ *and its applications*, to appear in Essential number theory

[9] N. Bergeron, A. Venkatesh, *The asymptotic growth of torsion homology for arithmetic groups,* J.Inst. Math. Jussieu 12(2) (2013), 391–447.

[10] J. Bergström, C. Faber, G. van der Geer *Siegel modular forms of degree three and the cohomology of local systems,* Selecta Math. (N.S.) 20 (2014), 83-124.

[11] J. Bergström; N. Dummigan; T.Mégarbané; T. Ibukiyama; H. Katsurada, *Eisenstein congruences for SO(4,3), SO(4,4), spinor, and triple product L-values,* Exp. Math. 27 (2018), 230-250.

[12] A. Borel, *Sur la cohomologie des espaces fibrés principaux et des espaces homogènes de groupes de Lie compacts,* Ann. of Math. (2)57(1953), 115–207.

[13] A. Borel *Stable real cohomology of arithmetic groups I*, Annales scientifiques de l'É.N.S. 4e série, 7, (1974), 235-272.

G. Harder, *Cohomology of Arithmetic Groups*,
Springer Monographs in Mathematics, https://doi.org/10.1007/978-3-032-10378-9

[14] A. Borel, *Stable real cohomology of arithmetic groups II,* In: Manifolds and Lie groups (Notre Dame, Ind., 1980), pp. 21–55, Progr. Math., 14, Birkhäuser, Boston, Mass., 1981.

[15] A. Borel, *Regularization theorems in Lie algebra cohomology. Applications,* Duke Math. J. 50 (1983), 605–623.

[16] A. Borel and W. Casselman, *L^2-cohomology of locally symmetric manifolds of finite volume,* Duke Math. J. 50 (1983), 625-647Raghu.

[17] A. Borel and H. Garland, *Laplacian and the discrete spectrum of an arithmetic group,* Amer. J. Math. 105 (1983), 309–335.

[18] A. Borel and Harish-Chandra, *Arithmetic subgroups of algebraic groups,* Ann. of Math. (2) 75 (1962), 485–535.

[19] A. Borel and J. Tits *Groupes réductifs,* Inst. Hautes Études Sci. Publ. Math. 27 (1965), 55-150.

[20] A. Borel and N. Wallach, Continuous Cohomology, Discrete Subgroups, and Representations of Reductive Groups, 2nd edition. Mathematical Surveys and Monographs, 67. American Mathematical Society, Providence, RI, 2000.

[21] A. Borel and J-P. Serre, *Corners and arithmetic groups,* Comment. Math. Helvetici 48 (1973), 436-491.

[22] N. Bourbaki, Groupes et Algèbres de Lie, Chapitres 4,5, et 6, Hermann, Paris, 1968

[23] R. Branchereau, *An upper bound on the denominator of Eisenstein classes in Bianchi manifolds*, Preprint

[24] J. H. Bruinier et al., *The 1-2-3 of Modular Forms*, Lectures from the Summer School on Modular Forms and their Applications, held in Nordfjordeid, June 2004. Edited by Kristian Ranestad. Universitext. Springer-Verlag, Berlin, 2008.

[25] J.P. Buhler, D. Harvey, *Irregular primes to 163 million*, Math. Comp. 80 (2011), 2435-2444.

[26] H. Busemann, *The Geometry of Geodesics.* Academic Press Inc., New York, N. Y., 1955

[27] L. Carlitz, *A generalization of Maillet's determinant and a bound for the first factor of the class number.* Proc. Amer. Math. Soc. 12 (1961), 256–261.

[28] G. Chenevier, J. Lannes, Automorphic Forms and Even Unimodular Lattices, Ergebnisse der Mathematik und ihrer Grenzgebiete Vol. 69, Springer, Berlin-Heidelberg, 2019.

[29] L. Clozel, *Motifs et formes automorphes: applications du principe de fonctorialité.* In: Automorphic Forms, Shimura Varieties, and L-functions, Vol. I (Ann Arbor, MI, 1988), 77–159, Perspect. Math., 10, Academic Press, Boston, MA, 1990.

[30] R. M. Damerell, *L-functions of elliptic curves with complex multiplication. I,II*, Acta Arith.17(1970), 287–301. Acta Arith. 19 (1971), 311–317.

[31] P. Deligne, *Formes modulaires et représentations ℓ-adiques*, Séminaire Bourbaki. Vol. 1968/69: Exposés 347-363, Exp. No. 355, 139-172, Lecture Notes in Math., 175, Springer, Berlin, 1971.

[32] P. Deligne, *Valeurs de fonctions L et périodes d'intégrales*, With an appendix by N. Koblitz and A. Ogus. In: Proc. Sympos. Pure Math., XXXIII, Automorphic forms, representations and L-functions, Part 2, pp. 313–346, Amer. Math. Soc., Providence, R.I., 1979.

[33] M. Demazure, A. Grothendieck, *Schémas en groupes* SGA 3. Lect. Notes Math., Vols. 151, 152, 153. Springer-Verlag (1970).

[34] T. Deng, *Torsion in Cohomology of* $\mathrm{Sl}_2(\mathbb{Z})$ *and Congruence of Modular Forms*, arXiv:1905.05260.

[35] T. Dokchitser, *Computing special values of motivic L-functions*, Experiment. Math. 13 (2004), 137–149.

[36] J. Elstrodt, F. Grunewald, J. Mennicke, Groups Acting on Hyperbolic Space, Springer Monographs in Mathematics, Berlin-Heidelberg, 1998.

[37] C. Faber, G. van der Geer, *Sur la cohomologie des systèmes locaux sur les espaces de modules des courbes de genre 2 et des surfaces abéliennes.I*, C. R. Math. Acad. Sci. Paris 338 (2004), no. 5, 381–384. II, no. 6, 467–470.

[38] J. Franke, *Harmonic analysis in weighted* L_2*-spaces.* Ann. Sci. École Norm. Sup. (4) 31 (1998), 181–279.

[39] H. Garland, W.-C.Hsiang *A square integrability criterion for the cohomology of arithmetic groups*, Proc. Nat. Acad. Sci. U.S.A.59(1968), 354–360.

[40] R. Godement, *Notes on Jacquet-Langlands theory*, IAS Princeton, 1970.

[41] D. Grayson, *Reduction theory using semistability*, Comment. Math. Helv. 59 (1984), 600-634.

[42] K. Haberland, *Perioden von Modulformen einer Variabler and Gruppencohomologie. I, II, III,* Math. Nachr. 112 (1983), 245-282, 283-295, 297-315.

[43] G. Harder, *General aspects in the theory of modular symbols,* In: Seminar on Number Theory, Paris 1981-1982, pp. 73-88, Progr. Math., 38, Birkhäuser, Boston 1983.

[44] G. Harder, *Eisenstein cohomology of arithmetic groups. The case* GL_2, Invent. Math. 89 (1987), 37–118.

[45] G. Harder, Eisensteinkohomologie und die Konstruktion gemischter Motive, Lect. Notes in Maths., 1562, Springer:Berlin-Heidelberg, 1993.

[46] G. Harder - N.Schappacher, *Special values of Hecke L-functions and abelian integrals.* Workshop Bonn 1984 (Bonn, 1984), 17-49, Lect. Notes in Maths., 1111, Springer:Berlin, 1985.

[47] G. Harder, *Eisenstein cohomology of arithmetic groups and its applications to number theory.* In: Proceedings of the International Congress of Mathematicians, Vol. I, II (Kyoto, 1990), 779-790, Math. Soc. Japan, Tokyo, 1991.

[48] G. Harder, Lectures on Algebraic Geometry I, Aspects of Mathematics, vol. 35 (second edition), Vieweg + Teubner: Wiesbaden, 2011.

[49] G. Harder, Lectures on Algebraic Geometry II, vol. 39, Vieweg + Teubner: Wiesbaden, 2011.

[50] G. Harder, *Interpolating coefficient systems and p-ordinary cohomology of arithmetic groups,* Groups Geom. Dyn., 5 (2011), 393–444.

[51] G. Harder *A congruence between a Siegel and an Elliptic Modular Form,* In: The 1-2-3 of Modular Forms, Springer: Berlin-Heidelberg 2008.

[52] G. Harder *The topological trace formula* `https://archive.mpim-bonn.mpg.de/id/eprint/5217/4/TopologicalTraceFormular.pdf`

[53] G. Harder, *Eisenstein cohomology for* $SL_2(\mathbb{Z}[i])$ *and special values of L-functions,* In: Cohomology of arithmetic groups, 51-82, Springer Proc. Math. Stat., 245, Springer, Cham, 2018.

[54] G. Harder, *Can we compute the denominators of the Eisenstein classes?,* Advanced Studies: Euro-Tbilisi Mathematical Journal, Special issue 9 (2021), 131-165.

[55] G. Harder, *Arithmetic aspects of rank one Eisenstein cohomology,* In: Proceedings of the International Colloquium on Cycles, Motives and Shimura Varieties, Mumbai 2008, pp. 131-190, Studies in Mathematics, vol. 21, Tata Institute, Narose Publ. House, 2010.

[56] G. Harder, *Eisenstein Cohomology and the Construction of mixed Motives,* see `https://archive.mpim-bonn.mpg.de/id/eprint/5217/1/Mix-Mot-2017.pdf`

[57] G. Harder, *Eisenstein cohomology,arithmetic applications* , preprint 2023, see `https://archive.mpim-bonn.mpg.de/id/eprint/5217/2/Talk-Lille.pdf`

[58] G. Harder, $d_2^{p,q} : E_2^{p,q} \to E_2^{p+2,q-1}$, preprint 2023, see `https://archive.mpim-bonn.mpg.de/id/eprint/5217/5/PAD.pdf`

[59] G. Harder, R. Pink, *Modular konstruierte unverzweigte abelsche p-Erweiterungen von $\mathbb{Q}(\zeta_p)$ und die Struktur ihrer Galoisgruppen,* Math. Nachr. 159 (1992), 83–99.

[60] G.Harder and A. Raghuram, *Eisenstein cohomology for GL_n and the Special Values of Rankin–Selberg L-functions,* Annals of Mathematics Studies, vol. 203, Princeton: University Press. 2020.

[61] Harish-Chandra, *Automorphic Forms on Semisimple Lie Groups*, Lect. Notes in Maths., 62, Springer:Heidelberg 1968.

[62] E. Hecke, *Ueber Modulfunktionen und die Dirichletschen Reihen mit Eulerscher Produktentwicklung. I,* Math. Ann. 114 (1937), no. 1, 1–28.

[63] S. Helgason, Differential Geometry and Symmetric Spaces, Pure and Applied Mathematics, Vol. XII. Academic Press, New York-London 1962.

[64] H. Hida, *Congruence of cusp forms and special values of their zeta functions.* Invent. Math. 63 (1981), no. 2, 225-261.

[65] H. Hida,*Galois representations into $GL_2(Z_p[[X]])$ attached to ordinary cusp forms*, Invent. Math. 85 (1986), no. 3, 545–613. MR 848685

[66] H. Hida,*On p-adic Hecke algebras for GL2 over totally real fields,*Ann. of Math. (2) 128 (1988), no. 2, 295–384. MR 960949

[67] F. Hosseinijafari *Eisenstein cohomology and the critical values of certain L-functions: The case G_2*, arXiv:2501.03546

[68] A. Hurwitz, *Über die Entwicklungskoeffizienten der lemniskatischen Funktionen;*
first communication in: Nachr. k. Ges. Wiss. Göttingen, Math. Phys. Kl. 1897, 273–276 = Math. Werke II, no LXVI, 338–341. Published in extenso: Math. Ann. 51 (1899), 196–226 = Math. Werke II, no LXVII, 342–373.

[69] K. Iwasawa, Lectures on p-adic L-functions, Annals of Mathematics Studies, 74. Princeton: University Press, Princeton, N.J., 1972.

[70] H. Jacquet, R.P. Langlands, *Automorphic Forms on Gl_2*, Lect. Notes in Maths., 114, Springer:Heidelberg 1970.

[71] C. Kaiser *Die Nenner von Eisensteinklassen für gewisse Kongruenzgruppen* Diplomarbeit Bonn 1990 (Homepage Harder folder Buch)

[72] M. Kneser, *Strong approximation*, In: Algebraic Groups and Discontinuous Subgroups (Proc. Sympos. Pure Math., Boulder, Colo., 1965) pp. 187-196, Amer. Math. Soc., Providence, R.I., 1966.

[73] B. Kostant, *Lie algebra cohomology and the generalized Borel-Weil theorem*, Ann. of Math., Vol 74 (1961), 329–387.

[74] T. Kubota, H-W. Leopoldt, *Eine p-adische Theorie der Zetawerte. I. Einführung der p-adischen Dirichletschen L-Funktionen.* J. Reine Angew. Math. 214(215) (1964), 328–339.

[75] J.C. Lagarias, H.L. Montgomery, A.M. Odlyzko, *A bound for the least prime ideal in the Chebotarev density theorem*, Invent. Math. 54 (1979), 271–296.

[76] R. Langlands, Euler Products, Yale University Press, 1971.

[77] R. Langlands, *On the functional equations satisfied by Eisenstein series.* Lect. Notes in Maths., Vol. 544, Springer: Berlin-New York, 1976.

[78] R. Langlands, *Automorphic representations, Shimura varieties, and motives. Ein Märchen.* In: Automorphic forms, representations and L-functions (Proc. Sympos. Pure Math., Oregon State Univ., Corvallis, Ore., 1977), Part 2, pp. 205-246, Proc. Sympos. Pure Math., XXXIII, Amer. Math. Soc., Providence, R.I., 1979.

[79] J.-P. Labesse, R. P. Langlands, *L-indistinguishability for SL(2)*, Canadian J. Math. 31 (1979), 726–785.

[80] R. Lee, R. H Szczarba, *On the homology and cohomology of congruence subgroups*, Invent. Math. 33 (1976), 15–53.

[81] J.-S. Li, J.Schwermer, *On the Eisenstein cohomology of arithmetic groups,* Duke Math. J. 123 (2004), 141-169.

[82] H, Maennel *Nenner von Eisensteinklassen: auf Hilbertschen Modulvarietäten und die p-adischen KLassenzahlformeln*, Bonner Math. Schriften, Nr. 247

[83] B. Mazur, A.Wiles, *Class fields of abelian extensions of* $\mathbb{Q}$*,* Invent. Math. 76 (1984), 179–330.

[84] J. Milnor and J. Moore, *On the structure of Hopf algebras*, Ann. of Math. (2) 81 (1965), 211–264.

[85] L. J. Mordell, *On Mr. Ramanujan's empirical expansions of modular functions,* Proc. Cambridge Philos. Soc. 19 (1917) 117-124.

[86] G.D. Mostow, T. Tamagawa, *On the compactness of arithmetically defined homogeneous spaces.* Ann. of Math. (2)76(1962), 446–463.

[87] W. Müller, F. Rochon, *Exponential growth of torsion in the cohomology of arithmetic hyperbolic manifolds*, Math. Zeitschrift 298 (2021), 79-106.

[88] J. Neukirch, Algebraic Number Theory, Grund. der Math. Wissenschaften, vol. 322 Springer: Berlin-Heidelberg 1999.

[89] T. Ono, *Arithmetic of algebraic tori*, Ann. of Math. (2) 74 (1961), 101-139.

[90] V. Platonov-A. Rapinchuk, Algebraic Groups and Number Theory, Pure and Applied Maths., vol. 139, Boston: Academic Press 1994.

[91] M. S. Raghunathan, *A note on quotients of real algebraic groups by arithmetic subgroups*, Invent. Math. 4 (1967/68), 318–335.

[92] S. Ramanujan, *On certain arithmetical functions,* Trans. Cambridge Philos. Soc. 22 (1916) 159–184. [Collected Papers of Srinivasa Ramanujan, AMS Chelsea Publishing, American Mathematical Society, Providence, RI, 2000, pp. 136-162.]

[93] G. de Rham, *Differentiable Manifolds. Forms, Currents, Harmonic Forms,* Grund. der Math. Wissenschaften, 266, Springer: Berlin-Heidelberg, 1984.

[94] K.A. Ribet, *A modular construction of unramified p-extensions of* $\mathbb{Q}(\mu_p)$, Invent. Math. 34 (1976), 151-162.

[95] J. Rosenberg, Algebraic K-theory and its Applications, Graduate Texts Maths., 147, Springer: Berlin-Heidelberg 1994.

[96] T. Scholl, *Motives for modular forms,* Invent. Math. 100 (1990), 419–430.

[97] H.-M. Sengün, *On the integral cohomology of Bianchi groups,* Exp. Math.20 (2011), 487–505.

[98] J. Schwermer, Kohomologie arithmetisch definierter Gruppen und Eisensteinreihen, Lect. Notes in Maths., 988, Springer-Verlag, Berlin, 1983.

[99] J. Schwermer, *Cohomology of arithmetic groups, automorphic forms and L-functions,* In: Cohomology of Arithmetic Groups and Automorphic Forms (Luminy-Marseille, 1989), pp. 1–29, Lect. Notes in Maths., 1447, Springer, Berlin, 1990.

[100] J-P. Serre, *Le probléme des groupes de congruence pour* SL_2, Ann. Math.92 (1970), 489–527.

[101] J-P. Serre, *Une interprétation des congruences relatives à la fonction* τ *de Ramanujan,* 1969 Séminaire Delange-Pisot-Poitou: 1967/68, Théorie des Nombres, Fasc. 1-2, Exp. 14, 17 pp.

[102] J-P. Serre, *A Course in Arithmetic,* Graduate Texts Maths., 7, Springer: Berlin-Heidelberg, 1973.

[103] J-P. Serre, *Cohomologie Galoisienne,* Lect. Notes in Maths. 5, Springer: Berlin-Heidelberg, 1964.

[104] G. Shimura, *The special values of the zeta functions associated with Hilbert modular forms,* Duke Math. J. 45 (1978), 637–679.

[105] C.L. Siegel, *Gesammelte Abhandlungen, I-III,* Springer: Berlin-Heidelberg 1966.

[106] T.A. Springer, Linear Algebraic Groups, Progress in Maths. vol. 9, Sec. Ed., Birkhäuser: Boston 1998.

[107] R. Steinberg, *Lectures on Chevalley Groups,* Yale University 1967.

[108] U. Stuhler, *Eine Bemerkung zur Reduktionstheorie quadratischer Formen,* Arch. Math. 27(1976), 604-610.

[109] H.P.F. Swinnerton-Dyer, *On ℓ-adic representations and congruences for coefficients of modular forms,* In: Modular Functions of One Variable III, Lect. Notes in Maths., 350, pp. 1-55, Springer: Berlin-Heidelberg.

[110] J. T. Tate, *Fourier analysis in number fields, and Hecke's zeta-functions,* In: Algebraic Number Theory (Proc. Instructional Conf., Brighton, 1965) ed. J.W.S. Cassels, A. Fröhlich, pp. 305-347, Academic Press: London, 1967.

[111] W.T. van Est, *A generalisation of the Cartan-Leray spectral sequence. I, II.* Nederl. Akad. Wetensch. Proc. Ser. A (Indag. Math.) 20, (1958), 399 – 419.

[112] V. Vatsal, *Congruences for the special values of modular L-functions,* C. R. Math. Acad. Sci. Soc. R. Can. 20 (1998), 77–82.

[113] D. Vogan, G. Zuckerman, *Unitary representations with nonzero cohomology,* Compos. Math. 53 (1984), 51–90.

[114] S. Wagstaff, *Zeros of p-adic L-functions.* Math. Comp. 29 (1975), no. 132, 1138–1143.

[115] N. Wallach, *On the constant term of a square integrable automorphic form,* In: Arsene, G. et al. (eds.) Operator Algebras and Group Representations. II. (Monogr. Stud. Maths., vol. 18, pp. 227- 237) London: Pitman 1984.

[116] L. Washington, Introduction into Cyclotomic Fields, Graduate Texts Maths., 83, Springer: Berlin-Heidelberg, 1997.

[117] U. Weselmann, *Siegel Modular varieties and the Eisenstein cohomology of* PGL_{2g+1}, Manuscripta Math. 145 (2014), 175–220.

[118] E. T. Whittaker, G.N. Watson *A course of Modern Analysis* Fourth edition Cambridge University Press

[119] A. Wiles, *Modular elliptic curves and Fermat's last theorem,* Ann. of Math. (2) 141 (1995), 443–551.

Index

G. Harder, *Cohomology of Arithmetic Groups*,
Springer Monographs in Mathematics, https://doi.org/10.1007/978-3-032-10378-9

MIX
Papier aus verantwortungsvollen Quellen
Paper from responsible sources
FSC® C105338

If you have any concerns about our products,
you can contact us on
ProductSafety@springernature.com

In case Publisher is established outside the EU,
the EU authorized representative is:
Springer Nature Customer Service Center GmbH
Europaplatz 3, 69115 Heidelberg, Germany

Printed by Libri Plureos GmbH
in Hamburg, Germany